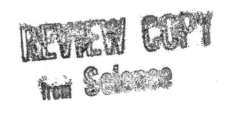

CONFORMATIONAL THEORY OF LARGE MOLECULES

CONFORMATIONAL THEORY OF LARGE MOLECULES

The Rotational Isomeric State Model in Macromolecular Systems

Wayne L. Mattice

Institute of Polymer Science, The University of Akron, Akron, Ohio

and

Ulrich W. Suter

Institute für Polymere, Eidgenössische Technische Hochschule, ETH-Zentrum, Zürich, Switzerland

A WILEY INTERSCIENCE PUBLICATION

JOHN WILEY & SONS, INC.

New York • Chichester • Brisbane • Toronto • Singapore

Library of Congress Cataloging in Publication Data:

Mattice, Wayne L.
    Conformational theory of large molecules: the rotational isomeric
state model in macromolecular systems / Wayne L. Mattice and Ulrich
W. Suter.
        p.   cm.
    "A Wiley-Interscience Publication."
    Includes indexes.
    ISBN 0-471-84338-5
    1. Polymers—Mathematical models.   I. Suter, U., 1935–   .
II. Title.
QD381.9.M3M39   1994
547.7′0442—dc20   93-40718

Printed in the United States of America

10 9 8 7 6 5 4 3 2 1

CONTENTS

PREFACE

One way in which *macromolecules* differ from small molecules is in the *enormous* number of conformations that they populate at ordinary temperatures. A variety of models have been developed for averaging conformation-dependent physical properties over the enormous number of conformations accessible to a macromolecule. The *rotational isomeric state model in macromolecular systems* is unique among these models in that it describes the properties of the macromolecule in terms of the *real covalent structure* (or in terms of the properties of small molecules that reflect sections of the chain), and does so in a computational *tractable* (indeed, computationally *efficient*!) manner. The formalism allows the incorporation into the calculation of as much structural detail as any chemist might wish, without compromise in computational efficiency.

The application of rotational isomeric state techniques to polymers is not new. Their earliest use on polymers was reported by Volkenstein over four decades ago,[1] and the generator matrix technique employed in rotational isomeric state calculations was invented a decade earlier,[2] for another purpose. Applications to macromolecules began to appear in the literature with some regularity in about 1960.[3] Flory's classic book on rotational isomeric state theory was published in 1969.[4]

The intervening quarter of a century since the first appearance of Flory's book has seen important improvements. Formulations for the quantitative description of several properties have been simplified and generalized, new formulations for the treatment of additional properties have been created, and an explosion in applications to different types of polymers has occurred. The technique has become popular as chemists have synthesized polymers and copolymers with an increasing variety of topologies, compositions, constitutions, and configurations, often using monomer units of increasing complexity. Their curiosity about the manner in which the conformation-dependent physical properties of these macromolecules are determined by the monomer units that occur in the chains leads naturally to the use of rotational isomeric state the-

---

[1] Volkenstein, M. V. *Dokl. Akad. Nauk SSSR* **1951**, *78*, 879.

[2] Kramers, H. A.; Wannier, G. H. *Phys. Rev.* **1941**, *60*, 252.

[3] Gotlib, Yu. Ya. *Zh. Tekhn. Fiz.* **1959**, *29*, 523. Birshtein, T. M.; Ptitsyn, O. B. *Zh. Tekhn. Fiz* **1959**, *29*, 1048. Lifson, S. *J. Chem. Phys.* **1959**, *30*, 964. Nagai, K. *J. Chem. Phys.* **1959**, *31*, 1169. Hoeve, C. A. J. *J. Chem. Phys.* **1960**, *32*, 888.

[4] Flory, P. J. *Statistical Mechanics of Chain Molecules*, Wiley-Interscience, New York, **1969**; reprinted with the same title by Hanser, München, (Munich), **1989**.

ory for the answers to these questions. Another factor contributing to the use of rotational isomeric state theory is the availability to any researcher (often at the desk site) of a computer that delivers all of the computational resources required for the calculations. The recent advent of commercial software packages for the visualization and manipulation of atom-based models of polymers also contributes to the use of rotational isomeric state techniques. For all of these reasons, it seems that a book describing the current status of rotational isomeric state theory might be of use to those who wish to understand the manner in which the conformation-dependent physical properties of a polymer are determined by its covalent structure.

We suspect that many of the readers of this book will be using it to understand how to perform rotational isomeric state calculations for polymers that are of interest to them. Among these readers there may be many who are using the book for self-instruction, without the benefits that arise from an instructor and a classroom setting. We hope that such readers will be assisted by the level at which the material is introduced, by the style in which it is presented, by the 174 problems that appear at the ends of the chapters (along with answers to 72% of the problems, in Appendix B), and by the inclusion (in Appendix C) of a FORTRAN program that will perform calculations for the most important property, the characteristic ratio (or mean-square unperturbed end-to-end distance).

The book contains 15 chapters that can conceptually be arranged into four groups:

1. The first two chapters are introductory material for the theoretical description of chain molecules (Chapter I) and the experimental observables that are most frequently addressed by rotational isomeric state methods (Chapter II).
2. Chapters III–VI develop the rotational isomeric state model, starting from the well-known conformational properties of very simple alkanes. In order to emphasize the fundamental features of the model, these chapters are restricted to simple chains, and to a limited subset of the properties that can be addressed by rotational isomeric state techniques.
3. Chapters VII–XII expand the scope of the types of polymers that can be treated with rotational isomeric state techniques.
4. Chapters XIII–XV expand the scope of the types of conformation-dependent physical properties that can be treated with rotational isomeric state techniques.

When we began this project, it was with the intent of including in the book a reference to every published application of rotational isomeric state techniques to polymers. We soon realized that the number of publications that would need to be cited and discussed, if we were to adhere to the original objective, would require a book much larger than desirable. Although a complete review of the known RIS models and their parameters would have exceeded the scope of this book, a comprehensive listing will appear elsewhere.[5] The selection of applications for inclusion in this book has been based primarily on our perception of the needs of a reader who uses the book for self-instruction in performing rotational isomeric state calculations. It attempts a balance between early work that provides the foundation for the rotational isomeric model, and more recent developments of the model and applications to polymers.[6] Over 500 names are listed in the author index, but we could not describe work by every author who has used the rotational isomeric state model, nor could we discuss all of the publications by those authors that were cited. Our apologies to the many authors of publications that we did not cite in the text.

Both authors acknowledge a special debt to Professor Flory for fostering their interest in rotational isomeric state techniques, through his writing and through personal interaction. Their research in this area has been supported over the years by a number of government-supported funding agencies and private foundations, notable among which are the Army Research Office, Edison Polymer Innovation Corporation, German Bundesministerium für Forschung and Technologie, Guggenheim Foundation, National Institutes of Health, National Science Foundation, Office of Naval Research, Petroleum Research Fund, Schweizerischer Nationalfonds, and United States Department of Agriculture. Both authors would like to acknowledge the competent and generous help of Ms. Lera Tomasic in the preparation of the figures in this book. Valuable feedback on earlier drafts of the manuscript was received from Professors Akihiro Abe, Ivet Bahar, Burak Erman, James Mark, Wilma Olson, Enrique Saiz, and Alan Tonelli, from Dr. Do Yoon, and from members of the research groups of WLM (Mr. R. Balaji, Mr. M. Cotah, Mr. M. Fuller, Dr. T. Haliloğlu, Dr. C. Helfer, Mr. E.-G. Kim, Mr. M. Matties, Dr. S. Misra, Ms. M. Nguyen-Misra, Mr. Q. Ou, Mr. J.-S. Sun, Dr. Y. Wang, Prof. Y. Zhan, Dr. R. Zhang) and UWS (Dr. B. M. Forrest, Dr. A. A. Gusev, Dr. M. Laso, Dr. E. Leontidis, Mr. R. Rapold, Mr. M. Tomaselli, Mr. A. H. Widmann, Dr. B. Watzke-Lerebours, and Mr. M. Zehnder). WLM extends warm thanks to UWS and to ETH for their generous hospitality during his leave in 1993, at which time the manuscript for the book was completed.

WAYNE L. MATTICE
ULRICH W. SUTER

Akron, Ohio
Zürich, Switzerland
May 1994

---

[5]Rehahn, M.; Mattice, W. L.; Suter, U. W. *Adv. Polym. Sci.* **1995**.
[6]The distribution of the dates of publication of the citations is 12% pre-1960, 21% from the 1960s, 27% from the 1970s, 29% from the 1980s, and 10% from the 1990s, through 1994.

CONFORMATIONAL THEORY OF LARGE MOLECULES

1 Aren't There Enough Models to Treat Chain Statistics? A Brief Review of Simple Models

A polymer consists of *macromolecules*, molecules that often comprise tens of thousands of atoms. There are more possible constitutional isomers than can be individually considered (or comprehended), and the number of *configurations* and *conformations*[1] is similarly overwhelming. Naturally, a comprehensive description of the molecular structure requires methods designed to handle the multiplicity of possible topologies and spatial arrangements of molecular moieties.

A number of different models and theoretical devices has been created since the early days of polymer science for the treatment of macromolecules. A model can be constructed for chains of almost any constitutional complexity, but the vast majority of all efforts have addressed the behavior of the simplest macromolecular topology, the single-strand chain, and have been limited to the simplest linear constitution, a uniform chain that is a concatenation of identical entities ("segments").

An even more drastic simplification for almost all models lies in the assumption of "ideality," which in this context means the assumption that the effects of excluded volume are not affecting the modeled observables. In other words, it may be assumed that the segments of the macromolecule do not exclude other segments (of the same or another chain) from the volume they themselves occupy. This assumption of being in the Θ *state*[2] is a most severe limitation,

---

[1] The terms *constitution, configuration* (and associated with it *diastereoisomerism*), and *conformation* are employed here as is the custom in organic chemistry: a *conformation* is a specified mutual spatial arrangement of atoms or groups of the (chain) molecule, inasmuch as this arrangement can be determined by specification of torsion angles—bond lengths and bond angles are part of the specification of a configuration. The *conformation about a bond* (or *conformation of a bond*) refers to the relative position of groups directly bonded to two atoms that are connected by a (single) bond—the conformation of any larger unit can be described by the conformation of all bonds of which it consists. We follow in large part the principles of nomenclature as stated by the International Union of Pure and Applied Chemistry's (IUPAC) Commission on Macromolecular Nomenclature in *Pure Appl. Chem.* **1981**, *53*, 733, which are largely consistent with the nomenclature used in organic chemistry as defined by the IUPAC Commission on Nomenclature of Organic Chemistry in *Pure Appl. Chem.* **1975**, *45*, 11.

[2] Flory, P. J. *J. Chem. Phys.* **1949**, *17*, 303. Fox, T. G.; Flory, P. J. *J. Phys. Chem.* **1949**, *53*, 197. Flory, P. J. *Principles of Polymer Chemistry*, Cornell University Press, Ithaca, NY, **1953**.

since it is rare that a polymer is in a Θ state.[3] Under ordinary conditions there are significant long-range (intramolecular) correlations that "perturb" the conformation of the chains. But under the special conditions at the Θ state, the effects of the excluded volume of a segment vanish and the dimensions of the macromolecule take on their unperturbed values. The perturbations incurred under other than Θ conditions are difficult to quantify, since the interactions between pairs (or higher multiplets) of chain segments that are far apart in sequence along the macromolecule are mainly responsible for the changes in spatial dimensions. On the other hand, it is usually possible, albeit with some effort, to find Θ conditions and thereby situations where the experimental determination of unperturbed values is possible. For the most pertinent observables, the unperturbed mean square values of the end-to-end distance and the radius of gyration, there exist in addition procedures for their deduction from experiments taken under other than Θ conditions. Hence, the Θ state is the situation favorably addressed first.

We will review five of the most important models briefly below to set the stage for the introduction of the rotational isomeric state techniques. These models describe or predict properties of the individual chains (or of ensembles of independently acting chains) from molecular structural features. Of most immediate interest is usually a description of the spatial arrangement of the atoms of the macromolecule, or, since the number of distinguishable possible arrangements is usually enormously high and since only an ensemble average is usually of interest, a description of the probabilistic distribution for the *positions* r_i of all atoms i of the macromolecule, $W(r_i)$. Frequently, only the distributions of the end-to-end distance, $W(r)$, of a linear chain or of the radius of gyration, $W(s)$, are required. Very often it is even sufficient to know only the second moment of a distribution, that is, the average value of r^2 or s^2. These averages are denoted by $\langle r^2 \rangle$ or $\langle s^2 \rangle$.[4] The subscript 0 denotes the average for an ensemble in the Θ state, as in $\langle r^2 \rangle_0$ and $\langle s^2 \rangle_0$.

Following the brief exposé on ideal chains, a few paragraphs will serve as introduction to and overview of the effects of excluded volume on polymer dimensions. Even though most of this text will focus on ideal chain molecules, the subject of nonideal chains is sufficiently important and the possibility to apply realistic ideal chain models to nonideal situations (non-Θ state) so promising that a very short summary seems appropriate.

1. THE END-TO-END DISTANCE AND RADIUS OF GYRATION

Although our interest here is in flexible polymers that occupy an enormous number of conformations, the preferred starting point for a discussion of such

---

[3]Symbols are defined in the text when first used. If they are used in different places in the book, they are also defined in Appendix A.

[4]The angle brackets $\langle \ \rangle$ denote a conformational ensemble average of the enclosed property.

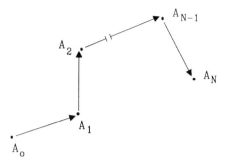

Figure I-1. A representative conformation of a chain of N steps, connecting $N + 1$ atoms.

chains is an instantaneous snapshot of one of them in an arbitrary conformation. We temporarily suppress the important fact that the chain actually has access to a large number of conformations, in order to devise a useful device for the description of a single conformation. The averaging over all conformations will be deferred until the second section of this chapter (Section I-2).

Any one of the many conformations of the chain can be pictured as a succession of N steps (bonds) $\mathbf{L}_i$,[5] joining $N+1$ atoms $(A_0, A_1, \ldots, A_N)$, as depicted in Fig. I-1. The *end-to-end vector* $\mathbf{r}$ is

$$\mathbf{r} = \sum_{i=1}^{N} \mathbf{L}_i \tag{I-1}$$

The direction of $\mathbf{r}$ is affected by our arbitrary decision as to the direction for indexing the A_i along the chain. The length of $\mathbf{r}$, however, is independent of an arbitrary selection of the direction for indexing. The square of the length of $\mathbf{r}$ is easily generated as[6]

$$r^2 \equiv \mathbf{r} \cdot \mathbf{r} \equiv \mathbf{r}^T \mathbf{r} = \left(\sum_i \mathbf{L}_i\right) \cdot \left(\sum_j \mathbf{L}_j\right)$$

$$= \sum_i \mathbf{L}_i \cdot \mathbf{L}_i + \sum_{i \neq j} \mathbf{L}_i \cdot \mathbf{L}_j \tag{I-2}$$

---

[5]The length of step i is denoted by L_i, which conveys no information about its direction. Both length and direction are specified by the vector, denoted by $\mathbf{L}_i$.

[6]In this text, vectors are always represented by column matrices, such as the end-to-end vector with the components r_x, r_y, and r_z is written as

$$\mathbf{r} = \begin{bmatrix} r_x \\ r_y \\ r_z \end{bmatrix}$$

The transpose is $\mathbf{r}^T = [r_x \quad r_y \quad r_z]$, and the scalar product is $\mathbf{r} \cdot \mathbf{r} \equiv \mathbf{r}^T \mathbf{r} = r_x^2 + r_y^2 + r_z^2$.

The *squared radius of gyration* is the mean of all $\mathbf{s}_i \equiv \mathbf{r}_i - \mathbf{g}$, where $\mathbf{g}$ is the center of mass vector, $(N + 1)^{-1} \sum_i \mathbf{r}_i$:[7]

$$s^2 = \frac{1}{N + 1} \sum_{i=0}^{N} (\mathbf{r}_i - \mathbf{g}) \cdot (\mathbf{r}_i - \mathbf{g}) \tag{I-3}$$

It is possible to show that[8]

$$s^2 = \frac{1}{(N + 1)^2} \sum_{i=0}^{N-1} \sum_{j=i+1}^{N} r_{ij}^2 \tag{I-4}$$

where r_{ij} is the distance between atoms i and j, and the double sum is over all pairs $i < j$.

The relationships in Eqs. (I-1) through (I-4) are independent of any model for the conformational ensemble. Equations (I-3) and (I-4) merely assume a set of $N + 1$ points of identical mass, fixed in space. Equations (I-1) and (I-2) assume also the specification of the sequence in which the points are connected, so that we can identify the two points in the set that define $\mathbf{r}$ and r^2. (We really need labels on only A_0 and A_n in order to determine $\mathbf{r}$, and A_0 need not be distinguished from A_n if we seek r^2.)

The selection of a model is necessary to average each of these properties over all the conformations in an ensemble. The model describes the relationship between the conformations.

2. IDEAL POLYMER CHAINS[9]

In this section, a brief and necessarily cursory overview of five idealized models for chain molecules is presented. Since the purpose of this book lies in the description of "flexible" chains, specifically, of macromolecules that can take on many conformations, we will limit ourselves mostly to model chains that are best suited for random-coiling polymers.

---

[7] Actually, s^2 is defined more generally by

$$s^2 = \frac{\sum_{i=0}^{N} m_i s_i^2}{\sum_{i=0}^{N} m_i}$$

where m_i is the mass of A_i, and s_i its distance from the center of mass. If all A_i have the same mass, this expression is identical with Eq. (I-3). For homopolymers and random copolymers, inclusion of the m_i usually has a trivial effect on the computed value of s^2. Hence the definition expressed in Eq. (I-3) will be adopted here.

[8] See, for example, Appendix A of Flory, P. J. *Statistical Mechanics of Chain Molecules*, Wiley-Interscience, New York, **1969**: reprinted by Hanser, München, **1989**.

[9] More thorough and detailed treatments of these matters as well as proofs of the relationships given here can be found in Volkenstein, M. V. *Configurational Statistics of Polymer Chains*, Wiley-

A. The Random Walk (or Random Flight) Chain

The *random-flight* chain has the simplest mathematical properties, but, unfortunately, also has the smallest degree of structural fidelity to the real macromolecule it is designed to represent. It is based on a random walk that departs at the origin of the coordinate system and in which each successive step can take all orientations in space with equal probability, independently of its predecessors. The *step length* L varies according to a probability distribution $W(L)$, the step length distribution, which may be of almost any form. The unperturbed *mean square end-to-end distance* $\langle r^2 \rangle_0$ is simply[10]

$$\langle r^2 \rangle_0 = N \langle L^2 \rangle \tag{I-5}$$

where $\langle L^2 \rangle$ signifies the mean square length of all bonds. This relationship is simple and significant. It indicates a proportionality between $\langle r^2 \rangle_0$ and N that is universally true for all ideal flexible chains in the limit of large N and also gives physical meaning to the constant of proportionality. The subscript 0 denotes an *unperturbed average*, that is, one taken under conditions where effects of the excluded volume are compensated and where no external forces act on the macromolecule. This state corresponds to the Θ state of a real chain.

The *mean square radius of gyration* is the single most important observable characterizing a macromolecule's dimensions. From Eq. (I-4)

$$\langle s^2 \rangle = \frac{1}{(N+1)^2} \sum_{i=0}^{N-1} \sum_{j=i+1}^{N} \langle r_{ij}^2 \rangle \tag{I-6}$$

which applies equally to perturbed and to unperturbed chains. From combi-

---

Interscience, New York, **1963**. Birshtein, T. M.; Ptitsyn, O. B. *Conformations of Macromolecules*, Wiley-Interscience, New York, **1966**. Flory, P. J. *Statistical Mechanics of Chain Molecules*, Wiley-Interscience, New York, **1969**: reprinted by Hanser, München, **1989**. Yamakawa, H. *Modern Theory of Polymer Solutions*, Harper & Row, New York, **1971**. Doi, M.; Edwards, S. F. *The Theory of Polymer Dynamics*, Clarendon Press, Oxford, **1986**. Freed, K. R. *Renormalization Group Theory of Macromolecules*, Wiley-Interscience, New York, **1987**. des Cloizeaux, J.; Jannink, G. *Polymers in Solution. Their Modelling and Structure*, Clarendon Press, Oxford, **1990**.

[10]The first sum in Eq. (I-2) is $N \langle L^2 \rangle$, independent of any assumption about the conformation. The second sum concerns $L_i L_j \cos \theta$, where θ is the angle between $\mathbf{L}_i$ and $\mathbf{L}_j$. Bond orientations in which $\cos \theta = x$ are of the same probability as bond orientations where $\cos \theta = -x$. Also, L_i is independent of all other bond lengths in the chain, in this model, and is also uncorrelated with θ. Hence all $\mathbf{L}_i \cdot \mathbf{L}_j$ will, in the conformational average, be zero. The average value of $\mathbf{r} \cdot \mathbf{r}$ is, therefore, simply

$$\langle \mathbf{r} \cdot \mathbf{r} \rangle_0 = N \langle L^2 \rangle + \left\langle \sum_{i \neq j} \mathbf{L}_i \cdot \mathbf{L}_j \right\rangle = N \langle L^2 \rangle$$

which is Eq. (I-5).

nation of this relationship with Eq. (I-5), where an unperturbed chain is now assumed, one derives

$$\lim_{N\to\infty} \langle s^2 \rangle_0 = \frac{\langle r^2 \rangle_0}{6} \tag{I-7}$$

This relationship is true for almost any function chosen for $W(L)$ ($\langle L^2 \rangle$ must exist), and hence for all polymer chains of sufficient length.[11]

Often a Gaussian distribution is chosen for the consequent mathematical simplifications:

$$W(\mathbf{L}_i) = \left(\frac{3}{2\pi\langle L_i^2 \rangle} \right)^{3/2} \exp\left(\frac{-3L_i^2}{2\langle L_i^2 \rangle} \right) \tag{I-8}$$

Then, the distribution of $\mathbf{r}$ is particularly simple:

$$W(\mathbf{r}) = \left(\frac{3}{2\pi\langle r^2 \rangle_0} \right)^{3/2} \exp\left(\frac{-3r^2}{2\langle r^2 \rangle_0} \right) \tag{I-9}$$

and the distribution of the segments about the center of gravity also:[12]

$$W(\mathbf{s}_i) \approx \left(\frac{3}{2\pi\langle s^2 \rangle_0} \right)^{3/2} \exp\left(\frac{-3s_i^2}{2\langle s^2 \rangle_0} \right) \tag{I-10}$$

The chain with Gaussian step distribution, Eq. (I-8), is often called the *Gaussian chain*. Equations (I-9) and (I-10) hold asymptotically for all flexible linear chains as $N \to \infty$.

Averages of higher even powers of r and of s (the "higher moments") are important for several properties addressed in other chapters. The dimensionless ratios formed from the "higher moments" and the second moments depend on N, but for all linear chains in the limit $N \to \infty$ the dimensionless ratios approach constant values, several of which are collected in Table I-1. Note that $\langle r^4 \rangle_0 / \langle r^2 \rangle_0^2 > \langle s^4 \rangle_0 / \langle s^2 \rangle_0^2$, which signifies that the distribution for r^2 is broader than the distribution for s^2.

Averages of odd powers of $\mathbf{r}$ are coordinate system dependent, and vanish in the limits of high N.

---

[11] The border between "sufficient" and "not sufficient" can be hard to define.
[12] Every conformation has one $\mathbf{r}$, and $N + 1$ values of s_i. The atoms that determine $\mathbf{r}$ are always A_0 and A_N. Every atom has its own s_i. For a very short chain (such as $N = 2$), atoms at the ends, and atoms near the middle of the chain, may have different $W(\mathbf{s}_i)$.

TABLE I-1. Limiting Values of $\langle r^{2p}\rangle_0/\langle r^2\rangle_0^p$ and $\langle s^{2p}\rangle_0/\langle s^2\rangle_0^p$, as $N \to \infty$

	r		s	
p	Exact	Approximate	Exact	Approximate
2	$\frac{5}{3}$	1.7	$\frac{19}{15}$	1.267
3	$\frac{35}{9}$	3.9	$\frac{631}{315}$	2.003
4	$\frac{35}{3}$	11.7	$\frac{1219}{315}$	3.870
5	$\frac{385}{9}$	42.8		

B. The Freely Jointed Chain

The *freely jointed* (or hinged) chain is a random-walk chain (Section I.1.A) with fixed bond length. Usually all N bonds have the same length L for convenience.[13] Equation (I-5) then reads

$$\langle r^2\rangle_0 = NL^2 \tag{I-11}$$

and Eq. (I-9) and the $\langle r^{2p}\rangle_0/\langle r^2\rangle_0^p$ in Table I-1 hold also asymptotically as $N \to \infty$. The value of $\langle s^2\rangle_0$ at any N is

$$\langle s^2\rangle_0 = \left(\frac{N+2}{N+1}\right)\frac{\langle r^2\rangle_0}{6} \tag{I-12}$$

which recovers Eq. (I-7). The fourth moments at any N have values of $\langle r^4\rangle_0/\langle r^2\rangle_0^2$ given by[14]

$$\frac{\langle r^4\rangle_0}{\langle r^2\rangle_0^2} = \frac{5N-2}{3N} \tag{I-13}$$

and values of $\langle s^4\rangle_0/\langle s^2\rangle_0^2$ given by[15]

$$\frac{\langle s^4\rangle_0}{\langle s^2\rangle_0^2} = \frac{19N^3 + 45N^2 + 32N - 6}{15N(N+1)(N+2)} \tag{I-14}$$

which recover the first line of entries in Table I-1. One can also derive ratios

[13] $W(L)$ is a δ function centered on this value of L.
[14] Jernigan, R. L.; Flory, P. J. *J. Chem. Phys.* **1969**, *50*, 4178.
[15] Mattice, W. L.; Sienicki, K. *J. Chem. Phys.* **1989**, *90*, 1956.

involving mixed moments, such as[16]

$$\frac{\langle r^2 s^2 \rangle_0}{\langle r^2 \rangle_0 \langle s^2 \rangle_0} = \frac{4N - 1}{3N} \tag{I-15}$$

The mixed moments are required for the calculation of correlation coefficients, as illustrated by Problem I-4.

The Equivalent Chain. The fact that the freely jointed chain is a model with only one structural parameter, L (besides the chain length parameter, N), has made it particularly attractive for the representation of real macromolecules by a simple model. A real linear chain of sufficient length, composed of n *skeletal bonds* of *length* l,[17,18] may be represented by a freely jointed chain consisting of N "bonds" of length L. The two parameters N and L are determined by the conditions that $\langle r^2 \rangle_0$ and the length of the fully extended chain must be the same for the two models, $NL^2 = \langle r^2 \rangle_0$ and $NL = r_{max}$, and hence

$$L = \frac{\langle r^2 \rangle_0}{r_{max}} \tag{I-16}$$

$$N = \frac{r_{max}^2}{\langle r^2 \rangle_0} \tag{I-17}$$

The *length of the fully extended chain*, or *contour length*, r_{max}, is taken to be the maximum achievable length without unphysical distortion of the molecular frame. For a simple polymer such as polyethylene, for instance, the variable internal degrees of freedom are usually taken to be the *torsion angles*, ϕ, at the $n - 2$ internal C—C bonds, but the C—C—C bond angle is restricted to 112° and l is restricted to 154 pm. The restrictions on θ and l lead to $r_{max} = nl \sin(\theta/2) \approx 0.83nl$, where θ denotes the *bond angle*. The application to polyethylene is developed further in Problem I-5. The freely jointed chain that is equivalent to a real chain through Eq. (I-16) and (I-17) is often called the *equivalent chain*.[19]

---

[16]Equation (I-15) applies to a freely jointed chain, but the formalism of rotational isomeric state theory was used in its derivation.[15]

[17]We will distinguish between "real" bonds, that is, bonds that are part of the actual chemical structure of a macromolecule, and "hypothetical" bonds, that is, bonds from which the hypothetical chain models (ideal or not) are constructed; real bonds are designated by $\mathbf{l}$ and their lengths by l, hypothetical bonds by $\mathbf{L}$ and their lengths by L.

[18]For chains where not all skeletal bonds are of equal length, the root-mean-square value of all n of the l_i is used, $l \equiv \sqrt{\overline{l^2}} = \sqrt{n^{-1} \sum_i l_i^2}$.

[19]Kuhn, W. *Kolloid Z.* **1936**, *76*, 258; *Kolloid Z.* **1939**, *87*, 3.

The Characteristic Ratio. A convenient comparison between the real chain and the equivalent chain is the *characteristic ratio*, C_n, defined by

$$C_n = \frac{\langle r^2 \rangle_0}{nl^2} \tag{I-18}$$

and usually given as the limiting value $C_\infty \equiv \lim_{n \to \infty} C_n$. The equivalent chain has a value of unity for the ratio $\langle r^2 \rangle_0 / NL^2$ for all values of N. Therefore C_n is a measure for how much larger the actual chain dimensions are in the real chain than they would be in the equivalent freely jointed one with the same bond lengths, that is, if all correlations between bond directions would cease to exist.

The value C_∞ is proportional to n/N (or L/l), the number of real skeletal bonds per segment of the freely jointed chain (or the ratio of the lengths of the step length in the freely jointed chain and the real chemical bond). For chains with large values of C_∞, the step length required in the freely jointed chain model may be several multiples of the length of the chemical bond.

C. The Freely Rotating Chain

The *freely rotating chain* is identical to the freely jointed chain (Section I.2.B) in all respects except that the angle between two adjoining bonds is held fixed at a predetermined value. The torsion angles of the bonds can take on all values with equal probability. It is the result of an attempt to introduce the reality of chemical structure into the ideal chain description: The bond angles of real chains are nearly always confined to certain values within narrow limits. Note that the model does not require that rotation is easy or fast, only that all values of the torsion angles of each bond occur with equal frequency. If the bond angles are all taken to be the same, the characteristic ratio is[20]

$$C_n = \frac{1 - \cos\theta}{1 + \cos\theta} + \frac{2\cos\theta[1 - (-\cos\theta)^n]}{n(1 + \cos\theta)^2} \tag{I-19}$$

$$C_\infty = \frac{1 - \cos\theta}{1 + \cos\theta} \tag{I-20}$$

(Alternative expressions for C_n and C_∞, in terms of the complement of the bond angle, are the subject of Problem I-6.) The corresponding expression for the radius of gyration is more complex than the expression for C_n, but Eq. (I-7) holds, as for all chains. For many simple chains such as polyethylene,

---

[20] C_n is undefined when $\theta = 180°$ because $\langle r^2 \rangle_0$ is then equal to $N^2 L^2$, which causes $\langle r^2 \rangle_0 / NL^2$ to increase without limit as $N \to \infty$. Free rotation about bonds has no influence on $\langle r^2 \rangle_0$ when $\theta = 180°$.

assumption of one value for all skeletal bond angles is often approximately correct, and the angle is usually approximately tetrahedral. Since the angles between real bonds are almost always greater than 90°, the value of C_∞ is nearly always greater than 1. The fact that the bond angles are fixed has clearly an appreciable effect on the unperturbed dimensions of the macromolecule.

A characteristic measure for the dimensions of a long, real chain, often used in the older literature, is the *steric factor*, the ratio of the root mean square end-to-end distance of a real polymer chain to that of a freely rotating chain with the same skeletal structure, $\sigma^2 = \langle r^2 \rangle_0^{\text{real}} / \langle r^2 \rangle_0^{f.\text{rot.}}$. For a chain with approximately uniform bond angles,

$$\sigma = \left[\frac{C_\infty^{\text{real}}(1 + \cos \theta)}{1 - \cos \theta} \right]^{1/2} \tag{I-21}$$

a quantity that commonly assumes values between 1.5 and 2.5.

D. The Worm-like, Porod–Kratky, or Continuously Curved Chain[21]

This chain model can be viewed as a limiting case of the freely rotating chain (Section I.2.C). As a preliminary exercise, it is advantageous to introduce first the *persistence length a*.[22] Picture a freely rotating chain with N bonds of length L. The average projection of the end-to-end vector on the direction of the first bond $(\mathbf{L}_1)$ is

$$\frac{\langle \mathbf{r}^T \mathbf{L}_1 \rangle}{L} = L \left[\frac{1 - (-\cos \theta)^N}{1 + \cos \theta} \right] \tag{I-22}$$

The limiting value for an infinitely long chain $(NL \to \infty)$ is called the *persistence length* (this definition is *model-independent*!)

$$a \equiv \lim_{NL \to \infty} \frac{\langle \mathbf{r}^T \mathbf{L}_1 \rangle}{L} = \frac{L}{1 + \cos \theta} \tag{I-23}$$

The *worm-like chain* is obtained when the bond length of the finite length chain is reduced to zero while at the same time the number of segments is increased without limit $(N \to \infty)$ in order to keep the *contour length* constant, and the bond angle is opened to let $1 + \cos \theta$ vanish so that the persistence length retains a finite value. This procedure affords a chain of continuous cur-

---

[21]Porod, G. *Monatsh. Chem.* **1949**, *80*, 251. Kratky, O.; Porod, G. *Rec. Trav. Chim.* **1949**, *68*, 1106. Harris, R. A.; Hearst, J. E. *J. Chem. Phys.* **1966**, *44*, 2595.
[22]The persistence length is often identified in the literature with the term $1/2\lambda$; the two representations are identical.

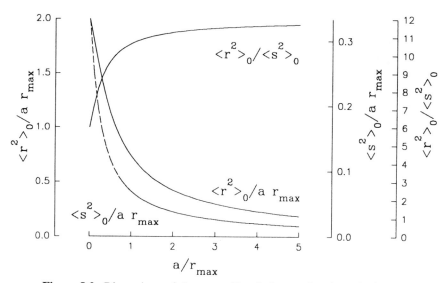

Figure I-2. Dimensions of the worm-like chain as a function of a/r_{max}.

vature, where the direction of curvature at any point is random. The resulting chain is described by two parameters only, the stiffness parameter a and the contour length r_{max}.

$$\langle r^2 \rangle_0 = 2ar_{max}\left\{1 - \frac{a}{r_{max}}\left[1 - \exp\left(\frac{-r_{max}}{a}\right)\right]\right\} \qquad (\text{I-24})$$

and[23]

$$\langle s^2 \rangle_0 = \frac{ar_{max}}{3}\left\{1 - \frac{3a}{r_{max}}\left[1 - \frac{2a}{r_{max}} + 2\left(\frac{a}{r_{max}}\right)^2\right.\right.$$
$$\left.\left. \cdot \left(1 - \exp\left(\frac{-r_{max}}{a}\right)\right)\right]\right\} \qquad (\text{I-25})$$

Figure I-2 depicts the dimensionless ratios $\langle r^2 \rangle_0/ar_{max}$ and $\langle s^2 \rangle_0/ar_{max}$. The dimensionless ratios approach a limit as $a/r_{max} \to 0$. This limit corresponds to a long *flexible polymer chain*. The limits yield

[23]Benoit, H.; Doty, P. *J. Phys. Chem.* **1953**, *57*, 958.

$$\frac{\langle r^2 \rangle_0}{r_{max}} = 2a\left(1 - \frac{a}{r_{max}}\right) \approx 2a \tag{I-26}$$

and

$$\frac{\langle s^2 \rangle_0}{r_{max}} = \frac{a}{3}\left(1 - \frac{3a}{r_{max}}\right) \approx \frac{a}{3} \tag{I-27}$$

and the relationships depicted in Table I-1.

For a truly *stiff* macromolecule (where r_{max}/a is small) $\langle r^2 \rangle_0$ grows with r_{max}^2. This behavior is what one would expect for a rigid rod. The rigid rod is not addressed in this book, but an interesting property of the worm-like chain model is its ability to describe simple chains of very different local character, from the very flexible to the very stiff ones, with only one structural parameter.

The contour length in the model is the geometric length of the continuous space curve that represents the macromolecule. The term r_{max} for it is chosen here to make comparison with real chains easy. If the characteristic ratio needs to be evaluated, for instance, for the purpose of comparison with other chain models, the value of r_{max} must be established. The same procedure as above [see text after Eq. (I-17)] may be applied.

E. The Simple Chain with Symmetric Hindered Rotation[24]

Real macromolecules almost always exhibit values for C_∞ significantly larger than a freely rotating chain with the same bond angles. This result is obtained because bonds in the real chain cannot assume all values of their torsion angles with equal probability, but instead have a preference for some bond conformations over others. Such preferences are well known for low molecular weight compounds,[25] and it is only natural to assume that they are also manifest in macromolecules.

A model that leads to simple results is based on a chain with rigid bonds, all of equal length, and with constant bond angles, all of equal magnitude. The conformations of the bonds do not occur with equal probability but are determined by a torsion angle-dependent potential energy function $E(\phi)$ that is *symmetric*[26] *with respect to* $\phi = 0$ (and, of course, with respect to $\phi = \pi$).[27] Thus $E(\phi) = E(-\phi)$, for all ϕ. At equilibrium, the bond conformations are distributed

---

[24]Oka, S. *Phys. Math. Soc. Japan* **1942**, *24*, 657. Kuhn, H. *J. Chem. Phys.* **1947**, *15*, 843. Taylor, W. J. *J. Chem. Phys.* **1947**, *15*, 412, *J. Chem. Phys.* **1948**, *16*, 257. Benoit, H. *J. Chem. Phys.* **1947**, *44*, 18.

[25]Hindered rotation is exhibited in molecules as simple as hydrogen peroxide, H_2O_2.

[26]Often we will use the word *symmetric* to define a torsion potential energy function that causes each conformation of the chain to have the same probability as its mirror image. A symmetric torsion potential energy function has $\langle \sin \phi \rangle = 0$.

[27]We use the convention where $\phi = 0$ for a *cis* (synperiplanar) placement.

according to a *Boltzmann distribution*, in which the frequency of occurrence is $\sim \exp(-E(\phi)/kT)$, where k is the *Boltzmann constant* and T the *absolute temperature*. The characteristic ratio of the chain is determined by the (temperature dependent!) average value of the cosine of the torsion angle. This average can be readily obtained from the assumed potential energy function by[28]

$$\langle \cos \phi \rangle = \frac{\int_{-\pi}^{\pi} \cos \phi \exp[-E(\phi)/kT]d\phi}{\int_{-\pi}^{\pi} \exp[-E(\phi)/kT]d\phi} \tag{I-28}$$

The characteristic ratio of the infinitely long chain can be expressed by

$$C_\infty = \left(\frac{1 - \cos \theta}{1 + \cos \theta} \right) \left(\frac{1 - \langle \cos \phi \rangle}{1 + \langle \cos \phi \rangle} \right) \tag{I-29}$$

The presence of T in Eq. (I-28) implies $\langle \cos \phi \rangle$ may be temperature dependent, and hence the value of C_∞ evaluated from Eq. (I-29) may also depend on temperature. The temperature dependence of C_∞ is usually presented as $\partial \ln C_\infty / \partial T$, which is numerically equal to $(\partial \ln \langle r^2 \rangle_0 / \partial T)_\infty$. The unit is $\deg^{-1}$ (reciprocal degree), and a typical order of magnitude for many flexible chains is $\sim \pm 10^{-3}$. For the simple chain with symmetric hindered rotation

$$\frac{\partial \ln C_\infty}{\partial T} = -\left(\frac{2}{(1 - \langle \cos \phi \rangle)^2} \right) \frac{\partial \langle \cos \phi \rangle}{\partial T} \tag{I-30}$$

If all torsion angles are equally likely, that is, either $E(\phi)$ is independent of ϕ or $T \to \infty$, $\langle \cos \phi \rangle = 0$ and Eq. (I-29) becomes identical to the equation for C_∞ of a freely rotating chain, Eq. (I-20). If ϕ shows a preference for values close to zero (synperiplanar conformation), $\langle \cos \phi \rangle$ is positive and C_∞ is smaller than it would be for the otherwise identical freely rotating chain. Conversely, if the preferred values of ϕ are close to π (antiperiplanar conformation), $\langle \cos \phi \rangle$ is negative and C_∞ larger than it would be for the corresponding freely rotating chain. Finally, it is conceivable that $E(\phi)$ might be variable, but with a form that will produce $\langle \cos \phi \rangle = 0$ at finite T.[29] Then the chain will have the C_∞ expected for a freely rotating chain, even though rotation is hindered.

If all pairs of consecutive bonds in a real chain have a single value for θ, a value of $\langle \cos \phi \rangle$ can be found that will provide agreement with a measured C_∞ and the prediction from Eq. (I-29). However, if the real chain does *not* have

---

[28]Given the assumed symmetry of the torsion potential energy function, the upper or lower limits for both integrals could be changed to 0 without affecting the result.
[29]For example, the symmetric torsion potential-energy function will have $\langle \cos \phi \rangle = 0$ if $E(\phi) = E(\pi - \phi)$.

independent bonds that are subject to a symmetric torsion potential-energy function, the model that successfully explains C_∞ via Eq. (I-29) may yield nonsense when applied to $\partial \ln C_\infty / \partial T$ via Eq. (I-30). This behavior is obtained for a chain as simple as polyethylene, as shown in Problem I-12. In the case of polyethylene, the failure occurs because the bonds are *not* independent.

3. MODELS THAT TAKE INTO ACCOUNT THE EFFECTS OF EXCLUDED VOLUME[30]

For ideal chain models many well developed and well-analyzed variants have been created (five of these variants have been introduced above), and it would seem natural that a similar statement could be made about models that do not assume perfect compensation of the effects of excluded volume. Unfortunately, this assumption is not true. The difficulties associated with the introduction of the excluded volume are so formidable that despite the expense of extraordinary effort only relatively few quantitative statements can be made.

While many of the results from different theoretical endeavors are model-dependent, there are nevertheless observations that persist across models. These apparently model-independent features are probably correct, and some of them are reproduced here, partially as an introduction for later chapters that treat the application of rotational isomeric state methods to nonideal chains.

The effects of excluded volume in nonideal polymer solutions are most readily observed in the chain dimensions. Under the influence of excluded volume interactions the chain expands or contracts from the dimensions it assumes at a Θ state. A measure of the change in linear dimensions is the *expansion factor* α

$$\alpha^2 = \frac{\langle r^2 \rangle}{\langle r^2 \rangle_0} \tag{I-31}$$

The expansion factor depends on the chain length and the state of the polymer system, specifically, the interactions of chain segments with each other and the solvent. For long chains, $\langle r^2 \rangle_0$ is proportional to n, as shown in Eqs. (I-5) and (I-11). The chain length dependence of $\langle r^2 \rangle$ is not known with generality, but for long chains it is possible to show that a simple power law must be observed:

---

[30]More detailed treatments of these matters as well as proofs of the relationships given here can be found in Yamakawa, H. *Modern Theory of Polymer Solutions*, Harper & Row, New York, **1971**. De Gennes, P. G. *Scaling Concepts in Polymer Physics*, Cornell University Press, Ithaca, NY, **1979**. Doi, M.; Edwards, S. F. *The Theory of Polymer Dynamics*, Clarendon Press, Oxford, **1986**. Freed, K. F. *Renormalization Group Theory of Macromolecules*, Wiley-Interscience, New York, **1987**.

$$\langle r^2 \rangle \propto n^{2\nu} \qquad \text{and, therefore} \qquad \alpha \propto n^{(2\nu-1)/2} \qquad \text{(I-32)}$$

Obviously, the value of ν is important. For flexible chains under Θ conditions, ν is $1/2$ [see Eqs. (I-5) and (I-11)]. It can be significantly lower (a completely collapsed chain would be characterized by $\nu \to 1/3$) or higher (for a very "stiff" chain, the proportionality between $\langle r^2 \rangle$ and L^2, under all conditions, would require a value of $\nu \to 1$). For very good solvents one finds that[31]

$$\alpha^5 - \alpha^3 \approx 2.60z \qquad \text{(I-33)}$$

where z is a dimensionless parameter proportional to $n^{1/2}$ and to the "strength" of the excluded volume interaction.[32] This result leads to

$$\nu = \tfrac{3}{5} \qquad \text{and} \qquad \alpha \propto n^{1/10} \qquad \text{(I-34)}$$

for large z.

For rather poor solvents, such as for situations not very far from the Θ state and where z is small, the best current information comes from a perturbation calculation that gives[33]

$$\alpha^2 = 1 + \tfrac{3}{4}z - 2.075z^2 + 6.297z^3 - 25.057z^4$$
$$+ 116.135z^5 - 594.717z^6 + \cdots \qquad \text{(I-35)}$$

Even though this series expansion yields coefficients for the terms of low power in z that are quite different from the ones derived from Eq. (I-33) (there the estimate for α^2 at small z is $1 + 2.6z + \cdots$), a similar value for ν for the case of the very good solvent (where z is large) can be estimated[34] from it:

[31] Flory, P. J. *J. Chem. Phys.* **1949**, *17*, 303. Flory, P. J.; Fox, T. G. *J. Am. Chem. Soc.* **1951**, *72*, 1904.

[32] For a model chain with N segments, the centers of which are linked by bonds of fixed length where the segments possess a "segmental excluded volume" β, z is given by

$$z = \left(\frac{3}{2\pi L^2} \right)^{3/2} \beta N^{1/2}$$

The parameter β is defined in terms of an effective pair interaction energy between two polymer segments, $E(x)$ (where x is the distance between the centers of segments i and j), as

$$\beta = \int_0^\infty \{1 - \exp[-E(x)/kT]\} \, dx$$

[33] Muthukumar, M.; Nickel, B. G. *J. Chem. Phys.* **1984**, *80*, 5839.
[34] Le Guillou, J. C.; Zinn-Justin, J. *Phys. Rev. Lett.* **1977**, *39*, 95. Oono, Y. *Adv. Chem. Phys.* **1985**, *61*, 301.

$$\nu = 0.588 \pm 0.001 \tag{I-36}$$

It is probably safe today to assume that the exponent ν for long chains and good solvents is very slightly smaller than 0.6, and that for the situations not far from Θ conditions Eq. (I-35) applies.

PROBLEMS

Appendix B contains answers for all problems in this chapter.

I-1. Which dimensionless ratio experiences the slower change as one proceeds from a freely jointed chain with very large N to a freely jointed chain of smaller N, $\langle r^4 \rangle_0 / \langle r^2 \rangle_0^2$ or $\langle s^4 \rangle_0 / \langle s^2 \rangle_0^2$?

I-2. Which distribution function is broader for an infinitely long unperturbed chain, the distribution for r^2 or for s^2?

I-3. For freely jointed chains, are the distribution functions for r^2 and s^2 broader or narrower at small N than at large N? How can this result be rationalized?

I-4. The correlation coefficient for two variables, x and y, is defined as

$$\rho = \frac{\dfrac{\langle xy \rangle}{\langle x \rangle \langle y \rangle} - 1}{\left[\left(\dfrac{\langle x^2 \rangle}{\langle x \rangle^2} - 1 \right) \left(\dfrac{\langle y^2 \rangle}{\langle y \rangle^2} - 1 \right) \right]^{1/2}} \tag{I-37}$$

where the range is $-1 \le \rho \le 1$. Using relationships for the freely jointed chain, show that $\rho \to (5/8)^{1/2}$ when $x = r^2$, $y = s^2$, and $N \to \infty$. How should this result be interpreted? (Consider "typical" conformations, as well as "atypical" ones such as rings with $\mathbf{r} = \mathbf{0}$ and conformations where $r \to r_{max}$.)

I-5. Given $C_\infty = 6.7$ for polyethylene at $140°C$,[35] find N and L for the equivalent chain for a polyethylene of $M = 10^6$. The length of the C—C bond is 153 pm, and θ_{CCC} is $112°$. Also estimate the persistence length.

I-6. If the complement of the bond angle, $\theta_c = \pi - \theta$, is used instead of the bond angle itself, what are the simplest forms of the expressions for C_n and C_∞ for a freely rotating chain?

---

[35] Chiang, R. *J. Phys. Chem.* **1965**, *69*, 1645.

I-7. Bonds in several small molecules have a symmetric torsion potential energy function that can be written as

$$E(\phi) = \left(\frac{\Delta E}{2} \right) (1 + \cos 3\phi) \tag{I-38}$$

The torsion potential energy function for the C—C bond in ethane can be described by this expression with $\Delta E \approx 12$ kJ mol^{-1}. Is $\langle \cos \phi \rangle$ positive or negative? How does it change with temperature? What are the implications for C_∞, and the sign of its temperature coefficient, if a long chain could be constructed from bonds subject only to this torsion potential energy function?

I-8. How might the qualitative answers to the questions posed in Problem I-7 be modified if the effective potential energy function were changed so that the minimum at $\phi = 180°$ were lower in energy than the others?

I-9. Find the second virial coefficient, A_2,

$$\frac{PV}{nRT} = 1 + A_2 \left(\frac{n}{V} \right) + A_3 \left(\frac{n}{V} \right)^2 + \cdots \tag{I-39}$$

when the equation of state is

$$\frac{PV}{nRT} = \frac{V}{V - nb} - \frac{an}{VRT} \tag{I-40}$$

Then provide an interpretation of the behavior of this gas when $T = a/bR$.

I-10. Design a chain with $C_\infty < 1$ and $(\partial \ln \langle r^2 \rangle_0 / \partial T)_\infty < 0$ at 300 K, subject to the following constraints.

(a) All bond lengths are identical and independent of T.

(b) All bond angles are identical and independent of T.

(c) Bonds are independent, with a symmetric torsion potential energy function.

I-11. Find C_∞ and $(\partial \ln \langle r^2 \rangle_0 / \partial T)_\infty$ at 300 K for the following chains.

Chain	θ, deg	ϕ_i, deg	$(E_{\phi_i},$ kJ mol$^{-1})$	
A	112	90(0)	−90(0)	
B	112	180(0)	−60(0)	60(0)
C	112	0(0)	−120(0)	120(0)
D	112	180(0)	−60(2)	60(2)
E	112	180(2)	−60(0)	60(0)
F	105	180(0)	−60(2)	60(2)
G	120	180(0)	−60(2)	60(2)

I-12. Polyethylene has θ_{CCC} = 112°, C_∞ = 6.7 at 140°C,[35] and $(\partial \ln \langle r^2 \rangle_0 / \partial T)_\infty$ = −1.1 × 10^{-3} deg^{-1}.[36] If the bonds are confined to three states with ϕ = 180° and ±60°, with $E(180°)$ = 0 and $E(60°) = E(-60°) = E_g$, can a single value of E_g be assigned such that Eqs. (I-29) and (I-30) will account for both C_∞ at 140°C and its temperature coefficient?

[36]Ciferri, A.; Hoeve, C. A. J.; Flory, P. J. *J. Am. Chem. Soc.* **1961**, *83*, 1015. Flory, P. J.; Ciferri, A.; Chiang, R. *J. Am. Chem. Soc.* **1961**, *83*, 1023.

II A Brief Review of Common Observables

Rotational isomeric state theory provides the link between the detailed covalent structure of a macromolecule and measurable conformation-dependent physical properties that depend on the structure. This chapter summarizes several of the observables that are susceptible to a rotational isomeric state analysis. Subsequent chapters will develop the rotational isomeric state treatment for these observables. We begin with properties that depend only on the dimensions, through an appropriate manipulation of the bond vectors. Then we turn to other observables that depend on additional properties (polarizability, dipole moment, magnetic moment) of the macromolecule.

1. PROPERTIES THAT DEPEND ON $\langle r^2 \rangle_0$

The peculiar fixation on the mean square end-to-end distance and the related radius of gyration, apparent in the previous chapter and in much of the literature on the conformation of macromolecules, originates in the fact that the most important macromolecular observables are intimately connected to these quantities. The calculation of $\langle r^2 \rangle_0$ and $\langle s^2 \rangle_0$ from the rotational isomeric state model will be described in Chapter VI, after the introduction of the conformational partition function and methods for calculation of the conformations of individual bonds and sequences of bonds.

A. The Proportionality between Chain Length and $\langle r^2 \rangle_0$ from Small Chains to Macromolecules

In the previous chapter it has been stated that the unperturbed mean square end-to-end distance is exactly proportional to chain length for the random-walk chain and the freely jointed chain with bonds of constant length, and that this proportionality is true for all *very long* chains. It was not indicated, however, how rapidly this limit was approached.

One can write[1] the characteristic ratio of a general chain of n bonds as

---

[1] Nagai, K. *J. Chem. Phys.* **1966**, *45*, 838. Flory, P. J. *Statistical Mechanics of Chain Molecules*, Wiley-Interscience, New York, **1969**: reprinted by Hanser, München, **1989**.

$$C_n = C_\infty + \frac{A}{n} + \frac{B}{n^2} + \cdots \tag{II-1}$$

where the terms A and B are dependent on the details of the model involved. Both terms may vary with n, particularly at small values of n. However, it is possible to show that B and the quantities multiplying the terms of even higher negative powers of n [not shown in Eq. (II-1)], become irrelevant for large values of n. As a practical matter, it is usually possible to perform rotational isomeric state calculation of C_n at values of n where Eq. (II-1) can be truncated after the term in n^{-1}, and where A has assumed a constant value. Hence, C_∞ *can be obtained from C_n by linear extrapolation against* $1/n$.

The freely rotating chain shall serve as a simple example. Equation (I-19) gives the exact expression for C_n, and $A = 2\cos\theta[1 - (-\cos\theta)^n]/(1 + \cos\theta)^2$. For large values of n the term $[(1 - (-\cos\theta)^n] \rightarrow 1$ (except for the truly rigid rod, where $\theta = \pi$, a case which is not considered here) and $A \rightarrow 2\cos\theta/(1 + \cos\theta)^2$. For tetrahedral bond angles the limit for A has a value of $-3/2$. The infinite chain value is approached very fast, and for $n = 5$, A is already $-1.494\ldots$.

In practice one often performs numeric calculations for chains of successively doubling degree of polymerization. The postulated linear dependence between C_n and $1/n$ makes very simple extrapolations possible. For example, two contiguous entries of a list of values for C_n of successively double chain length, C_n and C_{2n}, give an estimate $C_\infty \approx 2C_{2n} - C_n$. Experience shows that this extrapolated value is very precise already for chains of only several hundred bonds. Application to a freely rotating chain with a tetrahedral bond angle is presented in Table II-1. This method for calculation of C_∞ for rotational isomeric state chains is used in the FORTRAN program listed in Appendix C.

TABLE II-1. Estimation of C_∞ from C_n for a Freely Rotating Chain with a Tetrahedral Bond Angle

n	C_n	$2C_{2n} - C_n$
1	1	1.66667
2	1.33333	1.92593
4	1.62963	1.99543
8	1.81253	1.99997
16	1.90625	2.00000
32	1.95312	2.00000
64	1.97656	
$\vdots$		
∞	2	2

B. The Distribution of r^2 and Its Impact on Observables

The properties that are largely dependent on the distribution of the end-to-end distance, or, more generally, on the distribution of mass in the domain of a macromolecule (and hence on s^2 also) reflect the configuration and conformation of the entire chain.

Dilute Solution Viscosity. The viscosity of dilute polymer solutions is commonly explained on the basis of a very simple concept. For the purpose of hydrodynamic modeling one represents the isolated coiled macromolecule by a *hydrodynamically equivalent sphere*, that is, a hypothetical sphere, impenetrable to the solvent, displaying in a hydrodynamic field the same frictional effects as the actual chain. The radius of that sphere is usually measured in terms of the radius of gyration of the real macromolecule

$$R_E = \gamma \langle s^2 \rangle^{1/2} \approx \left(\frac{\gamma}{\sqrt{6}} \right) \langle r^2 \rangle^{1/2} \qquad \text{(II-2)}$$

where the relationship with $\langle r^2 \rangle$ holds only under strictly Θ conditions and for very long chains. One would expect the value of γ, that is, the size of the hydrodynamically equivalent sphere, to be different for different types of motions of the macromolecule in the fluid, such as for diffusion and for viscous flow. It is usually taken to be in the range 0.7–1.0. We will employ the mean value of the range, $\gamma \approx 0.85$.

Einstein's law for the viscosity of a dilute suspension of solid spheres is[2]

$$\frac{\eta - \eta_s}{\phi \eta_s} = \frac{5}{2} \qquad \text{(II-3)}$$

where η represents the zero-shear viscosity of the suspension, η_s the zero-shear viscosity of the pure solvent, and ϕ the volume fraction of spheres. Taking the equivalent sphere for a chain, one substitutes $\phi \approx c V_E \mathscr{L} / M$, with c being the mass concentration of spheres, V_E the equivalent sphere's volume ($4 \pi R_E^3 / 3$), and $\mathscr{L}$ Avogadro's number. Further, one assumes that Eq. (I-7) holds, that is, that the approximate relationship in Eq. (II-2) holds. This assumption is valid at the Θ state for very long chains. These assumptions lead to

$$\frac{\eta - \eta_s}{c \eta_s} = \left(\frac{10 \pi \mathscr{L} \gamma^3}{3 \cdot 6^{3/2}} \right) \frac{\langle r^2 \rangle_0^{3/2}}{M} \qquad \text{(II-4)}$$

---

[2]Einstein, A. *Ann. Physik.* **1906**, *19*, 289; *Ann. Physik.* **1911**, *34*, 591.

It is customary to replace the factor in parentheses by the so-called viscosity function, Φ.[3] Since we know from Eqs. (I-5) and (I-18) that $\langle r^2 \rangle_0$ is proportional to M, it is convenient to write the resulting formula as

$$\frac{\eta - \eta_s}{c\eta_s} = \Phi \left(\frac{\langle r^2 \rangle_0}{M} \right)^{3/2} M^{1/2} = KM^{1/2} \qquad \text{(II-5)}$$

The corresponding experimental quantity must, in accordance with the conditions valid for Einstein's model, be determined at high dilution and zero shear. It is termed the "intrinsic viscosity" and is defined by

$$[\eta] = \lim_{c \to 0} \left(\frac{\eta - \eta_s}{c\eta_s} \right) \qquad \text{(II-6)}$$

and should be measured at (or corrected to) Θ conditions. Then

$$[\eta]_\Theta = K_\Theta M^{1/2} = \Phi \left(\frac{\langle r^2 \rangle_0}{M} \right)^{3/2} M^{1/2} \qquad \text{(II-7)}$$

This expression is a special case of the heuristic equation that is known as the Mark–Houwink relationship, $[\eta] = KM^a$. At the Θ state it assumes the form of Eq. (II-7).

If the intrinsic viscosity of a polymer with uniform chains is experimentally known in a Θ state, knowledge of the molar mass of the chain can yield

$$\langle r^2 \rangle_0 = \left(\frac{K_\Theta}{\Phi} \right)^{2/3} M \qquad \text{(II-8)}$$

In reality there is experimental error, of course, and the polymer samples are not uniform with respect to molecular weight, and special procedures are required to correct reliably for the imperfections of reality,[4] but dilute-solution viscosity is a convenient and trustworthy source of information on $\langle r^2 \rangle_0$ [or $\langle s^2 \rangle_0$].

Parenthetically we want to note that it is not necessarily appropriate to assume spherical symmetry of the polymer chains's shape. In a frame of reference external to the chain, as in laboratory coordinates, the distribution of any measure of the average shape and size of the macromolecule will be spherically

---

[3]Currently available experimental data indicate a value of $\gamma \approx 0.85$, and hence $\Phi \approx 2.6 \times 10^{23}$ mol^{-1}, at least for polymers of reasonably narrow molecular weight distribution.

[4]Bareiss, R. E. in *Polymer Handbook*, 3rd ed., Brandrup, J.; Immergut, E. H., eds., Wiley-Interscience, New York, **1989**.

symmetric and located on the center of gravity of the chain, as a result of orientational averaging. If the chain is observed in an "internal" frame of reference, for example, the principal-axis system of the instantaneous radius of gyration tensor, and care is taken that orientational averaging is avoided, the shape of the coil resembles a "cake of soap"[5] much more than a sphere, and the deviation from spherical shape may be enhanced with short chains.[6] Close inspection even reveals that the density has two equal maxima rather than one maximum at the center of mass.[7] Nevertheless, identification of the average chain conformation's physical habit with a simple hard sphere of radius $\gamma \langle s^2 \rangle_0^{1/2}$ works remarkably well in practice.

Local Concentration in Dilute Solution. In a polymer solution at high concentrations the domains that contain the coils overlap strongly and the segment concentration is uniform throughout the space of the solution. Changing the macroscopic concentration will cause a proportional change in local segment density. If the concentration is successively lowered, one reaches a "crossover concentration," c^*, at which the sum of the volumes of the domains occupied by the solute molecules is approximately equal to the total volume of the solution. Concentration c^* marks the upper limit of the dilute-solution range. At concentrations significantly below c^* the coils are essentially isolated, and the segment concentration is no longer uniform. Because of the intramolecular connectivities, the chain segments are clustered and a local segment density, independent of the macroscopic concentration, is established. The actual concentration at some point in space (inside the domain of the macromolecule selected for analysis) varies because of reorientation of the entire chain and also conformational fluctuations. On the average, the segment density is largest in the center of the coil (at the center of gravity) and decreases with increasing distance. Equation (I-10) gives a reasonable approximation to the distribution of mass at the Θ state. We are interested here in a simple estimate of the average concentration in the coil's domain.

We approximate the real macromolecule with an equivalent chain, specifically, with a freely jointed chain of N bonds ($N + 1$ atoms) of length L, and identify the bonds with "chain segments." The mean local segment concentration is estimated, adopting the equivalent sphere concept, by assuming that the macromolecule resides entirely inside the boundaries of the equivalent sphere. This approach yields an average concentration of segments—close to the core of the real coil the segment concentration is higher than the one modeled here, and outside it is lower. The average local concentration of segments inside a chain's domain, c_E, is $c_E = N/V_E$. On replacing V_E with $4\pi R_E^3/3$, R_E with $\gamma(\langle r^2 \rangle_0/6)^{1/2}$, and $\langle r^2 \rangle_0$ with NL^2, we obtain

---

[5]Solc, K.; Stockmayer, W. H. *J. Chem. Phys.* **1971**, *54*, 2756.
[6]Mattice, W. L. *Macromolecules* **1980**, *13*, 506.
[7]Theodorou, D. N.; Suter, U. W. *Macromolecules* **1985**, *18*, 1206.

TABLE II-2. Average Segment Concentrations for Polystyrene Under Θ Conditions

M	N	Segment Conc. (mol liter^{-1})	Styryl Conc. (mol liter^{-1})	%v/v
7×10^4	100	0.16	1.14	≈ 10
7×10^5	1,000	0.051	0.36	≈ 4
7×10^6	10,000	0.016	0.11	≈ 1

$$c_E = \frac{N}{V_E} = \frac{3 \cdot 6^{3/2}}{4\pi\gamma^3 L^3 N^{1/2}} \tag{II-9}$$

The segment (or bond) length L is a measure for the real chain's geometric persistence (or "stiffness"). According to Eq. (II-9), the segment density decreases rapidly with increasing stiffness of the chain, and also decreases with increasing chain length. Typical values for a flexible polymer chain can be obtained for atactic polystyrene, where $L \approx 18$ Å and there are ~7 styryl units per segment. Representative results with $\gamma \approx 0.85$ are listed in Table II-2.

An average concentration of monomeric units of a few percent, as found here for polystyrene, is typical for many polymers. It is remarkably high, considering that this is the *lowest* possible concentration inside the coiled chain that can be achieved no matter what macroscopic dilution is employed. Obviously, the common method of strongly diluting to obtain situations where simple "ideal-solution laws" are valid cannot yield results in polymer systems comparable to those obtained for mixtures of compounds of low molecular mass only. On the other hand, if one estimates the fraction of segment–segment contacts a chain segment experiences on the average (of all contacts, with segments and solvent) in these Θ-state coils, the number is remarkably low. The frequency of such contacts can be estimated as the square of the volume-fraction concentrations in Table II-2 and is, for the monomeric units (styryl units) in the three cases listed there, about 1%, about 0.2%, and 0.01% in order of increasing molecular weight. Clearly, in long flexible chains in very dilute solution, the average number of contacts between monomeric units is low.

Macrocyclization in Condensation Polymers.[8] Step reaction polymerization almost invariably leads to formation of cyclics of various sizes besides linear chains, and a considerable portion of the material of high molecular weight

[8] Jacobson, H.; Stockmayer, W. H. *J. Chem. Phys.* **1950**, *18*, 1600. Flory, P. J. *Principles of Polymer Chemistry*, Cornell University Press, Ithaca, NY, **1953**, p. 622. Flory, P. J. *Statistical Mechanics of Chain Molecules*, Interscience, New York, **1969**: reprinted by Hanser, München, **1989**. Semlyen, J. A. *Adv. Polym. Sci.* **1976**, *21*, 41. Semlyen, J. A. in *Cyclic Polymers*, Semlyen, J. A., ed., Elsevier, London, **1986**, Chapter 1, as well as chapters by other authors in Semlyen's book.

material may consist of rings. One distinguishes situations where the formation of cyclics is kinetically controlled and those where ring–chain equilibria are established. The equilibrium case will be addressed in detail in Chapter XIII. Here we will focus on the central fact common to both situations, kinetically and thermodynamically controlled ring formation, that the probability distribution for the chain end of the linear macromolecule determines the probability of ring closure.

Ring formation had baffled researchers until, in 1950, Jacobson and Stockmayer showed that for the equilibrium case the absolute concentration of each cyclic species is not dependent on the total concentration of polymeric material in the system (this result implies that a certain concentration exists for each macrocyclization equilibrium system below which essentially only cyclics are present), but rather on the density of a chain end of the linear chain at the locus of the other end of the same chain, $W(\mathbf{r} = \mathbf{0})[W(\mathbf{0})\,d\mathbf{r}$ is the probability for the two chains ends to approach each other to within a displacement difference $d\mathbf{r}]$. If the distribution of the end-to-end vector can be assumed to be Gaussian, that is, to follow Eq. (I-9), then

$$W(\mathbf{0}) = \left(\frac{3}{2\pi \langle r^2 \rangle_0} \right)^{3/2} \qquad \text{(II-10)}$$

and since $\langle r^2 \rangle_0$ is proportional to the chain length, $W(\mathbf{0})$ is an easily determined quantity. Just as the Gaussian chain is an unrealistic detailed model of a real macromolecule and can yield functional forms only for the limiting case $N \to \infty$ [there is still $\langle r^2 \rangle_0$ to be determined before anything about $W(\mathbf{r})$ can be said], so is the distribution in Eq. (I-9) unrealistic in detail, and Eq. (II-10) will be valid only for very long chains.

Detailed analysis of the ring-forming system in a Θ state unequivocally gives the chain length dependence of the concentrations of rings with x monomeric units for the *kinetically* controlled ring closure as

$$[\text{cyclic } x\text{-mer}] \sim x^{-3/2}, \qquad x \to \infty \qquad \text{(II-11)}$$

The absolute values of the ring yields depend also on the details of the chemical reaction that leads to rings. The *equilibrium* concentrations of the cyclic x-mer, on the other hand, can be estimated quantitatively from the statistical properties of the x-meric chain. If the distribution of the end-to-end vector is the same in the preceding considerations, specifically, Gaussian, the chain length dependence of the concentrations of rings with x monomeric units for the thermodynamically controlled ring closure is

$$[\text{cyclic } x\text{-mer}] \sim x^{-5/2}, \qquad x \to \infty \qquad \text{(II-12)}$$

The higher negative power of x, compared to the kinetically controlled case, is due to the fact that the rate of the ring-opening reaction increases with x also.

For very large rings, the ring–chain equilibrium is fully understood once $\langle r^2 \rangle_0$ is known. For smaller rings additional considerations, susceptible to evaluation from a structurally detailed model, are necessary (see Chapter XIII).

Scattering of Radiation. Scattering of radiation is the most versatile class of existing experiments. Depending on the design of the experiment, a wide variety of phenomena can be observed. In all of the experiments, a beam of radiation traverses a medium and may be attenuated and partially scattered. Here, a small subset of experimental setups are considered, namely, those cases in which the attenuation of the incident beam is due only to scattering, the energy of the scattered quanta is the same as that of quanta in the primary beam, and phase relationships between independent scatterers are retained, that is, we focus on elastic coherent scattering only. Furthermore, only situations where the beam of radiation can be viewed as a plane wave and where the sample is macroscopically isotropic shall be described. The scattered particles are typically photons, electrons, or neutrons.

Why Does a Medium Scatter Radiation? Perfect continua, where the "scattering power" is equal everywhere, do not scatter radiation—a contrast in scattering power, occurring on a length scale commensurate with the wavelength of the incident radiation, is necessary. This contrast can be caused by density fluctuations (due to local differences in temperature or pressure) or composition fluctuations. Here, the differences in local composition on a molecular scale are of interest. For *small, infinitely dilute* scatterers, let the scattered fraction of the radiation impinging on a molecular entity be called the scattering factor f (f_s for the solvent, f_x for the monomeric unit, and f_{pol} for the entire macromolecule). The amplitude of the scattered radiation is then proportional to the contrast factor

$$\Delta f_{pol} \equiv f_{pol} - \left(\frac{\overline{V}_{pol}}{\overline{V}_s} \right) f_s \tag{II-13}$$

where $\overline{V}_{pol}$ is the partial molar volume of the polymer and $\overline{V}_s$ that of the solvent. Since the intensity of radiation is proportional to the square of the amplitude, the scattered intensity is proportional to $(\Delta f_{pol})^2$. If one assumes that the scattering power of a polymer is just that of its monomeric units times the degree of polymerization, and the same relationship holds also for the partial molar volume, one obtains for the scattered intensity at some angle θ

$$i_\theta \propto x^2 \left[f_x - \frac{\overline{V_x}}{\overline{V_s}} f_s \right]^2 \qquad \text{(II-14)}$$

The fact that the scattered intensity depends on the square of the degree of polymerization (i.e., on M^2) is the basis of the well-known molecular weight determination by light scattering.[9]

If a Molecule Is So Large That It Acts as a Collection of Several Subscatterers. Two scatterers, i and j, with scattering factors f_i and f_j, *fixed in space*, each scatter radiation independently. The incident waves, impinging on both scatterers in the direction of the unit vector $\mathbf{e}_0$, are exactly in phase, but the scattered waves, arriving at the detector in the direction of the unit vector $\mathbf{e}_\theta$, are not since they have traveled different distances. If the vector from i to j is termed $\mathbf{r}_{ij}$, the difference in path length for the rays arriving on the detector is $\mathbf{r}_{ij} \cdot \mathbf{e}_0 + (-\mathbf{r}_{ij} \cdot \mathbf{e}_\theta)$. The difference in phase of the two rays at the detector must be $(2\pi/\lambda) \cdot$ (path difference). For convenience, a *scattering vector* $\mathbf{Q}$ shall be defined by[10]

$$\mathbf{Q} \equiv \frac{2\pi}{\lambda}(\mathbf{e}_0 - \mathbf{e}_\theta) \qquad \text{hence} \qquad Q = |\mathbf{Q}| = \frac{4\pi}{\lambda} \sin\left(\frac{\theta}{2}\right) \qquad \text{(II-15)}$$

The phase difference at the detector of the two rays scattered by i and j is then $\mathbf{r}_{ij} \cdot \mathbf{Q}$. The phase difference causes the rays to interfere, and the resultant intensity at the detector is[11]

---

[9]It is convenient to express the scattered intensity in terms of the so-called excess Rayleigh ratio ΔR_θ, which is corrected for the effects of the distance d between scatterer and detector, for the size of the scattering volume, V_{sc}, for scattering of the pure solvent, i_s, and for the intensity of the incident radiation, I_0

$$\Delta R_\theta = \frac{(i_\theta - i_{\theta,s})d^2}{I_0 V_{sc} p}$$

where p is a polarization factor (see below). The excess Rayleigh ratio has the dimensions of length and is usually expressed in reciprocal centimeters (cm^{-1}). The equation for the radiation scattering of solutions of an infinitely dilute, small solute is then

$$\frac{\mathscr{K} c}{\Delta R_\theta} = \frac{1}{M_w}$$

where $\mathscr{K}$ is a system dependent constant of dimensions (length^2 mol mass^{-2}). For the most common situations $\mathscr{K}$ and p are given in Table II-3.

[10]Many different symbols have been used for this quantity, most commonly μ, $\mathbf{s}$ and $\mathbf{S}$, $\mathbf{q}$ and $\mathbf{Q}$, $\mathbf{h}$ and $\mathbf{H}$.

[11]We use here the convenient exponential expressions. One could, of course, replace $\exp(\tilde{i}\mathbf{r}_{ij} \cdot \mathbf{Q})$ by $\cos(\mathbf{r}_{ij} \cdot \mathbf{Q}) + \tilde{i}\sin(\mathbf{r}_{ij} \cdot \mathbf{Q})$. ($\tilde{i}$ denotes $\sqrt{-1}$.)

TABLE II-3. $\mathcal{H}$ and p for Common Scattering Experiments

Radiation	p	$\mathcal{H}$	Terms
Light	1 if H $\cos^2\theta$ if V $(1+\cos^2\theta)/2$ if U	$(4\pi^2\tilde{n}_s^2/\mathcal{L}\lambda^4)(\partial\tilde{n}/\partial c)_{c=0}^2$	$\tilde{n}(\tilde{n}_s)$: refractive index of solution (solvent) λ: wavelength Polarization of incident light: H: horizontal, V: vertical U: unpolarized
X-rays	≈ 1 if $\theta \leq 5°$	$\mathcal{L}i_e(\partial\rho_e/\partial c)_{c=0}^2$ $\approx (\mathcal{L}/M_x^2)[z_x - (\bar{V}_x/\bar{V}_s)z_s]^2$	i_e: scattering factor for one electron ρ_e: electron density z_x (z_s): number of electrons per monomeric unit (solvent)
Electrons	≈ 1 if $\theta \leq 5°$	$(\mathcal{L}/M_x^2)[\xi_x - (\bar{V}_x/\bar{V}_s)\xi_s]^2$	ξ_x (ξ_s): scattering factor of a monomeric unit (solvent)
Neutrons	≈ 1 if $\theta \leq 5°$	$(\mathcal{L}/M_x^2)[B_x - (\bar{V}_x/\bar{V}_s)B_s]^2$	B_x (B_s): total coherent scattering amplitude of a monomeric unit (solvent)

$$i_\theta \propto f_i f_j \exp(\tilde{i}\mathbf{r}_{ij}\cdot\mathbf{Q}) \tag{II-16}$$

If $\mathbf{Q} = 0$ or $\mathbf{r}_{ij} = 0$, $\exp(\tilde{i}\mathbf{r}_{ij}\cdot\mathbf{Q})$ assumes its maximum value of unity since interference is constructive. Hence i_0 is the maximum value of i_θ [if $\theta > 0$, $\exp(\tilde{i}\mathbf{r}_{ij}\cdot\mathbf{Q}) < 1$ since interference is necessarily less than constructive and i_θ is less than i_0].

If K scatterers instead of two exist, all *fixed in space*, the additivity of amplitudes of all rays arriving at the detector allows for a straightforward extension of the above relationships. One obtains[12]

$$i_\theta \propto \sum_{i=1}^{K}\sum_{j=1}^{K} f_i f_j \exp(\tilde{i}\mathbf{r}_{ij}\cdot\mathbf{Q}) \tag{II-17}$$

[12] At $\theta = 0$ (i.e., at $\mathbf{Q} = 0$)

$$i_0 \propto \sum_{i=1}^{K}\sum_{j=1}^{K} f_i f_j = \left[\sum_{i=1}^{K} f_i\right]^2 = f_{av}^2 K^2$$

The K scatterers could be identified for instance with monomeric units or with atoms. Since we will only be interested in the angular dependence (i.e., the Q dependence) of the scattered intensity, it is convenient to divide by the "forwardscattered" intensity

$$\frac{i_\theta}{i_0} = \frac{1}{K^2} \sum_{i=1}^{K} \sum_{j=1}^{K} \left(\frac{f_i f_j}{f_{av}^2} \right) \exp\left(\tilde{i} \mathbf{r}_{ij} \cdot \mathbf{Q} \right) \tag{II-18}$$

For our purposes it is sufficient to focus on chains with uniform scatterers where all f_i are equal, and

$$\frac{i_\theta}{i_0} = \frac{1}{K^2} \sum_{i=1}^{K} \sum_{j=1}^{K} \exp\left(\tilde{i} \mathbf{r}_{ij} \cdot \mathbf{Q} \right) \tag{II-19}$$

Since the macromolecule is in reality not fixed in space but is able to *assume all orientations*, the appropriate average of i_θ/i_0 is required. Rotational averaging yields the *Debye scattering function* $P(Q)$ of a rigid arrangement of K identical scattering centers:

$$P(Q) \equiv \left(\frac{i_\theta}{i_0} \right)_{\text{orientation average}} = \frac{1}{K^2} \sum_{i=1}^{K} \sum_{j=1}^{K} \frac{\sin\left(Q r_{ij} \right)}{Q r_{ij}} \tag{II-20}$$

where $P(Q)$ can assume values between 0 and 1. In particular, $P(0) = 1$ and $P(\infty) = 0$. For values to be significantly less than unity, that is, for sizable destructive interference to occur, $Q r_{ij}$ must be of order unity.

For *flexible chains* with K scattering centers the conformational ensemble average must be computed:

$$P(Q) = \frac{1}{K^2} \sum_{i=1}^{K} \sum_{j=1}^{K} \left\langle \frac{\sin\left(Q r_{ij} \right)}{Q r_{ij}} \right\rangle \tag{II-21}$$

This Debye scattering function fully describes the angular dependence of radiation scattered at a flexible macromolecule. It is a readily determinable experimental quantity, since it is free of difficult system-dependent constants and calibration procedures. It can also be computed directly if a model for the conformation of a chain is assumed. More easily applied, however, is the procedure outlined below.

Information Contained in the Debye Scattering Function. Replacing the sine function in Eq. (II-21) with a power series expansion yields

$$P(Q) = 1 - \left(\frac{Q^2}{3!}\right)\frac{1}{K^2}\sum_{i=1}^{K}\sum_{j=1}^{K}\langle r_{ij}^2\rangle$$

$$+ \left(\frac{Q^4}{5!}\right)\frac{1}{K^2}\sum_{i=1}^{K}\sum_{j=1}^{K}\langle r_{ij}^4\rangle - \cdots \qquad \text{(II-22)}$$

Since $\mathbf{r}_{ij} = \mathbf{r}_i - \mathbf{r}_j$, application of Eq. (I-6) leads to

$$P(Q) = 1 - \left(\frac{2}{3!}\right)Q^2\langle s^2\rangle + \cdots \qquad \text{(II-23)}$$

In other words, the initial slope of a plot of $P(Q)$ versus Q^2 is $\langle s^2\rangle/3$. This relationship is clearly a direct and important method of determining the radius of gyration without limitations and, via Eq. (I-7), $\langle r^2\rangle$ for long flexible chains at or near the Θ state. This property is the origin of the angle dependent part of the so-called Zimm plot.[13]

The Debye Scattering Function of Some Simple Chain Models. The scattering function $P(Q)$ can be analytically determined for several simple models. A *homogeneous sphere* of radius R generates

$$P(Q)_{\text{sphere}} = \frac{9}{Q^6 R^6}[\sin(QR) - QR\cos(QR)]^2 \qquad \text{(II-24)}$$

with $\langle s^2\rangle = 3R^2/5$, hence $Q^2\langle s^2\rangle = 3Q^2R^2/5$. The possibly most extremely different shape, a *long thin rod* of length L is associated with

$$P(Q)_{\text{thin rod}} = \frac{2}{QL}\int_{0}^{QL}\frac{\sin x}{x}dx - \left[\frac{\sin(\frac{1}{2}QL)}{\frac{1}{2}QL}\right]^2 \qquad \text{(II-25)}$$

---

[13]In the procedure commonly termed "Zimm plot," due to Bruno H. Zimm, the corrected and normalized scattering intensity ΔR_θ is plotted according to

$$\left(\frac{\mathcal{K}c}{\Delta R_\theta}\right) = \frac{1}{M_w P(Q)} + 2A_2 c - \cdots$$

where A_2 is the second virial coefficient of the chemical potential of the solvent (a measure of the interactions between polymer and solvent). The method consists of a double extrapolation, $Q \to 0$ and $c \to 0$, and yields *at once* M_w, A_2, and $\langle s^2\rangle$.

with $\langle s^2 \rangle = L^2/12$, hence $Q^2 \langle s^2 \rangle = Q^2 L^2/12$. For long, flexible, ideal chains, specifically, for chains where $W(\mathbf{r})$ satisfies Eq. (I-9), the terms in Eq. (II-22) of the form $\langle r_{ij}^{2p} \rangle$ can be computed exactly. Some manipulation yields

$$P(Q)_{\text{ideal coil}} = \frac{2}{v^2} \left[v - 1 + \exp(-v) \right], \qquad v = Q^2 \langle s^2 \rangle_0 \qquad \text{(II-26)}$$

The scattering functions for homogeneous spheres, long thin rods, and ideal flexible coils are plotted against Q^2 in Fig. II-1.

A convenient way of displaying the angular dependence of radiation scattering for large values of Q is the *Kratky plot*. One plots $Q^2 P(Q)$ against Q

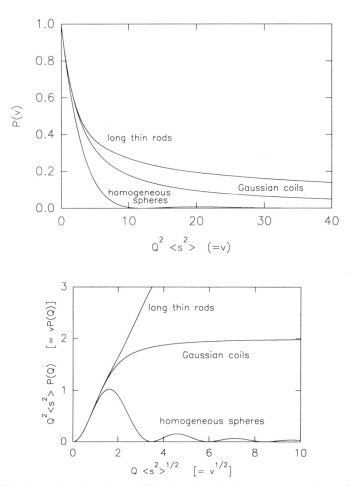

Figure II-1. Scattering functions and Kratky plots for homogeneous spheres, long thin rods, and ideal flexible coils.

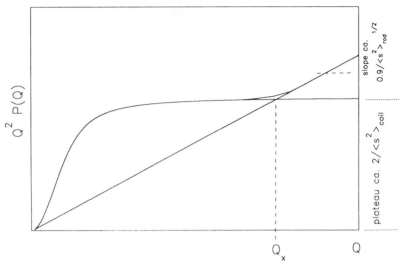

Figure II-2. Kratky plot for a flexible chain, illustrating the cross-over between coil and rod behavior.

(see Fig. II-1) to obtain a line that departs from the origin with an approximate Q^2 dependence and then approaches a straight line with slope $\approx 0.9Q/\langle s^2 \rangle^{1/2}$. Different models for the conformation then show different characteristics. The line for the homogeneous sphere oscillates with consecutively smaller amplitude, the first zero being at $Q \approx 4/\langle s^2 \rangle^{1/2}$. The long thin rod's curve continues on its straight course with the same slope. The ideal coil levels off to a plateau of height $Q^2 P(Q) \approx 2/\langle s^2 \rangle_0$. Obviously, the Kratky plot gives a very suggestive rendering of the chain's conformational characteristics.

The long polymer chains scatter radiation in agreement with the flexible-coil model. As the length of the scattering vector, Q, increases, however, there must be a point for any real coil where it becomes "stiff" on the decreasing scale of observation ($\approx 2\pi/Q$). At some critical value of Q, at Q_x, the flexible coil curve in the Kratky plot will turn up again to follow a thin rod-like line. It is possible to deduce from the Q value at this point an estimate of the macromolecule's persistence length:[14]

$$a \approx \frac{2.3}{Q_x} \tag{II-27}$$

This result is a "global" measure of conformational characteristics deduced from a local observation, namely, from an experiment that depends largely on a small segment of the macromolecule. Consequently, the reliability of values of a obtained in this way should not be overestimated.

[14]Durchschlag, H.; Kratky, O.; Breitenbach, J. W.; Wolf, B. A. *Monatsh. Chem.* **1970**, *101*, 1462.

2. PROPERTIES THAT DEPEND NOT ONLY ON r^2

Most macromolecular observables are determined by a great variety of physical causes, not only by the distribution of the end-to-end vector. We give here a very brief overview of some of the most commonly encountered ones that yield to the rotational isomeric state method. (The actual treatment in the rotational isomeric state method is postponed, however, to later chapters.)

A. Nuclear Magnetic Resonance Observables: Chemical Shift and Coupling Constants

The techniques collectively termed *nuclear magnetic resonance* (nmr) undoubtedly are among the most powerful in analyzing molecular constitution, configuration, and conformation. Their particular attraction for the purposes of conformational analysis lies in the fact that they provide insight into *local* spatial arrangements and the dynamics of their interchange. Here, attention shall be focused only on the static (average) properties.

Chemical Shift. In a homogeneous magnetic field, atomic nuclei possessing magnetic moments can enter into resonance with external electromagnetic fields. Nuclei of the same kind (i.e., ^1H, ^{13}C, ^{19}F, ...) that reside in different chemical environments are in resonance at different frequencies. This "chemical shift" [expressed in parts per million (ppm) relative to a reference signal and represented by the symbol δ] is sensitive even to subtle differences in the environment. Configurational differences can often be deduced from nmr spectra, and conformation affects the chemical shift in measurable ways.

The mechanisms effective in causing a chemical shift are not well understood. It is simply known that the different (highly populated) conformations of chemical compounds experience different values for δ. The ^{19}F spectra of fluoroalkanes[15] shall serve as an example. A fluorine atom in the central CF_2 group in the linear sequence $-CF_2-CH_2-CF_2-CH_2-CF_2-$ experiences a shift of 11 ppm if the bond drawn with a long dash is *gauche* ($\pm$ synclinal) relative to one in the situation where this bond is *trans* (antiperiplanar). This phenomenon is called a "γ effect." The values for the shift parameters come from model-supported inspection of spectra. Once a set of shift parameters has been determined for a nucleus, all one needs to know to estimate a spectrum is the probability for a bond to be in each possible conformation.

Coupling Constants. An observable of even more local sensitivity than the chemical shift is the "vicinal spin–spin coupling constant" ^3J, usually

[15] Tonelli, A. E.; Schilling, F. C.; Cais, R. E. *Macromolecules*, **1982**, *15*, 849.

between two ^{1}H nuclei. Simple relationships of long standing[16] connect the conformation, that is, the dihedral angle between the coupled nuclei, to the magnitude of the coupling. The coupling between the amide proton and the C^α proton in a peptide group shall serve as an illustration. The coupling constant between the two bolded hydrogen atoms in the structure **H–N—C^α–H**, when the N—C bond assumes a dihedral angle β, is

$$^3J_{HH} = 8.9\cos^2 \beta - 0.9\cos \beta + 0.9\sin^2 \beta \qquad \text{(in hertz)} \qquad \text{(II-28)}$$

Experimentally determined values can be compared to a conformationally averaged value of $8.9\langle\cos^2 \beta\rangle - 0.9\langle\cos \beta\rangle + 0.9\langle\sin^2 \beta\rangle$.

If vicinal coupling constants can be observed and if their dependence on conformation is similarly well established, as for the example just displayed, they are observables of unparalleled utility in conformational investigations.

B. Dipole Moment

The (permanent) charge distribution in neutral molecules can be characterized by (permanent) dipole moments.[17] If a long chain molecule is assumed to consist only of unpolarizable segments, it is possible to obtain the total molecular dipole moment μ of the macromolecule as the sum of all local dipole moments $\mathbf{m}_i$:

$$\mu = \sum_i \mathbf{m}_i \qquad \text{(II-29)}$$

This equation is the analog of defining the end-to-end vector as a sum of all bond vectors. Even though this assumption of field-independent charge distributions is probably not very realistic, the approach is very successful in practice.

The quantity readily available from experiment is the magnitude $\mu = (\mu^T\mu)^{1/2}$, or, in a flexible molecule, the mean-square conformational average $\langle\mu^2\rangle$. It is determined from measurement of the static dielectric constant

---

[16]Karplus, M. *J. Chem. Phys.* **1959**, *30*, 11; *J. Chem. Phys.* **1960**, *33*, 1842; *J. Am. Chem. Soc.* **1963**, *35*, 2870. Bovey, F. A.; Brewster, A. I.; Patel, D. J.; Tonelli, A. E.; Torchia, D. A. *Acct. Chem. Res.* **1972**, *5*, 193.

[17]Actually, the dipole moment characterizes one term in a multipole expansion for the description of the charge distribution. The potential energy of an assembly of charges q_i in an electric field $\mathscr{E}$ is

$$E_{pot} = |\mathscr{E}| \sum_i q_i + \mathscr{E}\cdot \sum_i q_i s_i + \cdots$$

where the first term defines the net charge and the second the *molecular dipole moment* $\mu = \sum_i q_i s_i$, s_i being the vector from the center of charge to charge i.

ϵ and the refractive index of a solution in a nondipolar solvent at different concentrations. Extrapolation to zero solute concentration yields the necessary quantities.[18]

$$\frac{\langle \mu^2 \rangle}{x} = \frac{27kTM_x}{4\pi \mathcal{L} \rho_s} \left[\frac{(\partial \epsilon / \partial w)_{w=0}}{(\epsilon_s + 2)^2} + \frac{(\partial \tilde{n} / \partial w)_{w=0}}{(\tilde{n}_s^2 + 2)^2} \right] \qquad \text{(II-30)}$$

where w is the weight fraction of polymer in the solution. Note that it is not necessary to know the molar mass or its distribution precisely. The dipole moment per monomeric unit is molecular weight dependent only for relatively short chains. It usually has approached its limit already for the macromolecules of interest.

It is often convenient to define a characteristic dipole moment ratio, D, in analogy to the characteristic ratio as defined by Eq. (I-18).

$$D_x = \frac{\langle \mu^2 \rangle_0}{x m_x^2} \qquad \text{(II-31)}$$

where D_x measures the degree of correlation between unit dipole moment vectors in the chain of x monomeric units, just as C_n measures the correlation between n skeletal bonds. With increasing chain length, D_x rapidly converges towards its limiting value, $D_\infty \equiv \lim_{x \to \infty} D_x$. The calculation of $\langle \mu^2 \rangle_0$ from the rotational isomeric state model will be presented in Chapter XIV.

The mean square dipole moment has been a favored measure of unperturbed coil conformation for many years, because of theoretical arguments[19] that lead to the conclusion that $\langle \mu^2 \rangle$ must be insensitive to the "quality of the solvent," that is, the same regardless of whether the system is in a Θ state, for all polymer structures that conform to certain (nor very restrictive) symmetry conditions. Furthermore, Nagai and Ishikawa[20] were able to show that the chain expansion factor, defined by Eq. (I-31), and a similar factor α_μ, $\alpha_\mu^2 \equiv \langle \mu^2 \rangle / \langle \mu^2 \rangle_0$, are related by

$$\lim_{n \to \infty} \frac{\alpha_\mu^2 - 1}{\alpha^2 - 1} = \frac{\langle (\mathbf{r}^T \boldsymbol{\mu})^2 \rangle_0}{\langle r^2 \rangle_0 \langle \mu^2 \rangle_0} \qquad \text{(II-32)}$$

This equation points to an important role for a property of the *unperturbed*

[18] Guggenheim, E. A. *Trans. Faraday Soc.* **1949**, *45*, 714. Smith, J. W. *Trans. Faraday Soc.* **1950**, *46*, 394. Guggenheim, E. A. *Trans. Faraday Soc.* **1951**, *47*, 573. Smith, J. W. *Electric Dipole Moments*, Butterworths, London, **1955**, p. 60.

[19] Marchal, J.; Benoit, H. *J. Chim. Phys. Phys.-Chim. Biol.* **1955**, *52*, 818. *J. Polym. Sci.* **1957**, *23*, 223. Stockmayer, W. H. *Pure Appl. Chem.* **1967**, *15*, 539.

[20] Nagai, K.; Ishikawa, T. *Polym. J.* **1971**, *2*, 416.

chain, $\langle \mathbf{r}^T \boldsymbol{\mu} \rangle_0$, in determining the manner in which $\langle \mu^2 \rangle$ will respond to the chain expansion produced by the excluded volume effect. Since $\langle (\mathbf{r}^T \boldsymbol{\mu})^2 \rangle_0 \geq 0$, Eq. (II-32) refutes the possibility that $\langle \mu^2 \rangle$ might decrease when the chain expands, if the chain is infinitely long. Furthermore, $\alpha_\mu^2 = \alpha^2$ is predicted if $\langle (\mathbf{r}^T \boldsymbol{\mu})^2 \rangle_0 = \langle r^2 \rangle_0 \langle \mu^2 \rangle_0$, so that under these circumstances a theory for α^2 would also yield α_μ^2, provided the chain is infinitely long. Finally, there is the important case where $\langle \mathbf{r}^T \boldsymbol{\mu} \rangle_0 = 0$. This case is important because the symmetry conditions in many polymers produce $\langle \mathbf{r}^T \boldsymbol{\mu} \rangle_0 = 0$, and (II-32) states that for these polymers $\langle \mu^2 \rangle$ is insensitive to excluded volume, in the limit where $n \to \infty$. The term $\langle \mathbf{r}^T \boldsymbol{\mu} \rangle_0$ is equal to zero in the limit of infinitely long chains for structures where all local dipole moments (e.g., those assignable to independent monomeric units) are perpendicular to the chain contour.[21] The restriction $n \to \infty$ must be emphasized in application of Eq. (II-32). It is easy to design *finite* chains that have $\alpha_\mu^2 < 1 < \alpha^2$, and chains that have $\alpha_\mu^2 \neq 1$ even though $\langle \mathbf{r}^T \boldsymbol{\mu} \rangle_0 = 0$.[22] The behavior predicted by Eq. (II-32) is recovered for these chains as $n \to \infty$.[23] Chapter XIV presents the calculation of $\langle \mathbf{r}^T \boldsymbol{\mu} \rangle_0$ for the rotational isomeric state model of a chain.

C. Optical Anisotropy and Its Consequences: Stress-Optical Effect, Kerr Effect, Cotton–Mouton Effect, and Depolarized Light Scattering

Polarizability. Electric fields polarize molecules and induce dipole moments. An induced dipole moment in an isotropic[24] molecule is always parallel to the inducing field. If the field strength is small, one can write

$$\mathbf{p} = \alpha \mathscr{E} \qquad \text{(II-33)}$$

where $\mathbf{p}$ is the induced dipole moment, and α is the molecular polarizability.[25] In an anisotropic molecule the polarization is not equal in different directions and the induced dipole's direction and strength depend on the molecule's orientation in the field. The molecular polarizability is, therefore, a second order tensor, $\boldsymbol{\alpha}$.

$$\mathbf{p} = \boldsymbol{\alpha} \mathscr{E} \qquad \text{(II-34)}$$

Tensor $\boldsymbol{\alpha}$ can be represented in the form of a 3×3 matrix

[21] Doi, M. *Polym. J.* **1972**, *3*, 252.

[22] Mattice, W. L.; Carpenter, D. K. *Macromolecules* **1984**, *16*, 625. Mattice, W. L. *Macromolecules* **1988**, *21*, 3320.

[23] Mattice, W. L. *J. Chem. Phys.* **1987**, *87*, 5512.

[24] *Isotropic* means here of cubic or higher molecular symmetry, for example, belonging to symmetry classes T, T_h, T_d, O, O_h, I, I_h, K, K_h.

[25] A factor $4\pi\epsilon_0$ is required if the SI (Système International) system of units is to be used. In order to retain formal simplicity, we employ the Gaussian cgs (centimeter-gram-second) system here.

$$\boldsymbol{\alpha} = \begin{bmatrix} \alpha_{xx} & \alpha_{yx} & \alpha_{zx} \\ \alpha_{xy} & \alpha_{yy} & \alpha_{zy} \\ \alpha_{xz} & \alpha_{yz} & \alpha_{zz} \end{bmatrix} \tag{II-35}$$

which is symmetric, $\alpha_{xy} = \alpha_{yx}, \alpha_{xz} = \alpha_{zx}, \alpha_{yz} = \alpha_{zy}$. The first subscript indicates the direction of the component of the inducing electric field; the second subscript, the direction of the corresponding component of the induced dipole moment.

It is convenient to split $\boldsymbol{\alpha}$ into an *isotropic* and an *anisotropic* part. The *average polarizability*, $\bar{\alpha}$, is[26]

$$\bar{\alpha} = \frac{1}{3} \operatorname{tr} \boldsymbol{\alpha} = \frac{\alpha_{xx} + \alpha_{yy} + \alpha_{zz}}{3} \tag{II-36}$$

The *polarizability anisotropy*, $\hat{\boldsymbol{\alpha}}$, is obtained by subtracting the isotropic tensor[27] $\bar{\alpha}\mathbf{I}_3$ from $\boldsymbol{\alpha}$:

$$\hat{\boldsymbol{\alpha}} = \boldsymbol{\alpha} - \bar{\alpha}\mathbf{I}_3 \tag{II-37}$$

Traditionally, the scalar quantity[28]

$$\langle \gamma^2 \rangle = \tfrac{3}{2} \langle \operatorname{tr} \hat{\boldsymbol{\alpha}}^2 \rangle \tag{II-38}$$

is called the *molecular anisotropy*.

The usefulness of the polarizability in the conformational analysis of macromolecules rests on the assumption, usually termed the *valence optical principle*,[29] of additivity of local polarizability. It applies to the polarizability tensor as well as to its anisotropy and $\bar{\alpha}$

$$\boldsymbol{\alpha} = \sum_i \boldsymbol{\alpha}_i \qquad \hat{\boldsymbol{\alpha}} = \sum_i \hat{\boldsymbol{\alpha}}_i \qquad \bar{\alpha} = \sum_i \bar{\alpha}_i \tag{II-39}$$

[26]The trace of a square matrix $\boldsymbol{\alpha}$, denoted by "tr," is the sum of all elements on the main diagonal, $\sum_i \alpha_{ii}$. It is a scalar and an invariant of that matrix; that is, it does not depend on the orientation of the tensor in the coordinate frame. Here the property of invariance states (the obvious) that the average polarizability does not depend on the orientation of the molecule with respect to the axes $x, y,$ and z.

[27]Here $\mathbf{I}_3$ is the unit (or identity) tensor of order 3, and is in matrix representation

$$\mathbf{I}_3 = \begin{bmatrix} 1 & 0 & 0 \\ 0 & 1 & 0 \\ 0 & 0 & 1 \end{bmatrix}$$

[28]This is an invariant, like the average polarizability.

[29]Meyer, E. L.; Otterbein, G. *Phys. Z.* **1931**, *32*, 290. Volkenstein, M. V. *Configurational Statistics of Polymeric Chains*, Wiley-Interscience, New York, **1963**, Chapter 7.

where the subunits i over which the sums extend can be monomeric units, chain segments, and so on. Orientation of parts of the chain causes orientation of the polarizability tensors and, hence, different total polarizability tensors of the entire chain.

Linear Birefringence Methods. Several experimental observables are very sensitive to polarizability. The most prominent class among the experiments form methods that rely on the linear birefringence induced in a polymer solution by an external field. For the examples mentioned below in greater detail, the fields are mechanical ($\rightarrow$ strain birefringence), electrical ($\rightarrow$ electric birefringence or Kerr effect), and magnetic ($\rightarrow$ magnetic birefringence or Cotton–Mouton effect). The principle is always the same.

If an isotropic ensemble of anisotropic molecules (e.g., a solution) is subjected to an external, directionally orienting field, the molecules will favor one spatial direction over others. This field will tend to align the tensors of optical anisotropy that are fixed in the molecular frame, counteracted, of course, by Brownian motion. Nevertheless, the average polarizabilities parallel ($\alpha_\parallel$) and perpendicular ($\alpha_\perp$) to the external field direction assume different values and $\Delta\alpha \equiv \alpha_\parallel - \alpha_\perp$. The corresponding observable quantity is the difference in refractive index parallel and perpendicular to the straining direction, $\Delta\tilde{n} \equiv \tilde{n}_\parallel - \tilde{n}_\perp$.[30]

$$\Delta\tilde{n} = \frac{(\tilde{n}^2 + 2)^2}{6\tilde{n}} \left(\frac{4\pi\mathscr{L}c}{3M} \right) \Delta\alpha \tag{II-40}$$

where $\Delta\tilde{n}$ is readily observable.[31]

Strain Birefringence; the Stress-Optical Coefficient. If a crosslinked, non-glassy, amorphous polymeric sample (a "rubber") is strained, the average polarizabilities parallel and perpendicular to the straining direction are at the focus of interest, and one measures the difference in refractive index parallel and perpendicular to the straining direction. Since $\Delta\tilde{n}$ is proportional to the applied stress,[32] data are conveniently reduced to a system constant, the *stress-optical coefficient B*, by

---

[30]The relationship between polarizability and refractive index is assumed to obey the simple Lorentz–Lorentz equation

$$\frac{\tilde{n} - 1}{\tilde{n} + 2} = \frac{4\pi}{3}\rho\alpha$$

where ρ is the number density of molecules of polarizability α.

[31]Le Fèvre, C. G.; Le Fèvre, R. J. W. *Rev. Pure Appl. Chem.* **1955**, *5*, 269.

[32]Kuhn, W.; Grün, F. *Kolloid Z.* **1942**, *101*, 248. Treloar, L. R. G. *The Physics of Rubber Elasticity*, 2nd ed., Clarendon Press, Oxford, **1958**.

$$B \equiv \frac{\Delta \tilde{n}}{\tau} \tag{II-41}$$

where τ is the axial stress on the sample. (If the sample were to be stressed either biaxially or in a geometry different from the simple tension implied here, the difference in the stresses parallel and perpendicular to the straining direction, $\Delta \tau \equiv \tau_{\parallel} - \tau_{\perp}$, would have to be used.) The stress can be measured, and B is a reliably obtainable quantity.

The molecular interpretation of the stress-optical coefficient for small stresses and strains and long polymer chains between network junction points rests on the assumption that the chain is composed of segments that, in their own local coordinate systems, have segment polarizability tensors that are unaffected by the stress. The stress merely changes the orientation of these segment polarizability tensors. Conformations within a segment do not change, because such changes would alter the segments polarizability tensor when expressed in its own local coordinate system. For a *freely jointed chain*, one can define the segmental polarizability parallel ($a_{\parallel}$) and perpendicular ($a_{\perp}$) to the "bond," and the segmental polarizability anisotropy

$$\Delta a \equiv a_{\parallel} - a_{\perp} \tag{II-42}$$

For purely historical reasons, one usually employs

$$\Gamma_2 \equiv \frac{3\Delta a}{5} \tag{II-43}$$

The stress-optical coefficient can then be shown to depend on Γ_2 as

$$B = \left(\frac{2\pi\Gamma_2}{27kT} \right) \frac{(\tilde{n}^2 + 2)^2}{\tilde{n}} \tag{II-44}$$

For real chains, the connection between the monomeric unit's polarizability tensor, constitution, configuration, and conformation, and Γ_2, must be established. It is possible to demonstrate[33] that for small strains and long chains between network junction points, and under Θ conditions (which are approximately established in the amorphous bulk), Γ_2 can be expressed as

$$\Gamma_2 = \frac{9}{10} \frac{\langle \mathbf{r}^T \hat{\boldsymbol{\alpha}} \mathbf{r} \rangle_0}{\langle r^2 \rangle_0} \tag{II-45}$$

[33]Gotlib, Yu. Ya.; Volkenstein, M. V.; Byutner, E. K. *Dokl. Akad. Nauk SSSR* **1954**, *99*, 935. Volkenstein, M. V. *Configurational Statistics of Polymer Chains*, Wiley-Interscience, New York, **1963**. Flory, P. J.; Jernigan, R. L.; Tonelli, A. E. *J. Chem. Phys.* **1968**, *48*, 3822.

This relationship is a reflection of the fact that the orientation of the polarizability anisotropy (the cause for the birefringence) is performed by the orientation of the macromolecules via the end-to-end vectors (the mechanical field acts on the chains through the points at which they are attached to the network, at the network junction points). Assumption of the valence-optical principle (see above) reduces the problem of a molecular definition of the stress-optical coefficient to one of appropriately determining the average of the quantity $\mathbf{r}^T \hat{\boldsymbol{\alpha}} \mathbf{r}$, which will be discussed in Chapter XIV.

Electric Birefringence; the Kerr Constant. An electric field induces birefringence in a homogeneous solution; in other words, the average polarizabilities of the solution parallel and perpendicular to the field direction assume different values. The difference in refractive index parallel and perpendicular to the field direction is measured. For small electric fields the birefringence is proportional to the square of the strength of the field.[34] In solution one defines a *molar Kerr constant* by relating $\Delta\alpha$ to the square of the "local field strength," $E_{local} = (\epsilon + 2)\mathscr{E}/3$, and to one molecule of solute.

$$\langle {}_m K \rangle_0 = \left(\frac{4\pi}{3} \right) \frac{\mathscr{L}\Delta\alpha}{E_{local}^2} \tag{II-46}$$

This term is experimentally accessible. It is the result of an extrapolation to zero field strength and zero concentration of the birefringence suffered by the solution:[35]

$$\frac{\langle {}_m K \rangle_0}{x} = \frac{54\tilde{n}_s}{(\tilde{n}_s^2 + 2)^2(\epsilon_s + 2)^2}$$
$$\cdot \left[\lim_{\mathscr{E},m_x \to 0} \left(\frac{\Delta\tilde{n} - \Delta\tilde{n}_s}{m_x \mathscr{E}^2} \right) + \overline{V}_x \lim_{\mathscr{E} \to 0} \frac{\tilde{n}_s}{\mathscr{E}^2} \right] \tag{II-47}$$

Here m_x is the molarity of the polymer and $\overline{V}_x$ its average molar volume. The subscript s denotes pure solvent. Similarly to the situation with the molecular dipole moment, it is not necessary to know the molar mass or its distribution precisely. The molar Kerr constant per monomeric unit is dependent on molecular weight only for relatively short chains and usually has approached its limit already for the macromolecules of interest.

The molecular origin of electric birefringence is in part similar to that of strain birefringence. Here, however, two mechanisms of orientation are acting: (1) the macromolecule is oriented with respect to the electric field by a tendency to minimize the energy of the *permanent molecular dipole* in the field,

[34] Kerr, J. *Phil. Mag. Series* **1880**, *9*, 157.
[35] Saiz, E.; Suter, U. W.; Flory, P. J. *Trans. Faraday Soc. II* **1977**, *73*, 1538.

which is $-\boldsymbol{\mu}^T\mathscr{E}$ (the orientation of minimum potential energy is with $\boldsymbol{\mu}\|\mathscr{E}$); (2) the macromolecule is also oriented by a tendency to minimize the energy of the *field-induced dipole* in the field, which is $-(\frac{1}{2})\mathscr{E}^T\hat{\boldsymbol{\alpha}}\mathscr{E}$ (the orientation of minimum potential energy is with the axis of the largest principal component of $\hat{\boldsymbol{\alpha}}\|\mathscr{E}$). As a result, the interpretation of the molar Kerr constant assumes the form[36]

$$\langle {}_mK \rangle_0 = \frac{2\pi\mathscr{L}}{15}\left(\frac{\langle \boldsymbol{\mu}^T\hat{\boldsymbol{\alpha}}\boldsymbol{\mu}\rangle_0}{(kT)^2} + \frac{(\epsilon_s - 1)\langle \mathrm{tr}\,\hat{\boldsymbol{\alpha}}^2\rangle_0}{(\tilde{n}_s^2 - 1)kT} \right) \qquad \text{(II-48)}$$

Of the two mechanisms, the one through the permanent dipole moment and giving rise to the first term in Eq. (II-48), and the one based on polarizability anisotropy and causing the second term, the first one is the more important one for all cases where the dipole moment is substantial.[37] Assuming that $\boldsymbol{\mu} = \sum_i \mathbf{m}_i$ and that the valence optical principle (see above) is valid reduces Eq. (II-48) to a problem of determining the averages of the quantities $\boldsymbol{\mu}^T\hat{\boldsymbol{\alpha}}\boldsymbol{\mu}$ and $\mathrm{tr}\,\hat{\boldsymbol{\alpha}}^2$. Evaluation of these averages will be addressed in Chapter XIV.

There are no direct experimental results or analytical theories that describe the manner in which $\langle {}_mK \rangle$ is affected by the chain expansion produced by the excluded volume effect. Simulations suggest that there are at least some chains for which the Kerr constant is affected by this chains expansion.[38]

Magnetic Birefringence; the Cotton–Mouton Constant. A magnetic field causes orientation much as an electric field does and hence induces birefringence in a homogenous solution.[39] The effect is quadratic in magnetic field, and a *molar Cotton–Mouton* constant can be defined in close analogy to the molar Kerr constant. Again, extrapolation to zero field strength and zero concentration is required. The mechanism of orientation is similar to the one orienting molecules in an electric field by the molecular anisotropy. The magnetic field induces a magnetic dipole in the molecule, whose energy in the field $(-\mathbf{H}^T\boldsymbol{\chi}\mathbf{H})$ the system attempts to minimize by orientation. Permanent magnetic dipoles are not evident in macromolecules, and the second mechanism effective in electric fields has no counterpart here. In close analogy to the formulation for the Kerr constant, the molecular interpretation of the

[36]Buckingham, A. D.; Pople, J. A. *Phys. Soc. A* **1955**, *68*, 905. Nagai, K.; Ishikawa, T. *J. Chem. Phys.* **1965**, *43*, 4508. Buckingham, A. D.; Orr, B. J. *Quart. Rev.* **1967**, *21*, 195. Flory, P. J.; Jernigan, R. L. *J. Chem. Phys.* **1968**, *48*, 3822.
[37]Rule of thumb: if $\mu > 1$ Debye.
[38]Mattice, W. L. *Macromolecules* **1988**, *21*, 2320.
[39]Buckingham, A. D.; Pople, J. A. *Proc. Phys. Soc. B* **1956**, *69*, 1133.

Cotton–Mouton constant is[40]

$$_mC = \frac{2\pi}{15kT} \langle \operatorname{tr} \hat{\alpha}\hat{\chi} \rangle_0 \tag{II-49}$$

If we extend the valence optical assumption to magnetic susceptibilities also (where it actually is more appropriate than for polarizabilities!) so that $\hat{\chi} = \sum_i \hat{\chi}_i$, Eq. (II-49) reduces to a purely mathematical problem of properly evaluating the average of the trace of $\hat{\alpha}\hat{\chi}$. It is advisable to measure the Cotton–Mouton constant in a Θ state.

Depolarized Rayleigh Scattering. In an anisotropic molecule the induced electric dipole moment is in general not parallel to the inducing field vector. Hence, if an isotropic solution of anisotropic molecules is irradiated with *vertically* polarized light, the light scattered at 90° is not only *vertically* polarized but contains a horizontal component. The *horizontally* polarized part of the light scattered at 90° is directly proportional to the polarization anisotropy. It contains, however, an adventitious component that must be removed before experimental results can be utilized.[41] The normalized and corrected scattering intensity, the depolarized Rayleigh ratio R_{HV}, determines the experimental molecular anisotropy $\langle \gamma^2 \rangle$ (see above).

$$\frac{\langle \gamma^2 \rangle}{x} = \left(\frac{90M_x}{\mathscr{L}\rho_s} \right) \left(\frac{\lambda_v}{2\pi} \right)^4 \lim_{w \to 0} \left(\frac{R_{HV}}{w(\tilde{n}^2 + 2)^2} \right) \tag{II-50}$$

Here λ_v is the wavelength of the incident beam in vacuo.

Molecular interpretation yields a theoretical value of[42]

$$\langle \gamma^2 \rangle_0 = \tfrac{3}{2} \langle \operatorname{tr} \hat{\alpha}^2 \rangle \tag{II-51}$$

which is Eq. (II-38) for an unperturbed polymer chain. The right-hand side of

[40]Buckingham, A. D.; Pritschard, W. H.; Whiffen, D. H. *Trans. Faraday Soc.* **1967**, *63*, 1057. Le Fèvre, R. J. W.; Murthy, D. S. N. *Austral. J. Chem.* **1969**, *22*, 1415. Suter, U. W.; Flory, P. J. *Trans. Faraday Soc.* **1977**, *73*, 1521. Erman, B.; Wu, D.; Irvine, P. A.; Marvin, D. C.; Flory, P. J. *Macromolecules* **1982**, *15*, 670.

[41]The optical anisotropy observed in condensed phases contains contributions from collision-induced perturbations in addition to the intrinsic molecular anisotropy and the solute–solute correlations for which dilution to zero concentration corrects. The collision-induced components of the depolarized scattering must be removed by spectroscopic (e.g., optical filtering) means. The linear birefringence methods are linear in $\hat{\alpha}$ and the collision-induced fluctuations cancel. Depolarized scattering, however, is quadratic in $\hat{\alpha}$ and must therefore be corrected.

[42]Ingwall, R. T.; Flory, P. J. *Biopolymers* **1972**, *11*, 1527. Patterson, G. D.; Flory, P. J. *Trans. Faraday Soc.* **1972**, *68*, 1098. *Trans. Faraday Soc. II* **1972**, *68*, 1111. Carlson, C. W.; Flory, P. J. *Trans. Faraday Soc. II* **1977**, *73*, 1005.

Eq. (II-51) is identical (except for a numeric factor) to the second term in the expression for the Kerr constant, Eq. (II-48). Its evaluation will be postponed until Chapter XIV.

D. Optical Activity

Plane-polarized light passing through a medium consisting of chiral molecules can suffer a rotation of the plane of polarization. This phenomenon is well known and is tightly coupled to the preferential absorption of photons of one of the two senses of chiral polarization (R and S) in the medium. While the preferential absorption of R or S light cannot be quantitatively explained in a form applicable to the purpose of this book, there is a simple model for the rotation of plane polarized light as a function of structure and conformation.

The experimentally observable quantity is a rotation angle α after light of wavelength λ has passed through a polymer solution (in an inactive solvent), and the *average molar rotation* per monomeric unit (or segment) is defined by

$$\frac{[M]_\lambda^T}{x} = \frac{M_1\alpha}{10 dw\rho_s} \tag{II-52}$$

where d is the path length in the solution. Units are usually given in deg cm^2 mol^{-1} (degrees per square centimeter per mole). Experimental measurement is extremely simple and precise. For polymeric alkanes, for example, values for $[M]_\lambda^T/x$ between 0 and ± 300 deg cm^2 mol^{-1} are common; the precision of the values is typically better than 1%.

In contrast to the marvelous simplicity of the experiment, there is no simple and convincing theory. Polarizability theory has laid a solid foundation,[43] but the methods are currently not applicable to flexible molecules. For a special class of substances, however, an effective and simple model,[44] originally introduced as a purely empirical correlation, has proved to be surprisingly accurate and reliable.[45]

Brewster's uniform conductor model[46] starts from the concept that chemical bonds can be modeled as consisting of uniformly conducting material, polarizable according to the appropriate bond refractivities, and that every succession of three bonds can be viewed as a helix segment (a *skew conformational unit*) that contributes an additive increment to the average chiral effect of the molecule. The contribution of a *skew* conformational unit with approximately tetrahedral bond angles to the total molar rotation is

---

[43] DeVoe, H. *J. Chem. Phys.* **1965**, *43*, 3199. Applequist, J. *J. Chem. Phys.* **1973**, *58*, 4251.
[44] Whiffen, D. H. *Chem. Ind.* **1956**, *964*. Brewster, J. H. *J. Am. Chem. Soc.* **1959**, *81*, 5475.
[45] Pucci, S.; Aglietto, M.; Luisi, P. L.; Pino, P. *Org. Chem.* (Chiurdoglu, G., ed.) **1971**, *21*, 203.
[46] Brewster, J. H. *Topics Stereochem.* (Eliel, E. L.; Allinger, N. L., eds.) **1967**, *2*, 1.

$$[\Delta\Phi]_\lambda = C_{ij} \sin \beta \qquad \text{(II-53)}$$

where β is the dihedral angle in the unit and C_{ij} is a material constant

$$C_{ij} \approx \frac{10^{10}}{\lambda^2} \left(\frac{l_1 l_2 l_3}{l_1 + l_2 + l_3} \right) \frac{(\tilde{n}^2 + 2)^2}{9\tilde{n}} \sum \Delta R_\lambda \qquad \text{(II-54)}$$

Here the l_i are the bond lengths of the three bonds that form the *skew* conformational unit and the ΔR_λ are the corresponding three bond or group refractivities at the observation wavelength. For commonly used wavelengths such as the sodium D-line, bond refractivities for many chemical structures are tabulated.[47] In a molecule consisting of several *skew* conformational units [e.g., a $C(sp^3)$—$C(sp^3)$ bond contains nine such units because each of the two carbon atoms has three substituents], they add up to yield the incremental molar rotation of that bond, $[\Delta M]_{\lambda,i}$, which can be conveniently written as

$$[\Delta M]_{\lambda,i} = K_{1,j} \sin \phi_i + K_{2,i} \cos \phi_i \qquad \text{(II-55)}$$

where ϕ_i is the traditional torsion angle of bond i and the two constants $K_{1,i}$ and $K_{2,i}$ are obtained from the appropriate C_{ij}. The value $[M]_\lambda^T$ of a given chain conformation is simply the sum of the contributions $[\Delta M]_{\lambda,i}$ of all bonds that make up the molecule. The conformational averages require only knowledge of the average values $\langle \sin \phi_i \rangle$ and $\langle \cos \phi_i \rangle$ for all bonds.

Optical rotations, like nmr coupling constants, are very local probes for conformation and, therefore, not sensitive to the thermodynamic state of a polymer solution. Molar rotation does not have to be measured at a Θ state. However, the method requires the availability of optically active material.

E. Electronic Excitation: Excimers, Excitation Transfer[48]

The *singlet electronic excited state* produced by excitation of a *chromophore*, Eq. (II-56), can relax back to the singlet ground state via a variety of mechanisms, including the emission of a photon of slightly lower energy than the one absorbed during the excitation, Eq. (II-57), $h\nu_A > h\nu'_A$.

---

[47]Vogel, A. I.; Cresswell, W. T.; Jeffrey, G. H.; Leicester, J. *J. Chem. Soc.* **1952**, *514*.

[48]Several excellent books provide more detail on the application of fluorescence methods to the study of the conformations of macromolecules, including the following: Lakowicz, J. *Principles of Fluorescence Spectroscopy*, Plenum, New York, **1983**. Winnik, M. A., ed. *Photophysical and Photochemical Tools in Polymer Science*, Reidel, Dordrecht, **1985**. Guillet, J. *Polymer Photophysics and Photochemistry: An Introduction to the Study of Photoprocesses in Macromolecules*, Cambridge University Press, Cambridge, **1985**. Phillips, D. ed. *Polymer Photophysics: Luminescence, Energy Migration and Molecular Motion in Synthetic Polymers*, Chapman and Hall, London, **1985**. Kalyanasundaram, K. *Photochemistry in Microheterogenous Systems*, Academic Press, Orlando, **1987**.

$$A + h\nu_A \rightarrow A^*$$ (II-56)

$$A^* \rightarrow A + h\nu'_A$$ (II-57)

This process is called *fluorescence*. The differences in the energies of the two photons arises from the different vibrational levels involved in the two transitions. Typically the transition in Eq. (II-56) originates with A in its vibrational ground state, because the spacing of vibrational energy levels is usually larger than kT, and hence the ground state is much more highly populated than vibrational excited states for ordinary molecules at ordinary temperatures. The absorption, however, produces an electronic excited state that is also vibrationally excited. Therefore $h\nu_A$ must supply the energy for the vibrational excitation of A^*, as well as the electronic excitation. The vibrational excitation is quickly dissipated (usually within $\sim 10^{-12}$ s) in condensed media (Kasha's rule), which is the environment in which polymers must be found. Isolated small molecules, which can exist in vacuo, may retain the vibrational excitation for much longer times. Fluorescence occurs on a much longer time scale (typically $\sim 10^{-9}$ s) than the time scale for dissipation of excess vibrational energy in condensed media. Therefore the transition in Eq. (II-57) originates with A^* in its vibrational ground state, but it can lead initially to a vibrational excited electronic ground state. Hence $h\nu'_A$ carries away the energy released in the electronic transition, less whatever vibrational excitation exists on A at the end of the electronic transition. Since $h\nu_A$ and $h\nu'_A$ have different energies, experiments are easily designed for the study of either of these transitions.

The efficiency of the overall process is described as the *quantum yield for fluorescence*, Q, which is the ratio of the number of photons emitted as fluorescence (with energy $h\nu'_A$) to the number of photons absorbed (with energy $h\nu_A$) to produce the singlet excited state responsible for the fluorescence. The quantum yield for fluorescence will be less than one if other processes, either radiative or nonradiative, can compete with the one depicted in Eq. (II-57) for the deactivation of the singlet excited state. Substances that reduce the quantum yield for fluorescence are termed *quenchers*. Two types of quenching mechanisms are important in the context of this book, because they provide information on the distance between the chromophore and the quencher, and this information is related to the conformation of a macromolecule if the chromophore and quencher are covalently attached to the same chain.

Excimers. The word *excimer* is a compact jargon for "excited-state dimer." It describes a dimer of two identical chromophores that is stable in the presence of electronic excitation, but unstable in the electronic ground state. (The excited complex is usually called an *exciplex* if the two chromophores are not chemically identical.) The processes involved in the formation and radiative dissociation of an excimer can be represented as

$$A^* + A \rightarrow (AA)^* \tag{II-58}$$

$$(AA)^* \rightarrow A + A + h\nu'_{AA} \tag{II-59}$$

The first process depicts the formation of a dimer by two chemically identical chromophores, the one in a singlet excited state and the other in the electronic ground state, and the sharing of the excitation in this dimer. The second process depicts fluorescence from the excited dimer, accompanied by dissociation of the dimer to two identical ground-state species. Since the excited dimer is stable relative to separated A^* and A, the process depicted in Eq. (II-58) is accompanied by a reduction in energy. For this reason, and because the ground state dimer is dissociative, the existence of the excimer is apparent in the emission spectrum, $h\nu'_{AA} < h\nu'_A$. Experimentally, one can determine whether an emitted photon came from A^*, Eq. (II-57), or from $(AA)^*$, Eq. (II-59).

The present interest in excimers can be understood by consideration of the full implications of Eq. (II-59). Since the dimer dissociates in the absence of electronic excitation, it follows that the formation of the excimer requires that one A must approach another A to such a close distance that the conformation is unstable when both species are in their electronic ground states. In other words, the two species are in "contact," and that contact is over such a short distance that it is repulsive. Consider now the case where the two species of A are covalently attached to opposite ends of a single chain molecule. Then the only conformations of the chain that can produce excimer emission are those conformations that are cyclic, bringing the two ends very close together.[49] More generally, if the two chromophores are attached at positions i and j, excimer emission will probe that portion of the distribution function, $W(r_{ij})$, where r_{ij} is very small. Issues of orientation, as well as distance, are also involved. The "As" are usually planar aromatic ring systems, such as naphthalene, anthracene, carbazole, and pyrene. Two angles (the angles between the $\mathbf{r}_{ij}$ and the normal to each planar ring system), along with r_{ij}, are involved. The preferred geometry for the singlet excimer has a parallel arrangement for the normals to the two ring systems such that the two ring systems produce a "sandwich."[50] However, absence of precise knowledge of the detailed relationship between I_E and the geometric constraints often causes one to seek no better than a semiquantitative interpretation of I_E/I_M.[51]

Another issue that affects the interpretation is related to the finite lifetime of A^*. The excited molecule emits with a rate constant $k_e = 1/\tau$, where τ is the fluorescence lifetime. A typical order of magnitude for τ is 10^{-9} s. If the system is mobile, A and A^* may diffuse during the time the excitation is present,

[49]Winnik, M. A. *Acct. Chem. Res.* **1985**, *18*, 73.
[50]Braun, H.; Förster, Th. *Ber. Bunsenges. Phys. Chem.* **1966**, *70*, 1091.
[51]This ratio is often denoted by I_E/I_M or I_D/I_M. The numerator is the intensity of the fluorescence from the excimer, and the denominator is the intensity of the fluorescence from the monomer.

and form an excimer as a result of an encounter produced by this diffusion, even though they were separated by a larger distance at the moment A^* was initially produced by absorption of the electronic excitation, Eq. (II-56). Hence the interpretation of the I_E/I_M is tied to the value of τ/η, where η denotes the viscosity of the medium. In the limit where $\tau/\eta \rightarrow 0$, the system is nearly immobile on the time scale of the experiment, and an excimer can be formed only by two species of A that were already very close to one another before the excitation occurred. In this limit, I_E/I_M should be interpreted using the equilibrium distribution function, provided that the excitation cannot hop from one chromophore to another, and thereby find an excimer forming site as a consequence of energy migration. If a series of macromolecules are labeled with A at sites i and j, such that $|i-j|$ differs for members of the series, the dependence of measured values of I_E/I_M on $|i-j|$ probes the fraction of $W(r_{ij})$ where $r_{ij} \sim$ 0.33–0.4 nm. At larger τ/η the values of I_E/I_M may increase because of the increase in the length of the path traversed by A^* during its lifetime.

Nonradiative Singlet Energy Transfer. Another fluorescence technique is capable of measuring distances on a longer scale than those probed by excimer formation. This technique is often called *nonradiative* singlet energy transfer or *Förster transfer.*[52] Consider a system containing two fluorescent groups, A and B, where A is subject to the transitions in Eqs. (II-56) and (II-57), and in addition we have

$$B + h\nu_B \rightarrow B^* \tag{II-60}$$

$$B^* \rightarrow B + h\nu'_B \tag{II-61}$$

$$A^* + B \rightarrow A + B^* \tag{II-62}$$

Equations (II-60) and (II-61) for B are the counterparts of Eqs. (II-56) and (II-57) for A. The key equation for Förster transfer is Eq. (II-62), which depicts the nonradiative transfer of the singlet excitation from A to B. If the transfer of excitation in Eq. (II-62) can compete with the direct emission from A^* in Eq. (II-57), we will observe (1) emission from B^*, with energy $h\nu'_B$, as a consequence of excitation of A with energy $h\nu_A$ and (2) quenching of the fluorescence from A, at $h\nu'_A$, due to the presence of B. It is the distance dependence of the competition of these processes that makes this phenomenon of importance in studies of macromolecules.

The efficiency of Förster transfer for a static system is usually interpreted with the relationship

[52]Förster, Th. *Ann. Physik.* **1948**, *2*, 55.

$$E = \frac{1}{1 + (r_{ij}/R_0)^6} \tag{II-63}$$

where r_{ij} is the distance between the two chromophores, and R_0 is a spectroscopic constant, often called the *Förster radius*. Equation (II-63) shows that one interpretation of R_0 is the value of r_{ij} at which $E = \frac{1}{2}$. The form of the efficiency specified by Eq. (II-63) is depicted in Fig. II-3. The efficiency specified by Eq. (II-63) changes rapidly from 1 to 0 as r_{ij} increases from ~$0.5R_0$ to ~$1.5R_0$, and is equal to $\frac{1}{2}$ when $r_{ij} = R_0$.

The value of R_0 can be established for any pair of chromophores by independent measurements.

$$R_0^6 = \frac{8.8 \times 10^{-25} \kappa^2 Q_A}{\tilde{n}_s^4} \int_0^\infty F_A(\nu) \epsilon_B(\nu) \frac{d\nu}{\nu^4} \tag{II-64}$$

Here the integral measures the overlap of the normalized fluorescence of A and the excitation of B, and κ^2 is a factor whose value depends on the angular orientation of the transition moments in the two chromophores. The range is $0 \le \kappa^2 \le 4$, and for systems where the orientation is completely random, $\kappa^2 = \frac{2}{3}$. There are extensive tabulations of the values of R_0 for pairs of

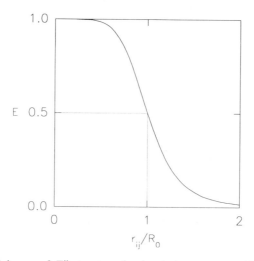

Figure II-3. Efficiency of Förster transfer in static systems with random orientation ($\kappa^2 = \frac{2}{3}$).

chromophores.[53] Pairs of probes are available for which R_0 is in the range 1–8 nm, and therefore measurements of E can explore values of r_{ij} over a rather broad range, provided suitably labeled samples are available. Large values of R_0 are favored by $h\nu'_A = h\nu_B$ (so that the overlap integral will be large) and large values for the quantum yield for fluorescence of A and the extinction coefficient for absorption by B.

With static systems, the most direct connection between E and rotational isomeric state theory lies in the use of theory to calculate the distribution function, $W(r_{ij})$, and then evaluation of the expected efficiency as

$$\langle E \rangle = \left\langle \frac{1}{1 + (r_{ij}/R_0)^6} \right\rangle \tag{II-65}$$

If the rotational isomeric state theory suggests that the orientation of the two chromophores is not isotropic, the influence of the orientation can be incorporated by evaluation of κ^2 for each conformation. Usually this extra effort does not yield a significant improvement in the result for synthetic polymers.

As with the interpretation of I_E/I_M for excimer fluorescence, the situation becomes more complicated when the chromophores can undergo significant diffusion on the time scale defined by τ. Then the value of r_{ij} may change significantly while A^* exists, and a conformation with $r_{ij} > R_0$ at the moment of excitation may convert to a conformation with a $r_{ij} < R_0$ within the time τ. Interpretation of E with a static model is not appropriate under these conditions.

3. IS A ROTATIONAL ISOMERIC STATE APPROACH REALLY NECESSARY? WHAT ARE ITS BIGGEST DRAWBACKS?

Chapter I was used to briefly review the state of the art in modeling macromolecules by general approaches, that is, by methods that do not take account of the structural details of the chains. Mathematically they are comparatively approachable. However, they can only address properties that depend on the segment distribution in the coil, that is, properties that were mentioned in Section II.1 (of this chapter). Even for these properties the general models are only useful to explain; they have the correct functional forms, but do not produce quantitative predictions. Observables that are sensitive to other characteristics, such as the nmr chemical shift and the coupling constants, or any of the other properties mentioned in the second part of this chapter, usually cannot be explained or predicted by the general chain models. There is clearly a need for a methodology that provides *quantitative predictions* and *quantita-*

[53]Berlman, I. B. *Energy Transfer Parameters of Aromatic Compounds*, Academic Press, New York, **1973**.

tive explanations from the known chemical structure of a flexible macromolecule.

Rotational isomeric state (RIS) theory can do that. It provides methods that allow predictions and explanations based on the chemical detail of the structures investigated—bond lengths, bond angles, torsion angles, and much more are considered and crucially influence the results (just as these details determine many properties of actual polymers). The properties mentioned in Section II.2 of this chapter, for instance, all yield to analysis with RIS methods. This ability alone suffices, we feel, to justify an investment of time and effort into the RIS approach.

There are also drawbacks. For one, RIS theory normally does not yield analytical expressions that can be evaluated by hand—a computer is usually necessary. This requirement largely prevents the manipulation of algebraic functions that can produce much insight into a phenomenon and prevents one from obtaining certain information. One can, however, easily numerically vary parameters of interest and at least estimate the required partial derivatives. And the widespread availability of computers, coupled with the computational efficiency of most of the RIS formalism, means that the requirement for the resort to a computer does not really impose a very large barrier.

The RIS method is also set up to primarily handle polymer chains in their unperturbed state, that is, in a Θ solution or the amorphous bulk. The effects of excluded volume are not easily treated, and if they are addressed, they can be investigated only numerically. However, as will be shown in Chapter XV, it is possible to focus on the effects of excluded volume on the single-chain behavior for any property that yields to RIS analysis.

Last, the RIS method does not concern itself with solution thermodynamics (beyond the default implication of an unperturbed state; this limitation can, however, be avoided). It is, therefore, best suited for dilute solutions, although it is possible to include a concentration dependence into the computations much as the effects of excluded volume on the conformations of a single chain can be treated.

PROBLEMS

Appendix B contains answers for all problems in this chapter.

II-1. What ranges of sizes for θ will cause a chain of 10 freely rotating bonds to have a value of C_n that is within 1% of the value of C_∞?

II-2. The value of the Mark–Houwink constant K_Θ for polydimethylsiloxane in butanone at 20°C is 7.8×10^{-4}, for $[\eta]$ in dl g^{-1} (decaliters per gram) and M in g mol^{-1}.[54] What is the value of C_∞? The length of an Si—O bond is 163 pm.

[54]Crescenzi, V.; Flory, P. J. *J. Am. Chem. Soc.* **1964**, *86*, 141.

II-3. What property of polymer chains causes the average segment concentration within an isolated molecule to decrease as M increases?

II-4. If a coupling constant for a rigid molecule is given by Eq. (II-28), does measurement of $^3J_{HH}$ uniquely define the conformation at the N—C^α bond?

II-5. After considering the limiting behavior of α^2 as $n \to \infty$, give the most conservative interpretation of the α_μ^2 predicted by Eq. (II-32) for a chain with $\langle \mathbf{r}^T \boldsymbol{\mu} \rangle_0 = 0$.

II-6. If the polarizability tensor is diag$(2x, x, x,)$, find $\bar{\alpha}$ and $\hat{\boldsymbol{\alpha}}$

II-7. The scalars $\bar{\alpha}$, Γ_2, $_mK$, $_mC$, and $\langle \gamma^2 \rangle$ all depend on $\hat{\boldsymbol{\alpha}}$. Which of these five properties must always be positive, and which can be positive or negative?

II-8. Which molecule exhibits the most intense excimer emission in dilute solution: 2-phenylpropane, *racemic*-2,4-diphenylpentane, or *meso*-2,4-diphenylpentane?

III The Rotational Isomeric State Model

The widespread use of rotational isomeric state concepts for the study of the conformation-dependent physical properties of polymers did not occur until scientists had easy access to computers. The concepts themselves are not new, however. The earliest application of rotational isomeric state concepts in polymers was reported over 40 years ago.[1] This chapter summarizes the relationship between the classical partition function, Q, and the *conformational partition function*, Z, that assumes the central role in rotational isomeric state (RIS) theory. It also develops the basis for the RIS model for small molecules, with frequent reference to the well-known conformational properties of three simple alkanes: ethane, *n*-butane, and *n*-pentane.

1. BOND LENGTHS, BOND ANGLES, AND ROTATIONS ABOUT BONDS

Macromolecules differ from molecules of low molar mass only by the much larger number of constituent atoms. While this large number of atoms may aggravate some of the problems encountered in the detailed analysis of small flexible molecules, it is not a fundamental difficulty. In contrast to most small molecules, however, macromolecules cannot be obtained in an isolated condition. The interaction of the macromolecule with its environment is an unavoidable fact for polymer chains. In dilute solution it is the solvent that provides interactions with many degrees of freedom. In concentrated solution it is, in addition, the other chains in the system that also need to be considered, and in the bulk (amorphous, liquid crystalline, or crystalline) it is the other chains that provide interactions with an environment of random or more or less ordered character. The unavoidable and extraordinarily numerous intermolecular interactions require special measures, namely, judiciously applied simplifications, in the analysis of macromolecules.

The process of constructing expressions for properties of macromolecular systems begins, following accepted practice with molecules of low molecular weight, by formulating a partition function for the system. A choice must be

[1] Volkenstein, M. V. *Dokl. Akad. Nauk SSSR* **1951**, *78*, 879.

made between a quantum-mechanical and a classical mechanical approach. The decision is not obvious, but careful consideration[2] has shown that the *classical partition function* is sufficient to describe the thermodynamics of flexible macromolecules in the random coil conformations. Following Flory, we will describe a simple example.

Consider a system consisting of one flexible (macro)molecule and N_s solvent molecules. The solvent molecules have spherical appearance and no internal degrees of freedom, but the solute (macro)molecule is composed of many "centers" (atoms). The total system contains N centers, N_s solvent centers, and $N - N_s$ centers in the flexible solute molecule. The system's total energy is composed of kinetic energy, associated with the momenta of all centers, and the total potential energy, $U(\mathbf{R}^N)$, which is a function of the configuration, $\mathbf{R}^N$ (which symbolizes all values necessary to specify the exact instantaneous molecular arrangement of the single flexible solute and all solvent molecules). All centers (atoms of the flexible solute molecule and the centers of the simple solvent) are expressed in Cartesian coordinates, and the momenta and coordinates are therefore separable (i.e., independent). Following standard texts of thermodynamics or statistical mechanics, one obtains after integration over the momenta the following expression for the classical partition function

$$Q = \left[\frac{(2\pi m k T)^{1/2}}{h} \right]^{3N} Z_N \qquad \text{(III-1)}$$

where the first factor contains the contributions of the momenta (m is the "average mass"), and Z_N is the part of the partition function that contains all coordinate information, the *configuration partition function*; Z_N is given by

$$Z_N = \int_V^{\text{ext}} \int^{\text{int}} \exp\left[\frac{-U(\mathbf{R}^N)}{kT} \right] d\mathbf{R}^N \qquad \text{(III-2)}$$

Integration is over $3N$ variables that are conveniently grouped into $3N_s + 6$ "external" coordinates and $3(N - N_s) - 6$ "internal" degrees of freedom. The external coordinates describe the location in space of all solvent molecules ($3N_s$ coordinates), of the flexible solute molecule (three coordinates), and the orientation of the flexible molecule (three values). The $3(N - N_s) - 6$ "internal" degrees of freedom, $\mathbf{R}_{\text{int}}$, specify the molecular constitution, chemical configuration, and conformation, without regard for spatial location or orientation. It is usually sufficient to consider only the average solvent configuration and the solute's location and orientation. One integrates over the coordinates of the centers of the molecules and the solute's orientation as if they were independent of the internal degrees of freedom of the solute, and corrects for the error com-

---

[2]Flory, P. J. *Macromolecules* **1974**, *7*, 381.

mitted by replacing the "true" potential energy $U(\mathbf{R}^N)$ by a *potential of mean force with respect to the solvent*, $\overline{U}(\mathbf{R}_{int})$.

$$Z_N = C_{int} \int^{int} \exp \left[\frac{-\overline{U}(\mathbf{R}_{int})}{kT} \right] d\mathbf{R}_{int} \tag{III-3}$$

where C_{int} is a constant of dimensions V^{N_s+2} since this integration was over $3N_s + 6$ degrees of freedom.[3]

Usually one wants to express the "internal" degrees of freedom in *internal coordinates*, (i.e., in bond lengths, bond angles, and torsion angles). Investigations of small molecules carried out over many decades have shown that bond lengths are usually confined to a narrow range about a mean value, and that they can therefore often be considered to be of constant length for analysis of flexible molecules. One has found that bond angles are not quite as rigidly defined, but it is often acceptable to consider them to adopt values not far from the mean and to keep them fixed at given values for conformational analysis. Therefore one integrates over the set of bond lengths $\{l\}$ and the set of bond angles $\{\theta\}$, assuming them to be independent of conformation, and correcting for the error thus committed by replacing the potential of mean force with respect to the solvent by a (new) *potential of mean force with respect to the solvent and chemical structure*, $\hat{U}(\{\phi\})$:

$$Z_N = C_{\{\phi\}} \int \exp \left[\frac{-\hat{U}(\{\phi\})}{kT} \right] d\{\phi\} \tag{III-4}$$

where $C_{\{\phi\}}$ is of dimension V^N.[4]

The integral in Eq. (III-4) represents that part of the configuration partition function that depends on the torsion angles. In the situations addressed in this book, the conformational variables are by far the most relevant degrees of freedom and the torsion angles are the variables to consider. The integral in Eq.

[3] The potential of mean force $\overline{U}(\mathbf{R}_{int})$ is defined as the function that will make Eq. (III-3) correct, that is, by

$$C_{int} \exp \left[\frac{-\overline{U}(\mathbf{R}_{int})}{kT} \right] = \int_V^{ext} d\mathbf{R}_{ext} \exp \left[\frac{-U(\mathbf{R}^N)}{kT} \right]$$

[4] Again, the potential of mean force $\hat{U}(\{\phi\})$ is defined as the function that will make Eq. (III-4) correct, that is, by

$$C_{\{\phi\}} \exp \left[\frac{-\hat{U}(\{\phi\})}{kT} \right] = \int_V^{ext} d\mathbf{R}_{ext} \int \exp \left[\frac{-U(\mathbf{R}^N)}{kT} \right] d\{l\} \, d\{\theta\}$$

$$= C_{int} \int \exp \left[\frac{-\overline{U}(\mathbf{R}_{int})}{kT} \right] d\{l\} \, d\{\theta\}$$

(III-4) is the most relevant quantity. It is convenient to term it a partition function $Z_{\{\phi\}}$,

$$Z_{\{\phi\}} = \int \exp\left[\frac{-\hat{U}(\{\phi\})}{kT}\right] d\{\phi\} \qquad (III-5)$$

the *classical configuration partition function of the flexible (macro)molecule in terms of conformation*, often simply called the *conformational partition function*, so that $Z_N = C_{\{\phi\}}Z_{\{\phi\}}$. Frequently we will write the conformational partition function more simply as Z.

If one is interested, as we are, in conformation-dependent physical properties, all relevant information of the classical partition function Q is found in $Z_{\{\phi\}}$, since the dynamic degrees of freedom contribute only trivially through the first factor on the right-hand side of Eq. (III-1) and $C_{\{\phi\}}$ is a constant. At constant T and V, there are actually no changes in Q except those in the conformational partition function $Z_{\{\phi\}}$.

It is important to realize that in computing conformational characteristics from atomistic models, if, as is usually done, only torsional degrees of freedom are analyzed and integrated, the potential of mean force with respect to the solvent and chemical structure, $\hat{U}(\{\phi\})$, is the appropriate potential energy function, and not the "true" potential energy $U(\mathbf{R}^N)$. The approximation used for $\hat{U}(\{\phi\})$ must comprise solvent contributions as well as terms correcting for the averaging of bond lengths and bond angles. Sometimes this is done implicitly in work reported in the literature, but more frequently the error is committed that very accurate "true" potential energies, often from extensive and accurate quantum-chemical calculations, are directly compared to conformational energies that would require a potential of mean force, not the true potential provided.

In the following section, we consider conformational variability in small molecules. Whenever a potential energy is given as a function of torsion angles only, it is implicit that the potential of mean force with respect to the solvent and chemical structure is used.

2. TORSION POTENTIAL ENERGY FUNCTIONS FOR SIMPLE MOLECULES

A. The Rotational Isomeric State Approximation: Well-Defined Minima

Ethane. Consider the conformations of ethane, $H_3C—CH_3$. If bonds of fixed length and bond angles of fixed magnitude are assumed, the molecule consists of two methyl groups each of which has C_{3v} symmetry. They are joined by a C—C bond with torsion angle ϕ. The torsional potential energy function

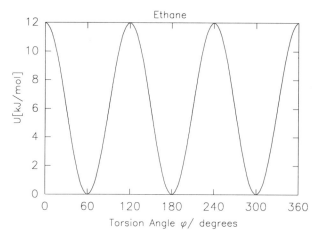

Figure III-1. Torsion potential energy function for the C—C bond in ethane.

$\hat{U}(\phi)$ must have threefold periodic symmetry. Experimental and theoretical information could in principle be employed to deduce the exact shape of $\hat{U}(\phi)$, but little is known beyond the fact that there is a threefold barrier of energy $U_b \approx 12$ kJ mol^{-1}.[5] One of the simplest threefold periodic functions is of the form

$$\hat{U}(\phi) = \frac{U_b}{2} (1 + \cos 3\phi) \qquad \text{(III-6)}$$

This function has found widespread use for the purpose of conformational analysis of ethane. In Fig. III-1 we plot $\hat{U}(\phi)$ versus ϕ, and Fig. III-2 depicts the population density of each conformation, $p(\phi)$,[6] for three temperatures. One notices immediately that the population of conformations is strong around the staggered positions, $\phi = \pm 60°$ and $180°$, and weak everywhere else, even at the comparatively high temperature of 500 K. As an expedient, therefore, one could replace the torsion angle continuum by three "states" only, centered at the three staggered positions.

n-Butane. The torsion potential energy function for the internal C—C bond in *n*-butane, CH_3CH_2—CH_2CH_3, has a related functional form. The form given

[5] Smith, L. G. *J. Chem. Phys.* **1949**, *17*, 139. Herschbach, D. R. *Intl. Symp. Molec. Struct. Spectry.,* *Tokyo 1962*, Butterworths, London, **1963**.
[6] The frequency of occurrence of each torsion angle is proportional to the Boltzmann factor, namely, to $\exp[-\hat{U}(\phi)/kT]$, and must, of course, be normalized so that $\int_0^{2\pi} p(\phi)\,d\phi = 1$. $p(\phi)\,d\phi$ is the probability of occurrence of the molecule in the range between ϕ and $\phi + d\phi$.

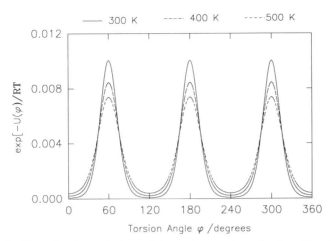

Figure III-2. Normalized populations of the torsion angle at the C—C bond in ethane at 300, 400, and 500 K.

by Tang[7] is displayed in Fig. III-3. The population density deduced from this potential energy function at three temperatures is depicted in Fig. III-4. Also here it is noteworthy how strongly the positions close to the staggered positions at $\pm\sim60°$ and 180° are favored. However, here the three staggered positions are not equally populated, as they were in ethane. The two g conformations[8] are equally populated because the torsion potential energy function is *symmetric*, $\hat{U}(\phi) = \hat{U}(-\phi)$, but the t conformation is more populated than the two g conformations.[9,10]

In these two and in countless other cases investigated but not reviewed here, one has found the population of conformations to be very distinctly grouped

---

[7]We follow Tang, A.-C. *J. Chinese Chem. Soc.* **1952**, *19*, 33 and *Sci. Sinica (Peking)* **1954**, *3*, 279 in using

$$\hat{U}(\phi) = 11.8 + 7.66\cos\phi + 4.64\cos 2\phi + 8.8\cos 3\phi \qquad (\text{kJ mol}^{-1})$$

Sometimes Au-Chin Tang is cited as Au Chin-Tang.

[8]The three minima in the conformational energy surface at $\phi \approx 180°, 60°$, and $-60°$ are denoted by t, g^+, and g^-, which are compact notations for *trans*, *gauche$^+$*, and *gauche$^-$*, respectively; g refers to the *gauche* minima collectively.

[9]A mnemonic device for the convention for "+" and "−" rotations is as follows: Imagine the bond whose conformation is discussed to be a *right-handed* thredded screw. The two atoms at either end of the bond and their substituents are nuts that have been thredded onto the screw. Starting from the zero position, a torsion angle is *positive* if the corresponding rotation *increases* the distance between the nuts. We use g^+ to denote the g conformation produced by a positive rotation of $\sim60°$ from the reference (c) conformation.

[10]In the older convention, where ϕ is defined as $0°$ in the t placement, g^+ denoted the g conformation produced by a positive rotation of $\sim120°$, starting from t. This g^+ is the mirror image of the g^+ defined here with c as a reference.

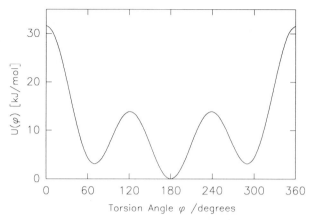

Figure III-3. Torsion potential energy function for the central C—C bond in *n*-butane.

about a few torsion angle values. The RIS approach now adopts a thermodynamically equivalent conformational model in which a few *discrete hypothetical isomers*, so-called *rotamers* or *conformers*, with torsion angles similar to the
most favored conformations are representing the molecule's very many conformations. (These "isomers" can almost never be isolated, in reality, because the
conformational transitions between rotational isomers are usually fast on the
human being time scale.) In other words, one chooses discrete rotational states
to replace the continuum over certain domains of "conformation space." The

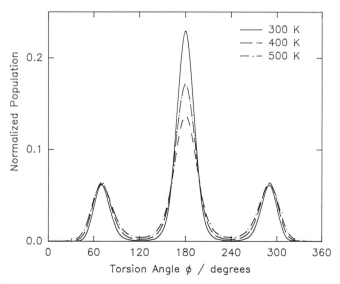

Figure III-4. Normalized populations of the torsion angle at the central C—C bond in
n-butane at 300, 400, and 500 K.

set of domains must be chosen so as to (1) not overlap, but to (2) fully represent the accessible conformation space.

The (hypothetical) rotational isomers are thought of as distinct chemical species that have associated thermodynamic state functions just like any compound. The word "state" is frequently applied to a rotational isomer, as in "*trans* state" or "*gauche* state." This use of the word should not be equated to its use in quantum mechanics. A multitude of quantum mechanical states contribute to each rotational isomeric state.

There is a one-to-one correspondence between concepts in the continuum representation of conformation and the RIS model, as summarized in Table III-1. Note that ϕ_i must not be identified with the value of ϕ at the energy minimum of the domain i, but rather with $\langle \phi \rangle_i$, since for an asymmetric potential energy well $\phi_{\min E} \neq \langle \phi \rangle$.

For convenience, one usually takes one state as reference (often the one with lowest energy), and, if state j is the reference state, $E_j = S_j = 0, u_j = 1$.

n-Butane shall serve as a simple example. An intuitively pleasing choice of rotational isomers is evident from the population of rotations (see Fig. III-4). They are t, g^+, and g^-.[8] The two g rotamers are mirror images and have therefore identical state functions in an achiral environment. Consequently, several of the properties of the rotational isomers are those summarized in Table III-2. The continuous distribution of torsion angles of *n*-butane has been replaced by only three isomers. The population of rotational isomers is given by only one statistical weight, σ. The statistical weight can be related to the thermodynamic state functions of the rotational isomers by

TABLE III-1. Corresponding Concepts between the Continuum and RIS Representation

	Continuum	RIS
Conformational Partition Function	$z = \int_0^{2\pi} \exp(-\hat{U}/kT)\,d\phi$	$z = \sum u_i$ $u_i = \exp(-G_i/RT)$ $= u_0 \exp(-E_i/RT)$ with $u_0 = \exp(S_i/R)$
Population	$p_i = z^{-1} \int_{\text{domain}} \exp(-\hat{U}/kT)\,d\phi$	$p_i = z^{-1} u_i$
Representative energy	$\langle \hat{U} \rangle_i = z^{-1} \int_{\text{domain}} \hat{U} \exp(-\hat{U}/kT)\,d\phi$	$E_i \approx H_i^a$
Representative entropy	$\langle S \rangle_i = z^{-1} \int_{\text{domain}} (\hat{U}/T) \exp(-\hat{U}/kT)\,d\phi$	S_i
Representative angle	$\langle \phi \rangle_i = z^{-1} \int_{\text{domain}} \phi \exp(-\hat{U}/kT)\,d\phi$	ϕ_i

[a]For the consideration of conformations one usually assumes that the (partial) molar volumes of all conformers are identical, that is, that for the most frequent situation of constant pressure, the term PV is irrelevant and enthalpy can be safely approximated by internal energy.

TABLE III-2. Rotational Isomeric States of *n*-Butane

	trans	*gauche*[+]	*gauche*[−]
Energy	0	E_σ	E_σ
Entropy	0	S_σ	S_σ
Statistical weight[a]	1	σ	σ
Population	$(1 + 2\sigma)^{-1}$	$\sigma(1 + 2\sigma)^{-1}$	$\sigma(1 + 2\sigma)^{-1}$
Population at 300 K[b]	0.56	0.22	0.22
Average ϕ at 300 K[b]	180°	~70°	~−70°

[a] The conformational partition function is the sum of the statistical weights for all conformations, $Z = 1 + 2\sigma$.
[b] Using the torsional potential energy function described by Tang.[7]

$$\sigma = \sigma_0 \exp\left(\frac{-E_\sigma}{RT}\right) \tag{III-7}$$

with

$$\sigma_0 = \exp\left(\frac{S_\sigma}{R}\right) \tag{III-8}$$

where it is assumed that the (partial) molar volume of all conformers is identical, that is, that for the conformational equilibria, enthalpy can be approximated by internal energy and Gibbs energy by Helmholtz energy.

The quantities E_σ and S_σ are, by construction, *relative* quantities giving the amount of energy and entropy of the g conformer in excess of those of the t conformer. The energy difference E_σ has a straightforward interpretation as the difference in mean energy between the g and the t state, $\langle \hat{U} \rangle_g - \langle \hat{U} \rangle_t$. A value of $E_\sigma \approx 2$ kJ mol^{-1} is satisfactory. The entropy difference S_σ, on the other hand, measures the relative population of the two states after differences in mean energy have been accounted for (see Table III-1). It is sensitive to the relative *shape* of the potential energy well. From the curves in Fig. III-4 it is probably safe to take S_σ to be zero and $\sigma_0 = 1$.

***n*-Pentane.** Now consider *n*-pentane, CH_3CH_2—CH_2—CH_2CH_3. A map of an approximation to the potential of mean force as functions of torsion angles ϕ_2 and ϕ_3, $\hat{U}(\phi_2, \phi_3)$, is plotted in Fig. III-5, where for convenience the methyl groups have been replaced by appropriate spherical pseudoatoms (the torsional behavior of the terminal methyl groups is of little importance here and irrelevant for polymers).[11] The global minimum is indicated by × and isoenergy contour lines are drawn at 4, 8, 12, 16, and 20 kJ mol^{-1} above the minimum value

[11] Following Abe, A.; Jernigan, R. L.; Flory, P. J. *J. Am. Chem. Soc.* **1966**, *88*, 631.

of $\hat{U}(\phi_2, \phi_3)$ at $\phi_2 = \phi_3 = 180°$.[12] At first blush it seems as if states located independently at ϕ_2, $\phi_3 = 180°$, $\sim \pm 70°$ would be most natural. Indeed, analogy with ethane and n-butane suggests that nine rotational isomers, termed tt, tg^+, tg^-, g^+t, g^+g^+, g^+g^-, g^-t, g^-g^+, and g^-g^-,[13] should be used for analysis of n-pentane. This analogy with ethane implies that $\langle\phi_2\rangle$ and $\langle\phi_3\rangle$ are identical for all rotamers. Closer inspection of Fig. III-5 shows that the independence between the two relevant torsion angles is only approximate, and that in particular the states g^+g^- and g^-g^+ might be split in two. Considering the fact that the states in question are of relatively high energy and, hence, not very strongly populated, together with the need for simplicity leads us, however, to presently assume that only one conformer is necessary to represent the domains $(30° < \phi_2 < 120°, -30° > \phi_3 > -120°)$ and $(-30° > \phi_2 > -120°, 30° < \phi_3 < 120°)$ and to approximate the torsion angle behavior by the "independent" values mentioned above.

Comparison of the conformational energy map for n-pentane in Fig. III-5 and the conformational energy curve for n-butane in Fig. III-3 suggests that in large part n-pentane can be viewed simply as "double n-butane." If one of the two bonds, 2 and 3, is in an exact t conformation, the dependence of the energy $\hat{U}$ on the other torsion angle mirrors the dependence of the energy of n-butane on its sole torsion angle. Furthermore, the energy of the states g^+g^+ and g^-g^- appears to be simply the sum of two n-butane g bonds. Seven of the nine rotational isomeric states can be represented in perfect analogy to the situation in n-butane, and one chooses the same energy E_σ as for n-butane. Only the two states where g conformations of opposite sign occur, g^+g^- and g^-g^+, have clearly different appearance and energy than what would be anticipated from n-butane's conformational energy alone.

A look at a model (see Fig. III-6) quickly reveals that the two conformers with unexpectedly high energy have their methyl groups very close in space. Traditionally this close interaction has been termed the "pentane effect," the "1,4-interaction" (counting bonds between interacting groups), the "1,5-interaction" (counting carbon atoms from H_3C- to $-CH_3$ along the chain), and the "boat effect" (the shape of n-pentane in this conformation resembles the bow of a boat). The energy associated with this special interaction is called E_ω. From the potential energy map for n-pentane one would estimate the value of $E_\omega \approx 7.5$ kJ mol^{-1} (roughly the difference between $\hat{U}^{g^+g^-}$ and $\hat{U}^{g^+g^+}$). The rotational isomeric states of n-pentane have the characteristics presented in Table III-3.

---

[12]The symmetry of the graph is determined by the molecular symmetry. Two mirror symmetries exist: $\phi_3 = \phi_2$ and $\phi_3 = -\phi_2$.

[13]The symmetry in the molecular frame and energy plot renders several states equivalent:

$$tg^+ \approx tg^- \approx g^+t \approx g^-t$$
$$g^+g^+ \approx g^-g^-$$
$$g^+g^- \approx g^-g^+$$

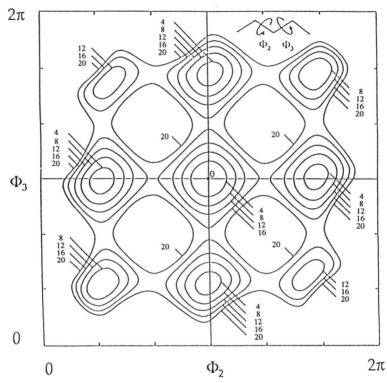

Figure III-5. Conformational energy surface for the two internal C—C bonds in *n*-pentane.

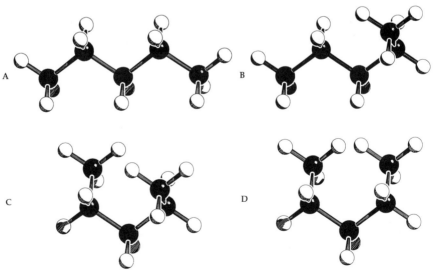

Figure III-6. *n*-Pentane in the *tt*, *tg*$^+$, *g*$^+$*g*$^+$, and *g*$^+$*g*$^-$ conformations.

TABLE III-3. Rotational Isomeric States of *n*-Pentane

State	ϕ_2	ϕ_3	E	u_i^a	u_i/Z^b
tt	$180°$	$180°$	0	1	0.36
tg^+	$180°$	$\sim 70°$	E_σ	σ	0.13
tg^-	$180°$	$\sim -70°$	E_σ	σ	0.13
g^+t	$\sim 70°$	$180°$	E_σ	σ	0.13
g^+g^+	$\sim 70°$	$\sim 70°$	$2E_\sigma$	σ^2	0.06
g^+g^-	$\sim 70°$	$\sim -70°$	$2E_\sigma + E_\omega$	$\sigma^2\omega$	0.003
g^-t	$\sim -70°$	$180°$	E_σ	σ	0.13
g^-g^+	$\sim -70°$	$\sim 70°$	$2E_\sigma + E_\omega$	$\sigma^2\omega$	0.003
g^-g^-	$\sim -70°$	$\sim -70°$	$2E_\sigma$	σ^2	0.06

[a] All S_i are taken to be zero; Z is the sum of the u_i, $1 + 4\sigma + 2\sigma^2 + 2\sigma^2\omega$.
[b] Fraction assuming $\sigma = 0.43$ and $\omega = 0.049$. These values of the statistical weights are appropriate for 300 K with $E_\sigma = 2.1$ kJ mol^{-1} and $E_\omega = 7.5$ kJ mol^{-1}.

It has become customary to distinguish between interactions that depend on the state of a single bond only and those that take specification of more than one bond for their description. Those interactions that can be described by the rotational isomeric state of *one* bond, say, E_σ, are called *first-order interactions*. Interactions that require specification of the states of *two* bonds, say, E_ω, are termed *second order interactions*, and so on.

Comparison of Tables III-2 and III-3 is illuminating. Even though the *tt* conformation in *n*-pentane is favored over its nearest competitors by the same amount of energy, E_σ, as the *t* conformation in *n*-butane, its predominance is significantly smaller in *n*-pentane than in *n*-butane. This difference is due to the fact that the less favored conformations grow faster in number than the "best" conformation and "outnumber" it, as suggested in Fig. III-7. The *t* conformation of *n*-butane must compete with only two conformations, each of energy E_σ, but the *tt* conformation of *n*-pentane must compete with four conformations of energy E_σ, and also four other conformations of higher energy. Generalizing, one can state that, as the number of bonds increases in a macromolecule, the importance of the most favored state usually decreases. The concentration of the all-*t* state is already vanishingly small for very modest *n*-alkane chain lengths

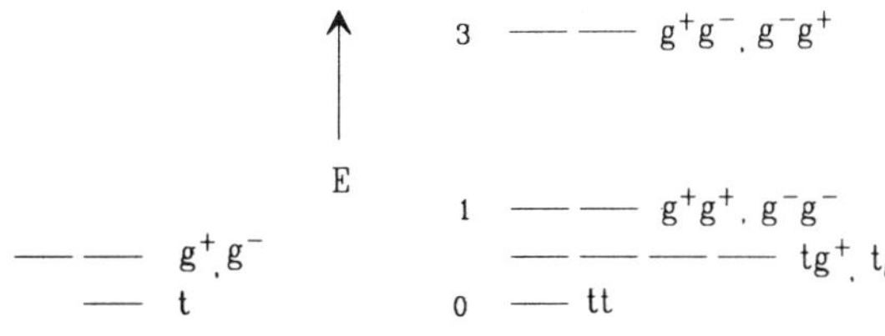

Figure III-7. Depiction of the conformational energies of the 3 and 9 rotational isomers for *n*-butane and *n*-pentane, respectively, relative to the energy of the all-*trans* state.

(although it is still the conformer of highest individual concentration, it does not dominate any property anymore). This point is developed in Problem III-9.

The rapidly growing number of conformations does not only make the "most favorable" conformation irrelevant, it also overpowers any conceivable method of analysis in which every conformer is analyzed individually. For noncyclic alkanes with n skeletal carbon–carbon bonds, for example, the number of conformations grows with 3^{n-2}. This number is 27 for hexane, 81 for heptane, and reaches 10^{90} for the modest chain length of 191 carbon atoms. This fantastic number is P. A. M. Dirac's estimate of the number of protons in the universe—a formidable number indeed.[14] Obviously individual analysis of conformers must cease long before this point, but polymers usually have much higher numbers of conformationally relevant skeletal bonds. This situation has precluded the detailed conformational analysis of macromolecules until the advent of RIS theory, which provides a powerful methodology for realistic simplification of the problem. The introduction of these methods will begin in the next chapter.

The temperature coefficients of the properties can be addressed in an RIS model. In general, the temperature coefficients depend on changes in the statistical weights as well as the geometry (i.e., θ_i, ϕ_i, and l_i). Often it is sufficient to only consider changes in the statistical weights, assuming the geometry to be temperature-independent. In principle, changes in geometry with temperature could be incorporated in the RIS scheme; to date, this seems not yet to have been necessary.

B. Adaptation to Cases without Well-Defined Minima

The construction of a rotational isomeric state model for ethane, n-butane, and n-pentane in the preceding section was based on the strong preference of these molecules for some few conformations that were chosen as loci for the rotational isomeric states. What if a molecule has small rotation barriers and large areas of conformation space are significantly populated even at moderate temperatures?

From Table III-1 it is clear that, mathematically speaking, the RIS model is simply replacing integration over the conformation hypersurface by a summation. This replacement process can only then lead to reliable results if all areas of relatively low $\hat{U}$ are represented by a term in the sum. This requirement means that not only valleys of potential energy but also plateaus of low energy are important. It is also not sufficient to represent strongly asymmetric or banana-shaped valleys by a single rotational isomeric state. A satisfactory RIS model can be obtained by application of the following rules:

1. The entire energy hypersurface must be inspected and all valleys must be

---

[14]More recent estimates suggest Dirac's estimate might be a bit too large.

found. As a first approximation, every valley of not too excessive energy is assigned a conformer.

2. Around the minima one charts domains whose boundaries either touch the periphery of other domains or areas of such high energy that their population is completely irrelevant. Boundaries between domains shall preferentially be located on ridges between valleys.

3. If domains become too large, that is, if the edge lengths of boundaries in one torsion angle exceed ~60–80° (usually due to large areas of low energy but without minima), it is advisable to split domains into subdomains of similar size.

4. Characteristics of the domains are assigned to the conformers that represent them. Table III-1 shows how for a domain ζ of the local conformational partition function, z_ζ, the internal energy E_ζ, the entropy S_ζ, and the torsion angles $\phi_{i,\zeta}$ are obtained.

From the formulae in Table III-1 it is clear that the process of averaging always includes a weighting by the averaged variable. It is therefore possible that a RIS model that performs satisfactorily for one property does not suffice for another one. In that case, a refinement can be obtained by subdivision of domains (see rule 3 above) over which the property in question varies too rapidly.

3. HARD AND SOFT DEGREES OF FREEDOM

Development of the formulas for the potential of mean force at the beginning of this chapter might have left the impression that it is always advisable to fix the values of bond lengths and bond angles in conformational analysis and only consider variation of torsion angles. This point of view is often overly simplified. It is more appropriate to conceptually divide all degrees of freedom accessible to the macromolecule into *hard* and *soft* ones, that is, those that do not vary significantly (or that do not by their variation influence the partition function or the relevant observables significantly) and those that do. A first approximation is to assume that bond lengths and bond angles belong to the hard degrees of freedom, while torsion angles belong to the soft ones.

It is, however, possible to divide in any way one desires between hard and soft variables. In some structures, for instance, are certain bond angles that vary strongly from conformer to conformer, or within conformers. An example is θ_{SiOSi} in poly(dimethylsiloxane), which has an unusually small force constant.[15] One must, however, be careful to assess the statistical mechanical effect of such a manipulation. In principle a different potential of mean force should

[15]Lord, R. C.; Robinson, D. W.; Schumb, W. C. *J. Am. Chem. Soc.* **1957**, *78*, 1327. Aronson, J. R.; Lord, R. C.; Robinson, D. W. *J. Chem. Phys.* **1960**, *33*, 1004. Durig, J. R.; Flanagan, M. J.; Kalasinsky, V. F. *J. Chem. Phys.* **1977**, *66*, 2775.

be employed if a degree of freedom is changed from being viewed as hard to soft.

PROBLEMS

Appendix B contains answers for all problems in this chapter.

III-1. In which of the following alkanes is the population of rotational isomeric states independent of temperature: ethane, n-butane, n-pentane?

III-2. Compare the rotational isomeric state analysis of 1,2-dichloroethane with the analysis of n-butane in Table III-2. How much of the formalism is common to both molecules? Are new terms needed for 1,2-dichloroethane, or can it be treated by simply assigning different values to the terms defined for n-butane?

III-3. Compare the RIS analysis of 1,3-dichloropropane with the analysis of n-pentane in Table III-3. How much of the formalism is common to both molecules? Are new terms needed for 1,3-dichloropropane, or can it be treated by simply assigning different values to the terms defined for n-pentane?

III-4. Estimate the ratio of the populations of g^+ and t at 300, 400, and 500 K, using Fig. III-4. How closely are your estimates reproduced by the expressions for the populations in Table III-2?

III-5. Convince yourself that the sum of the fraction of bonds in the g^+ and g^- states in both n-butane and n-pentane is given by

$$\frac{1}{n-2} \frac{\partial \ln Z}{\partial \ln \sigma} \tag{III-9}$$

using the expressions for Z in Tables III-2 and III-3.

III-6. How should we differentiate Z for n-pentane if we want to extract the fraction of the molecules that are in the g^+g^- and g^-g^+ states?

III-7. How should we differentiate Z for n-pentane if we want to extract the fraction of the molecules that are in the g^+g^+ and g^-g^- states? (Consider the introduction into the expression for Z of a dummy variable, ψ, which can be assigned the value $\psi = 1$ after evaluation of $\partial \ln Z / \partial \ln \psi$.)

III-8. Construct a RIS model for n-butane that uses six states, three of which are the t, g^+, and g^- states described in this chapter, and the other three of which are c, s^+, and s^- (*cis*, *skew*$^+$, *skew*$^-$) states with ϕ at the *maxima* in the potential energy function depicted in Fig. III-3. For this six-state model, construct the populations of the six states at 300 K, and compare them with the results in Table III-2 for the three-state model. Which model is preferred?

III-9. Add the enumeration of the 27 conformations of *n*-hexane to the diagram in Fig. III-7. Assume that the only pertinent energies are E_σ for each *g*, and E_ω for each $g^\pm g^\mp$ pair. What does the modified figure imply about the population of the most stable state in *n*-hexane, *ttt*, as compared with the population of *tt* in *n*-pentane?

III-10. There is a strongly repulsive third-order interaction (call it $E_{\omega'}$) between the terminal methyl groups in the $g^\pm g^\mp g^\pm$ conformations of *n*-hexane. How would inclusion of this third-order interaction change the modification of Fig. III-7 that was produced in Problem III-9? With the aid of your modified Fig. III-7, can you present a rationalization for why neglect of $E_{\omega'}$ might be less worrisome than neglect of E_ω, even if $E_{\omega'} > E_\omega$?

IV The Conformation Partition Function for a Simple Chain

Chapter III developed the rationale for the rotational isomeric state (RIS) analysis of the conformations of flexible molecules, drawing on the well-known conformational properties of three simple alkanes: ethane, n-butane, and n-pentane. It also described the important roles of the conformational partition function, denoted by Z_N, and the potential of mean force with respect to the chemical structure and solvent, $\hat{U}(\{\phi\})$. Here we will reexamine the conformational partition function in the RIS approximation, denoted simply by Z, and also adopt a simpler expression, E, for the appropriate potential of mean force. The objective is to establish a general method by which the discrete enumeration of rotational isomers for the three small alkanes considered in Chapter III can be made useful for chain molecules that contain *any* number of bonds.

1. EXPRESSION OF Z BY INTEGRALS AND SUMMATIONS

A large number of variables are required for specification of the internal conformation of a macromolecule. If bond lengths and bond angles are held constant, a linear chain of n bonds contains $n - 2$ internal bonds about which rotation produces a change in internal conformation. The expression for Z could be written as $n - 2$ integrals, one integral for each variable internal torsion angle, ϕ_2 through ϕ_{n-1}.

$$Z = \int_0^{2\pi} \int_0^{2\pi} \cdots \int_0^{2\pi} \exp\left(\frac{-E_{\phi_2, \phi_3 \cdots \phi_{n-1}}}{kT} \right) d\phi_2 \, d\phi_3 \cdots d\phi_{n-1} \qquad \text{(IV-1)}$$

Occasionally there will be a special case where E can be approximated as a sum of independent contributions that are determined by individual torsion angles:

$$E_{\phi_2, \phi_3, \cdots, \phi_{n-1}} = \sum_{i=2}^{n-1} E_i \qquad \text{(IV-2)}$$

68

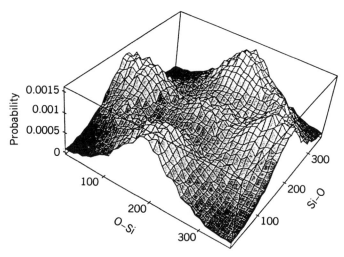

Figure IV-1. Profile of the joint probability for rotational states at two successive Si—O bonds meeting at a silicon atom in an isolated chain of poly(diethylsiloxane), illustrating a case of interdependent rotations. Reprinted with permission from Neuburger, N.; Bahar, I.; Mattice, W. L. *Macromolecules* **1992**, *25*, 2447. Copyright 1992 American Chemical Society.

Under these circumstances, the expression for Z simplifies to

$$Z = \prod_{i=2}^{n-1} z_i \qquad (\text{IV-3})$$

$$z_i = \int_0^{2\pi} \exp\left(\frac{-E_i}{kT}\right) d\phi_i \qquad (\text{IV-4})$$

where z_i is the conformational partition function for bond i. A chain is described by Eqs. (IV-2), (IV-3), and (IV-4) if the bonds are *independent*.

More typical is a chain in which the bonds are *interdependent*. In such a chain, the change in E produced by an alteration in conformation at bond i depends not only on ϕ_i but also on the conformations adopted at one or more neighboring bonds. Figure IV-1 depicts a probability profile for a polymer where nearest-neighbor bonds are interdependent.[1] The positions of high probability for bond i depend on the selection of ϕ_{i-1}. If the interdependence of the bonds in a chain is confined to nearest neighbors, we obtain

[1] Neuburger, N.; Bahar, I.; Mattice, W. L. *Macromolecules* **1992**, *25*, 2447.

$$E_{\phi_2,\phi_3,\cdots,\phi_{n-1}} = E_2 + \sum_{i=3}^{n-1} E_{i-1,i} \tag{IV-5}$$

When the bonds are interdependent, Eq. (IV-2) is not an acceptable replacement for Eq. (IV-5).

In many chains, the integrations that appear in Eq. (IV-1) and (IV-4) can be replaced by summations with little effect on most conformation-dependent physical properties obtained from a derivative of ln Z:

$$Z = \sum_{\phi_2} \sum_{\phi_3} \cdots \sum_{\phi_{n-1}} \exp\left(\frac{-E_{\phi_2,\phi_3,\cdots\phi_{n-1}}}{kT}\right) \tag{IV-6}$$

$$z_i = \sum_{\eta} \exp\left(\frac{-E_{\eta;i}}{kT}\right) \tag{IV-7}$$

In Eq. (IV-7), η indexes the states at bond i. The justification for this replacement of integrals by summations is most easily seen when the rotation potential energy function for each bond is characterized by a small number of narrow minima separated by regions of extremely high energy.[2] In such cases, the regions of high energy make a vanishingly small contribution to the integral. Attention can be confined to the narrow minima, and each minimum can be associated with a single value of the torsion angle if the minimum is sufficiently sharp. Each of the $\exp(-E_{\eta;i}/kT)$ in Eq. (IV-7) is the statistical weight for the ηth sharp minimum, and z_i is the sum of these statistical weights. The probability for occupation of state η at bond i for a chain with independent bonds is

$$p_{\eta;i} = z_i^{-1} \exp\left(\frac{-E_{\eta;i}}{kT}\right) \tag{IV-8}$$

Broader minima, or minima of unequal widths, may require the use of several points, or a factor before the exponential that incorporates information about the breadth of the minimum, for an adequate description, as will be seen in the examples to be discussed in Section IV.3. Even when the barriers between minima are relatively low, and the rotational isomers are poorly defined, it may still be an acceptable approximation to replace the integrals by summations that contain an adequate number of terms. Of course, the magnitude of Z will depend

[2]Volkenstein, M. V. in *Configurational Statistics of Polymeric Chains*, Timasheff, S. N.; Timasheff, M. J., eds., Wiley-Interscience, New York, **1963**.

on the range of the index in each sum. The consequences of this dependence are more important for properties that are evaluated directly from $\ln Z$ (such as entropy) than for properties that are extracted from a derivative of $\ln Z$. One might partially correct for the discrepancy introduced by replacing the integral form for Z by a summation of exponential terms: Multiplying the sum with a factor of $2\pi/\nu_i$ for each torsion angle $\phi_2 \cdots \phi_{n-1}$, where ν_i is the number of states for ϕ_i, would yield the integral form as all $\nu_i \to \infty$. The precision of the partition functions thus obtained is still not sufficient for many applications, however, and one usually refrains from corrections of this kind.

2. EXTRACTION OF INFORMATION FROM Z

The usual manipulations that evaluate a derivative of $\ln Z$ can be employed for extraction of information from Z. The amount by which the average conformational energy exceeds the arbitrarily selected zero is

$$\langle E \rangle - E_0 = kT^2 \left(\frac{\partial \ln Z}{\partial T} \right) \tag{IV-9}$$

Care must be taken in treating properties that depend directly on $\ln Z$, as illustrated by the conformational entropy. The magnitude of Z is influenced by the number of terms used in each of the summations in Eqs. (IV-6) and (IV-7), and consequently a conformational entropy evaluated from

$$S = \frac{\langle E \rangle - E_0}{T} + k \ln Z \tag{IV-10}$$

is at the mercy of the number of terms in this sum. Most of the conformational information derived from Z is obtained from a derivative of $\ln Z$, and consequently it is less sensitive than S to the conversion from the integral form of Z to the form based on discrete summations.

Figure IV-2 provides a schematic description of the dependence of $\ln Z$ on T. One curve (for present purposes, it does not matter which one) represents the behavior of $\ln Z$ when the evaluation is via the integrals in Eq. (IV-1). The other curve represents a successful replacement (for the purposes of properties that depend on a derivative of $\ln Z$) of the integrals by summations with a finite number of terms, Eq. (IV-6). The replacement is deemed "successful" because it faithfully reproduces the behavior of $\partial \ln Z/\partial T$. It does not, however, give the correct value of $\ln Z$ at any T. Consequently the replacement is adequate for evaluation of $\langle E \rangle - E_0$ via Eq. (IV-9), but it cannot be used for evaluation of S via Eq. (IV-10).

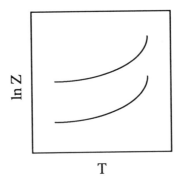

Figure IV-2. Schematic behavior of ln Z for two formulations of the conformational partition function.

3. Z FOR CHAINS WITH INDEPENDENT BONDS

A. First-Order Interactions

Figure IV-3 depicts three consecutive bonds in a generalized chain molecule. The bonds are indexed successively from 1 to 3. Atoms at either end of bond i are denoted by A_{i-1} and A_i. The distance between A_{i-1} and A_i is determined by the length of bond i, and the distance between atoms A_{i-2} and A_i is determined by the lengths of bonds $i - 1$ and i, as well as the angle, θ_{i-1}, between these bonds. In many applications of Z, the bond lengths and bond angles can be considered to be constant. The interactions between A_{i-1} and A_i, and between A_{i-2} and A_i, are then independent of the conformation of the chain.

Constancy of bond lengths and bond angles does not enforce a constant distance between A_{i-2} and A_{i+1}. If all bond lengths and bond angles are identical, the distance between these two atoms is (in Figure IV-3, take $i = 2$)

$$r_{i-2;i+1} = (3 - 4\cos\theta + 2\cos^2\theta - 2\sin^2\theta\cos\phi_i)^{1/2}l \qquad (IV-11)$$

which depends on a variable, the torsion angle, ϕ_i, that describes rotation about

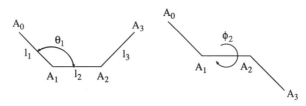

Figure IV-3. Three successive bonds in a chain. The value of ϕ_i is $0°$ in the drawing at the left and $180°$ in the drawing at the right.

bond i. Since rotation about internal bonds is the primary means by which a macromolecule changes its internal conformation, the torsion angle is a variable that must be considered in the construction of Z. The interaction of A_{i-2} and A_{i+1} is termed a *first-order interaction* because it is an interaction that is controlled by a single variable, ϕ_i.

If the bonds are independent, the conformational partition function for this bond, z_i, will be determined by the manner in which the energy of interaction of A_{i-2} and A_{i+1} depends on ϕ_i. A consideration of the first-order interactions is then sufficient for the formulation of Z, as shown by Eq. (IV-3). Only in very unusual circumstances will it be necessary to separately evaluate $n-2$ different z_i. If the macromolecule is a homopolymer, the z_i will repeat in a sequence that is determined by the number of backbone bonds in a monomer unit. In a simple chain with independent bonds, all of which are identical, the expression for Z becomes

$$Z = z^{n-2} \tag{IV-12}$$

B. Evaluation of z_i

Either Eq. (IV-4) or (IV-7) is a candidate for the evaluation of z_i. The selection should be guided by the manner in which the conformational energy depends on ϕ_i and the nature of the conformation-dependent physical property that is the target of the calculation. The basis for this statement will be illustrated by consideration of four profiles of E_i γs. ϕ_i.

Free Rotation. Consider first the very simple profile depicted in Fig. IV-4a, where E_i is independent of ϕ_i, being zero everywhere. Bond i is subject to free rotation. Implementation of Eq. (IV-4) immediately yields $z_i = 2\pi$. Utilization of Eq. (IV-7) invokes the approximation of a continuously accessible range of ϕ_i by a finite number of suitably chosen torsion angles. For example, the three angles $180°$, $-60°$, and $60°$ sample the available conformational space at equally spaced intervals. If the statistical weight for each of the three angles is formulated as $\exp(-E_i/kT)$, summation of the three statistical weights as required in Eq. (IV-7) yields $z_i = 3$. This interpretation of z_i correctly captures the fact that all three assigned states for ϕ_i are equally probable, but it does not reflect the continuous nature of the accessible ϕ_i. The consequences of the approximation will depend on the property that is ultimately calculated. Both expressions for z_i will lead to exactly the same result for the mean square unperturbed end-to-end distance, but they will produce different distribution functions for the end-to-end distance, $W(r)$, one distribution function being continuous and the other being discrete.

Energy Profiles Represented by Square Wells. The energy profile in Fig. IV-4b has three narrow square wells, each of width $\Delta\phi$, centered at $\phi_i = 180°$, $-60°$, and $60°$. Energies outside these square wells are so high that they can

74

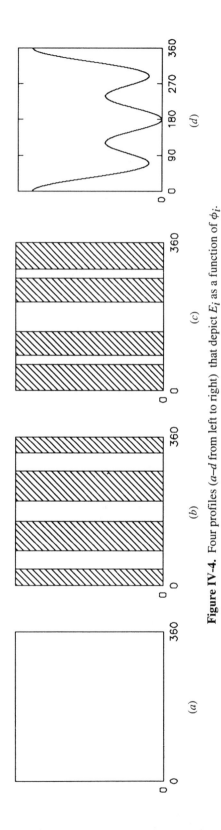

Figure IV-4. Four profiles (*a–d* from left to right) that depict E_i as a function of ϕ_i.

be considered to be infinite. The result for z_i from Eq. (IV-4) is $3\Delta\phi$. The probability for occupancy of a specified well is $z_i^{-1} \int \exp(-E_i/kT) d\phi_i$, with the integration extending over the well in question. That probability is $1/3$. The discrete summation via Eq. (IV-7) yields $z_i = 3$. The probability of occupancy of a specified well is $z_i^{-1} \exp(-E_i/kT)$, where the energy is that found at the well in question. This probability is also $1/3$.

The energy profile in Fig. IV-4c has three square wells, but the well centered at $180°$ has twice the width of the other wells. The expression for z_i from Eq. (IV-4) is $4\Delta\phi$. Simple utilization of three angles, $180°$, $60°$, and $-60°$, in Eq. (IV-7) will yield $z_i = 3$, a result that is identical with the one obtained from the energy profile in Fig. IV-4b. The identity of these two z_i is undesirable because it does not capture the important fact that the t well in Fig. IV-4c should be weighted as heavily as the two g wells combined. Figures IV-4b and IV-4c demonstrate that it may be misleading to simply assign one term in Eq. (IV-7) to each minimum in the energy profile, with $\exp(-E_i/kT)$ evaluated from the energy at the minimum. The energy profile in Fig. IV-4c is better represented by either of two alternatives.

The first alternative assigns four, rather than three, states to the energy profile in Fig. IV-4c. These states are assigned torsion angles of $180° + \Delta\phi/2$, $180° - \Delta\phi/2$, $60°$, and $-60°$. Each assigned state is in the middle of nonoverlapping regions of ϕ_i, of width $\Delta\phi$, in which E_i is zero. (This approach clearly requires that all square wells have widths that are integral multiples of $\Delta\phi$.) Equation (IV-7) yields $z_i = 4$, which signifies that the four states are weighted equally. The second alternative retains three states, but weights each state as the integral of $\exp(-E_i/kT)$ over the entire well. This alternative is of more general application, because it does not require widths for the square wells that are integral multiples of $\Delta\phi$. The integration permits a greater contribution to z_i from the well centered at $\phi_i = 180°$. The result for z_i will be $4\Delta\phi$, which is numerically the same as that obtained from Eq. (IV-4). However, the detailed models differ. The model from Eq. (IV-4) views each ϕ_i as being continuously variable over each well. The second alternative from Eq. (IV-7), on the other hand, envisions only three discrete values of ϕ_i, with $\phi_i = 180°$ being weighted twice as heavily as either $\phi_i = -60°$ or $\phi_i = 60°$. And the first alternative from Eq. (IV-7) envisions four discrete values of ϕ_i that are equally weighted.

Which model with $z_i = 4$ is preferred? The question loses significance as $\Delta\phi \to 0$. The three models for the chain become indistinguishable from one another if the wells are sufficiently narrow. When the wells are of finite width, there is generally a preference for the simplest model that can successfully account for measured conformation-dependent physical properties. The simplest model is often the one with the fewest states. Whether this model can describe experimental data in a satisfactory manner must be investigated on a case-by-case basis.

Continuous Energy Profiles. Figure IV-d depicts a continuous energy profile in which there are three minima. This profile is more realistic than those that contain square wells. It is similar to the energy profile for the internal C—C bond in n-butane (Fig. III-3). Utilization of Eq. (IV-7) will produce a z_i of the form $1 + 2\sigma$, which is the same as the result reported in Table III-2.[3] This form is easily understood if one state is assigned to each low-energy region. The ϕ_i characteristic of each state can be taken to be the value that produces the minimum in the conformational energy if the profile is symmetric (as it is in the t state), but may be displaced from the minimum if the profile is not symmetric (as in the two g states). Then σ is $\exp[-(E_g - E_t)/kT]$, where E_g and E_t are the energies at the g and t minima, respectively. An alternative approach, which becomes more desirable as the shapes near the minima become more diverse, is interpretation of σ as

$$\sigma = \frac{\int_g \exp[-(E_{\phi_g} - E_0)/kT]\,d\phi}{\int_t \exp[-(E_{\phi_t} - E_0)/kT]\,d\phi} \tag{IV-13}$$

where the first integration runs over a g well, the second integration runs over the t well, and E_0 is an arbitrarily chosen zero for the energy. In both cases, σ represents the statistical weight of a g state relative to the t state. The t state is arbitrarily assigned a statistical weight of 1. The value of σ can be greater or less than 1, depending on whether the chain prefers g or t states.[4] The probability for the t state at this bond is

$$p_{t;i} = \frac{1}{z_i} \tag{IV-14}$$

and the probability for a particular g state at this bond is

$$p_{g^+;i} = p_{g^-;i} = \frac{\sigma}{z_i} \tag{IV-15}$$

In a simple chain with n identical independent bonds, Eqs. (IV-14) and (IV-15) also give the fractions of all internal bonds that are in the three states.

The examples in Figs. IV-4b, IV-4c, and IV-4d depict energy profiles with only three minima. Of course, applications are not restricted to energy profiles with three minima. Furthermore, the number of terms in Eq. (IV-7) need not be identical with the number of minima, as shown by the possible description of free rotation with three states and the description of Fig. IV-4c with four states.

[3] Abe, A.; Jernigan, R. L.; Flory, P. J. *J. Am. Chem. Soc.* **1966,** *88,* 631.
[4] Polyethylene and polyoxymethylene are examples of chains with $\sigma < 1$ and $\sigma > 1$, respectively.

4. Z FOR CHAINS WITH PAIRWISE INTERDEPENDENT BONDS

A. Second-Order Interactions

Figure IV-5 depicts four different conformations of four consecutive bonds in a simple chain molecule. The values of ϕ_i and ϕ_{i-1} are both 180° in Fig. IV-5a. In Fig. IV-5b ϕ_{i-1} is 60°. This conformational change alters the distance between A_i and A_{i-3}. Since the separation of these two atoms is controlled by a single variable, ϕ_{i-1}, there is a change in a first-order interaction. If the weighting scheme were the one discussed in connection with n-butane, the statistical weight contributed by bond $i - 1$ would change from 1 in Fig. IV-5a to σ in Fig. IV-5b.

The change in internal conformation from Fig. IV-5b to Figs. IV-5c and IV-5d is accomplished by a rotation of 120° about bond i. These rotations are in opposite directions, with the value of ϕ_i being 60° in Fig. IV-5c and −60° in Fig. IV-5d. The separation of A_{i-2} and A_{i+1} in Figs. IV-4c and IV-4d is the same as the separation of A_{i-3} and A_i. A weighting scheme based exclusively on first-order interactions will not discriminate between the conformations depicted in Figs. IV-5c and IV-5d. Each conformation is assigned a statistical weight of σ^2; one factor of σ is contributed by each of the two internal bonds. This weighting scheme would be sufficient if the bonds were independent.

Independence of the bonds implies that the distance-dependent interaction between A_{i-3} and A_{i+1} makes no contribution to Z. This assumption will be invalid if the interaction of these two atoms provides different contributions to the conformational energies in the four structures depicted in Fig. IV-5. At

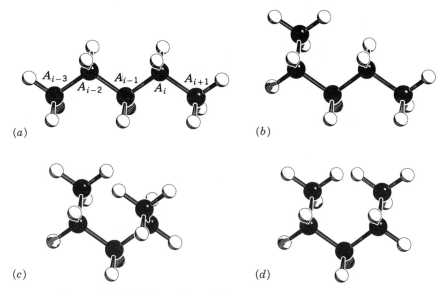

Figure IV-5. Conformations of four successive bonds in a chain.

TABLE IV-1. Decrease in Distances Controlling Second-Order Interactions, $r = r_{i-3,i+1}$, When All Bond Lengths and Bond Angles Are Identical, and $\phi = 180°, \pm 60°$.

Transition at bonds $i-1$ and i	$[r(180°, 180°) - r(\phi_{i-1}, \phi_i)]/l$	$[r(180°, 180°) - r(\phi_{i-1}, \phi_i)]/l$ (Tetrahedral Bond Angles)
$tt \rightarrow g^+t$	$3^{1/2}\sin\theta$	$(8/3)^{1/2} \sim 1.63$
$tt \rightarrow g^+g^+$	$3^{1/2}[(3/2)(1-\cos\theta)]^{1/2}\sin\theta$	$(16/3)^{1/2} \sim 2.31$
$tt \rightarrow g^+g^-$	$3^{1/2}[(1/2)(5-3\cos\theta)]^{1/2}\sin\theta$	$8^{1/2} \sim 2.83$

issue is a conformational energy surface that depicts only the energy of the interaction of A_{i-3} with A_{i+1}, as functions of ϕ_{i-1} and ϕ_i. If the undulations on this surface are small compared to kT, the *second-order interaction* can be neglected in good approximation. Conversely, the second-order interaction must be incorporated into Z if this energy surface has strong variations in regions of ϕ_{i-1}, ϕ_i that are allowed by first-order interactions.

The basis for the emphasis on the g^+g^- and g^-g^+ conformations in the introduction of second-order interactions is shown in Table IV-1. This table contains the reduction in the separation of A_{i-3} and A_{i+1} when ϕ_{i-1}, ϕ_i change from the tt state to other states, for the case where all bond lengths and bond angles are identical. The reduction in the separation of A_{i-3} and A_{i+1} is largest when the transition is $tt \rightarrow g^+g^-$ (or, equivalently, $tt \rightarrow g^-g^+$). Frequently the separation of A_{i-3} and A_{i+1} becomes so small that their interaction is repulsive in the conformation depicted in Fig. IV-5d, where there are two successive g placements of opposite sign.[3] In the other three conformations, the larger separation of these two atoms (Table IV-1) often causes their interaction to be so weak that it can be ignored in the formulation of Z.[5] A realistic weighting scheme must usually produce a substantially smaller statistical weight for the conformation in Fig. IV-5d than for the conformation in Fig. IV-5c. This objective is obtained by the introduction of a new statistical weight, ω, for a second-order interaction. The interaction is of *second-order* because the separation of the interacting atoms, A_{i-3} and A_{i+1}, is controlled by two variables, ϕ_{i-1} and ϕ_i. The presence of second-order interactions means the bonds are interdependent. The energy of a specified conformation is given by Eq. (IV-5).

B. Special Case of a Symmetric Threefold Torsion Potential Energy Function

Internal bonds in a linear polyethylene chain are subject to a symmetric threefold torsion potential energy function. The first-order interactions do not

---

[5] A more general matrix, to be introduced in Problem IV-10, can accommodate cases where important second-order interactions occur when two successive bonds adopt g placements of the same sign (Fig. IV-5c) or t placements (Fig. IV-5a).

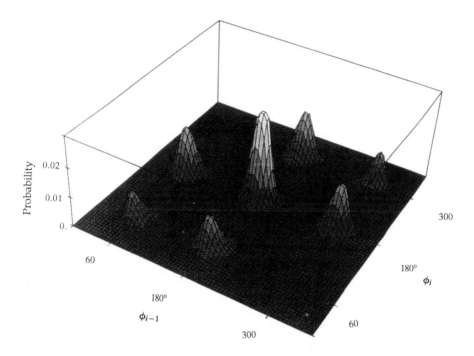

Figure IV-6. Probability profile for two consecutive C—C bonds in an isolated polyethylene chain, extracted from a long molecular dynamics trajectory at 300 K. Reprinted with permission from Bahar, I.; Zúñiga, I.; Dodge, R.; Mattice, W. L. *Macromolecules* **1991**, *24*, 2986. Copyright 1991 American Chemical Society.

discriminate between g^+ and g^-. The ratio of the statistical weight of either a g^+ or g^- state in n-butane to the statistical weight of the t state is denoted by σ (recall Table III-2). If the bonds were independent, an unperturbed chain would have $Z = (1 + 2\sigma)^{n-2}$. However, the bonds in polyethylene are not independent. There is a highly repulsive interaction of A_{i-3} and A_{i+1} in the conformation depicted in Fig. IV-5d.[3,6] The magnitude of this repulsion can be appreciated from inspection of Fig. IV-6, which depicts the probability profile for occupancy of ϕ_{i-1} and ϕ_i at internal bonds in an isolated polyethylene chain.[7] Peaks are easily observed in the two corners that correspond to g placements of the same sign at these two bonds, but peaks are barely perceptible in the two corners where the g placements are of opposite sign.[8] The formalism for the evaluation

[6]Tsuzuki, S.; Schäfer, L.; Gotō, H.; Jemmis, E. D.; Hosoya, H.; Siam, K.; Tanabe, K.; Ōsawa, E. *J. Am. Chem. Soc.* **1991**, *119*, 4665.
[7]Bahar, I.; Zúñiga, I.; Dodge, R.; Mattice, W. L. *Macromolecules* **1991**, *24*, 2986.
[8]The regions of low energy in the conformational energy surface for n-pentane, depicted as Fig. III-5, appear as peaks in the probability profile for a pair of successive internal C—C bonds in a long random-coil polyethylene (Fig. IV-6).

of Z must be modified so that the interdependence of the bonds is recognized.[3] This formalism will produce $Z < (1 + 2\sigma)^{n-2}$ if $\omega < 1$ and $n > 3$.

Assuming that the interdependence of the bonds is restricted to nearest neighbors, Eq. (IV-5) shows that the conformational energy of the chain can be broken down into local contributions in a manner that ignores the interdependence only when the first term, E_2, is considered. The interdependence is incorporated on consideration of subsequent terms, where $2 < i < n$. Consider first the statistical weights that might arise from E_2 in Eq. (IV-5). These statistical weights account for the first-order interaction of A_0 and A_3. They are 1, σ, and σ for t, g^+, and g^- states, respectively, at bond 2 . It will prove convenient to write these statistical weights as a row vector denoted $\mathbf{R}_2$. Each element in $\mathbf{R}_2$ is the statistical weight arising from E_2 when ϕ_2 occupies the specified rotational isomer. In general, $\mathbf{R}_2$ contains ν_2 elements, where ν_i denotes the number of rotational isomers at bond i:

$$\begin{array}{ccc} t & g^+ & g^- \end{array}$$
$$\mathbf{R}_2 = \begin{bmatrix} 1 & \sigma & \sigma \end{bmatrix}. \tag{IV-16}$$

In the following development, it will be important to understand a broad interpretation of each element in $\mathbf{R}_2$ and of the sum of these elements. *Simply stated*, the first element in $\mathbf{R}_2$ is the contribution to the statistical weight by bond 2 when this bond is in the t state. *With equal justification*, the row vector can be described as $\mathbf{R}_i$, with a first element that is the sum of the contributions made to the statistical weight of the chain by bonds 2 through bond i when the last bond considered (bond i) is in state t. This element from $\mathbf{R}_i$ can be denoted by $\sum w_{\ldots t;i}$. Similarly, the second and third elements are the sums of the contributions made to the statistical weight by bonds 2 through bond i when the last bond considered is in state g^+ or g^-, respectively.[9]

$$\mathbf{R}_i = \begin{bmatrix} \sum w_{\ldots t;i} & \sum w_{\ldots g^+;i} & \sum w_{\ldots g^-;i} \end{bmatrix} \tag{IV-17}$$

The sum of the elements in $\mathbf{R}_2$ gives Z in the special case where $n = 3$. A broader interpretation is that the sum of the elements in $\mathbf{R}_i$ is the contribution to Z made by rotation about bonds 2 through i. This sum is also equal to Z for a chain of $i + 1$ bonds; Z for a chain of n bonds is

$$Z = \mathbf{R}_{n-1} \begin{bmatrix} 1 \\ 1 \\ 1 \end{bmatrix} \tag{IV-18}$$

The $E_{i-1,i}$ in Eq. (IV-5) provide a contribution to the conformational energy that depends on the states of two successive bonds. If each bond has three

---

[9]This interpretation is cumbersome when $n = 3$, but has advantages at larger n.

accessible states, all possible contributions from $E_{i-1,i}$ to the statistical weight of the chain can be arranged as the nine elements in a 3×3 matrix. It will be convenient to index the rows and columns in the same sequence as was done for the row vector in Eq. (IV-16). The columns index the state of bond i, and the rows index the state of bond $i - 1$. The elements of the matrix are the statistical weights that arise from the first-order interaction of A_{i-2} and A_{i+1} and the second-order interaction of A_{i-3} and A_{i+1}.[10] The matrix is denoted by $\mathbf{U}_i$.

$$\downarrow \text{State of bond } i$$

$$\mathbf{U}_i \;=\; \begin{array}{c} \\ t \\ g^+ \\ g^- \end{array} \begin{array}{c} t \quad g^+ \quad g^- \\ \left[\begin{array}{ccc} 1 & \sigma & \sigma \\ 1 & \sigma & \sigma\omega \\ 1 & \sigma\omega & \sigma \end{array} \right]_i \end{array} \tag{IV-19}$$

State of bond $i - 1 \uparrow$

It is called the *statistical weight matrix* for bond i. The statistical weight for the second-order interaction in the g^+g^- and g^-g^+ conformations is denoted by ω. A statistical weight for a second-order interaction cannot occur in $\mathbf{R}_2$, because it would imply an interaction between chain atom A_3 and nonexistent "atom A_{-1}." The second-order interactions are incorporated into the weighting scheme as soon as it is physically sensible to do so, and that position is at the statistical weight matrix for bond 3. In general, jth-order interactions can first occur in $\mathbf{U}_{j+1}$.

If a chain were to contain only four bonds, Z would be the sum of the statistical weights for the nine conformations denoted by tt, g^+t, g^-t, tg^+, g^+g^+, g^-g^+, tg^-, g^+g^-, and g^-g^- (recall Table III-3). The statistical weights for all of the first- and second-order interactions are present in Eqs. (IV-16) and (IV-19). Consider the consequences of processing this information by formation of the product $\mathbf{R}_2\mathbf{U}_3$. The product is a new row that will be denoted by $\mathbf{R}_3$.

$$\mathbf{R}_3 \equiv \mathbf{R}_2\mathbf{U}_3 = \begin{bmatrix} \overset{tt+g^+t+g^-t}{1+\sigma+\sigma} & \overset{tg^++g^+g^++g^-g^+}{\sigma+\sigma^2+\sigma^2\omega} & \overset{tg^-+g^+g^-+g^-g^-}{\sigma+\sigma^2\omega+\sigma^2} \end{bmatrix}$$

$$\tag{IV-20}$$

The first element in $\mathbf{R}_3$ is obtained by multiplication of $\mathbf{R}_2$ onto the first col-

---

[10]The statistical weight matrix for bond i does not include a statistical weight for the first-order interaction of A_i and A_{i-3} because that first-order interaction is counted in the consideration of the bond in which the value of the index i is smaller by one. For example, when $i = 3$, the statistical weights that arise from the interaction of A_4 with A_1 and A_0 are found in $\mathbf{U}_3$, but the statistical weights that arise from the interaction of A_3 with A_0 are in $\mathbf{R}_2$.

umn in U_3. Inspection of the indexing for the elements in R_2, and the rows and columns in U_3 shows that the combined indices describe the conformations tt, g^+t, and g^-t. In each case, bond 3 is indexed as t because all elements from U_3 come from the first column. The statistical weights for these three conformations are 1, σ, and σ, respectively. Their sum appears in the first element in R_3. Similarly, the second element of R_3 contains the statistical weights for tg^+, g^+g^+, and g^-g^+, which consider all conformations in which bond 3 is g^+. The remaining three conformations, in which bond 3 is in state g^-, contribute the statistical weights found in the third element of R_3.

The elements in R_3 conform to the broad interpretation of R_i that was introduced in the paragraph immediately after Eq. (IV-16). For example, the first element is the sum of the statistical weights contributed by bonds 2 and 3 when bond 3 is in the t state. Hence Z for a chain of four bonds is the sum of the elements in R_3. More importantly, the procedure used to generate R_3 from R_2 can be repeated to give new rows, R_4, R_5, and so on, in which the elements retain the significance in the broad interpretation of R_i. The repetitive multiplication of R_{i-1} onto U_i can be continued without restraint on the upper limit for i. All first- and second-order interactions are incorporated. Interactions of longer range can be included by using larger matrices, as will be demonstrated in subsequent chapters.

A more concise equation can be developed by recognition that R_2 is often[11] the first row of U:

$$R_2 = [1 \quad 0 \quad 0] \, U_2 \tag{IV-21}$$

and the summation of the three elements in R_i is obtained by multiplication of R_i onto a column consisting of three elements, each of which is 1. Therefore Z is given by[12]

$$Z = U_1 U_2 U_3 \cdots U_n = \prod_{i=1}^{n} U_i \tag{IV-22}$$

where U_i, $1 < i < n$, are given by Eq. (IV-19), and the first and last statistical weight matrices are

$$U_1 = [1 \quad 0 \quad 0] \tag{IV-23}$$

---

[11] An exception is presented in Problem IV-4. For some chains U is formulated such that the first row contains a statistical weight for a second-, or higher-, order interaction. In general, R_i and U_i can contain statistical weights for interactions of order no higher than $i - 1$.
[12] An alternative notation for Eq. (IV-22) is $Z = U_1^{(n)}$.

$$\mathbf{U}_n = \begin{bmatrix} 1 \\ 1 \\ 1 \end{bmatrix} \tag{IV-24}$$

When all of the $\mathbf{U}_i$ are identical for $1 < i < n$, Eq. (IV-22) can be simply written as

$$Z = \mathbf{U}_1 \mathbf{U}^{n-2} \mathbf{U}_n \tag{IV-25}$$

An equivalent route to the conformational partition function is to write the statistical weights for first- and second-order interactions in different matrices, $\mathbf{D}_i$ and $\mathbf{V}_i$, respectively. The matrix of statistical weights for the first-order interactions is diagonal, and the matrix of statistical weights for the second-order interactions is symmetric:

$$\mathbf{D}_i = \text{diag}(1, \sigma, \sigma)_i \tag{IV-26}$$

$$\mathbf{V}_i = \begin{bmatrix} 1 & 1 & 1 \\ 1 & 1 & \omega \\ 1 & \omega & 1 \end{bmatrix}_i \tag{IV-27}$$

Here Z is generated as the sum of all elements of the serial product $\mathbf{D}_2 \mathbf{V}_3 \mathbf{D}_3 \cdots \mathbf{V}_{n-1} \mathbf{D}_{n-1}$, in which every matrix is either diagonal or symmetric, and $\mathbf{V}_i$ appears between $\mathbf{D}_{i-1}$ and $\mathbf{D}_i$. This expression is made identical with Eq. (IV-22) by appropriate combination of the matrices. The statistical weight matrix $\mathbf{U}_i$ is $\mathbf{V}_i \mathbf{D}_i$, and $[1 \quad 1 \quad 1]\mathbf{D}_2 = [1 \quad 0 \quad 0]\mathbf{U}_2$.

Although ω is much less than 1 for polyethylene chains, it is nevertheless worthwhile to consider the consequences of an assignment of the value of 1 to this statistical weight. The rows of $\mathbf{U}_i$ are then identical, each being $[1 \quad \sigma \quad \sigma]$. Identity of the rows means that the contribution to the statistical weight introduced via $\mathbf{U}_i$ is independent of the state adopted at bond $i-1$. Serial multiplication of the statistical weight matrices as specified in Eq. (IV-25) will now yield $Z = (1 + 2\sigma)^{n-2}$. This result is correct for the case where $\omega = 1$, but it could have been obtained with less labor by recognition that the identity of the rows in all of the $\mathbf{U}_i$ signals that the bonds are independent. With the independence of bonds, Z is given by Eq. (IV-12), which is a product of scalars.

5. REDUCTION WHEN TORSION POTENTIAL ENERGY FUNCTIONS ARE SYMMETRIC

Inspection of Eq. (IV-16) and (IV-20) shows that numerically identical elements appear in the second and third columns of $\mathbf{R}_i$. The identity of these two elements follows from the fact that the statistical weight of any conformation is equal to the statistical weight of its mirror image when the torsion potential energy function is *symmetric*. The second element of $\mathbf{R}_i$ contains the sums of the statistical weights of all conformations of bonds 2 through i in which bond i is g^+, and the third element contains an equivalent sum for all conformations of bonds 2 through i in which bond i is g^-. The two elements would not be identical if the torsion potential energy function were asymmetric, that is, if the first-order interaction for g^+ were different from the first-order interaction for g^-, or if the second-order interaction for g^+g^- were different from the second-order interaction for g^-g^+.

The symmetry of the torsion potential energy function permits computation of Z via statistical weight matrices that are more compact than the 3×3 matrix defined in Eq. (IV-19),[13] because we really do not need to evaluate separately the second and third elements of $\mathbf{R}_i$. The improvement in the efficiency of the computations is not important for many applications, because the calculations with the unreduced matrices require CPU (central processing unit) and memory that are easily supplied by the current generation of computers. An exception is provided by some of the higher moments that will be discussed in Chapter XIII. For the purposes of the present chapter, the condensation is important because it provides an easy entry into the expression of Z by eigenvalues.

Consider the 2×2 statistical weight matrix defined in Eq. (IV-28).

$$
\downarrow \text{State of bond } i
$$

$$
\begin{array}{cc} & t \quad\; g^+ + g^- \end{array}
$$

$$
\mathbf{U}_i = \begin{array}{c} t \\ g^+ + g^- \end{array} \left[\begin{array}{cc} 1 & 2\sigma \\ 1 & \sigma(1+\omega) \end{array} \right]_i \qquad \text{(IV-28)}
$$

$$
\text{State of bond } i-1 \uparrow
$$

The index for the first row and first column is t, as it was in Eq. (IV-19). The remaining index is $g^+ + g^-$, which combines the two symmetric g states. The value of $u_{1,1}$ is 1, as it was in Eq. (IV-19). Element $u_{1,2}$ describes the contribution from bond i when it is in either g state, given that there is a t state at bond $i-1$. This contribution is σ for either g state. The factor of two arises because this element now incorporates both types of g placements. Element $u_{2,1}$ is 1 because the statistical weight contributed by a t placement is always 1. The remaining element combines the contributions when both bonds adopt g placements that are of the same sign (σ) and opposite signs ($\sigma\omega$). Insertion of Eq. (IV-28) in Eq. (IV-25), along with $\mathbf{U}_1 = [1 \quad 0]$ and $\mathbf{U}_n^T = [1 \quad 1]$, yields exactly the same result as Eqs. (IV-19), (IV-23), (IV-24), and (IV-25). The sum of the two elements in Eq. (IV-30) is the same as the sum of the three elements

[13]Nagai, K. *J. Chem. Phys.* **1965**, *42*, 516; *J. Chem. Phys.* **1968**, *48*, 5646. Flory, P. J.; Abe, Y. *J. Chem. Phys.* **1971**, *54*, 1351.

in Eq. (IV-20).

$$\mathbf{R}_2 = [1 \quad 0]\mathbf{U}_2 = [1 \quad 2\sigma] \tag{IV-29}$$

$$\mathbf{R}_3 = \mathbf{R}_2\mathbf{U}_3 = [1 + 2\sigma \quad 2\sigma + 2\sigma^2(1 + \omega)] \tag{IV-30}$$

The general procedure that is applied in the condensation of $\mathbf{U}$ for chains with symmetric torsion potential energy functions has been summarized by Flory.[14] The general description employs a separation of the rotational isomeric states into two types, based on whether they are, or are not, distinguishable from their mirror image. Those rotational isomeric states that are distinguishable from their mirror image are denoted by Greek letters, with superscripts $+$ and $-$ denoting the members of each pair. The minimal set of rotational isomers for implementation of the reduction is α^+ and α^-. These two rotational isomers must be equally probable, because we have assumed the rotation potential energy function is symmetric. Other pairs, denoted by β^+ and β^-, and so on, are permissible. The description may also contain any number (including zero) of rotational isomers that are indistinguishable from the mirror images. These rotational isomers are denoted by $0, 1, \ldots$.

In the present case, the stated $0, \alpha^+, \alpha^-$ are identified with t, g^+, g^-. Two rectangular matrices are defined in general as

$$\mathbf{X}^0 = \begin{array}{c} \\ 0 \\ \alpha^{\pm} \\ \beta^{\pm} \\ \vdots \end{array} \begin{array}{ccccc} 0 & \alpha^+ & \alpha^- & \beta^+ & \beta^- \quad \cdots \\ \left[\begin{array}{ccccc} 1 & 0 & 0 & 0 & 0 \quad \cdots \\ 0 & \frac{1}{2} & \frac{1}{2} & 0 & 0 \quad \cdots \\ 0 & 0 & 0 & \frac{1}{2} & \frac{1}{2} \quad \cdots \\ \vdots & \vdots & \vdots & \vdots & \vdots \quad \ddots \end{array}\right] \end{array} \tag{IV-31}$$

$$\mathbf{Y}^0 = \begin{array}{c} \\ 0 \\ \alpha^+ \\ \alpha^- \\ \beta^+ \\ \beta^- \\ \vdots \end{array} \begin{array}{ccc} 0 & \alpha^{\pm} & \beta^{\pm} \quad \cdots \\ \left[\begin{array}{ccc} 1 & 0 & 0 \quad \cdots \\ 0 & 1 & 0 \quad \cdots \\ 0 & 1 & 0 \quad \cdots \\ 0 & 0 & 1 \quad \cdots \\ 0 & 0 & 1 \quad \cdots \\ \vdots & \vdots & \vdots \quad \ddots \end{array}\right] \end{array} \tag{IV-32}$$

which, in the present case, become

[14]Flory, P. J. *Macromolecules* **1974**, 8, 381.

$$\mathbf{X}^0 = \begin{array}{c} \\ 0 \\ \alpha^{\pm} \end{array} \begin{array}{ccc} 0 & \alpha^+ & \alpha^- \\ \left[\begin{array}{ccc} 1 & 0 & 0 \\ 0 & \frac{1}{2} & \frac{1}{2} \end{array} \right] \end{array} \qquad \text{(IV-33)}$$

$$\mathbf{Y}^0 = \begin{array}{c} \\ 0 \\ \alpha^+ \\ \alpha^- \end{array} \begin{array}{cc} 0 & \alpha^{\pm} \\ \left[\begin{array}{cc} 1 & 0 \\ 0 & 1 \\ 0 & 1 \end{array} \right] \end{array} \qquad \text{(IV-34)}$$

In general, the dimensions of $\mathbf{X}^0$ are $(\nu + 1)/2 \times \nu$, and $\mathbf{Y}^0$ is $\nu \times (\nu + 1)/2$. For the present case, these dimensions are 2×3 and 3×2, respectively. The elements in $\mathbf{X}^0$, denoted by x_{jk}^0, are of three types:

$x_{jk}^0 = 1$ if row j and column k are indexed by the same integer.
$x_{jk}^0 = \frac{1}{2}$ if row j and column k are indexed by the same Greek letter.
All remaining $x_{jk}^0 = 0$.

For $\mathbf{Y}^0$ the elements are

$y_{jk}^0 = 1$ if row j and column k are indexed by the same integer or Greek letter.
All remaining $y_{jk}^0 = 0$.

If these rules are applied to a chain that has no pair of symmetric states (no states indexed by Greek letters), both $\mathbf{X}^0$ and $\mathbf{Y}^0$ are the identity matrix, $\mathbf{I}_\nu$.

The product $\mathbf{Y}^0\mathbf{X}^0$ is of dimensions $\nu \times \nu$, and therefore is of the same dimensions as $\mathbf{U}_i$, $1 < i < n$. If the rows and columns of $\mathbf{U}_i$ are indexed in the order 0, α^+, α^-, β^+, β^-, and so on, then

$$\mathbf{U}_i\mathbf{Y}^0\mathbf{X}^0 = \mathbf{U}_i \qquad \text{(IV-35)}$$

and, for the two special cases of the row represented by $\mathbf{U}_1$ and the column represented by $\mathbf{U}_n$

$$\mathbf{U}_1\mathbf{Y}^0\mathbf{X}^0 = \mathbf{U}_1 \qquad \text{(IV-36)}$$

$$\mathbf{Y}^0\mathbf{X}^0\mathbf{U}_n = \mathbf{U}_n \qquad \text{(IV-37)}$$

Interdigitation of $\mathbf{Y}^0\mathbf{X}^0$ between successive $\mathbf{U}_i$ in Eq. (IV-25) will not affect the final result for Z. This interdigitation yields

$$Z = \mathbf{U}_1\mathbf{Y}^0\mathbf{X}^0\mathbf{U}_2\mathbf{Y}^0\mathbf{X}^0\mathbf{U}_3\mathbf{Y}^0\mathbf{X}^0 \cdots \mathbf{Y}^0\mathbf{X}^0\mathbf{U}_{n-1}\mathbf{Y}^0\mathbf{X}^0\mathbf{U}_n \qquad \text{(IV-38)}$$

where the serial indices have been placed on the internal $\mathbf{U}_i$. Conceptually the matrices in the serial product can be grouped (without changing the order of their appearance in the product) as

$$Z = (\mathbf{U}_1\mathbf{Y}^0)(\mathbf{X}^0\mathbf{U}_2\mathbf{Y}^0)(\mathbf{X}^0\mathbf{U}_3\mathbf{Y}^0)\mathbf{X}^0 \cdots \mathbf{Y}^0(\mathbf{X}^0\mathbf{U}_{n-1}\mathbf{Y}^0)(\mathbf{X}^0\mathbf{U}_n) \qquad \text{(IV-39)}$$

where $\mathbf{U}_1\mathbf{Y}^0$ is a row of $(\nu + 1)/2$ elements, $\mathbf{X}^0\mathbf{U}_n$ is a column of $(\nu + 1)/2$ elements, and the internal $\mathbf{X}^0\mathbf{U}_i\mathbf{Y}^0$ are square matrices of dimensions $(\nu + 1)/2 \times (\nu + 1)/2$. Hence the square matrices that appear as $\mathbf{U}^{n-2}$ in Eq. (IV-25) have been reduced in dimensions from $\nu \times \nu$ to $(\nu+1)/2 \times (\nu+1)/2$. Application of this procedure to the 3×3 matrix in Eq. (IV-19) yields the 2×2 matrix in Eq. (IV-28).

6. IMPLEMENTATION OF MATRIX DIAGONALIZATION

When applicable,[15] the implementation of matrix diagonalization simplifies the numeric evaluation of Z because the product of matrices is replaced by an expression using scalars. This procedure employs a diagonal matrix, $\mathbf{\Lambda}$, of the same dimensions as $\mathbf{U}$. The elements on the diagonal are the eigenvalues, λ_i, indexed such that λ_1 is the largest. The procedure leads to an expression for Z in terms of the λ_i. This expression may take on a very simple form as $n \to \infty$, and therefore it greatly reduces the labor required for evaluation of the properties of very long chains.[16] A simple illustration is presented in this section. Further details on diagonalization are presented elsewhere.[17]

When the $\mathbf{U}_i$ are identical for $1 < i < n$, the substitution of $\mathbf{U} = \mathbf{A}^{-1}\mathbf{\Lambda}\mathbf{A}$ into Eq. (IV-25), and cancellation of each $\mathbf{A}\mathbf{A}^{-1}$, yields[18]

$$Z = \mathbf{U}_1\mathbf{A}^{-1}\mathbf{\Lambda}^{n-2}\mathbf{A}\mathbf{U}_n \qquad \text{(IV-40)}$$

---

[15] The procedure is applicable whenever the $\mathbf{U}_i$ in Z appear in a pattern that repeats every j bonds. The illustration here uses $j = 1$.

[16] By "greatly reduces the labor" we do not imply that Eq. (IV-25) is unusable as it stands. On the contrary, it is easily evaluated by any microcomputer, even for large n. But it would be extremely tedious to evaluate by hand. The expression obtained via diagonalization can easily be evaluated by hand.

[17] See, for example, Perrin, C. L. *Mathematics for Chemists*, Wiley-Interscience, New York, **1970**, Chapter 8, and Kahn, P. B. *Mathematical Methods for Scientists and Engineers*, Wiley-Interscience, New York, **1990**, Chapter 1.

[18] $\mathbf{A}$ is the matrix that achieves the diagonalization of $\mathbf{U}$. The product of a matrix and its inverse, whether written as $\mathbf{A}\mathbf{A}^{-1}$ or as $\mathbf{A}^{-1}\mathbf{A}$, yields the identity matrix, $\mathbf{I}_\nu$.

Here Λ is a diagonal matrix with nonzero elements λ_i.[19] If U is a 2×2 matrix, as in Eq. (IV-28):

$$\Lambda = \begin{bmatrix} \lambda_1 & 0 \\ 0 & \lambda_2 \end{bmatrix} \qquad \text{(IV-41)}$$

Completion of the matrix multiplication required by Eq. (IV-25) leads to

$$Z = \frac{(1 - \lambda_2)\lambda_1^{n-1} + (\lambda_1 - 1)\lambda_2^{n-1}}{\lambda_1 - \lambda_2} \qquad \text{(IV-42)}$$

where, when U_i is given by Eq. (IV-28),

$$\lambda_1, \lambda_2 = \frac{1 + \sigma(1 + \omega) \pm \sqrt{(1 - \sigma(1 + \omega))^2 + 8\sigma}}{2} \qquad \text{(IV-43)}$$

with $\lambda_1 > \lambda_2$. In the limit as $n \to \infty$, Z approaches a simpler form

$$Z \to \left(\frac{1 - \lambda_2}{\lambda_1 - \lambda_2} \right) \lambda_1^{n-1} \qquad n \to \infty \qquad \text{(IV-44)}$$

which is often approximated as $Z \approx c\lambda_1^n$, or

$$\ln Z \approx n \ln \lambda_1 + \text{constant} \qquad n \to \infty \qquad \text{(IV-45)}$$

This approximation is especially useful when the objective is a property of an infinitely long chain that can be obtained from a derivative of $\ln Z$, because the problem is quickly converted via Eq. (IV-45) into a derivative of $\ln \lambda_1$, as illustrated by Problem IV-8. An example of the approach to the behavior predicted by Eq. (IV-45) is presented in Table IV-2. The values of σ and ω are those appropriate for polyethylene at 420 K.

Computation of Z for a simple chain via Eq. (IV-42) is rapid. However, the diagonalization procedure for computation of Z is not universally applicable. It is of little use for a chain in which the U_i do not describe a repeating pattern.

---

[19]The eigenvalues are the solution to the equation

$$\det |U - \lambda I| = 0$$

For the special case where U is a 2×2 matrix, as in Eq. (IV-28), the determinant can be solved for λ_1 and λ_2 by using the quadratic equation.

TABLE IV-2. Z and Its Approximation for Unperturbed Polyethylene ($\sigma = 0.54$, $\omega = 0.088$, $\lambda_1 = 1.8532575\ldots$, $\lambda_2 = -0.2657374\ldots$)

n	Z	$\ln Z$	$\ln Z - n \ln \lambda_1$
4	3.7945	1.3336	-1.1342
10	154.04	5.0372	-1.1322
30	3.5184×10^7	17.3761	-1.1322
100	2.0038×10^{26}	60.5623	-1.1322
300	7.7448×10^{79}	183.9512	-1.1322

Furthermore, as will be seen in a later chapter, it is convenient to leave Z in the matrix form, Eq. (IV-25), because this form facilitates incorporation of the chain geometry required for the computation of $\langle r^2 \rangle_0$ and other conformation-dependent physical properties. The chain geometry will be introduced in Chapter VI.

PROBLEMS

Appendix B contains answers for Problems IV-1, IV-2, and IV-4–IV-12.

IV-1. A chain has independent bonds subject to the threefold torsion potential energy function

$$\frac{\Delta E}{2} (1 + \cos 3\phi) \qquad\qquad \text{(IV-46)}$$

How might the evaluation and interpretation of z_i be affected by

(a) $\Delta E \ll kT$?   (b) $\Delta E \sim kT$?   (c) $\Delta E \gg kT$?

IV-2. Enumerate the statistical weights for the 27 conformations of n-hexane. Multiply $\mathbf{R}_3$ from Eq. (IV-20) onto the $\mathbf{U}$ in Eq. (IV-19) to generate $\mathbf{R}_4$. Verify that the sum of the three elements in $\mathbf{R}_4$ is Z for this chain.

IV-3. Verify that the statistical weight matrices in Eqs. (IV-19) and (IV-28) produce the same Z.

IV-4. Construct Z for a long chain that has a symmetric threefold torsion potential energy function in which the first-order interactions penalize g states relative to t states. Include two types of second-order interactions, one of which incorporates an attraction between A_{i-3} and A_{i+1} when there are t placements at bond $i - 1$ and i, and the other of which incorporates a repulsion between the same two atoms when bonds $i - 1$ and i adopt g placements of the same sign. Use a 3×3

statistical weight matrix. What property of this chain causes the first row of $\mathbf{U}_2$ to be different from the first row of $\mathbf{U}_i$, $2 < i < n$?

IV-5. The chain described in Problem IV-4 can also be treated using a 2×2 statistical weight matrix. What are the expressions for $\mathbf{U}_2$ and $\mathbf{U}_i$, $2 < i < n$?

IV-6. Estimate the average conformational energy at 420 K for a chain of 100 bonds described by Eqs. (IV-19), (IV-23), (IV-24), and (IV-25), $\sigma = \exp(-E_\sigma/RT)$, $\omega = \exp(-E_\omega/RT)$, $E_\sigma = 2.1$ kJ $\cdot$ mol^{-1}, and $E_\omega = 8.4$ kJ $\cdot$ mol^{-1}. Is it easier to obtain the solution analytically [from a derivative of Eq. (IV-25), as required by Eq. (IV-9), followed by substitution of appropriate values for σ, ω, and n] or numerically (by estimation of $\partial \ln Z/\partial T$ from numeric results computed at $T = 420 \pm \Delta T$)?

IV-7. Compare values of Z calculated by the exact relationship in Eq. (IV-42) and the approximate relationships in Eqs. (IV-44) and (IV-45), for $n = 5$, 10, and 100, using

(a) $\sigma = 1$, $\omega = 0.1$ (b) $\sigma = 0.1$, $\omega = 0.1$ (c) $\sigma = 0.01$, $\omega = 0.1$

(d) $\sigma = 0.1$, $\omega = 1$ (e) $\sigma = 0.1$, $\omega = 0.01$

IV-8. Convince yourself that the sum of the fraction of bonds in the g^+ and g^- states in the RIS model for unperturbed polyethylene is, in general

$$p_g = \frac{1}{n-2} \frac{\partial \ln Z}{\partial \ln \sigma} \tag{IV-47}$$

Also convince yourself that as $n \to \infty$, this result approaches $\partial \ln \lambda_1/\partial \ln \sigma$, which in turn is equal to $(\lambda_1 - 1)(\lambda_1 - \lambda_2)^{-1}$. [It may be helpful to realize that $\sigma(1 + \omega) = \lambda_1 + \lambda_2 - 1$ and $2\sigma = (\lambda_1 - 1)(1 - \lambda_2)$, both of which follow immediately from the definitions of λ_1 and λ_2.]

IV-9. Convince yourself that, for any n, the all-*trans* conformation of an unperturbed n-alkane is the conformation of largest statistical weight. Then calculate the dependence on n of the fraction of the population in the all-*trans* conformation at 420 K. In what range of n does it become unlikely that a 1-kg sample of an unperturbed n-alkane will contain *any* molecules in the all-*trans* conformation?

IV-10. A chain with a symmetric threefold torsion potential energy function, a special second-order interaction when two successive bonds occupy t states, and a special second-order interaction in g^+g^+ and g^-g^- states, and in the g^+g^- and g^-g^+ states, might have a statistical weight matrix given by

$$\mathbf{U} = \begin{bmatrix} \tau & \sigma & \sigma \\ 1 & \sigma\psi & \sigma\omega \\ 1 & \sigma\omega & \sigma\psi \end{bmatrix} \qquad \text{(IV-48)}$$

Reduce this 3×3 matrix to a 2×2 matrix.

IV-11. For the chain described in Problem IV-10, find the probability that two successive bonds will both be in t states, in the limit as $n \to \infty$.

IV-12. A famous model for a helix $\rightleftharpoons$ coil transition in a homopolymer uses a 2×2 statistical weight matrix[20]

$$\mathbf{U} = \begin{array}{c} \\ c \\ h \end{array} \begin{array}{c} c \quad h \\ \begin{bmatrix} 1 & \hat{\sigma}'s \\ 1 & s \end{bmatrix} \end{array} \qquad \text{(IV-49)}$$

where c denotes random coil and h denotes helix. The statistical weights are an initiation parameter (denoted here by $\hat{\sigma}'$, and typically on the order of 10^{-4}), and a temperature- and solvent-dependent propagation parameter (denoted by s, and typically in the range 0.5–2.0). This statistical weight matrix for the helix $\rightleftharpoons$ coil transition and the reduced form of the statistical weight matrix for the chain with a symmetric torsion potential energy function and pairwise interdependent bonds, Eq. (IV-28), are both of the form

$$\mathbf{U} = \begin{bmatrix} 1 & AB \\ 1 & B \end{bmatrix} \qquad \text{(IV-50)}$$

Why can a statistical weight matrix of the form in Eq. (IV-50) be used for an unperturbed flexible chain *and* for a helix $\rightleftharpoons$ coil transition? (Focus on the consequences for $p_{g\pm}$ and p_h, in the limit as $n \to \infty$, of the differences in the values of A for the two systems.)

---

[20]Zimm, B. H.; Bragg, J. K. *J. Chem. Phys.* **1959**, *31*, 526. The helix $\rightleftharpoons$ coil transition will be presented in Chapter X. Zimm and Bragg denoted the initiation parameter by σ, which we have replaced by $\hat{\sigma}'$ in order to avoid confusion with the statistical weight for a first-order interaction.

V Average Bond Conformations

This chapter describes simple procedures for extracting from Z the probabilities for occupancy of specific rotational isomeric states at specified bonds, pairs of consecutive bonds, triplets of bonds, and so on. In the next chapter this information will be combined with the chain geometry for rapid computation of the characteristic ratio via generator matrices, and also for estimation of the distribution function for r^2 via an efficient Monte Carlo simulation. The last section of the present chapter also demonstrates that the statistical weights are not uniquely determined by the probabilities for pairs of rotational isomeric states (but the probabilities are uniquely determined by the statistical weights).

1. AVERAGE BOND CONFORMATIONS FOR CHAINS WITH INDEPENDENT BONDS

Evaluation of the probabilities is simple when the bonds are independent. If bond i is independent and has statistical weight $u_{\eta;i}$ when it occupies state η, the probability for state η at bond i, denoted $p_{\eta;i}$, is

$$p_{\eta;i} = \frac{u_{\eta;i}}{z_i} \tag{V-1}$$

If the $n - 2$ internal bonds are not only independent but are also identical, the probability any internal bond in the chain occupies state η, p_η, is identical with $p_{\eta;i}$. This chain is not subject to end effects, because $p_{\eta;2}$ and $p_{\eta;n-1}$ are both equal to $p_{\eta;i}$ for $2 < i < n - 1$. Eq. (IV-14) and (IV-15) are the special cases for a symmetric threefold torsion potential energy function, where $z_i = 1 + 2\sigma$. The probability that two successive bonds occupy states ξ and $\eta, p_{\xi\eta;i}$, is the product of $p_{\xi;i-1}$ and $p_{\eta;i}$ when the bonds are independent. The terms $p_{\eta;i}$ and $p_{\xi\eta;i}$ are examples of *bond-* and *pair probabilities*, respectively.

The probability the chain of n independent bonds will be in any particular state, denoted p_κ, is the product of $n - 2$ bond probabilities, one for each of the internal bonds,

$$p_\kappa = \prod_{i=2}^{n-1} p_{\eta;i} \tag{V-2}$$

with the selection at each i of the appropriate state, η, for the conformation in question. There are ν^{n-2} different values for the index κ. This index is determined by $n - 2$ values for the subscript η, one for each of the $n - 2$ internal bonds. Each η has ν possible values, where ν represents the number of rotational isomeric states at an internal bond.

2. AVERAGE BOND CONFORMATIONS FOR CHAINS WITH INTERDEPENDENT BONDS

A. Bond and Pair Probabilities at a Specific Location along the Chain

The evaluation of bond and pair probabilities for chains with interdependent bonds is more complicated than the simple expression in Eq. (V-1). No longer is $p_{\xi\eta;i}$ given by $p_{\xi;i-1}p_{\eta;i}$. A more general definition of $p_{\eta;i}$, which applies to all chains, irrespective of whether the bonds are independent or interdependent, is

$$p_{\eta;i} = \frac{Z_{\eta;i}}{Z} \qquad \text{(V-3)}$$

where $Z_{\eta;i}$ is the *sum of the statistical weights of all conformations in which bond i is in state η*. The $Z_{\eta;i}$ for n-hexane are presented in Table V-1. The general definition of $Z_{\eta;i}$ reduces to the ratio of a single statistical weight to a bond conformational partition function, Eq. (V-1), only when the bonds are independent. Evaluation of Z was described in the preceding chapter. The new information required for computation of $p_{\eta;i}$ is the sum denoted by $Z_{\eta;i}$. This information is obtained by a proper sorting of the terms that contribute to Z, because every term appearing in $Z_{\eta;i}$ also appears in Z.

Evaluation of $Z_{\eta;i}$ is facilitated by the imposition of a detailed notation for the statistical weights. In the event $\mathbf{U}_i$ is of dimensions 3×3, and the interdependence of the bonds is confined to nearest neighbors, let the statistical weights be indexed as

$$\mathbf{U}_i = \begin{bmatrix} \sigma_t\omega_{tt} & \sigma_{g^+}\omega_{tg^+} & \sigma_{g^-}\omega_{tg^-} \\ \sigma_t\omega_{g^+t} & \sigma_{g^+}\omega_{g^+g^+} & \sigma_{g^-}\omega_{g^+g^-} \\ \sigma_t\omega_{g^-t} & \sigma_{g^+}\omega_{g^-g^+} & \sigma_{g^-}\omega_{g^-g^-} \end{bmatrix}_i \qquad \text{(V-4)}$$

TABLE V-1. $Z_{\eta;i}$ for n-Hexane

i	$Z_{t;i}$	$Z_{g^+;i} = Z_{g^-;i}$
2, 4	$1 + 4\sigma + 2\sigma^2 + 2\sigma^2\omega$	$\sigma + 3\sigma^2 + \sigma^2\omega + \sigma^3 + 2\sigma^3\omega + \sigma^3\omega^2$
3	$1 + 4\sigma + 4\sigma^2$	$\sigma + 2\sigma^2 + 2\sigma^2\omega + \sigma^3 + 2\sigma^3\omega + \sigma^3\omega^2$

The first-order statistical weight for state η is σ_η, and the second-order statistical weight for successive states $\xi\eta$ is $\omega_{\xi\eta}$. The subscript i at the end of Eq. (V-4) applies to all the statistical weights in the matrix. The statistical weight matrix used for polyethylene, Eq. (IV-19), is the special case where $\sigma_{t;i} = 1, \sigma_{g^+;i} = \sigma_{g^-;i} = \sigma, \omega_{g^+g^-;i} = \omega_{g^-g^+;i} = \omega$, and all remaining ω's $= 1$.

Intuitive Evaluation of the Probabilities. The intuitive approach to the $p_{\eta;i}$ is to obtain $Z_{\eta;i}$ by a modification of Z that gives a statistical weight of zero to every conformation that does not have state η at bond i. This modification merely requires replacement by $\mathbf{0}$ of all columns in $\mathbf{U}_i$ that are not indexed by the desired η. If the internal bonds in the chain have three rotational isomeric states, the three $p_{\eta;i}$ are obtained as

$$p_{\eta;i} = Z^{-1}\mathbf{U}_1\mathbf{U}_2 \cdots \mathbf{U}_{i-1}\mathbf{U}'_{\eta;i}\mathbf{U}_{i+1} \cdots \mathbf{U}_n \tag{V-5}$$

where the three possible expressions for $\mathbf{U}'_{\eta;i}$ are

$$\mathbf{U}'_{t;i} = \begin{bmatrix} \sigma_t\omega_{tt} & 0 & 0 \\ \sigma_t\omega_{g^+t} & 0 & 0 \\ \sigma_t\omega_{g^-t} & 0 & 0 \end{bmatrix}_i \tag{V-6}$$

$$\mathbf{U}'_{g^+;i} = \begin{bmatrix} 0 & \sigma_{g^+}\omega_{tg^+} & 0 \\ 0 & \sigma_{g^+}\omega_{g^+g^+} & 0 \\ 0 & \sigma_{g^+}\omega_{g^-g^+} & 0 \end{bmatrix}_i \tag{V-7}$$

$$\mathbf{U}'_{g^-;i} = \begin{bmatrix} 0 & 0 & \sigma_{g^-}\omega_{tg^-} \\ 0 & 0 & \sigma_{g^-}\omega_{g^+g^-} \\ 0 & 0 & \sigma_{g^-}\omega_{g^-g^-} \end{bmatrix}_i \tag{V-8}$$

depending on which one of the three states at bond i is state η.

This intuitive approach to the $p_{\eta;i}$ has several obvious extensions. It immediately gives the pair probability, such as the probability that bond j is in state ξ and bond i is in state η:

$$p_{\xi;j;\eta;i} = Z^{-1}\mathbf{U}_1\mathbf{U}_2 \cdots \mathbf{U}_{j-1}\mathbf{U}'_{\xi;j}\mathbf{U}_{j+1} \cdots \mathbf{U}_{i-1}\mathbf{U}'_{\eta;i}\mathbf{U}_{i+1} \cdots \mathbf{U}_n \tag{V-9}$$

with an obvious extension to higher-order probabilities. In the special case of the pair probability with $i - j = 1$, we also can write

$$p_{\xi\eta;i} = Z^{-1}\mathbf{U}_1\mathbf{U}_2 \cdots \mathbf{U}_{i-1}\mathbf{U}'_{\xi\eta;i}\mathbf{U}_{i+1} \cdots \mathbf{U}_n \tag{V-10}$$

where $U'_{\xi\eta;i}$ is obtained from U_i by zeroing every element except the one element that is indexed by row ξ and column η.

$$U'_{\xi;i-1}U'_{\eta;i} = U_{i-1}U'_{\xi\eta;i} \qquad (V\text{-}11)$$

It also gives the probability for *not* having a particular rotational isomeric state at bond i. For example, the probability for not have a t state at bond i is

$$1 - p_{t;i} = Z^{-1}U_1U_2 \cdots U_{i-1}\begin{bmatrix} 0 & \sigma_{g^+}\omega_{tg^+} & \sigma_{g^-}\omega_{tg^-} \\ 0 & \sigma_{g^+}\omega_{g^+g^+} & \sigma_{g^-}\omega_{g^+g^-} \\ 0 & \sigma_{g^+}\omega_{g^-g^+} & \sigma_{g^-}\omega_{g^-g^-} \end{bmatrix}_i$$
$$\cdot U_{i+1} \cdots U_n \qquad (V\text{-}12)$$

A More Formal Evaluation of the Probabilities. All of the equations in the previous section can be justified by a more formal process. In the case of the bond probability denoted by $p_{g^-;i}$, we note that the statistical weight of a specified conformation of the chain will be either independent of $\sigma_{g^-;i}$ (if the conformation has state t or g^+ at bond i) or directly proportional to $\sigma_{g^-;i}$ (if the conformation has state g^- at bond i). The conformational partition function is of the form

$$Z = A + B\sigma_{g^-;i} \qquad (V\text{-}13)$$

where A denotes the sum of the statistical weights of all conformations ($2 \times 3^{n-3}$ in number if $\nu = 3$) in which bond i is in state t or g^+, and $B\sigma_{g^-;i}$ is the sum of the statistical weights of all conformations (3^{n-3} in number) where bond i is in state g^-. The desired sum of statistical weights for $Z_{g^-;i}$ is simply $B\sigma_{g^-;i}$. This term is extracted from Eq. (V-13) by differentiation of Z with respect to $\sigma_{g^-;i}$ followed by multiplication by $\sigma_{g^-;i}$.[1]

$$B\sigma_{g^-;i} = \sigma_{g^-;i}\left(\frac{\partial Z}{\partial \sigma_{g^-;i}}\right) \qquad (V\text{-}14)$$

Consequently $p_{g^-;i}$ is

$$p_{g^-;i} = \frac{\sigma_{g^-;i}}{Z}\left(\frac{\partial Z}{\partial \sigma_{g^-;i}}\right) \equiv \frac{\partial \ln Z}{\partial \ln \sigma_{g^-;i}} \qquad (V\text{-}15)$$

In other words, it is extracted by an appropriate differentiation of $\ln Z$ as is

---

[1] Jernigan, R. L.; Flory, P. J. *J. Chem. Phys.* **1969**, *50*, 4165.

frequently done in statistical mechanics (e.g., Eq. IV-9). This procedure can immediately be extended to pair probabilities; for example

$$p_{g^+g^-;i} = \frac{\partial \ln Z}{\partial \ln \omega_{g^+g^-;i}}$$

(V-16)

Since $\sigma_{g^-;i}$ occurs only in $\mathbf{U}_i$, the derivative required in Eq. (V-15) is

$$\frac{\partial Z}{\partial \sigma_{g^-;i}} = \mathbf{U}_1 \mathbf{U}_2 \cdots \mathbf{U}_{i-1} \left(\frac{\partial \mathbf{U}_i}{\partial \sigma_{g^-;i}} \right) \mathbf{U}_{i+1} \cdots \mathbf{U}_n$$

(V-17)

$$\frac{\partial \mathbf{U}_i}{\partial \sigma_{g^-;i}} = \begin{bmatrix} 0 & 0 & \omega_{tg^-} \\ 0 & 0 & \omega_{g^+g^-} \\ 0 & 0 & \omega_{g^-g^-} \end{bmatrix}_i$$

(V-18)

Multiplication of the derivative by $\sigma_{g^-;i}$, as required by Eq. (V-15), yields $\mathbf{U}'_{g^-;i}$.

$$\sigma_{g^-;i} \frac{\partial \mathbf{U}_i}{\partial \sigma_{g^-;i}} = \mathbf{U}'_{g^-;i} \equiv \begin{bmatrix} 0 & 0 & \sigma_{g^-}\omega_{tg^-} \\ 0 & 0 & \sigma_{g^-}\omega_{g^+g^-} \\ 0 & 0 & \sigma_{g^-}\omega_{g^-g^-} \end{bmatrix}_i$$

(V-19)

Comparison of Eqs. (V-4) and (V-19) shows $\mathbf{U}'_{\eta;i}$ is obtained from $\mathbf{U}_i$ by replacement with 0 of the elements in all columns other than column η. Similarly, $\mathbf{U}'_{\xi\eta;i}$ retains $u_{\xi\eta}$ from $\mathbf{U}_i$, but renders all other elements null. The expressions for the bond and pair probabilities are identical with the ones obtained by the intuitive approach, Eqs. (V-5) and (V-10).[1]

Behavior of $p_{\eta;i}$ and $p_{\xi\eta;i}$. If all bonds are identical in a long chain, the dependence of $p_{\eta;i}$ and $p_{\xi\eta;i}$ on i is determined by the importance of the second-order interactions. Both $p_{\eta;i}$ and $p_{\xi\eta;i}$ would be independent of i if Z were determined completely by first-order interactions, that is, if the bonds were independent. This behavior is illustrated in Fig. V-1. The calculations are performed for a long chain in which the statistical weight matrix for each bond is

$$\mathbf{U} = \begin{bmatrix} 1 & 0.543 & 0.543 \\ 1 & 0.543 & 0.543\omega \\ 1 & 0.543\omega & 0.543 \end{bmatrix}$$

(V-20)

This statistical weight matrix is of the form expected when bond i is subject to a symmetric threefold torsion potential energy function and nearest-neighbor

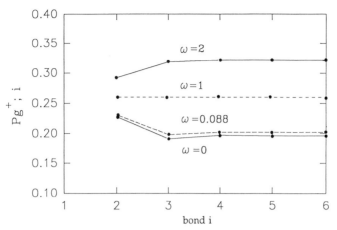

Figure V-1. Influence of ω on the behavior of $p_{g^+;i}$ near the end of a long chain. Since the torsion potential energy function is symmetric, $p_{g^+;i} = p_{g^-;i} = (1 - p_{t;i})/2$.

interdependence. The dependence of $p_{g^+;i}$ on i is confined to the region near the end of the chain, and this dependence disappears as $\omega \to 1$. The value of $p_{g^+;i}$ in the middle of a long chain decreases as ω decreases. Thus the bond probability, $p_{g^+;i}$ is determined by both the first- and second-order interactions.

When the elements of the statistical weight matrix are assigned the values appropriate for polyethylene at 140°C, the $p_{g^+;i}$ behave as depicted in Fig. V-2.[1] The statistical weight matrix used is the one specified by Eq. (V-20) with $\omega = 0.087$. Statistical weights are assigned as $-RT \ln \sigma = 2.1$ kJ mol^{-1} and $-RT \ln \omega = 8.4$ kJ mol^{-1}.[2] The end effects are rather small, and they are restricted to bonds near the ends of the chain. As $n \to \infty$ for an unperturbed polyethylene chain, the average $p_{g^+} \to p_{g^+;i}$, where $1 \ll i \ll n$, that is, where $p_{g^+;i}$ is evaluated for a bond far removed from either chain end.

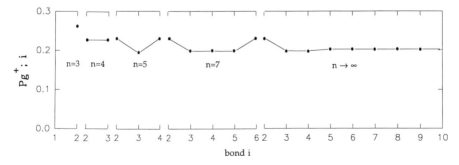

Figure V-2. Probability for a g^+ placement at bonds near the end of an unperturbed polyethylene chain at 140°C.

TABLE V-2. Relationships between the $p_{\eta;i}$ and $p_{\xi\eta;i}$ for a Chain with Identical Ends and Identical Pairwise Interdependent Bonds Subject to a Symmetric Threefold Rotation Potential Energy Function

Bond i	Bonds i and $i-1^a$	Bonds Equidistant from the Ends	Bond Pairs Equidistant from the Ends
$p_{g^+;i} = p_{g^-;i}$	$p_{g^+t;i} = p_{g^-t;i}$	$p_{\eta;i} = p_{\eta;n-i+1}$	$p_{\xi\eta;i} = p_{\eta\xi;n-i+2}$
	$p_{tg^+;i} = p_{tg^-;i}$		
	$p_{g^+g^+;i} = p_{g^-g^-;i}$		
	$p_{g^+g^-;i} = p_{g^-g^+;i}$		

^{a}If the bonds are sufficiently far removed from the ends so that end effects are no longer important, we also have $p_{g^+t;i} = p_{g^-t;i} = p_{tg^+;i} = p_{tg^-;i}$.

If all **U** are given by Eq. (V-20) with $\omega = 0.087$, a pair of bonds in the middle of a long chain has the $p_{\xi\eta;i}$ represented in matrix form as[1]

$$\mathbf{p}_{\xi\eta;i} = \begin{bmatrix} 0.321 & 0.138 & 0.138 \\ 0.138 & 0.0591 & 0.00516 \\ 0.138 & 0.00516 & 0.0591 \end{bmatrix}_i \tag{V-21}$$

and

$$p_{g^+;i} = p_{g^-;i} = \frac{1 - p_{t;i}}{2} = 0.202 \tag{V-22}$$

Both sets of probabilities are normalized, $\sum_\eta p_{\eta;i} = 1$ and $\sum_\xi \sum_\eta p_{\xi\eta;i} = 1$. Relationships between the probabilities for a simple chain are presented in Table V-2.

If the bonds were independent, a chain with a threefold symmetric torsion potential energy function would have $p_{tt;i} = p_{t;i-1}p_{t;i}, p_{tg^+;i} = p_{t;i-1}p_{g^+;i}$, and $p_{g^+g^+;i} = p_{g^+g^-;i} = p_{g^+;i-1}p_{g^+;i}$. None of these relationships between bond and pair probabilities is obeyed by the probabilities in Eqs. (V-21) and (V-22), however, because of the interdependence of the bonds.

B. Bond and Pair Probabilities Averaged over the Entire Chain

The average of $p_{\eta;i}$ over all $n - 2$ internal bonds always yields the fraction of the bonds in the chain that are in state η, irrespective of the value of n.

[2] Abe, A.; Jernigan, R. L.; Flory, P. J. *J. Am. Chem. Soc.* **1966,** *88,* 631.

$$p_\eta = \frac{1}{n-2} \sum_{i=2}^{n-1} p_{\eta;i} \tag{V-23}$$

The summation can be evaluated by a more efficient means than $n - 2$ independent executions of Eq. (V-5). The $n - 2$ terms in the sum are

$$p_{\eta;2} = Z^{-1} \mathbf{U}_1 \mathbf{U}'_{\eta;2} \mathbf{U}_3 \cdots \mathbf{U}_n \tag{V-24}$$

$$p_{\eta;3} = Z^{-1} \mathbf{U}_1 \mathbf{U}_2 \mathbf{U}'_{\eta;3} \mathbf{U}_4 \cdots \mathbf{U}_n \tag{V-25}$$

$$\vdots$$

$$p_{\eta;n-1} = Z^{-1} \mathbf{U}_1 \mathbf{U}_2 \cdots \mathbf{U}_{n-2} \mathbf{U}'_{\eta;n-1} \mathbf{U}_n \tag{V-26}$$

Each term contains one, and only one, primed statistical weight matrix. The bond that requires the primed statistical weight matrix has all values of the index i from 2 through $n - 1$. The desired sum is constructed by use of a series of row vectors that contain 2ν elements.[1] These elements are organized in two blocks of identical size. At bond 2, the elements are

$$\mathbf{R}_2 = [\mathbf{U}_1 \mathbf{U}_2 \quad \mathbf{U}_1 \mathbf{U}'_2] \tag{V-27}$$

The first block of ν elements in $\mathbf{R}_2$ is the product of the first two statistical weight matrices, and the last block replaces $\mathbf{U}_2$ with $\mathbf{U}'_2$. *More generally*, the first block in $\mathbf{R}_i$ is the product of the statistical weight matrices from $\mathbf{U}_1$ through $\mathbf{U}_i$, and the last block is the sum of all possible products of statistical weight matrices $\mathbf{U}_1$ through $\mathbf{U}_i$ where each matrix from $\mathbf{U}_2$ through $\mathbf{U}_i$ has its turn to bear the prime.

Consider the multiplication of $\mathbf{R}_2$ onto $\hat{\mathbf{U}}_{\eta;3}$, defined as the special case of $\hat{\mathbf{U}}_{\eta;i}$ for which $i = 3$:

$$\hat{\mathbf{U}}_{\eta;i} \equiv \begin{bmatrix} \mathbf{U} & \mathbf{U}'_\eta \\ \mathbf{0} & \mathbf{U} \end{bmatrix}_i \tag{V-28}$$

The matrix defined in Eq. (V-28) is of dimensions $2\nu \times 2\nu$, with twice as many rows, and twice as many columns, as $\mathbf{U}_i$. It is a matrix in which the elements are organized in four blocks. The two blocks on the main diagonal are $\mathbf{U}_i$, the block in the upper right is $\mathbf{U}'_{\eta;i}$ and the block in the lower left is null. The product of $\mathbf{R}_2$ and $\hat{\mathbf{U}}_{\eta;3}$ is

$$\mathbf{R}_3 \equiv \mathbf{R}_2 \hat{\mathbf{U}}_{\eta;3} = [\mathbf{U}_1 \mathbf{U}_2 \mathbf{U}_3 \quad \mathbf{U}_1 \mathbf{U}_2 \mathbf{U}'_{\eta;3} + \mathbf{U}_1 \mathbf{U}'_{\eta;2} \mathbf{U}_3] \tag{V-29}$$

The first block and the last block in $\mathbf{R}_3$ retain the general significance discussed in connection with $\mathbf{R}_2$. Occurrence of a primed matrix in the first block of $\mathbf{R}_3$ is rejected by the null block in $\hat{\mathbf{U}}_{\eta;3}$. The primed and unprimed matrices in the last ν columns of Eq. (V-28) are arranged so the last block of $\mathbf{R}_3$ will be the sum of $\mathbf{U}_1\mathbf{U}_2\mathbf{U}'_{\eta;3}$ and $\mathbf{U}_1\mathbf{U}'_{\eta;2}\mathbf{U}_3$. The procedure for generation of $\mathbf{R}_i$ from $\mathbf{R}_{i-1}$ can be continued through bond $n-1$.

$$\mathbf{R}_{n-1} = [\mathbf{U}_1\mathbf{U}_2 \cdots \mathbf{U}_{n-1} \quad \mathbf{U}_1\mathbf{U}'_{\eta;2}\mathbf{U}_3 \cdots \mathbf{U}_{n-1}$$
$$+ \mathbf{U}_1\mathbf{U}_2\mathbf{U}'_{\eta;3} \cdots \mathbf{U}_{n-1} + \cdots] \tag{V-30}$$

The sum of the probabilities in Eq. (V-23) is obtained by appending $\mathbf{U}_n$ to the end of the last ν elements in $\mathbf{R}_{n-1}$ and division of the resulting scalar by Z:

$$\sum_{i=2}^{n-1} p_{\eta;i} = \frac{1}{Z} [\mathbf{U}_1 \quad 0]\hat{\mathbf{U}}_{\eta;2}\hat{\mathbf{U}}_{\eta;3} \cdots \hat{\mathbf{U}}_{\eta;n-1} \begin{bmatrix} 0 \\ \mathbf{U}_n \end{bmatrix} \tag{V-31}$$

Division by the number of terms in the sum yields p_η, the fraction of bonds in the chain that are in state η:

$$p_\eta = \frac{1}{(n-2)Z} [\mathbf{U}_1 \quad 0]\hat{\mathbf{U}}_{\eta;2}\hat{\mathbf{U}}_{\eta;3} \cdots \hat{\mathbf{U}}_{\eta;n-1} \begin{bmatrix} 0 \\ \mathbf{U}_n \end{bmatrix} \tag{V-32}$$

If all bonds are identical, then

$$p_\eta = \frac{1}{(n-2)Z} [\mathbf{U}_1 \quad 0]\hat{\mathbf{U}}_\eta^{n-2} \begin{bmatrix} 0 \\ \mathbf{U}_n \end{bmatrix} \tag{V-33}$$

The average of the $n-3$ individual $p_{\xi\eta;i}, 2 < i < n$ is obtained in an analogous manner from the $\hat{\mathbf{U}}_{\xi\eta}$ constructed from the appropriate $\mathbf{U}'_{\xi\eta}$:

$$p_{\xi\eta} = \frac{1}{n-3} \sum_{i=3}^{n-1} p_{\xi\eta;i} \tag{V-34}$$

$$p_{\xi\eta} = \frac{1}{(n-3)Z} [\mathbf{U}_1\mathbf{U}_2 \quad 0]\hat{\mathbf{U}}_{\xi\eta}^{n-3} \begin{bmatrix} 0 \\ \mathbf{U}_n \end{bmatrix} \tag{V-35}$$

3. CONDITIONAL PROBABILITIES

When the bonds are interdependent, the probability for any single conformation of the chain cannot be extracted from Eq. (V-2) because the $p_{\eta;i}$ do not properly propagate the interdependence of the bonds. That interdependence is contained in the $p_{\xi\eta;i}$, and with proper manipulation they will provide a route to p_κ. The most convenient approach is via the *conditional probability*, $q_{\xi\eta;i}$, which is the probability bond i will be in state η, given we already know bond $i-1$ is in state ξ.[3]

$$q_{\xi\eta;i} = \frac{p_{\xi\eta;i}}{p_{\xi;i-1}} \tag{V-36}$$

The probability for a specific conformation of a long chain with pairwise interdependent bonds is given by $p_{\eta;2}$ and the $q_{\xi\eta;i}, 2 < i < n$, as

$$p_\kappa = p_{\alpha;2} q_{\alpha\beta;3} q_{\beta\gamma;4} \cdots q_{\psi\omega;n-1} \tag{V-37}$$

As an example, the $p_{\xi\eta;i}$ in Eq. (V-21) specify $q_{\xi\eta;i}$ represented in matrix form as

$$\mathbf{q}_{\xi\eta;i} = \begin{bmatrix} 0.538 & 0.231 & 0.231 \\ 0.682 & 0.292 & 0.026 \\ 0.682 & 0.026 & 0.292 \end{bmatrix}_i \tag{V-38}$$

This expression has simply renormalized each row of the matrix for $p_{\xi\eta;i}$ without changing the value of the ratio of any two elements in the row.

The distinction between the $p_{\xi\eta;i}$ and $q_{\xi\eta;i}$ can be drawn by contrasting two processes. If one were to select a chain at random, the probability bonds $i-1$ and i are in states ξ and η, respectively, is given by $p_{\xi\eta;i}$. However, if one looks at bond $i-1$, thereby gaining complete knowledge of its conformation, then the probabilities for the ν states at bond i are given by $q_{\xi\eta;i}$ using the known result for ξ. This latter process is pertinent in the stepwise assignment of the torsional angles for the $n-3$ internal bonds from bond 3 through bond $n-1$.

Rarely does one actually compute p_κ, which denotes the probability for a specified conformation of a long chain. However, it is useful to envision one could, in principle, compute all the p_κ, in order to average a conformation-dependent property over all conformations in an unperturbed ensemble. For some interesting properties, this incredibly tedious procedure can be replaced by a very efficient one, as we shall see in the next chapter.

---

[3] According to the definitions, $\sum_\eta p_{\xi\eta;i} = p_{\xi;i-1} \leq 1$, but $\sum_\eta q_{\xi\eta;i} = 1$. If bond $i-1$ populates more than one state, $\sum_\eta p_{\xi\eta;i} < 1$.

4. ESTIMATION OF STATISTICAL WEIGHTS FROM $p_{\xi\eta}$

This chapter has developed the relationships by which the probabilities for occupancy of different rotational isomeric states are extracted from the conformational partition function. Specifically, $p_{\eta;i}$ and $p_{\xi\eta;i}$ are obtained for chains with pairwise interdependent bonds via Eqs. (V-5) and (V-10), respectively. The methodology expressed by those equations permits computation of the probabilities from the statistical weights, which in turn are formulated from knowledge of the energies of the short-range interactions.

Now consider a situation where we do not have the energies of the short-range interactions, but instead have the probabilities at a particular temperature. In principle, one might obtain the probabilities from a sequence of experiments (perhaps Raman, nmr) that measure all of the p_η and $p_{\xi\eta}$. Let us assume the most favorable case, where experimental errors are nonexistent, and the theory for conversion from experiment to the probabilities is perfect.

The probabilities might also be obtained as the result of the computation of a molecular dynamics trajectory of suitable duration so bonds i and $i-1$ have attained an equilibrium sampling of their torsion angles. The data might appear as Fig. IV-6 for polyethylene, or as Figs. V-3 and V-4 for poly(dimethylsiloxane). The representation as probability surfaces has advantages for chains such as poly(dimethylsiloxane), which have other internal degrees of freedom coupled with ϕ_i and ϕ_{i-1}.

With polyethylene, the bond angles are constrained to values very close to the preferred value of 112°, and fixing θ at this value permits the calculation of an accurate conformational energy surface, ϕ_{i-1} and ϕ_i being the only two degrees of freedom. In contrast, in poly(dimethylsiloxane) θ_{SiOSi} is susceptible to extremely large fluctuations, and may actually attain a value of 180° with gain of only a fraction of a kilojoule per mole.[4] Arbitrary fixing of this bond angle at any particular value, in order to compute the conformational energy as a function of ϕ_{i-1} and ϕ_i, may yield a result that depends in important ways on the arbitrary selection of θ. However, a molecular dynamics trajectory, in which all internal degrees of freedom can be variables, will yield a probability surface for ϕ_{i-1} and ϕ_i that properly includes the coupling with all relevant internal degrees of freedom, including θ. Again, let us assume the most favorable case: the dynamics of the molecule obeys a known classical force fields, the integration of Newton's equations of motion is exact, and the trajectory is of sufficient duration to provide an equilibrium sampling of conformational space. If the probabilites were derived from a Monte Carlo simulation, we would also assume the most favorable case, where the number of iterations approaches ∞, and the rules for changing conformations are exact.

Under what circumstances can these perfect $p_{\eta;i}$ and $p_{\xi\eta;i}$, whether obtained from experiment, molecular dynamics, or a Monte Carlo simulation, be used for the formulation of Z? Can Z be recovered by the sequences

---

[4]Durig, J. R.; Flanagen, M. J.; Kalasinsky, V. F. *J. Chem. Phys.* **1977**, *66*, 2775.

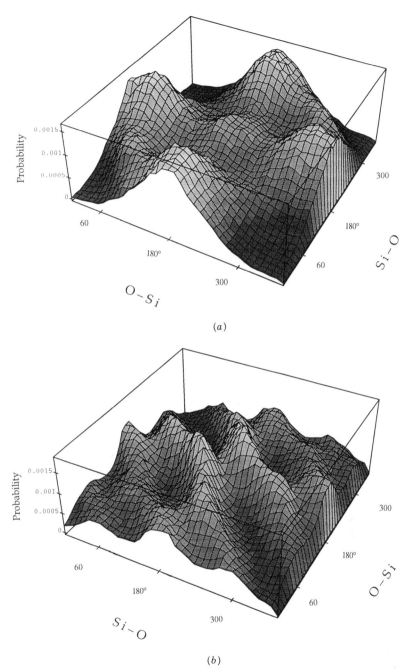

Figure V-3. Probability density at 300 K for two bonds that meet at a silicon atom (panel *a*) or an oxygen atom (panel *b*) in poly(dimethylsiloxane). Reprinted with permission from Bahar, I.; Zúñiga, I.; Dodge, R.; Mattice, W. L. *Macromolecules* **1991**, *24*, 2986. Copyright 1991 American Chemical Society.

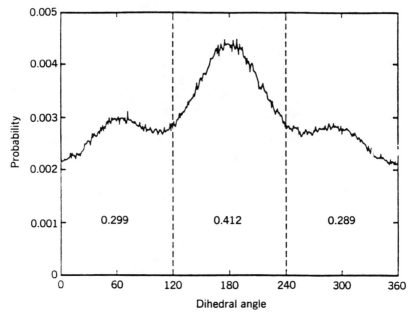

Figure V-4. Probability density at 300 K for an Si—O bond in poly(dimethylsiloxane). Reprinted with permission from Bahar, I.; Zúñiga, I.; Dodge, R.; Mattice, W. L. *Macromolecules* **1991**, *24*, 2986. Copyright 1991 American Chemical Society.

$$\left(\begin{array}{c} \text{Experiment} \\ \downarrow \\ p_{\eta;i} \quad \text{and} \quad p_{\xi\eta;i} \rightarrow \mathbf{U}_i \rightarrow Z \\ \uparrow \\ \text{Molecular dynamics or Monte Carlo} \end{array}\right) \qquad \text{(V-39)}$$

instead of the classic sequence

$$\text{Conformational energy surface} \;\rightarrow\; \mathbf{U}_i \;\rightarrow\; Z \qquad \text{(V-40)}$$

A. The Simple Chain

The formalism will be described in detail for the simplest case, poly(A), but it has been adapted to polymers such as poly(dimethylsiloxane), written as poly(A–B). Application of Eqs. (V-5) and (V-10) to a very long chain for which $\mathbf{U}_i$ for all internal bonds is given by Eq. (IV-48), which in condensed form (see Section IV.5) is

$$\mathbf{U} = \begin{bmatrix} \tau & 2\sigma \\ 1 & \sigma(\psi + \omega) \end{bmatrix} \tag{V-41}$$

yields, for a pair of bonds remote from either end

$$p_{g^+;i} + p_{g^-;i} = \frac{\partial \ln \lambda_1}{\partial \ln \sigma}, \tag{V-42}$$

$$p_{tt;i} = \frac{\partial \ln \lambda_1}{\partial \ln \tau} \tag{V-43}$$

$$p_{g^+g^+;i} + p_{g^-g^-;i} = \frac{\partial \ln \lambda_1}{\partial \ln \psi} \tag{V-44}$$

$$p_{g^+g^-;i} + p_{g^-g^+;i} = \frac{\partial \ln \lambda_1}{\partial \ln \omega} \tag{V-45}$$

$$\lambda_{1,2} = \frac{1}{2} \left\{ \tau + \sigma(\psi + \omega) \pm \sqrt{[\tau - \sigma(\psi + \omega)]^2 + 8\sigma} \right\} \tag{V-46}$$

and, since both sets of probabilities are normalized, $p_{t;i} = 1 - 2p_{g^+;i}$ and $p_{tg^+;i} = p_{tg^-;i} = p_{g^+t;i} = p_{g^-t;i} = \frac{1}{4}(1 - p_{tt;i} - 2p_{g^+g^+;i} - 2p_{g^+g^-;i})$. The four statistical weights can be written as four equations that can be solved iteratively, given the probabilities:

$$\sigma = (\tau - \lambda_2)^2 \frac{p_{g^+;i}}{p_{t;i}} \tag{V-47}$$

$$\psi^2 = \frac{1}{\sigma} \frac{p_{t;i}}{p_{g^+;i}} \left(\frac{p_{g^+g^+;i}}{p_{g^+t;i}} \right)^2 \tag{V-48}$$

$$\tau = \frac{1}{\psi} \frac{p_{tt;i}p_{g^+g^+;i}}{p_{tg^+;i}^2} \tag{V-49}$$

$$\omega = \psi \frac{p_{g^+g^-;i}}{p_{g^+g^+;i}} \tag{V-50}$$

These four equations specify a set of statistical weights that, when used in $\mathbf{U}$, produce a Z that recovers exactly the known $p_{\eta;i}$ and $p_{\xi\eta;i}$ for the bonds sufficiently far removed from either end of a long chain.

B. Lack of Uniqueness to the Solutions

Before becoming unduly enthusiastic about the likelihood of generating the statistical weights directly from the probabilities, it is instructive to examine the uniqueness of the solution.[5] Combination of Eqs. (V-42) and (V-46) with $p_{t;i} = 1 - 2p_{g^+;i}$ yields

$$p_{t;i} = \frac{\tau - \lambda_2}{\lambda_1 - \lambda_2} \tag{V-51}$$

The value of $p_{t;i}$ is unchanged if the numerator and denominator of Eq. (V-51) are multiplied by an arbitrary nonzero constant, c

$$p_{t;i} = \frac{c\tau - c\lambda_2}{c\lambda_1 - c\lambda_2} \equiv \frac{\tau_c - \lambda_{2,c}}{\lambda_{1,c} - \lambda_{2,c}} \tag{V-52}$$

where $\lambda_{1,2,c}$ are obtained from Eq. (V-46) with substitution of σ_c, τ_c, ψ_c, and ω_c for σ, τ, ψ, and ω, respectively.

$$\sigma_c = c^2\sigma, \qquad \tau_c = c\tau, \qquad \psi_c = \frac{\psi}{c}, \qquad \omega_c = \frac{\omega}{c} \tag{V-53}$$

Since c is arbitrary, there is an *infinite number of combinations* of σ_c, τ_c, ψ_c and ω_c that specify exactly the same set of $p_{\eta;i}$ and $p_{\xi\eta;i}$ as the "true" σ, τ, ψ, and ω. When one solves Eqs. (V-47)–(V-50) for the statistical weights, using the probabilities as input, it is not clear which member of this set will be obtained, that is, which value of c will correspond to the solution obtained. The $p_{\eta;i}$ and $p_{\xi\eta;i}$ for polyethylene at 140°C are reproduced just as well by any combination of σ_c, τ_c, ψ_c, and ω_c in Fig. V-5 as by the "true" $\sigma = 0.54$, $\tau = \psi = 1$, $\omega = 0.088$.

The "true" set of statistical weights can be recovered only if the true value of one of them ($\sigma, \tau, \psi, \omega$) has been established from other information. In practice, this objective is often achieved because the energy of at least one of these interactions is so weak that it can be equated to zero. The corresponding statistical weight can then be assigned a value of 1.

How severe are the consequences of the existence of an infinite set of statistical weights that determine the same $p_{\eta;i}$ and $p_{\xi\eta;i}$, $0 \ll i \ll n$? They are *irrelevant* if Z is to be used to compute $\langle r \rangle_0, \langle r^2 \rangle_0$ or any other conformation-dependent property for the unperturbed ensemble *at the temperature at which the $p_{\eta;i}$ and $p_{\xi\eta;i}$ themselves were calculated*. This statement follows from the fact that, while different choices of c will lead to numerically different results for Z, each of these Z will yield exactly the same set of $p_{\eta;i}$ and $p_{\xi\eta;i}$ and $\langle r \rangle_0, \langle r^2 \rangle_0$, and so on, are completely determined by those probabilities.

[5]Mattice, W. L.; Dodge, R.; Zúñiga, I.; Bahar, I. *Comput. Polym. Sci.* **1991**, *1*, 35.

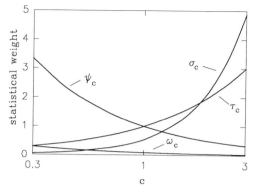

Figure V-5. σ_c, τ_c, ψ_c, and ω_c for polyethylene at 140°C.

The consequences are also nil for the temperature coefficients of the probabilities, and conformation-dependent properties determined by the temperature coefficients. The temperature coefficient for $p_{t;i}$ from Eq. (V-52), can be written as either

$$\frac{\partial \ln p_{t;i}}{\partial T} = \frac{\tau' - \lambda_2'}{\tau - \lambda_2} - \frac{\lambda_1' - \lambda_2'}{\lambda_1 - \lambda_2} \tag{V-54}$$

or

$$\left(\frac{\partial \ln p_{t;i}}{\partial T}\right)_c = \frac{\tau_c' - \lambda_{2,c}'}{\tau_c - \lambda_{2,c}} - \frac{\lambda_{1,c}' - \lambda_{2,c}'}{\lambda_{1,c} - \lambda_{2,c}} \tag{V-55}$$

where the prime, as a superscript, denotes the derivative with respect to T. Since $\tau_c', \lambda_{1,c}'$, and $\lambda_{2,c}'$ (as well as $\tau_c, \lambda_{1,c}$, and $\lambda_{2,c}$) are proportional to the arbitrarily chosen constant c, both equations specify the same temperature coefficient.

The energy of a conformation of the chain can be written in terms of the energies of the first- and second-order interactions as

$$E = n_g E_\sigma + n_{tt} E_\tau + n_{++} E_\psi + n_{+-} E_\omega \tag{V-56}$$

where n_g is the number of g states, without regard to sign, and n_{tt}, n_{++}, and n_{+-} are the number of pairs of consecutive bonds in tt states, g states of the same sign, and g states of opposite sign, respectively. The change in energy for the rotational isomerism denoted by $tg^+g^+ \rightarrow tg^-g^+$ is

$$\Delta E_{tg^+g^+ \rightarrow tg^-g^+} = E_\omega - E_\psi \tag{V-57}$$

which is unchanged if the same energy is added to both E_ω and E_ψ. In terms of Eq. (V-53), the ratio of ψ to ω is independent of c, specifically, $\psi/\omega = \psi_c/\omega_c$. However, if the rotational isomerism is $ttt \rightarrow tg^+t$ the change in energy is

$$\Delta E_{ttt \rightarrow tg^+t} = E_\sigma - 2E_\tau \qquad \text{(V-58)}$$

If an arbitrarily selected energy is added to E_τ, an energy twice that size must be added to E_σ if $\Delta E_{ttt \rightarrow tg^+t}$ is to be unaffected. In terms of Eq. (V-53), it is σ/τ^2 that must be independent of c, which, of course, requires that σ and τ individually depend on different powers of c.

The value of c is of importance if one wishes to identify a physically meaningful energy with a single statistical weight, via the relationship $E_w = -RT \ln w$. According to Eq. (V-53), the energy E_{w_c} that is extracted from w_c will differ from the true E_w by one or two terms of the size $RT \ln c$. Figure V-5 shows σ_c can be either less than 1 or greater than 1, depending on the size of c. The sign of the energy calculated as $-RT \ln \sigma_c$ becomes negative if c is larger than 1.36. The formulation of a Z from statistical weights evaluated via probabilities calculated (or measured) at a particular T is perfectly legitimate, but caution must be exercised in any attempt to extract physically meaningful energies from the statistical weights themselves, unless one has other information that provides the value of c. Similar considerations apply to poly(A–B).[5]

PROBLEMS

Appendix B contains answers for Problems V-2–V-10.

V-1. Figure V-2 depicts $p_{g^+;i}$ for several chains at 140°C, using the **U** in Eq. (IV-19), ω being assigned as 0.087. Calculate $p_{t,i}$ for these chains. Verify $p_{t,i} = 1 - 2p_{g^+,i}$.

V-2. When $\omega = 1$, verify Eq. (V-1) and Eq. (V-5) yield identical results for each bond probability at bond i.

V-3. For a short polyethylene chain (five bonds), calculate $p_{t;2}, p_{t;3}$ and $p_{t;4}$ from Eq. (V-5). Do the numeric results depend on the direction selected for indexing the bonds? Use the three $p_{t,i}$ to calculate p_t from Eq. (V-23), and verify the same result is obtained with Eq. (V-33).

V-4. Convince yourself that

$$\sum_\xi \sum_\eta p_{\xi\eta;i} = 1 \qquad \text{(V-59)}$$

$$\sum_{\xi} p_{\xi\eta;i} = p_{\eta;i} \qquad (V\text{-}60)$$

$$\sum_{\eta} p_{\xi\eta;i} = p_{\xi;i-1} \qquad (V\text{-}61)$$

$$\sum_{\eta} q_{\xi\eta;i} = 1 \qquad (V\text{-}62)$$

$$\sum_{\xi}\sum_{\eta} q_{\xi\eta;i} = \nu_{i-1} \qquad (V\text{-}63)$$

Is there a simple description of "?" in $\sum_{\xi} q_{\xi\eta;i} = ?$

V-5. Can you unambiguously generate the following information, given no data other than that specified in the question?

(a) Matrix of all $p_{\xi\eta;i}$ given $\mathbf{U}_i$?

(b) $\mathbf{U}_i$, given the matrix of all $p_{\xi\eta;i}$?

(c) Matrix of all $p_{\xi\eta;i}$, given the vector of all $p_{\eta;i}$?

(d) Vector of all $p_{\eta;i}$, given the matrix of all $p_{\xi\eta;i}$?

(e) Matrix of all $q_{\xi\eta;i}$, given the matrix of all $p_{\xi\eta;i}$?

(f) Matrix of all $p_{\xi\eta;i}$, given the matrix of all $q_{\xi\eta;i}$?

(g) Sign of $dp_{\xi\eta;i}/dT$, given the sign of $dq_{\xi\eta;i}/dT$?

(h) Sign of $dq_{\xi\eta;i}/dT$ given the sign of $dp_{\xi\eta;i}/dT$?

V-6. What is the average number of bonds in a sequence of t placements in unperturbed polyethylene over the temperature range 300–450 K?

V-7. Order the $p_{\xi\eta}$ for unperturbed polyethylene of very large n in the sequence of their sensitivity to temperature, as measured by the temperature coefficients $\partial \ln p_{\xi\eta;i}/\partial T$. Can you rationalize the signs and absolute values of the temperature coefficients?

V-8. If the bonds remote from either end of a long poly(A) have the $p_{\xi\eta;i}$ specified by Eq. (V-21), and $\sigma = 1$, what are the values of $\tau, \psi,$ and ω?

V-9. Assume bond i in the middle of a very long unperturbed polyethylene chain is in the g^+ state. How far must one proceed along the chain from bond i before $p_{g^-;i+j}$ becomes identical with p_{g^-}?

V-10. Let $\mathbf{p}_{(\xi;j)(\eta;i)}$ denote the matrix of probabilities that bond j is in state ξ and bond i is in state η. In this notation, the matrix in Eq. (V-21) is $\mathbf{p}_{(\xi;i-1)(\eta;i)}$. That matrix has elements with four distinct values, namely, 0.321, 0.138, 0.0591, and 0.00516. Explain why $\mathbf{p}_{(\xi;j)(\eta;i)}$ will contain elements with fewer distinct values at large $i - j$.

VI Mean-Square Unperturbed Dimensions for a Simple Chain

This chapter describes an efficient procedure, utilizing *generator matrices*, for exact computation of $\langle \mathbf{r} \rangle_0, \langle r^2 \rangle_0$, and $\langle s^2 \rangle_0$ for the simple chain for which Z was formulated in the preceding chapter. "Exact" means no mathematical approximations are used in the computation, and hence the evaluation of the model is exact. (The result is still at the mercy of the degree to which the model mimics reality.) The generator matrix technique was invented by Kramers and Wannier[1] for other purposes. Applications of this technique to polymer rotational isomeric states began in the late 1950s.[2]

We also describe a *Monte Carlo* method for estimation, with any degree of accuracy, of the distribution functions. The Monte Carlo calculations are less efficient than the generator matrix calculations, but they provide access to information that cannot be obtained using generator matrices.

Taken together, Chapters III–VI provide an introduction to applications of rotational isomeric state (RIS) theory that will be presented in subsequent chapters. One always needs a conformational partition function, and the procedure for its formulation was established in Chapters III and IV for a chain with pairwise interdependent bonds subject to a symmetric threefold rotation potential energy function. The basis for a description of the conformations of small molecules using RIS theory was presented in Chapter III, and the use of these concepts for the formulation of Z for a macromolecule was the topic of Chapter IV. Extraction from Z of the probabilities for various conformations of the chain was described in Chapter V. In the present chapter the information about the probabilities that is contained in Z is combined with geometric information for rapid computation of the averages.

1. LOCAL COORDINATE SYSTEMS AND TRANSFORMATION MATRICES

The end-to-end vector, $\mathbf{r}$, is the vector sum of n bond vectors connected head to tail. Execution of the sum requires that all the bond vectors be expressed in

[1] Kramers, H. A.; Wannier, G. H. *Phys. Rev.* **1941**, *60*, 252.
[2] Gotlib, Yu. Ya. *Zh. Tekhn. Fiz.* **1959**, *29*, 523. Birshtein, T. M.; Ptitsyn, O. B. *Zh. Tekhn. Fiz.* **1959**, *29*, 1048. Lifson, S. *J. Chem. Phys.* **1959**, *30*, 964. Nagai, K. *J. Chem. Phys.* **1959**, *31*, 1169. Hoeve, C. A. J. *J. Chem. Phys.* **1960**, *32*, 888.

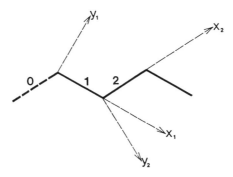

Figure VI-1. Local coordinate system defined by two consecutive bonds. From Abe, A.; Kennedy, J. W.; Flory, P. J. *J. Polym. Sci., Polym. Phys. Ed.* **1976**, *14*, 1337. Copyright © 1976 John Wiley & Sons, Inc. Reprinted by permission of John Wiley & Sons.

the same coordinate system. Temporarily using the superscript s to emphasize this fact, we see that the sum of the bond vectors is

$$\mathbf{r} = \sum_{i=1}^{n} \mathbf{l}_i^s \tag{VI-1}$$

Specification of the elements in $\mathbf{l}_i^s$ requires the length of the bond, the direction of the bond, and the orientation of the coordinate system in which $\mathbf{r}$ is to be expressed.

For present purposes, it is more convenient to define each bond vector so that it is completely specified by its length.[3] This objective is achieved by writing each $\mathbf{l}_i$ as it appears in a *local coordinate system* defined for bond i. The local coordinate system is centered on A_{i-1}, has an x axis that passes through A_i, a y axis in the plane of A_{i-2}, A_{i-1}, and A_i, and a z axis that completes a right-handed Cartesian coordinate system. The direction of the x axis is chosen so that the expression for $\mathbf{l}_i$ in this local coordinate system is

$$\mathbf{l}_i = \begin{bmatrix} l_i \\ 0 \\ 0 \end{bmatrix} \tag{VI-2}$$

The direction of the y axis is chosen so that the y coordinate of A_{i+1} is positive when $90° < \phi_i < 270°$. The first bond vector is a special case because there is no A_{i-2} for use in defining ϕ_1. The y axis for the initial coordinate system is taken to be in the plane of A_0, A_1, and A_2, and directed so that the y coordinate of A_2 is positive. Figure VI-1 depicts the local coordinate system.

[3]Flory, P. J. *Proc. Natl. Acad. Sci., USA* **1973**, *70*, 1819.

The orientation of all the l_i in a common coordinate system is handled by an operation that converts a vector expressed in the local coordinate system for bond $i + 1$ into the expression for the *same* vector in the local coordinate system for bond i. Repetitive application of this operation can express any l_i in the coordinate system defined for the initial bond. Using $\mathbf{v}^i$ and $\mathbf{v}^{i+1}$ as notations for the expression of the *same* vector in the local coordinate systems for bond i and $i + 1$, respectively, we obtain

$$\mathbf{v}^i = \mathbf{T}_i \mathbf{v}^{i+1} \tag{VI-3}$$

where $\mathbf{T}_i$ is a *transformation matrix*, of dimensions 3×3. The elements in $\mathbf{T}_i$ can be formulated from θ_i and ϕ_i. Recall that θ_i is the bond angle defined by A_{i-1}, A_i, and A_{i+1}. The torsion angle ϕ_i is zero for a *cis* placement of A_{i-2}, A_{i-1}, A_i, and A_{i+1}, and it increases when A_{i+1} rotates in a *clockwise* fashion (when viewed along the line from A_{i-1} to A_i). With these definitions for θ_i, ϕ_i, and the local coordinate systems[4]

$$\mathbf{T}_1 = \begin{bmatrix} -\cos\theta & \sin\theta & 0 \\ \sin\theta & \cos\theta & 0 \\ 0 & 0 & -1 \end{bmatrix}_1 \tag{VI-4}$$

$$\mathbf{T}_i = \begin{bmatrix} -\cos\theta & \sin\theta & 0 \\ -\sin\theta\cos\phi & -\cos\theta\cos\phi & -\sin\phi \\ -\sin\theta\sin\phi & -\cos\theta\sin\phi & \cos\phi \end{bmatrix}_i \qquad i > 1 \tag{VI-5}$$

The expression for the end-to-end vector in terms of l_i and $\mathbf{T}_i$ is

---

[4]The composition of the first column of $\mathbf{T}_i$ is a consequence of the fact that a unit vector along the x axis of the coordinate system for bond $i + 1$ has components in the coordinate system of bond i given by

$$\begin{bmatrix} -\cos\theta \\ -\sin\theta\cos\phi \\ -\sin\theta\sin\phi \end{bmatrix}$$

Similarly, unit vectors along the y and z axes of the coordinate system for bond $i + 1$ have components in the coordinate system of bond i given, respectively, by

$$\begin{bmatrix} \sin\theta \\ -\cos\theta\cos\phi \\ -\cos\theta\sin\phi \end{bmatrix} \quad \text{and} \quad \begin{bmatrix} 0 \\ -\sin\phi \\ \cos\phi \end{bmatrix}$$

$$\mathbf{r} = \mathbf{l}_1 + \mathbf{T}_1\mathbf{l}_2 + \mathbf{T}_1\mathbf{T}_2\mathbf{l}_3 + \cdots + \mathbf{T}_1\mathbf{T}_2 \cdots \mathbf{T}_{n-1}\mathbf{l}_n$$

$$= \sum_{i=1}^{n} \left(\prod_{j=0}^{i-1} \mathbf{T}_j \right) \mathbf{l}_i \qquad \text{(VI-6)}$$

with $\mathbf{T}_0 \equiv \mathbf{I}_3$. The serial product of $i - 1$ transformation matrices converts each $\mathbf{l}_i$ into its representation in the coordinate system defined for bond 1. It is in this coordinate system that $\mathbf{r}$ is expressed.

2. MATRIX GENERATION OF $\mathbf{r}$, r^2, AND s^2

In this section we temporarily set aside our interest in the average of conformation-dependent physical properties over an enormous number of conformations. Instead we focus on the evaluation of three conformation-dependent properties for a *single* arbitrarily chosen conformation, without regard for whether the conformation selected has a high or a low probability in the ensemble. Averaging over all conformations in the ensemble will be addressed in Section VI-3.

A. The End-to-End Vector

The sum in Eq. (VI-6) can be evaluated in a manner similar to the evaluation of p_η via the blocked row vector and matrix defined in Eqs. (V-27) and (V-28), respectively.[3,5] In the present application the $\mathbf{R}_i$ contain three rows and four columns. However, they can be written more concisely using a block form that gives the appearance of a row vector because there is one block of dimensions 3×3 followed by a second block of dimensions 3×1. The expression for $\mathbf{R}_i$ for the case where $i = 1$ is

$$\mathbf{R}_1 = [\mathbf{T}_1 \quad \mathbf{l}_1] \qquad \text{(VI-7)}$$

The first block in $\mathbf{R}_1$ is the transformation matrix for bond 1, and the second block is $\mathbf{l}_1$. *More generally*, the first block of $\mathbf{R}_i$ is the product $\mathbf{T}_1\mathbf{T}_2 \cdots \mathbf{T}_i$, and the second block is the vector sum of all $\mathbf{l}_j$ for j from 1 through i, all $\mathbf{l}_j$ being expressed in the coordinate system of bond 1. The second block in $\mathbf{R}_i$ is the end-to-end vector for the first i bonds, $\mathbf{r}_{0i}$, expressed in the local coordinate system for the first bond, as depicted in Fig. VI-2:

$$\mathbf{R}_i = [\mathbf{T}_1\mathbf{T}_2 \cdots \mathbf{T}_i \quad \mathbf{r}_{0i}] \qquad \text{(VI-8)}$$

[5]Flory, P. J. *Macromolecules* **1974**, *7*, 381.

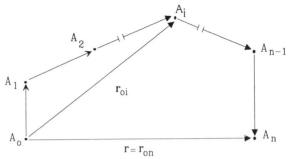

Figure VI-2. The r_{0i} for a short chain.

The counterpart of the block matrix in Eq. (V-28) is[3,5]

$$\mathbf{A}_i = \begin{bmatrix} \mathbf{T} & \mathbf{l} \\ \mathbf{0} & 1 \end{bmatrix}_i, \qquad 1 < i < n \qquad \text{(VI-9)}$$

where $\mathbf{A}_i$ is an example of a *generator matrix*. Other examples, designed for generation of other properties, will be encountered later in this chapter and in Chapters XIII and XIV. Generation of $\mathbf{R}_2$ as $\mathbf{R}_1 \mathbf{A}_2$ yields

$$\mathbf{R}_2 \equiv \mathbf{R}_1 \mathbf{A}_2 = [\mathbf{T}_1 \mathbf{T}_2 \quad \mathbf{T}_1 \mathbf{l}_2 + \mathbf{l}_1] \equiv [\mathbf{T}_1 \mathbf{T}_2 \quad \mathbf{r}_{02}] \qquad \text{(VI-10)}$$

Repetition of this procedure ultimately leads to

$$\mathbf{R}_n = [\mathbf{T}_1 \mathbf{T}_2 \cdots \mathbf{T}_n \quad \mathbf{T}_1 \mathbf{T}_2 \cdots \mathbf{T}_{n-1} \mathbf{l}_n + \cdots + \mathbf{T}_1 \mathbf{l}_2 + \mathbf{l}_1]$$

$$\equiv \left[\prod_{i=1}^{n} \mathbf{T}_i \quad \mathbf{r}_{0n} \right] \qquad \text{(VI-11)}$$

The second block is more compactly expressed as $\mathbf{r}$ since $\mathbf{T}_0 \equiv \mathbf{I}_3$. The first block in $\mathbf{R}_n$ is of no further interest because the transformation matrices have fulfilled their assignments when they generate $\mathbf{r}$. Since only the second block in $\mathbf{R}_n$ is needed, $\mathbf{A}_n$ can be written in a special form that is simply the last column of the general expression for $\mathbf{A}_i$:

$$\mathbf{r} = [\mathbf{T}_1 \quad \mathbf{l}_1] \mathbf{A}_2 \mathbf{A}_3 \cdots \mathbf{A}_{n-1} \begin{bmatrix} \mathbf{l}_n \\ 1 \end{bmatrix} \qquad \text{(VI-12)}$$

Alternatively, one can define $\mathbf{A}_1$ and $\mathbf{A}_n$ as the first row and last column of blocks, respectively, from the general form of $\mathbf{A}_i$:

$$\mathbf{r} = \mathbf{A}_1\mathbf{A}_2 \cdots \mathbf{A}_{n-1}\mathbf{A}_n \equiv \prod_{i=1}^{n} \mathbf{A}_i \tag{VI-13}$$

$$\mathbf{A}_1 = [\mathbf{T}_1 \quad \mathbf{l}_1] \tag{VI-14}$$

$$\mathbf{A}_n = \begin{bmatrix} \mathbf{l}_n \\ 1 \end{bmatrix} \tag{VI-15}$$

and $\mathbf{A}_i$ given by Eq. (VI-9) for $1 < i < n$. While written here in terms of the end-to-end vector and bond vectors, Eq. (VI-6) and the development that follows apply to any vector that is the sum of vectors rigidly attached to the local coordinate systems for the individual bonds. Application of this generalization will be deferred until Chapter XIV.

B. The Square of the End-to-End Distance

The square of the end-to-end distance is the self-dot product of $\mathbf{r}$, Eq. (I-2), written here as

$$r^2 \equiv \mathbf{r} \cdot \mathbf{r} = \left(\sum_{i=1}^{n} \mathbf{l}_i^s \right) \cdot \left(\sum_{i=i}^{n} \mathbf{l}_i^s \right) = 2 \sum_{k=1}^{n-1} \sum_{j=k+1}^{n} \mathbf{l}_k^s \cdot \mathbf{l}_j^s + \sum_{k=1}^{n} l_k^2 \tag{VI-16}$$

The factor of two before the double sum, and the lower limit of $k + 1$ for j, arise because $\mathbf{l}_k^s \cdot \mathbf{l}_j^s + \mathbf{l}_j^s \cdot \mathbf{l}_k^s$ is more simply evaluated as $2\mathbf{l}_k^s \cdot \mathbf{l}_j^s$. When $\mathbf{l}_k$ and $\mathbf{l}_j$ are expressed in their local coordinate systems, the dot product is

$$\mathbf{l}_k^s \cdot \mathbf{l}_j^s \equiv \mathbf{l}_k^T \mathbf{T}_k \mathbf{T}_{k+1} \cdots \mathbf{T}_{j-1}\mathbf{l}_j \tag{VI-17}$$

Here $\mathbf{T}_k\mathbf{T}_{k+1} \cdots \mathbf{T}_{j-1}\mathbf{l}_j$ expresses $\mathbf{l}_j$ in the local coordinate system for bond k. Combination of Eqs. (VI-16) and (VI-17) yields

$$r^2 = 2 \sum_{k=1}^{n-1} \sum_{j=k+1}^{n} \mathbf{l}_k^T \mathbf{T}_k \mathbf{T}_{k+1} \cdots \mathbf{T}_{j-1}\mathbf{l}_j + \sum_{k=1}^{n} l_k^2 \tag{VI-18}$$

The sums in Eq. (VI-18) are conveniently evaluated by an elaboration of the scheme that employs $\mathbf{R}_i$.[5,6] In the present case, $\mathbf{R}_i$ is a row vector of five elements written in three blocks. At bond 1, this row vector is

$$\mathbf{R}_1 = [1 \quad 2\mathbf{l}_1^T\mathbf{T}_1 \quad l_1^2] \tag{VI-19}$$

The last element in $\mathbf{R}_1$ is the square of the length of bond 1, and the middle block is the potential contribution by bond 1 to uncompleted terms in the double sum in Eq. (VI-18). *More generally*, the last element in $\mathbf{R}_i$ is the square of the distance from A_0 to A_i, r_{0i}^2. It includes the sum of l_1^2 through l_i^2 and all contributions from the double sum in Eq. (VI-18) in which j has an upper limit $j \leq i$. The double sum makes no contribution to the last element of $\mathbf{R}_i$ when $i = 1$, but it contributes at all larger i. The middle block in $\mathbf{R}_i$ is the sum of the potential contributions from bonds 1 through i to uncompleted terms in the double sum. When such terms are eventually completed, by appending $\mathbf{l}_j$, the indices in Eq. (VI-18) will be $1 \leq k \leq i$ and $j > i$. The first element in $\mathbf{R}_i$ is always 1 in the evaluation of r^2, as will generally be true for the $\mathbf{R}_i$ used for properties that include sums of terms such as $\mathbf{l}_i^T\mathbf{T}_i \cdots \mathbf{T}_{j-1}\mathbf{l}_j$. However, the first element in $\mathbf{R}_i$ is not 1 if the property requires sums of $\mathbf{T}_1\mathbf{T}_2 \cdots \mathbf{T}_{j-1}\mathbf{l}_j$, as illustrated by Eq. (VI-8):

$$\mathbf{R}_i = \left[1 \quad 2\sum_{j=1}^{i} \mathbf{l}_j^T\mathbf{T}_j\mathbf{T}_{j+1} \cdots \mathbf{T}_i \quad r_{0i}^2 \right] \tag{VI-20}$$

The square matrix used in conjunction with $\mathbf{R}_i$ is of dimensions 5×5, but it is more compactly written using nine blocks:[3,5]

$$\mathbf{G}_i = \begin{bmatrix} 1 & 2\mathbf{l}^T\mathbf{T} & l^2 \\ 0 & \mathbf{T} & \mathbf{l} \\ 0 & 0 & 1 \end{bmatrix}_i \qquad 1 < i < n \tag{VI-21}$$

The first row of $\mathbf{G}_i$ is of the same form as $\mathbf{R}_1$. Generation of $\mathbf{R}_2$ as $\mathbf{R}_1\mathbf{G}_2$ yields

$$\begin{aligned} \mathbf{R}_2 &= [1 \quad 2\mathbf{l}_1^T\mathbf{T}_1\mathbf{T}_2 + 2\mathbf{l}_2^T\mathbf{T}_2 \quad l_2^2 + 2\mathbf{l}_1^T\mathbf{T}_1\mathbf{l}_2 + l_1^2] \\ &\equiv [1 \quad 2\mathbf{l}_1^T\mathbf{T}_1\mathbf{T}_2 + 2\mathbf{l}_2^T\mathbf{T}_2 \quad r_{02}^2] \end{aligned} \tag{VI-22}$$

The disposition of the 1 and l_1^2 in $\mathbf{R}_1$, and the l_2^2 and 1 in the last column of $\mathbf{G}_2$, produces $l_2^2 + l_1^2$ in the last element in $\mathbf{R}_2$. More generally, the 1 in $\mathbf{R}_{i-1}$ and the 1 at the lower right of $\mathbf{G}_i$ produces the addition of l_i^2 to $r_{0,i-1}^2$. This procedure incorporates the contribution from l_i^2 in the last summation in Eq. (VI-18). The

[6]Flory, P. J. *J. Chem. Phys.* **1971**, *54*, 1351.

middle block in $\mathbf{R}_{i-1}$ and the middle block in the last column of $\mathbf{G}_i$ complete the contributions from the double sum in Eq. (VI-18) in which $j = i$, and place the result in the last element in $\mathbf{R}$, thereby completing r_{0i}^2. The interaction of the first two blocks in $\mathbf{R}_{i-1}$ with the $2\mathbf{l}_i^T\mathbf{T}_i$ and $\mathbf{T}_i$ in $\mathbf{G}_i$ appends a $\mathbf{T}_i$ to previously initiated incomplete terms in the double sum in Eq. (VI-18) and initiates new incomplete terms at bond i. Contributions from the last element in $\mathbf{R}_{i-1}$ are rejected by the null block at the end of the second column of blocks in $\mathbf{G}_i$. The first column in $\mathbf{G}_i$ is constructed so that the first element in $\mathbf{R}_i$ will be 1 for all i.

Continuation of this process will ultimately lead to $\mathbf{R}_n$, in which the last element is r^2. The first four elements (or first two blocks) are of no further interest. Consequently the square of the end-to-end distance is

$$r^2 = \begin{bmatrix} 1 & 2\mathbf{l}_1^T\mathbf{T}_1 & l_1^2 \end{bmatrix} \mathbf{G}_2\mathbf{G}_3 \cdots \mathbf{G}_{n-1} \begin{bmatrix} l_n^2 \\ \mathbf{l}_n \\ 1 \end{bmatrix} \qquad \text{(VI-23)}$$

Definition of $\mathbf{G}_1$ and $\mathbf{G}_n$ as being of the form of the first row, and last column, respectively, of the general $\mathbf{G}_i$ yields

$$r^2 = \mathbf{G}_1\mathbf{G}_2\mathbf{G}_3 \cdots \mathbf{G}_{n-1}\mathbf{G}_n \qquad \text{(VI-24)}$$

Alternative (but numerically equivalent) formulations for r^2 can be found in the earlier literature. These formulations utilize the same information ($\mathbf{T}_i$ and $\mathbf{l}_i$), but arrange that information differently in the block matrices.[7]

The same formalism can be employed for calculation of the square of the distance between chain atoms k and j, provided they are separated by more than one bond:

$$r_{kj}^2 = l_j^2 \qquad j - k = 1$$

$$r_{kj}^2 = \mathbf{G}_{[k+1}\mathbf{G}_{k+2}\mathbf{G}_{k+3} \cdots \mathbf{G}_{j-1}\mathbf{G}_{j]} \qquad j - k > 1 \qquad \text{(VI-25)}$$

---

[7]For example, if $\mathbf{G}_i$ of Eq. (VI-21) is redefined as

$$\mathbf{G}_i = \begin{bmatrix} 1 & \mathbf{l}^T\mathbf{T} & l^2/2 \\ 0 & \mathbf{T} & \mathbf{l} \\ 0 & 0 & 1 \end{bmatrix}_i \qquad 1 < i < n$$

and $\mathbf{G}_1$ and $\mathbf{G}_n$ are now defined as the first row and last column, respectively, of this matrix, then r^2 can be generated as

$$r^2 = 2\mathbf{G}_1\mathbf{G}_2\mathbf{G}_3 \cdots \mathbf{G}_{n-1}\mathbf{G}_n$$

The difference between this formulation and the one described in the text lies in the means of handling the factor of 2 that precedes the double sum in Eq. (VI-16).

We use the notation $\mathbf{G}_{[k+1}$ and $\mathbf{G}_{j]}$ for the first row and last column, respectively, of the 5×5 representations of $\mathbf{G}_{k+1}$ and $\mathbf{G}_j$. When the subscripts k and j are unspecified for r^2, as in Eq. (VI-23), it is understood that $k = 0$ and $j = n$. The ability to calculate the squared end-to-end distances for subchains via Eq. (VI-25) will be exploited in the next section, where we examine the square of the radius of gyration.

C. The Square of the Radius of Gyration

If the $n+1$ chain atoms are all of equal mass and these atoms are the only ones considered, the square of the radius of gyration, s^2, is

$$s^2 = \frac{1}{(n+1)^2} \sum_{k=0}^{n-1} \sum_{j=k+1}^{n} r_{kj}^2 \qquad \text{(VI-26)}$$

which is Eq. (I-4) for a rigid collection of n centers of the same mass. Incorporation of the expressions in Eq. (VI-25) for the r_{kj}^2 yields

$$s^2 = \frac{1}{(n+1)^2} \left(\sum_{k=0}^{n-2} \sum_{j=k+2}^{n} \mathbf{G}_{[k+1}\mathbf{G}_{k+2}\mathbf{G}_{k+3} \cdots \mathbf{G}_{j-1}\mathbf{G}_{j]} + \sum_{k=0}^{n-1} r_{k,k+1}^2 \right)$$

$$\text{(VI-27)}$$

The last term is just the sum of the squares of the bond lengths.

In construction of the generator matrix appropriate for evaluation of s^2, it is instructive to compare the structures of Eq. (VI-18) and Eq. (VI-27). Both equations contain a single sum of n scalars, and a double sum of terms, each of which starts with a row, ends with a column, and contains a variable number of square matrices in between (the number of square matrices is zero in some of the terms). The similarities in the structures of the equations imply a similarity in the structure of the generator matrices required for evaluation of the terms in the equations.

Simultaneous evaluation of the single sum and the double sum in Eq. (VI-27) can be performed systematically by blocked matrices that have a strong formal similarity to those employed for the computation of r^2.[5,6] The matrices required for s^2 are larger, with $\mathbf{R}_i$ having seven elements and the square matrix being of dimensions 7×7. However, the square matrix is more compactly written with nine blocks, and $\mathbf{R}_i$ can be written with three blocks. The expression for $\mathbf{R}_1$ is

$$\mathbf{R}_1 = [1 \quad \mathbf{G}_{[1} \quad l_1^2] \qquad \text{(VI-28)}$$

The last element in $\mathbf{R}_1$ is the square of the length of $\mathbf{l}_1$. More generally the last element in $\mathbf{R}_i$ is the contribution to the double sum in Eq. (VI-26) by those terms for which the upper limit for j is i. The middle block of $\mathbf{R}_i$ is the sum of the uncompleted portions of contributions to the double sum. These uncompleted terms have k that range from 0 through i, and j will be larger than i. The first element in $\mathbf{R}_i$ is 1. This element plays an equivalent role in the evaluation of r^2 and s^2:

$$
\mathbf{R}_i = \left[\begin{array}{cccc} 1 & \displaystyle\sum_{j=1}^{i} \mathbf{G}_{[j}\mathbf{G}_{j+1} & \cdots & \mathbf{G}_i \displaystyle\sum_{k=0}^{i-1}\sum_{j=k+1}^{i} r_{kj}^2 \end{array}\right] \tag{VI-29}
$$

The square matrix required for the evaluation of s^2 is[5,6]

$$
\mathbf{H}_i = \left[\begin{array}{ccc} 1 & \mathbf{G}_{[} & l^2 \\ 0 & \mathbf{G} & \mathbf{G}_{]} \\ 0 & 0 & 1 \end{array}\right]_i \qquad 1 < i < n \tag{VI-30}
$$

The product of $\mathbf{R}_1$ and $\mathbf{H}_2$ is

$$
\begin{aligned}
\mathbf{R}_2 &= \begin{bmatrix} 1 & \mathbf{G}_{[2} + \mathbf{G}_{[1}\mathbf{G}_2 & l_2^2 + \mathbf{G}_{[1}\mathbf{G}_{2]} + l_1^2 \end{bmatrix} \\
&\equiv \begin{bmatrix} 1 & \mathbf{G}_{[2} + \mathbf{G}_{[1}\mathbf{G}_2 & r_{12}^2 + r_{02}^2 + r_{01}^2 \end{bmatrix}
\end{aligned} \tag{VI-31}
$$

Repetition will lead eventually to an $\mathbf{R}_n$ in which the last element contains the double sum of Eq. (VI-26). The other elements in $\mathbf{R}_n$ are of no further interest. Consequently

$$
s^2 = \frac{1}{(n+1)^2} \begin{bmatrix} 1 & \mathbf{G}_{[1} & l_1^2 \end{bmatrix} \mathbf{H}_2\mathbf{H}_3 \cdots \mathbf{H}_{n-1} \begin{bmatrix} l_n^2 \\ \mathbf{G}_{n]} \\ 1 \end{bmatrix} \tag{VI-32}
$$

If $\mathbf{H}_1$ and $\mathbf{H}_n$ are defined as the first row and last column, respectively, of the general expression for $\mathbf{H}_i$, then

$$
s^2 = \frac{1}{(n+1)^2} \mathbf{H}_1\mathbf{H}_2\mathbf{H}_3 \cdots \mathbf{H}_{n-1}\mathbf{H}_n \tag{VI-33}
$$

Table VI-1 summarizes the generator matrices for **r**, r^2, and s^2.

TABLE VI-1. Generator Matrices for r, r^2, and s^2

Property	Bond 1	Bond i, $1 < i < n$	Bond n
r	$[\mathbf{T} \quad \mathbf{l}]$	$\begin{bmatrix} \mathbf{T} & \mathbf{l} \\ \mathbf{0} & 1 \end{bmatrix}$	$\begin{bmatrix} \mathbf{l} \\ 1 \end{bmatrix}$
r^2	$[1 \quad 2\mathbf{l}^T\mathbf{T} \quad l^2]$	$\begin{bmatrix} 1 & 2\mathbf{l}^T\mathbf{T} & l^2 \\ 0 & \mathbf{T} & \mathbf{l} \\ 0 & \mathbf{0} & 1 \end{bmatrix}$	$\begin{bmatrix} l^2 \\ \mathbf{l} \\ 1 \end{bmatrix}$
s^2	$[1 \quad 1 \quad 2\mathbf{l}^T\mathbf{T} \quad l^2 \quad l^2]$	$\begin{bmatrix} 1 & 1 & 2\mathbf{l}^T\mathbf{T} & l^2 & l^2 \\ 0 & 1 & 2\mathbf{l}^T\mathbf{T} & l^2 & l^2 \\ 0 & 0 & \mathbf{T} & \mathbf{l} & \mathbf{l} \\ 0 & 0 & \mathbf{0} & 1 & 1 \\ 0 & 0 & \mathbf{0} & 0 & 1 \end{bmatrix}$	$\begin{bmatrix} l^2 \\ l^2 \\ \mathbf{l} \\ 1 \\ 1 \end{bmatrix}$

If an application demands that the chain atoms be treated as having nonidentical masses, the mean square radius of gyration is obtained as[8]

$$s^2 = \left(\sum_{i=0}^{n} \sum_{j=0}^{n} m_i m_j \right)^{-1} \mathbf{H}_1 \mathbf{H}_2 \mathbf{H}_3 \cdots \mathbf{H}_{n-1} \mathbf{H}_n \tag{VI-34}$$

with

$$\mathbf{H}_i = \begin{bmatrix} 1 & m_{i-1}\mathbf{G}_{i[} & m_{i-1}m_i l_i^2 \\ 0 & \mathbf{G}_i & m_i \mathbf{G}_{i]} \\ 0 & 0 & 1 \end{bmatrix}_i \qquad 1 < i < n \tag{VI-35}$$

with $\mathbf{H}_1$ and $\mathbf{H}_n$ being now the first row, and last column, respectively, of the matrix in Eq. (VI-35).[6] This formulation of $\mathbf{H}_i$ might be useful, for instance, in calculation of the apparent s^2 measured by small angle neutron scattering for a polymer in which some units contain ^{1}H, while others contain ^{2}H. Then the m_i could be assigned to represent the scattering factor of each A_i and the hydrogen atoms bonded to it.

3. THE AVERAGE END-TO-END VECTOR

The end-to-end vector for a chain of three or more bonds depends on the torsion angle at one or more internal bonds. If the bond lengths are considered to be

[8]Mattice, W. L. *Macromolecules* **1976**, *9*, 48.

constant, the average of the $\mathbf{r}$ for the various conformations is obtained from Eq. (VI-6) as

$$\langle \mathbf{r} \rangle = \mathbf{l}_1 + \langle \mathbf{T}_1 \rangle \mathbf{l}_2 + \langle \mathbf{T}_1 \mathbf{T}_2 \rangle \mathbf{l}_3 + \cdots + \langle \mathbf{T}_1 \mathbf{T}_2 \cdots \mathbf{T}_{n-1} \rangle \mathbf{l}_n \qquad \text{(VI-36)}$$

When the bond angles are also constant, $\mathbf{T}_1$ can be removed from the angle brackets because it is determined entirely by the angle between the first two bonds [recall Eq. (VI-4)]:

$$\langle \mathbf{r} \rangle = \mathbf{l}_1 + \mathbf{T}_1 \mathbf{l}_2 + \mathbf{T}_1 \langle \mathbf{T}_2 \rangle \mathbf{l}_3 + \cdots + \mathbf{T}_1 \langle \mathbf{T}_2 \cdots \mathbf{T}_{n-1} \rangle \mathbf{l}_n \qquad \text{(VI-37)}$$

For a chain of only two bonds, the end-to-end vector is

$$\mathbf{r}_{02} = \begin{bmatrix} l_1 - l_2 \cos \theta_1 \\ l_2 \sin \theta_1 \\ 0 \end{bmatrix} \qquad \text{(VI-38)}$$

The expression for the average of $\mathbf{r}$ will usually change as the chain increases in length, but it will become independent of n at sufficiently large n if the chain is flexible. The third element will remain null if the torsion potential energy function is symmetric.

The path selected for further development is dictated by the treatment necessary for the remaining averages of the products of transformation matrices in Eq. (VI-37). The treatment is different for chains in which the bonds are independent or interdependent.

A. Independent Bonds

When the bonds are independent, the average of the product of transformation matrices can be replaced by the product of the averages:

$$\langle \mathbf{r} \rangle_0 = \mathbf{l}_1 + \mathbf{T}_1 \mathbf{l}_2 + \mathbf{T}_1 \langle \mathbf{T}_2 \rangle \mathbf{l}_3 + \cdots + \mathbf{T}_1 \langle \mathbf{T}_2 \rangle \cdots \langle \mathbf{T}_{n-1} \rangle \mathbf{l}_n \qquad \text{(VI-39)}$$

Following the line of development from Eqs. (VI-6)–(VI-13), with substitution of $\langle \mathbf{T}_i \rangle$ for $\mathbf{T}_i$, $1 < i < n$, ultimately leads to

$$\langle \mathbf{r} \rangle_0 = \mathbf{A}_1 \langle \mathbf{A}_2 \rangle \langle \mathbf{A}_3 \rangle \cdots \langle \mathbf{A}_{n-1} \rangle \mathbf{A}_n \qquad \text{(VI-40)}$$

where the internal $\langle \mathbf{A}_i \rangle$ are

$$\langle \mathbf{A}_i \rangle = \begin{bmatrix} \langle \mathbf{T} \rangle & \mathbf{l} \\ \mathbf{0} & 1 \end{bmatrix}_i \qquad 1 < i < n \qquad \text{(VI-41)}$$

If the internal bonds are identical

$$\langle \mathbf{r} \rangle_0 = \mathbf{A}_1 \langle \mathbf{A} \rangle^{n-2} \mathbf{A}_n \tag{VI-42}$$

which leads to[9]

$$\langle \mathbf{r} \rangle_0 = \mathbf{l} + \mathbf{T}_1 (\mathbf{I}_3 - \langle \mathbf{T} \rangle^n)(\mathbf{I}_3 - \langle \mathbf{T} \rangle)^{-1} \mathbf{l} \tag{VI-43}$$

For the special case of a symmetric torsion potential energy function, the limiting form as $n \to \infty$ is[6]

$$\lim_{n \to \infty} \langle \mathbf{r} \rangle_0 = \frac{l}{(1 + \cos \theta)(1 + \langle \cos \phi \rangle)} \begin{bmatrix} 1 + \cos \theta \langle \cos \phi \rangle \\ \sin \theta \\ 0 \end{bmatrix} \tag{VI-44}$$

The terms containing $\langle \cos \phi \rangle$ go to zero if rotation is free.

B. Interdependent Bonds

The average of the product of a series of transformation matrices cannot be separated into the product of the averages when the bonds are interdependent. A different strategy must be devised for finding the average of the $\mathbf{r}$ over all conformations. It will be convenient to express the probability for chain conformation κ as

$$p_\kappa = \frac{1}{Z} \prod_i w_{i,\kappa} \tag{VI-45}$$

where $w_{i,\kappa}$ is the statistical weight contributed by bond i to the statistical weight of chain conformation κ. For pairwise interdependent bonds, $w_{i,\kappa}$ is the element from $\mathbf{U}_i$ specified by the conformations adopted at bonds $i - 1$ and i when the chain is in conformation κ. For present purposes, we will not need to actually compute the large number of p_κ (which could be done in principle—but not in practice—for large n, because of the enormous size of ν^{n-2}) via Eq. (V-37) or Eq. (VI-45). Instead we will use the formalism in Eq. (VI-45) to devise a much simpler computational scheme for performing the averages over all conformations in the unperturbed ensemble ("simpler" because we will not have to compute *any* of the p_κ).

The average of the end-to-end vector in the unperturbed ensemble is

[9]Flory, P. J.; Yoon, D. Y. *J. Chem. Phys.* **1974**, *61*, 5358.

$$\langle \mathbf{r} \rangle_0 = \sum_{\kappa} p_{\kappa} \mathbf{r}_{\kappa} \tag{VI-46}$$

where $\mathbf{r}_{\kappa}$ is given by Eq. (VI-13). Insertion of Eqs. (VI-13) and (VI-45) into Eq. (VI-46) yields

$$\langle \mathbf{r} \rangle_0 = \frac{1}{Z} \sum_{\kappa} \left(\prod w_{i,\kappa} \right) (\mathbf{A}_1 \mathbf{A}_2 \cdots \mathbf{A}_n)_{\kappa} \tag{VI-47}$$

The $w_{i,\kappa}$ are scalars that can be interdigitated between the $\mathbf{A}_i$ in any useful order. The most useful order groups terms indexed by the same value of i:[5,6]

$$\langle \mathbf{r} \rangle_0 = \frac{1}{Z} \sum_{\kappa} (w_1 \mathbf{A}_1)_{\kappa} (w_2 \mathbf{A}_2)_{\kappa} \cdots (w_n \mathbf{A}_n)_{\kappa} = \frac{1}{Z} \sum_{\kappa} \left(\prod_{i=1}^{n} w_i \mathbf{A}_i \right)_{\kappa}$$

$$\tag{VI-48}$$

This sum can be compared with Z, which is the sum of the statistical weights of all conformations of the chain:

$$Z = \sum_{\kappa} (w_1 w_2 \cdots w_n)_{\kappa} = \sum_{\kappa} \left(\prod_{i=1}^{n} w_i \right)_{\kappa} \tag{VI-49}$$

Comparison of Eqs. (VI-48) and (VI-49) shows that an efficient procedure for evaluation of Z can be converted into an efficient procedure for evaluation of $\langle \mathbf{r} \rangle_0$ by merely expanding each scalar w_i to the matrix $w_i \mathbf{A}_i$. The summation in Eq. (VI-48) can be evaluated by a modification of Eqs. (IV-22) that simply *appends the appropriate expression for $\mathbf{A}_i$ onto each element of every $\mathbf{U}_i$*. If the chain is one subject to the symmetric threefold torsion potential energy function with nearest-neighbor interdependences used for the n-alkanes, the three types of statistical weight matrices are Eqs. (IV-23), (IV-19), and (IV-24), and the corresponding matrices for the generation of the summation in Eq. (VI-48) are[5,6]

$$\mathcal{A}_1 = [\mathbf{A}_1 \quad \mathbf{0} \quad \mathbf{0}] \tag{VI-50}$$

$$\mathscr{A}_i = \begin{bmatrix} \mathbf{A}_t & \sigma\mathbf{A}_{g^+} & \sigma\mathbf{A}_{g^-} \\ \mathbf{A}_t & \sigma\mathbf{A}_{g^+} & \sigma\omega\mathbf{A}_{g^-} \\ \mathbf{A}_t & \sigma\omega\mathbf{A}_{g^+} & \sigma\mathbf{A}_{g^-} \end{bmatrix}_i \qquad 1 < i < n \tag{VI-51}$$

$$\mathscr{A}_n = \begin{bmatrix} \mathbf{A}_n \\ \mathbf{A}_n \\ \mathbf{A}_n \end{bmatrix} \tag{VI-52}$$

The subscripts t, g^+, and g^- in Eq. (VI-51) denote the state assumed in the assignment of ϕ_i during the construction of $\mathbf{T}_i$ in each of $\mathbf{A}_i$. The summation in Eq. (VI-48) is generated as the serial product $\mathscr{A}_1 \cdots \mathscr{A}_n$. Division of the resulting vector by Z gives $\langle \mathbf{r} \rangle_0$.

$$\langle \mathbf{r} \rangle_0 = \frac{1}{Z} \mathscr{A}_1 \cdot \mathscr{A}_2 \cdots \mathscr{A}_n \tag{VI-53}$$

The dimensions of $\mathscr{A}_1$ are 3×12, $\mathscr{A}_n$ is a column vector of 12 elements, and the remaining $\mathscr{A}_i$ are square matrices of dimensions 12×12, if there are three rotational isomers at each internal bond. In general, for ν rotational isomers, the dimensions are $3 \times 4\nu, 4\nu \times 4\nu$, and $4\nu \times 1$ for $\mathscr{A}_1$, internal $\mathscr{A}_i$, and $\mathscr{A}_n$, respectively. If bonds $i - 1$ and i have different numbers of rotational isomers, the internal $\mathscr{A}_i$ are rectangular matrices of dimensions $4\nu_{i-1} \times 4\nu_i$.

The specific recipe usually used for generation of the internal $\mathscr{A}_i$ is often written as

$$\mathscr{A}_i = (\mathbf{U}_i \otimes \mathbf{I}_A) \begin{bmatrix} \mathbf{A}_t & 0 & 0 \\ 0 & \mathbf{A}_{g^+} & 0 \\ 0 & 0 & \mathbf{A}_{g^-} \end{bmatrix}_i \qquad 1 < i < n \tag{VI-54}$$

where $\mathbf{I}_A$ denotes the identity matrix of the same order as the generator matrix $\mathbf{A}$ (4×4 in the present case), and $\otimes$ denotes the direct product.[10] The block diagonal matrix, which is written more concisely as $\|\mathbf{A}\|_i$,

---

[10]If $\mathbf{U}_i$ is of dimensions $\nu_{i-1} \times \nu_i$ and $\mathbf{I}_A$ is of dimensions 4×4, the direct product, in the sequence $\mathbf{U}_i \otimes \mathbf{I}_A$, is of dimensions $4\nu_{i-1} \times 4\nu_i$, with the form

$$\mathbf{U}_i \otimes \mathbf{I}_A = \begin{bmatrix} u_{11}\mathbf{I}_A & u_{12}\mathbf{I}_A & \cdots \\ u_{21}\mathbf{I}_A & u_{22}\mathbf{I}_A & \cdots \\ \vdots & \vdots & \ddots \end{bmatrix}_i$$

where each element of $\mathbf{U}_i$ has been expanded to $u_{jk}\mathbf{I}_A$.

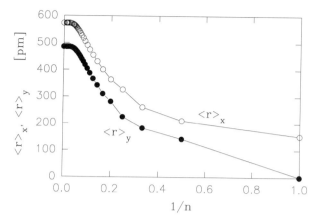

Figure VI-3. Components of $\langle \mathbf{r} \rangle_0$ for unperturbed linear polyethylene chains.

$$\|\mathbf{A}\|_i = \begin{bmatrix} \mathbf{A}_t & \mathbf{0} & \mathbf{0} \\ \mathbf{0} & \mathbf{A}_{g^+} & \mathbf{0} \\ \mathbf{0} & \mathbf{0} & \mathbf{A}_{g^-} \end{bmatrix}_i \qquad \text{(VI-55)}$$

has one $\mathbf{A}_{\eta;i}$ for each rotational isomer at bond i, and these $\mathbf{A}_{\eta;i}$ are written along the diagonal in the same sequence as the indexing of the columns in $\mathbf{U}_i$. This number will be different from three if the chain has $\nu \neq 3$.[11]

The method used for evaluation of $\langle \mathbf{r} \rangle_0$ deserves close inspection because it can be employed for the computation of the averages of a variety of conformation-dependent physical properties. The crucial step is the pairing in $\mathscr{A}_i$ of each element $u_{\xi\eta}$ in $\mathbf{U}_i$ with the form adopted by $\mathbf{A}_i$ when bond i is in state η.

Figure VI-3 depicts the x and y components, in the local coordinate system for bond 1, of $\langle \mathbf{r} \rangle_0$ for unperturbed polyethylene chains.[12] The symmetry of the torsion potential energy function for bonds in polyethylene causes the z component to be zero at all n. The y component is zero when $n = 1$, and the x component is then the length of the bond. At $n = 2$, the components are specified by the bond lengths and bond angle in the manner described in Eq. (VI-38). At all larger n, $\langle \mathbf{r} \rangle_0$ is computed from Eq. (VI-53). The x and y com-

---

[11] In a subtle way, Eq. (VI-54) is more restrictive than Eq. (VI-51). The block diagonal matrix in Eq. (VI-54) forces the use of exactly the same $\mathbf{A}_{\eta;i}$ throughout a column in the blocked form of $\mathscr{A}_i$, Eq. (VI-51). Hence $\mathbf{l}_i, \theta_i$, and ϕ_i must be identical in each of the $\mathbf{A}_{\eta;i}$ in this column. If we instead construct $\mathscr{A}_i$ directly from Eq. (VI-51), without making use of Eq. (VI-54), we have the option of using different $\mathbf{l}_i, \theta_i$, and ϕ_i in each $\mathbf{A}$ *within* a column. Then we may let the geometry of state η at bond i (namely, $\mathbf{l}_i, \theta_i, \phi_i$) depend not only on the state at bond i but also on the selection of the state, ξ, at bond $i - 1$. In practice, use is rarely made of this option. However, the formal structure of RIS theory will support this option in cases where it may be required.

[12] Yoon, D. Y.; Flory, P. J. *J. Chem. Phys.* **1974**, *61*, 5366.

ponents increase as n rises above 2, but they rapidly approach limiting values. The limiting expression for $\langle \mathbf{r} \rangle_0$ is $[575 \text{ pm} \quad 486 \text{ pm} \quad 0]^T$.[12]

The x component of the limiting expression for $\langle \mathbf{r} \rangle_0$ has often been identified with the persistence length of the worm-like chain model for a linear polymer, Eq. (I-23). This identification is not correct. A chain with atoms that alternate between A and B, poly(A–B), might be described by the worm-like chain model using a single value for the persistence length, independent of the structures of the terminal groups. However, if poly(A–B) is terminated at one end by A, and at the other end by B, and we also have $\theta_{ABA} \neq \theta_{BAB}$, the length of $\langle \mathbf{r}_0 \rangle$, as well as the sizes of the individual components, may depend on whether indexing of the bonds begins at the end terminated by A, or at the end terminated by B. Thus $\langle \mathbf{r} \rangle_0$ depends on the average projection of bond vectors in the local coordinate system defined for a bond, but the persistence length depends on average projections onto an axis that is colinear with the direction of the chain at the point of interest.

4. $\langle r^2 \rangle_0$ AND THE CHARACTERISTIC RATIO

The mean square unperturbed end-to-end distance for a freely jointed chain is the sum of the squares of the lengths of the bonds, Eq. (I-11). The constraints imposed by the limited range for bond angles, and the preference for some torsion angles over others as a result of short-range interactions, usually produce a different result for $\langle r^2 \rangle_0$. The characteristic ratio was defined, in Eq. (I-18), as the ratio of $\langle r^2 \rangle_0$ to the value expected for the freely jointed chain. In most real chains, C_n increases with n at small n and approaches a limiting value greater than one.

A. Independent Bonds

As was the case in the evaluation of $\langle \mathbf{r} \rangle_0$ for a chain with independent bonds, the computation of the unperturbed average of r^2 merely requires the substitution of $\langle \mathbf{T} \rangle$ for $\mathbf{T}$ in the expression for r^2, Eq. (VI-24):

$$\langle r^2 \rangle_0 = \mathbf{G}_1 \langle \mathbf{G}_2 \rangle \cdots \langle \mathbf{G}_{n-1} \rangle \mathbf{G}_n \qquad \text{(VI-56)}$$

where the internal $\langle \mathbf{G}_i \rangle$ are

$$\langle \mathbf{G}_i \rangle = \begin{bmatrix} 1 & 2\mathbf{l}^T \langle \mathbf{T} \rangle & l^2 \\ 0 & \langle \mathbf{T} \rangle & \mathbf{l} \\ 0 & 0 & 1 \end{bmatrix}_i \qquad 1 < i < n \qquad \text{(VI-57)}$$

The characteristic ratio is

$$C_n = \frac{1}{n \langle l^2 \rangle} \mathbf{G}_1 \langle \mathbf{G}_2 \rangle \cdots \langle \mathbf{G}_{n-1} \rangle \mathbf{G}_n \qquad \text{(VI-58)}$$

When all the bonds are of the same length, a slightly *simpler* form is

$$C_n = \frac{1}{n} \mathbf{G}_1 \langle \mathbf{G}_2 \rangle \cdots \langle \mathbf{G}_{n-1} \rangle \mathbf{G}_n \qquad \text{(VI-59)}$$

with the understanding that all l_i are to be assigned the value of 1 in the $\langle \mathbf{G}_1 \rangle, \langle \mathbf{G}_i \rangle$, and $\langle \mathbf{G}_n \rangle$.

$$\langle \mathbf{G}_i \rangle = \begin{bmatrix} 1 & 2[1 \quad 0 \quad 0]\langle \mathbf{T} \rangle & 1 \\ \mathbf{0} & \langle \mathbf{T} \rangle & [1 \quad 0 \quad 0]^T \\ 0 & \mathbf{0} & 1 \end{bmatrix}_i \qquad \text{(VI-60)}$$

A *much* simpler form can be written when all bond angles, torsion potential energy functions, and bond lengths are identical. When all bonds are identical and independent, the expression for $\langle r^2 \rangle_0$ obtained from Eq. (VI-18) is

$$\langle r^2 \rangle_0 = nl^2 + 2\mathbf{l}^T \left(\sum_{k=1}^{n-1} (n - k)\langle \mathbf{T} \rangle^k \right) \mathbf{l} \qquad \text{(VI-61)}$$

$$C_n = 1 + \frac{2}{n}[1 \quad 0 \quad 0] \left(\sum_{k=1}^{n-1} (n - k)\langle \mathbf{T} \rangle^k \right) \begin{bmatrix} 1 \\ 0 \\ 0 \end{bmatrix} \qquad \text{(VI-62)}$$

Replacement of the sums of $n \langle \mathbf{T} \rangle^k$ and $k \langle \mathbf{T} \rangle^k$ yields[13]

$$C_n = \left[(\mathbf{I}_3 + \langle \mathbf{T} \rangle)(\mathbf{I}_3 - \langle \mathbf{T} \rangle)^{-1} - \frac{2}{n}\langle \mathbf{T} \rangle(\mathbf{I}_3 - \langle \mathbf{T} \rangle^n)(\mathbf{I}_3 - \langle \mathbf{T} \rangle)^{-2} \right]_{11}$$
$$\text{(VI-63)}$$

[13] The replacements are

$$\sum_{k=1}^{n-1} \langle \mathbf{T} \rangle^k = (\langle \mathbf{T} \rangle - \langle \mathbf{T} \rangle^n)(\mathbf{I}_3 - \langle \mathbf{T} \rangle)^{-1}$$

$$\sum_{k=1}^{n-1} k \langle \mathbf{T} \rangle^k = \langle \mathbf{T} \rangle \left[\frac{\partial (\sum_{k=1}^{n-1} \langle \mathbf{T} \rangle^k)}{\partial \langle \mathbf{T} \rangle} \right]$$

These two expressions are used together to generate the substitution for $\sum_{k=1}^{n-1} k \langle \mathbf{T} \rangle^k$.

where the subscript 11 denotes the 1,1 element of the 3×3 matrix. The 1,1 element is selected by the vectors $[1 \quad 0 \quad 0]$ and $[1 \quad 0 \quad 0]^T$ in Eq. (VI-62). As n increases, C_n approaches the limiting value

$$C_\infty = [(\mathbf{I}_3 + \langle \mathbf{T} \rangle)(\mathbf{I}_3 - \langle \mathbf{T} \rangle)^{-1}]_{11} \qquad \text{(VI-64)}$$

B. Interdependent Bonds

The mean-square unperturbed end-to-end distance for chains with interdependent bonds is obtained by a method that has an exact parallel in the evaluation of $\langle \mathbf{r} \rangle_0$ for the same chain.[3,5] The replacement for Eq. (VI-46) is

$$\langle r^2 \rangle_0 = \sum_\kappa p_\kappa r_\kappa^2 \qquad \text{(VI-65)}$$

and a perfectly analogous development ultimately leads to

$$\langle r^2 \rangle_0 = \frac{1}{Z} \mathscr{G}_1 \mathscr{G}_2 \cdots \mathscr{G}_n \qquad \text{(VI-66)}$$

where the $\mathscr{G}_i$ are the matrices defined in Eqs. (VI-50)–(VI-52), with replacement of each $\mathbf{A}$ by the equivalent $\mathbf{G}$. If there are ν states for each internal bond, the dimensions of the $\mathscr{G}_i$ are $1 \times 5\nu$ for $\mathscr{G}_1$, $5\nu \times 5\nu$ for the internal $\mathscr{G}_i$, and $5\nu \times 1$ for $\mathscr{G}_n$. The program listed in Appendix C calculates C_n and C_∞ using Eq. (VI-66).

Much of the earlier literature uses another formulation for the computation of $\langle r^2 \rangle_0$ that appears different at first glance, but actually yields the same numeric results.[14] The formulation described here is preferred because it is more easily generalized to the computation of the averages of other conformation-dependent properties.

The characteristic ratios for unperturbed polyethylene chains were calculated by Abe et al.[15] using matrix methods equivalent[14] to those described here. The results obtained by these methods are depicted in Fig. VI-4. The charac-

---

[14]For example, $\langle r^2 \rangle_0$ for a chain with three rotational isomeric states per bond can be computed as

$$\langle r^2 \rangle_0 = \frac{2}{Z} [1 \quad 0 \quad 0][\mathbf{I}_3 \quad \mathbf{0} \quad \mathbf{0}]\mathscr{G}_1 \mathscr{G}_2 \cdots \mathscr{G}_{n-1} \begin{bmatrix} (l_n^2/2)\mathbf{I}_3 \\ \mathbf{I}_3 \otimes l_n \\ \mathbf{I}_3 \end{bmatrix} \begin{bmatrix} 1 \\ 1 \\ 1 \end{bmatrix}$$

where the $\mathscr{G}_i$ are 15×15 matrices written in block form as

$$\mathscr{G}_i = \begin{bmatrix} \mathbf{U} & (\mathbf{U} \otimes l^T)\|\mathbf{T}\| & \mathbf{0} \\ \mathbf{0} & (\mathbf{U} \otimes \mathbf{I}_3)\|\mathbf{T}\| & \mathbf{U} \otimes l \\ \mathbf{0} & \mathbf{0} & \mathbf{U} \end{bmatrix}$$

[15]Abe, A.; Jernigan, R. L.; Flory, P. J. *J. Am. Chem. Soc.* **1966**, *88*, 631.

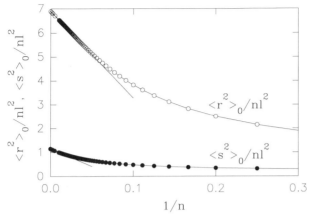

Figure VI-4. The dependence of C_n, Eq. (VI-66), and $C_{s,n}$, Eqs. (VI-70) and (VI-72), on $1/n$ for unperturbed polyethylene chains. The asymptotic limit at large n is the upper bound for the figure.

teristic ratio is larger than 1 whenever $n > 1$. It increases as n increases, and becomes linear in $1/n$ at sufficiently large n. If computations are performed for increasing n into the region where C_n is proportional to $1/n$, C_∞ can be obtained by a linear extrapolation to $1/n = 0$.[16]

The mean square end-to-end distance for a subchain within a chain that has interdependent bonds can be obtained from Eq. (VI-25). The conformational partition function must be combined with the $\mathscr{G}_i$ in a manner that properly incorporates the effect of bonds outside of the subchain on the conformations of the bonds within the subchain.

---

[16]The calculation required for Fig. VI-4 is for a system where $Z \to \infty$ as $n \to \infty$. At some n, numbers will be generated in $\Pi_i\, U_i$ and in $\Pi_i\, \mathscr{G}_i$ that exceed the largest number that can be stored in any computer. The calculation of Z and the serial product of the $\mathscr{G}_i$ can be kept within bounds, without changing the values of $\langle r^2 \rangle_0$ that result, by multiplying every U_i and $\mathscr{G}_i$ by a constant, c, such that Eq. (VI-66) becomes

$$\langle r^2 \rangle_0 = \frac{\mathscr{G}_1 [\Pi_{i=2}^{n-1} (c\mathscr{G}_i)] \mathscr{G}_n}{U_1 [\Pi_{i=2}^{n-1} (c U_i)] U_n}$$

Since numerator and denominator will both contain c^{n-2}, the value of $\langle r^2 \rangle_0$ is unaffected by any nonzero assignment for c. A good assignment, which will permit extension of the calculations to very large n, is $c = 1/\lambda_1$, where λ_1 is the largest eigenvalue of U_i. Alternatively, different scaling factors can be employed for U_i and $\mathscr{G}_i$, with the ratio of these factors carried along in the calculation.

$$\langle r^2 \rangle_0 = \left(\frac{c_U}{c_G} \right)^{n-2} \frac{\mathscr{G}_1 [\Pi_{i=2}^{n-1} (c_G \mathscr{G}_i)] \mathscr{G}_n}{U_1 [\Pi_{i=2}^{n-1} (c_U U_i)] U_n}$$

The latter procedure is adopted in the program listed in Appendix C.

$$\langle r_{kj}^2 \rangle_0 = \frac{1}{Z} \mathbf{U}_1 \cdots \mathbf{U}_{k-1} \mathscr{G}_{[k} \mathscr{G}_{k+1} \cdots \mathscr{G}_{j-1} \mathscr{G}_{j]} \mathbf{U}_{j+1} \cdots \mathbf{U}_n \qquad \text{(VI-67)}$$

$$\mathscr{G}_{[k} = \mathbf{U}_k \| \mathbf{G}_{[k} \| \qquad \text{(VI-68)}$$

$$\mathscr{G}_{j]} = (\mathbf{U}_j \otimes \mathbf{I}_5) \| \mathbf{G}_{[j} \| \qquad \text{(VI-69)}$$

5. THE MEAN SQUARE UNPERTURBED RADIUS OF GYRATION

For a flexible chain with a sufficiently large number of bonds, $\langle s^2 \rangle_0$ can be evaluated from C_∞, using Eq. (I-7). Since $s^2 = r^2/4$ when $n = 1$, it is obvious that $\langle s^2 \rangle_0$ may be different from $\langle r^2 \rangle_0/6$ for chains in which the number of bonds is small. The value of $\langle s^2 \rangle_0$ can be calculated directly for a chain of any n using the methods described above in conjunction with $\langle \mathbf{r} \rangle_0$ and $\langle r^2 \rangle_0$. The results are often presented as a dimensionless ratio defined as

$$C_{s,n} \equiv \frac{\langle s^2 \rangle_0}{n \langle l^2 \rangle} \qquad \text{(VI-70)}$$

The subscript s differentiates this ratio from C_n, which is the dimensionless ratio obtained with $\langle r^2 \rangle_0$ and defined in Eq. (I-18). For both C_n and $C_{s,n}$, the denominator is the value that would be assumed by $\langle r^2 \rangle_0$ if the chain were freely jointed. For a flexible chain, $C_{s,n}$ approaches $C_\infty/6$ as the number of bonds increases without limit.

A. Independent Bonds

Equation (VI-33) gives s^2 for a specified conformation. The transformation matrix for bond i is found in each $\mathbf{H}_i$. Replacement of each $\mathbf{T}_i$ by $\langle \mathbf{T}_i \rangle$ will yield $\langle s^2 \rangle_0$.

B. Interdependent Bonds

The replacement for Eq. (VI-46) is

$$\langle s^2 \rangle_0 = \sum p_\kappa s_\kappa^2 \qquad \text{(VI-71)}$$

The development described above in the evaluation of $\langle \mathbf{r} \rangle_0$ for chains with interdependent bonds ultimately leads to

$$\langle s^2 \rangle_0 = \frac{1}{(n+1)^2 Z} \mathscr{H}_1 \mathscr{H}_2 \cdots \mathscr{H}_n \qquad \text{(VI-72)}$$

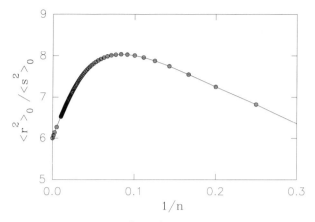

Figure VI-5. The dependence of $\langle r^2 \rangle_0 / \langle s^2 \rangle_0$ on $1/n$ for unperturbed polyethylene chains, using the data from Fig. VI-4.

where the appropriate **H** is substituted for **A** in each $\mathcal{H}_i$. The dimensions of the matrices are $1 \times 7\nu, 7\nu \times 7\nu$, and $7\nu \times 1$ for $\mathcal{H}_1$, the internal $\mathcal{H}_i$, and $\mathcal{H}_n$, respectively. As with $\langle r^2 \rangle_0$,[14] the earlier literature contains different (but numerically equivalent) formulations for the computation of $\langle s^2 \rangle_0$, which are not as easily generalized to the computation of the average values of conformation-dependent properties as the formulation described here.

Figure VI-4 depicts the chain length dependence of $C_{s,n}$ for polyethylene chains. As with C_n, this dimensionless ratio becomes linear in $1/n$ at sufficiently large n. However, the dimensionless ratio formulated from $\langle s^2 \rangle_0$ approaches its asymptotic limit at a rate different from that of C_n. For this reason, $\langle r^2 \rangle_0 / \langle s^2 \rangle_0$ is greater than six for chains of moderate size, but it approaches six as $n \to \infty$, as shown in Fig. VI-5.

6. SPECIAL FORMS FOR SYMMETRIC CHAINS WHEN U IS REDUCED

The $\mathcal{A}$, $\mathcal{G}$, and $\mathcal{H}$ matrices described in the previous sections of this chapter have not been designed so that they can take advantage of situations where **U** can be reduced, because of the symmetry of the torsion potential energy function. For example, the matrices constructed in Eqs. (VI-50), (VI-51), and (VI-52) envision a statistical weight matrix such as

$$\mathbf{U}_i = \begin{bmatrix} 1 & \sigma & \sigma \\ 1 & \sigma & \sigma\omega \\ 1 & \sigma\omega & \sigma \end{bmatrix}_i \qquad \text{(VI-73)}$$

such that the state for each column is uniquely defined, providing an unambiguous definition of the ϕ that should be used in the $\mathbf{A}, \mathbf{G}$, or $\mathbf{H}$ coupled with each element in the construction of $\mathscr{A}, \mathscr{G}$, or $\mathscr{H}$. However, since the torsion potential energy function is symmetric, we could also generate Z using a 2×2 statistical weight matrix, Eq. (IV-28). In that reduced statistical weight matrix, the indexing for the rows and columns was

$$
\mathbf{U}_i =
\begin{array}{c}
\\
t \\
g^+ + g^-
\end{array}
\begin{array}{cc}
t & g^+ + g^- \\
\left[\begin{array}{cc}
1 & 2\sigma \\
1 & \sigma(1 + \omega)
\end{array} \right]_i
\end{array}
\qquad \text{(IV-28)}
$$

How does one construct $\mathbf{A}, \mathbf{G}$, or $\mathbf{H}$ if this reduced statistical weight matrix is to be used for construction of the $\mathscr{A}, \mathscr{G}$, or $\mathscr{H}$ employed for computation of $\langle \mathbf{r} \rangle_0, \langle r^2 \rangle_0$, or $\langle s^2 \rangle_0$? Some comment is certainly in order, because a first glance would suggest that the torsion angle is not uniquely defined for the second column in the reduced form of the statistical weight matrix.

This section describes the forms of $\mathbf{A}, \mathbf{G}$, and $\mathbf{H}$ that are to be used in conjunction with the reduced forms of the statistical weight matrices, using the procedure described by Flory and Abe.[5,17] The material presented in this section will not permit computation of any results that could not also have been computed by the methods already presented earlier in this chapter, using the unreduced forms of the statistical weight matrices. Since the reduction in matrix size makes the computation more efficient, the methods described here will be faster. However, that increase in computational efficiency is not a strong motivation for their adoption in the computation of $\langle \mathbf{r} \rangle_0, \langle r^2 \rangle_0$, and $\langle s^2 \rangle_0$, because the computations using the unreduced matrices require trivial CPU time on the current generation of computers.

Readers whose primary interest lies in the computation of the conformation-dependent physical properties discussed earlier in this chapter may skip this section without sacrifice in their ability to understand the concepts presented in the remainder of this book. The greatest practical advantage to the generator matrices described in this section lies in their use for evaluation of higher moments, a topic that will be taken up in Chapter XIII. As the computation moves to higher and higher moments, the dimensions of the matrices increase without limit. Applications to higher moments can benefit from any means for achieving reductions in the dimensions of the matrices, including the one that is described here.

In the applications presented below, every $\mathbf{A}, \mathbf{G}$, or $\mathbf{H}$ in a column of $\mathscr{A}, \mathscr{G}$, or $\mathscr{H}$ must be identical; in other words, the conformation of state η at bond i is assumed to be independent of the states adopted by other bonds in the chain.

[17]Flory, P. J.; Abe, Y. *J. Chem. Phys.* **1971**, *54*, 1351.

A. The Mean End-to-End Vector

Let us write out, element by element, the forms adopted by the $\mathbf{A}_i$ defined by Eq. (VI-9) for two symmetrically disposed conformations, for one of which the torsion angle is ϕ, and for the other of which the torsion angle is $-\phi$. These two matrices will be denoted by $\mathbf{A}_+$ and $\mathbf{A}_-$, respectively. Both will be written with ϕ, with a change in sign before three elements in $\mathbf{T}_i$ in $\mathbf{A}_-$, since $\sin(-\phi) = -\sin\phi$.

$$
\mathbf{A}_+ = \begin{bmatrix} -\cos\theta & \sin\theta & 0 & l \\ -\sin\theta\cos\phi & -\cos\theta\cos\phi & -\sin\phi & 0 \\ -\sin\theta\sin\phi & -\cos\theta\sin\phi & \cos\phi & 0 \\ 0 & 0 & 0 & 1 \end{bmatrix}
\tag{VI-74}
$$

$$
\mathbf{A}_- = \begin{bmatrix} -\cos\theta & \sin\theta & 0 & l \\ -\sin\theta\cos\phi & -\cos\theta\cos\phi & \sin\phi & 0 \\ \sin\theta\sin\phi & \cos\theta\sin\phi & \cos\phi & 0 \\ 0 & 0 & 0 & 1 \end{bmatrix}
\tag{VI-75}
$$

An element by element comparison of these two matrices shows that all pairs of elements fall into one of three classes:

- 0: elements that are necessarily null in both matrices
- +: elements that might be nonzero, and identical in both matrices
- −: elements that might be nonzero, of opposite sign and identical magnitude

In this classification, the pair of matrices is represented by

$$
\begin{bmatrix} + & + & 0 & + \\ + & + & - & 0 \\ - & - & + & 0 \\ 0 & 0 & 0 & + \end{bmatrix}
\tag{VI-76}
$$

The representation of the elements by their class shows immediately that $\mathbf{A}_+$ and $\mathbf{A}_-$ are interconverted by the change in sign of all elements in the third column (or row), followed by change in sign of all elements in the third row (or column).

$$
\mathbf{A}_\pm = \mathbf{L}_-\mathbf{A}_\mp\mathbf{L}_-
\tag{VI-77}
$$

Here $\mathbf{L}_-$ is a diagonal matrix with elements -1 for those rows (columns) where the change in sign occurs, and elements $+1$ elsewhere on the diagonal. In the present case

$$\mathbf{L}_- = \text{diag}(1, 1, -1, 1) \tag{VI-78}$$

From the definition of $\mathbf{L}_-$, the identity matrix of the same order (written here as $\mathbf{I}_+$) is recovered as $\mathbf{I}_+ = \mathbf{I}_-^2$. Matrices denoted by $\mathbf{A}_0$ and $\mathbf{I}_0$ are the simple averages of the two corresponding matrices with subscripts $\pm$.

$$\mathbf{A}_0 = \tfrac{1}{2}(\mathbf{A}_+ + \mathbf{A}_-) \tag{VI-79}$$

$$\mathbf{I}_0 = \tfrac{1}{2}(\mathbf{I}_+ + \mathbf{I}_-) \tag{VI-80}$$

The matrices defined here can be employed for condensation of the $\mathscr{A}$ by coupling them with the elements of $\mathbf{X}^0$ and $\mathbf{Y}^0$, which were the rectangular matrices defined in Eqs. (IV-31) and (IV-32), and used for the condensation of $\mathbf{U}$. For the specific case at hand, these matrices take the forms

$$\mathbf{X}^0 = \begin{matrix} & \begin{matrix} 0 & \alpha^+ & \alpha^- \end{matrix} \\ \begin{matrix} 0 \\ \alpha^{\pm} \end{matrix} & \begin{bmatrix} 1 & 0 & 0 \\ 0 & \tfrac{1}{2} & \tfrac{1}{2} \end{bmatrix} \end{matrix} \tag{IV-33}$$

$$\mathbf{Y}^0 = \begin{matrix} & \begin{matrix} 0 & \alpha^{\pm} \end{matrix} \\ \begin{matrix} 0 \\ \alpha^+ \\ \alpha^- \end{matrix} & \begin{bmatrix} 1 & 0 \\ 0 & 1 \\ 0 & 1 \end{bmatrix} \end{matrix} \tag{IV-34}$$

In Chapter IV, the condensed forms of the statistical weight matrices are obtained from the uncondensed forms as $\mathbf{U}_1\mathbf{Y}^0, \mathbf{X}^0\mathbf{U}_i\mathbf{Y}^0$, and $\mathbf{X}^0\mathbf{U}_n$. The condensed forms of the $\mathscr{A}$ for use in conjunction with the reduced statistical weight matrices are obtained from the uncondensed forms as $\mathscr{A}_1\mathbf{Y}, \mathbf{X}\,\mathscr{A}_i\mathbf{Y}$, and $\mathbf{X}\,\mathscr{A}_n$, where the matrices $\mathbf{X}$ and $\mathbf{Y}$ are

$$\mathbf{X} = \begin{bmatrix} \mathbf{I}_0 & \mathbf{0} & \mathbf{0} \\ \mathbf{0} & \tfrac{1}{2}\mathbf{I}_+ & \tfrac{1}{2}\mathbf{I}_- \end{bmatrix} \tag{VI-81}$$

$$\mathbf{Y} = \begin{bmatrix} \mathbf{I}_0 & \mathbf{0} \\ \mathbf{0} & \mathbf{I}_+ \\ \mathbf{0} & \mathbf{I}_- \end{bmatrix} \tag{VI-82}$$

The matrices denoted by $\mathbf{X}$ and $\mathbf{Y}$ are obtained by expansion of each element in $\mathbf{X}^0$ and $\mathbf{Y}^0$, according to the following rules:

Each 0 in $\mathbf{X}^0$ and $\mathbf{Y}^0$ is replaced by $\mathbf{0}$.

Each nonzero element in $\mathbf{X}^0$ or $\mathbf{Y}^0$ is multiplied by

$\quad$ $\mathbf{I}_0$ if that column of $\mathbf{X}^0$, or row of $\mathbf{Y}^0$, was indexed by "0"

$\quad$ $\mathbf{I}_+$ if that column of $\mathbf{X}^0$, or row of $\mathbf{Y}^0$, was indexed by α^+

$\quad$ $\mathbf{I}_-$ if that column of $\mathbf{X}^0$, or row of $\mathbf{Y}^0$, was indexed by α^-

The final result for the reduced form of $\mathscr{A}_i$ is

$$\mathscr{A}_{i,\text{reduced}} = \begin{bmatrix} \mathbf{I}_0\mathbf{A}_0 & 2\sigma\mathbf{I}_0\mathbf{A}_+ \\ \mathbf{I}_0\mathbf{A}_0 & \sigma(\mathbf{I}_+ + \omega\mathbf{I}_-)\mathbf{A}_+ \end{bmatrix}_i, \qquad 1 < i < n \qquad \text{(VI-83)}$$

which is of dimensions 8×8. The unreduced matrix was of dimensions 12×12, with 2.25 times as many elements. The terminal matrices, of dimensions 3×8 and 8×1, respectively, are

$$\mathscr{A}_{1,\text{reduced}} = [\mathbf{A}_1\mathbf{I}_0 \quad \mathbf{0}] \qquad \text{(VI-84)}$$

$$\mathscr{A}_{n,\text{reduced}} = \begin{bmatrix} \mathbf{I}_0\mathbf{A}_n \\ \mathbf{I}_0\mathbf{A}_n \end{bmatrix} \qquad \text{(VI-85)}$$

B. Reduced $\mathscr{G}$ and $\mathscr{H}$ for $\langle r^2 \rangle_0$ and $\langle s^2 \rangle_0$

The reduced forms of $\mathscr{G}$ and $\mathscr{H}$ for computation of $\langle r^2 \rangle_0$ and $\langle s^2 \rangle_0$ are obtained in a completely analogous fashion, with the replacements for the expression in Eq. (VI-76) being

$$\begin{bmatrix} + & + & + & 0 & + \\ 0 & + & + & 0 & + \\ 0 & + & + & - & 0 \\ 0 & - & - & + & 0 \\ 0 & 0 & 0 & 0 & + \end{bmatrix} \qquad \text{(VI-86)}$$

$$\begin{bmatrix} + & + & + & + & 0 & + & + \\ 0 & + & + & + & 0 & + & + \\ 0 & 0 & + & + & 0 & + & + \\ 0 & 0 & + & + & - & 0 & 0 \\ 0 & 0 & - & - & + & 0 & 0 \\ 0 & 0 & 0 & 0 & 0 & + & + \\ 0 & 0 & 0 & 0 & 0 & 0 & + \end{bmatrix} \qquad \text{(VI-87)}$$

7. DISTRIBUTION FUNCTIONS FROM MONTE CARLO CALCULATIONS

Unfortunately, not every conformation-dependent physical property of a chain molecule is susceptible to evaluation by generator matrix methods. Sections VI-3–VI-5 dealt with the computations of average values extracted from the distribution functions for $\mathbf{r}, r^2$, and s^2 in an unperturbed ensemble. What of the distribution functions themselves? They cannot be calculated directly using generator matrices,[18] but they can be estimated by stochastic simulations that utilize the information found in Z. These simulations are often referred to with the words *Monte Carlo*, and we will use those words here, but they differ from the usual Monte Carlo simulations in that no use is made of a criterion (such as the Metropolis criterion[19]) for deciding whether to accept or reject a chain that has been generated. Instead an efficient procedure, which makes use of every chain that is generated, can be constructed by an appropriate manipulation of Z. A conformation-dependent physical property is extracted as the simple average over the $\mathcal{N}$ representative chains generated by this method. This procedure is not as computationally efficient as the generator matrix calculations that lead to $\langle \mathbf{r} \rangle_0, \langle r^2 \rangle_0$, and $\langle s^2 \rangle_0$, but it is more efficient than the usual Monte Carlo approach that rejects some of the chains that are generated.

One route to a distribution function, say, for $W(r^2)$, would be to randomly generate a large number of conformations, $\mathcal{N}$, measure r^2 for each, and weight each r^2 according to the p_κ computed for its conformation from $\prod_i w_i$, as specified by Eq. (VI-45). The distribution function evaluated over $\mathcal{N}$ such weighted r^2 should approach the desired $W(r^2)$ as $\mathcal{N} \to \infty$. The desired result can be obtained with less labor (with a smaller value of $\mathcal{N}$) by generating the chains in a nonrandom manner, such that the probability of generating a chain is proportional to its value of p_κ. This nonrandom method is preferred because it concentrates the numeric effort in chains that are more probable, with p_κ not drastically smaller than $p_{\kappa;\max}$, whereas the random approach tends to get bogged down in examination of many chains for which $p_\kappa \ll p_{\kappa;\max}$.

A. Monte Carlo Calculations Using Conditional Probabilities

The more efficient generation of a representative sample of $\mathcal{N}$ chains makes use of the information contained in Z in the selection of the chains. From Z one obtains the $p_{\eta;2}$ via Eq. (V-5) and the $q_{\xi\eta;i}, 2 < i < n$, via Eq. (V-36). Then a string of $n-2$ random numbers, each in the range $0 \leq random\,number \leq 1$, can

---

[18]The distribution functions can, in principle, be estimated to any desired degree of accuracy by more elaborate generator matrix calculations. For example, knowledge of all the $\langle r^{2p} \rangle_0 / \langle r^2 \rangle_0^p, p = 1, 2, \cdots, \infty$ completely describes the distribution function for r^2, and generator matrix expressions will be written for each of the $\langle r^{2p} \rangle_0$ in Chapter XIII. Unfortunately, these generator matrix expressions become unwieldy at large p.

[19]Metropolis, N.; Rosenbluth, A. W.; Rosenbluth, M. N.; Teller, A. H.; Teller, E. *J. Chem. Phys.* **1953**, *21*, 1087.

be used to select a representative conformation of the chain. The first random number is used in conjunction with the $p_{\eta;2}$ to select the conformation at bond 2. In the case of polyethylene, bond 2 will be assigned as

t if   *random number* $\leq p_{t;2}$

g^+ if $p_{t;2} < $ *random number* $\leq p_{t;2} + p_{g^+;2}$

g^- if $p_{t;2} + p_{g^+;2} < $ *random number*

For large $\mathcal{N}$ at 140°C, this procedure will assign t states at bond 2 in about 60% of the chains, and g states of each sign will equally populate the remaining 40% of the chains.

The assignment of the state at bond 3 will be made with knowledge of the prior assignment at bond 2. The appropriate information from Z is found in the $q_{\xi\eta;3}$. For example, in the case of polyethylene, if g^+ has been selected as the conformation at bond 2, then the state at bond 3 will be

t if   *random number* $\leq q_{g^+t;3}$

g^+ if $q_{g^+t;3} < $ *random number* $\leq q_{g^+t;3} + q_{g^+g^+;3}$

g^- if $q_{g^+t;i} + q_{g^+g^+;3} < $ *random number*

From the numeric values for $q_{\xi\eta}$ in Eq. (V-38), we see that this procedure will exert a strong preference, by more than a factor of 10, for $g^{\pm}g^{\pm}$ over $g^{\pm}g^{\mp}$. Propagation of this procedure along the chain permits assignment of the rest of the ϕ_i using the remaining random numbers and $q_{\xi\eta;i}$. After this procedure has been completed $\mathcal{N}$ times, using different sets of random numbers for each chain, any property of interest, such as $W(r^2)$, can be estimated for the unperturbed ensemble by *weighting equally* the values of r^2 for the $\mathcal{N}$ different chains. No differential weighting of the chains is appropriate for an unperturbed ensemble, because the relative weights of the different conformations (including the interdependence of these weights) was used in the generation of the chains. This procedure is more likely to generate a chain with p_κ not too different from $p_{\kappa;max}$ than it is to generate a chain with $p_\kappa \ll p_{\kappa;max}$. For this reason the property of interest, such as $W(r^2)$, is likely to converge with a smaller $\mathcal{N}$ than would have been the case if the chains were generated with equal probabilities for each of the ν states, and then weighted according to Eq. (VI-45).

The procedure can be used to evaluate the average, or the distribution function, for any property completely determined by the coordinates of the atoms in the unperturbed chain. Figure VI-6 depicts a distribution function computed

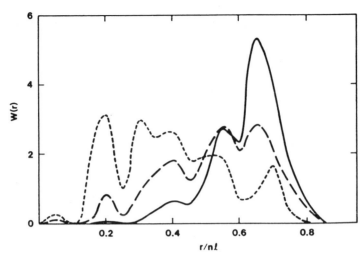

Figure VI-6. Distribution function for the end-to-end distance of polyethylene chains with $n = 10$ assuming free rotation (short-dashed curve), independent hindered rotations (long-dashed curve), and interdependent bonds (solid curve). Each Monte Carlo calculation employed 80,000 chains. Reproduced with permission from Mark, J. E.; Curro, J. G. *J. Chem. Phys.* **1984,** *81,* 6408. Copyright 1984 American Physical Society.

for the end-to-end distance of relatively short polyethylene chains. The distribution function is decidedly non-Gaussian for short chains. It cannot be calculated accurately by any simple closed-form expressions applicable to all short chains. However, it can be computed very accurately via Monte Carlo methods that use the $p_{\eta;2}$ and $q_{\xi\eta;i}, 2 < i < n$, extracted from an accurate RIS model for the chain.

B. Sample Size

The Monte Carlo procedure using the $p_{\eta;2}$ and $q_{\xi\eta;i}, 2 < i < n$ must reproduce the properties of the unperturbed ensemble in the limit $\mathcal{N} \to \infty$. In practice, it is possible to perform calculations for only a finite number of chains. How can one determine in advance the size of $\mathcal{N}$ needed for an estimate to the required degree of accuracy for the property of interest? When a property has been evaluated for a sample of very large $\mathcal{N}$, how can one estimate how closely it approaches the true value for the unperturbed ensemble?

One basis for assessing whether $\mathcal{N}$ was of adequate size is to compute, in addition to the property of interest, other relevant properties for which the exact answer is known. For example, the *exact* value of $\langle r^2 \rangle_0$ is rapidly calculated by the efficient generator matrix methods, Eq. (VI-66). Chapter XIII will present generator matrices that supply higher moments, $\langle r^{2p} \rangle_0, p > 1$. The *estimate* of $\langle r^{2p} \rangle_0$ can be obtained from the $\mathcal{N}$ Monte Carlo chains as

$$\langle r^{2p} \rangle_{0,\mathrm{MC}} = \frac{1}{\mathcal{N}} \sum r_{\mathrm{MC}}^{2p} \qquad (\text{VI-88})$$

If the purpose of the Monte Carlo calculations were to estimate $W(r^2)$, one should first check whether the $\langle r^{2p} \rangle_{0,\mathrm{MC}}$ are in good agreement with the $\langle r^{2p} \rangle_0$ that can be computed exactly by generator matrix methods. [The estimate of $W(r^2)$ cannot be very good if $\langle r^2 \rangle_{0,\mathrm{MC}}$ is in error.] If agreement is inadequate, a sample with more chains is required. Good agreement between $\langle r^2 \rangle_{0,\mathrm{MC}}$ and $\langle r^2 \rangle_0$ is a necessary, but not sufficient, condition for an accurate $W(r^2)$.

Another basis for assessing the accuracy of the Monte Carlo simulation is to realize that two independent simulations, each with an infinite number of chains, produce the same result, because the true properties of the unperturbed ensemble are recovered as $\mathcal{N} \to \infty$. One cannot actually produce M independent simulations, each containing an infinite number of chains, but one can produce M independent simulations with the same finite number of chains. Comparison of the results of the M different simulations is an excellent means of assessing the accuracy of the result. If X_i denotes a property X evaluated from the ith simulation, the information can be reported as $\langle X \rangle \pm (\langle X^2 \rangle - \langle X \rangle^2)^{1/2}$, where angle brackets here denote averages over the M independent simulations. If $\mathcal{N}$ for each of the independent simulations was sufficiently large for an accurate evaluation of X, the standard deviation should be small compared to $\langle X \rangle$.

Finally, one should expect that the value of $\mathcal{N}$ required for an accurate result will depend on n and on the property of interest. Usually $\mathcal{N}$ must increase as n increases, because of the larger number of conformations of the chain, and the likelihood that there is also a larger range of the property of interest for the chains in the ensemble. The data in Table I-1 show that the distribution function for r^2 is broader than the distribution function for s^2, and hence one might expect that $\langle s^2 \rangle_{0,\mathrm{MC}}$ will converge to $\langle s^2 \rangle_0$ at smaller $\mathcal{N}$ than required for similar convergence of $\langle r^2 \rangle_{0,\mathrm{MC}}$ to $\langle r^2 \rangle_0$.

The great attraction of the Monte Carlo method lies in its ability to provide access to conformation-dependent physical properties that are not susceptible to evaluation by generator matrices. Whenever either method can be applied, the generator matrix methods are preferable, because they yield the *exact* result for the ensemble with minimal computational effort.

PROBLEMS

Appendix B contains answers for Problems VI-1 through VI-6, VI-8 through VI-12, and VI-14 through VI-19.

VI-1. Find $\mathbf{v}^i$ of Eq. (VI-3) when $\mathbf{v}^{i+1}$ is $[1 \quad 0 \quad 0]^T, [0 \quad 1 \quad 0]^T$, and $[0 \quad 0 \quad 1]^T$. Then sketch $\mathbf{v}^i, \mathbf{v}^{i+1}$, and the two coordinate systems. Verify from the sketch that $\mathbf{v}^i = \mathbf{T}_i \mathbf{v}^{i+1}$.

VI-2. Sometimes $\mathbf{T}$ is written with $\cos \theta_c$ and $\sin \theta_c$, instead of $\cos \theta$ and $\sin \theta$, or setting $\phi = 0$ at a *trans* placement, instead of a *cis* placement. What are the expressions for $\mathbf{T}$ when these notations are used?[20]

VI-3. Explain why the product $\mathbf{T}_i\mathbf{T}_{i+1}$ has a simple form when all bond angles are equal and all torsion angles are $180°$.

VI-4. Draw the conformation of a chain when all bond angles are $90°$, the lengths of bonds 1 through 5 are 1, 2, 3, 2, and 1 nm, respectively, and the torsion angles at bonds 2, 3, and 4 are $180°, 180°$, and $0°$, respectively. Also draw the x and y axes of the local coordinate system for bond 1. Then verify that the vectors from A_0 to the various A_i are given by the last column of $\mathbf{A}_1\mathbf{A}_2 \cdots \mathbf{A}_i$, and $\mathbf{r}$ is given by Eq. (VI-13).

VI-5. For the chain described in Problem VI-4, verify that r^2 is given by Eq. (VI-23), and the other r_{ij}^2 are given by Eq. (VI-25). Use the r_{ij}^2 to calculate s^2 via Eq. (VI-26). Verify that the same value for s^2 can be obtained directly from Eq. (VI-33).

VI-6. What result is obtained from Eq. (VI-34) if we substitute $m_i = 0$ for $0 < i < n$, retaining nonzero values only for m_0 and m_n?

VI-7. Calculate $\langle \mathbf{r} \rangle_0$ as a function of n for a chain in which all bond lengths are 154 pm, all bond angles are $112°$, and each internal bond is subject to free rotation. (The result is depicted in Fig. 2 of Yoon and Flory.[12])

VI-8. Figure VI-3 depicts the components of $\langle \mathbf{r} \rangle_0$ for unperturbed polyethylene chains at $140°C$. How sensitive are these components to the interdependence of the bonds? Repeat the calculations for the cases where $\omega = 1$ and $\omega = 0$.

VI-9. Find $\langle \mathbf{T} \rangle$ for a freely jointed chain. (There is only one nonzero element.) Verify that the expected result for C_n is obtained from Eq. (VI-64) when all bonds are of the same length.

VI-10. Find $\langle \mathbf{T} \rangle$ for the freely rotating chain. (There are two nonzero elements.) Verify that the result obtained from Eq. (VI-64) for a freely rotating chain in which all bonds are of the same length, and all bond angles are identical, is identical with Eq. (I-20).

VI-11. Consider a chain in which the bond angle is tetrahedral, the bonds are independent, and the torsion potential energy function is given by the following square wells:

---

[20] A discussion of the form adopted by $\mathbf{T}$ when it is constructed from the supplement of the bond angle, and using $\phi = 0°$ for a *trans* placement, can be found in Appendix B of Flory, P. J. *Statistical Mechanics of Chain Molecules*, Wiley-Interscience, New York, **1969**: reprinted by Hanser, München, **1989**.

Energy = 0 at $\phi = 180° \pm \Delta\phi$ (t well).

Energy = 0 at $\phi = 60° \pm \Delta\phi$ (g^+ well).

Energy = 0 at $\phi = -60° \pm \Delta\phi$ (g^- well).

Energy is infinite at all ϕ outside these wells.

Find $\langle \mathbf{T} \rangle$ as a function of $\Delta\phi$ for $0° < \Delta\phi < 60°$. How sensitive are $\langle \mathbf{r} \rangle_\infty$ and C_∞ to the width of the square wells?

VI-12. Repeat Problem VI-11 with the t well weighted by 1, and the g wells each weighted by σ. At what value of $\Delta\phi$ does C_∞ differ by more than 5% from the C_∞ computed with the same σ, but using $\Delta\phi = 0$? If C_∞ can be measured with an accuracy of 5% in an experiment, at what value of $\Delta\phi$ does the finite width of the wells become large enough so it must be considered in the comparison of theory with experiment?

VI-13. Use Eqs. (VI-66) and (VI-72) to verify the behavior of C_n and $C_{s,n}$ depicted in Fig. VI-4.

VI-14. The usual RIS model for polyethylene provides only two values of r^2 for n-butane, and only four values of r^2 for n-pentane, which are easily calculated by direct enumeration of all the conformations in the ensembles when $n = 3$ or 4. At what value of n does the generation of the distribution function $W(r^2)$ by discrete enumeration of all conformations become so tedious that its replacement by a Monte Carlo simulation is an attractive alternative?

VI-15. Do the *odd*-indexed bonds or the *even*-indexed bonds make the larger contribution to the average end-to-end vector of polyethylene? Devise efficient evaluations of

$$\langle \mathbf{r} \rangle_{0,\text{odd}} = \mathbf{l}_1 + \mathbf{T}_1\langle \mathbf{T}_2\rangle\mathbf{l}_3 + \mathbf{T}_1\langle \mathbf{T}_2\mathbf{T}_3\mathbf{T}_4\rangle\mathbf{l}_5 + \cdots \qquad (\text{VI-89})$$

$$\langle \mathbf{r} \rangle_{0,\text{even}} = \mathbf{T}_1\mathbf{l}_2 + \mathbf{T}_1\langle \mathbf{T}_2\mathbf{T}_3\rangle\mathbf{l}_4 + \mathbf{T}_1\langle \mathbf{T}_2\mathbf{T}_3\mathbf{T}_4\mathbf{T}_5\rangle\mathbf{l}_6 + \cdots \qquad (\text{VI-90})$$

and compare the results.

VI-16. How different are the $\langle s^2 \rangle_0$ for an unperturbed polyethylene of 1000 bonds when calculated using

(a) All 1001 chain atoms, weighted equally?

(b) Only the 501 A_i with even i?

(c) Only the 101 A_i with $i = 0, 10, 20, \cdots, 1000$?

VI-17. In an unperturbed polyethylene with $n = 300$, how different are $\langle r_{0,100}^2 \rangle_0$ and $\langle r_{100,200}^2 \rangle_0$, that is, the mean square unperturbed end-to-end distances for the subchains consisting of the first 100 bonds and the central 100 bonds?

VI-18. Consider a chain with the structure A–(B–A)$_x$–B, where $\theta_{ABA} \neq \theta_{BAB}$. Will any of the following properties depend on which end of the chain is selected for the indexing of the bonds: $Z, \langle \mathbf{r} \rangle_0, \langle r^2 \rangle_0, \langle s^2 \rangle_0$?

VI-19. For a chain such as polyethylene, is it necessarily true that

$$\lim_{n \to \infty} \langle \mathbf{r}_{0n} \rangle_0 = \lim_{n \to \infty} \langle \mathbf{r}_{jn} \rangle_0 \qquad j > 0 \tag{VI-91}$$

$$\lim_{n \to \infty} \frac{\langle r_{0n}^2 \rangle_0}{nl^2} = \lim_{n \to \infty} \frac{\langle r_{jn}^2 \rangle_0}{(n-j)l^2} \qquad j > 0 \tag{VI-92}$$

VI-20. Use $p_{\eta;2}$ and $q_{\xi\eta;i}, 2 < i < n-1$, to construct representative chains of unperturbed polyethylene with 100 bonds. Calculate $\langle \mathbf{r} \rangle_0, \langle r^2 \rangle_0$, and $\langle s^2 \rangle_0$ from your sample, and compare with the exact values obtained using generator matrices. When evaluated by the Monte Carlo method, do all three properties converge to the exact values at the same value of $\mathcal{N}$?

VII Simple Chains with Symmetric Rotation Potential Energy Functions

Chapter IV introduced the conformational partition function in the rotational isomeric state (RIS) approximation, with attention focused on a simple chain in which all bonds are identical and subject to a pairwise interdependent symmetric rotation potential energy function with three rotational isomers per bond. That formulation of Z, with the statistical weight matrix in Eq. (IV-19), has been used successfully for treatment of the mean square dimensions of unperturbed n-alkanes.[1,2]

The methods developed in Chapters IV–VI are not restricted to linear chains with symmetric threefold potential energy functions and pairwise interdependent bonds. They can also be applied, with equal rigor, to chains with more complicated covalent structures, given the appropriate formulation of Z. This chapter and Chapters VIII–XII that follow describe the formulation of Z for chains that are more complicated than linear polyethylene. The focus in this chapter is on linear chains in which there are no side groups or branches that produce an asymmetry in the rotation potential energy function. Often these chains are treated with $\nu = 3$, as in polyethylene, but other chains are described with ν of 2, 4, 5, 6, or 9. The FORTRAN program listed in Appendix C can be used to calculate C_n and C_∞ for all of the chains described in this chapter.

Subsequent chapters will consider chains with *asymmetric* rotation potential energy functions, simplifications that arise from the presence of *rigid units* and the use of *virtual bonds*, the intramolecular formation of *helices* and *antiparallel sheets*, *star-branched polymers*, and polymers with *articulated side chains*, and special issues that may arise in the treatment of *copolymers*.

1. SYMMETRIC THREEFOLD ROTATION POTENTIAL ENERGY FUNCTIONS

Since many polymers are constructed with a chain of carbon atoms, and the C—C bond in ethane has a symmetric threefold rotation potential energy func-

[1] Abe, A.; Jernigan, R. L.; Flory, P. J. *J. Am. Chem. Soc.* **1966**, *88*, 631.
[2] Jernigan, R. L.; Flory, P. J. *J. Chem. Phys.* **1969**, *50*, 4165. Yoon, D. Y.; Flory, P. J. *Macromolecules* **1976**, *9*, 294.

tion, several important polymers have bonds subject to a potential energy function with this symmetry. The most famous example is polyethylene. Polyethylene is usually treated with two statistical weights, denoted by σ and ω, and the statistical weight matrix presented in Eq. (IV-19). A more general form for the statistical weight matrix for a simple chain subject to a symmetric threefold rotation potential energy function and pairwise interdependent bonds was presented as Eq. (IV-48), reproduced here as

$$\mathbf{U}_i = \begin{bmatrix} \tau & \sigma & \sigma \\ 1 & \sigma\psi & \sigma\omega \\ 1 & \sigma\omega & \sigma\psi \end{bmatrix}_i \tag{VII-1}$$

or, in reduced form

$$\mathbf{U}_i = \begin{bmatrix} \tau & 2\sigma \\ 1 & \sigma(\psi + \omega) \end{bmatrix}_i \tag{VII-2}$$

The reference states are t for the first-order interactions, and tg ($\equiv gt$) for the second-order interactions. This statistical weight matrix differs from the one presented in Eq. (IV-19) by the incorporation of two new statistical weights for second-order interactions. Two successive g placements of the *same* sign receive a weight of ψ, and two successive t states are weighted by τ. An excellent approximation for polyethylene is $\psi = \tau = 1$. However, as we shall see in this chapter, there are other chains in which these second-order statistical weights must be assigned values different from 1.

Often in formulating a statistical weight matrix for bond i it is helpful to write it as a product of a diagonal matrix that contains the statistical weights for the first-order interactions for ν_i rotational isomers, and a $\nu_{i-1} \times \nu_i$ matrix (which is symmetric if $\nu_{i-1} = \nu_i$ and the bonds are subject to symmetric rotation potential energy functions) that contains the statistical weights for the second-order interactions. The matrix in Eq. (VII-1) can be written as

$$\mathbf{U}_i = \mathbf{V}_i\mathbf{D}_i \equiv \begin{bmatrix} \tau & 1 & 1 \\ 1 & \psi & \omega \\ 1 & \omega & \psi \end{bmatrix}_i \operatorname{diag}(1, \sigma, \sigma)_i \tag{VII-3}$$

2. POLY(A)

By "poly(A)," we denote a chain of composition —A—A—A—A— and so on, in which all chain atoms are identical. The identity of all chain atoms means that a single statistical weight matrix is sufficient for description of Z. Also all bonds have the same l_i, θ_i, ν_i, and set of ϕ_i. The most famous example is polyethylene, with A = CH_2, but other possible assignments for A include CF_2, SiH_2, $Si(CH_3)_2$, S, and Se.

A. Polyethylene, Fluctuations, and Higher-Order Interactions

Linear *polyethylene* (or *polymethylene*) is the best investigated example of a chain with a symmetric rotation potential energy function and identical chain "atoms," defined here as CH_2 (which is not distinguished from the terminal CH_3). The conformational partition function must take account of the preferences contained in the conformational energy surface for *n*-pentane (see Fig. III-5) and in the probability profiles for two consecutive bonds in polyethylene (see Fig. IV-6). Both figures show well-defined rotational isomers, $\nu = 3$, a symmetric rotation potential energy function, the influence of a first-order interaction that penalizes *g* states, and the influence of a second-order interaction that strongly penalizes two successive *g* states of opposite sign. The important first- and second-order interactions generate statistical weights denoted by σ and ω. The energies associated with the short-range interactions have the relationship $E_\omega > E_\sigma > E_\psi \approx E_\tau \approx 0$.[3] Since E_τ and E_ψ are small compared to kT at the temperatures of interest, unperturbed polyethylene is usually treated using $\psi = \tau = 1$. For example, the calculated value of E_ψ corresponds to $\psi \sim 1.2$ at 140°C,[3] but C_∞ is affected very little by variation of ψ within the range $1 \leq \psi \leq 1.2$.[4] The common practice uses $\psi = 1$.

Several RIS models for polyethylene were described in the 1960s.[1,5] The models currently most widely used are two variants based on the work of Abe et al.[1] One model places the torsion angle for the *g* states exactly 120° from the torsion angle for *t*, and the other model displaces the *g* states slightly toward *t*, as suggested by Fig. III-4. The magnitude of the displacement, denoted by $\Delta\phi$, is typically ~7.5°. The small changes in ϕ_g require small changes in E_σ and E_ω, as shown in Table VII-1, in order to optimize agreement with measured values of C_∞ and its temperature coefficient,[1] which are 6.8 at 140° C[6] and -1.1×10^{-3} deg^{-1},[7] respectively. Often calculations for unperturbed polyethylene have used interactions energies of $E_\sigma = 0.5$ kcal mol^{-1} and $E_\omega = 2$ kcal mol^{-1}, which convert to 2.1 kJ mol^{-1} and 8.4 kJ mol^{-1}, respectively, although only one figure is significant for E_ω. Illustrative results were depicted in Fig. VI-4.

[3]Tsuzuki, S.; Schäfer, L.; Gotō, H.; Jemmis, E. D.; Hosoya, H.; Siam, K.; Tanabe, K.; Ōsawa, E. *J. Am. Chem. Soc.* **1991**, *119*, 4665.

[4]Mattice, W. L. *Comput. Polym. Sci.* **1991**, *1*, 173.

[5]Hoeve, C. A. J. *J. Chem. Phys.* **1961**, *35*, 1266. Nagai, K.; Ishikawa, T. *J. Chem. Phys.* **1962**, *37*, 496.

[6]Chiang, R. *J. Phys. Chem.* **1965**, *69*, 1645.

[7]Ciferri, A.; Hoeve, C. A. J.; Flory, P. J. *J. Am. Chem. Soc.* **1961**, *83*, 1015. Flory, P. J.; Ciferri, A.; Chiang, R. *J. Am. Chem. Soc.* **1961**, *83*, 1023.

TABLE VII-1. Selected RIS Models for Poly(A)[a]

A	l (pm)	θ (deg)	ν	U (Eq.)	ϕ (deg)	Nonzero Energies (kJ mol^{-1})	Ref.
CH_2	153	112	3	(VII-1)	180, ±67.5	$E_\sigma = 1.1 - 1.9$, $E_\omega = 5.4 - 6.7$	b
CH_2	153	112	3	(VII-1)	180, ±60	$E_\sigma = 1.8 - 2.5$, $E_\omega = 7.1 - 8.0$	b
CF_2	153	116	3	(VII-1)	180, ±65	$E_\sigma = 5, E_\omega = \infty$	c
CF_2	153	116	4	(VII-7)	±165, ±60	$E_\sigma = 5.9 \pm 1.7$, $E_\omega = 4.6 \pm 2.9$	c
SiH_2	234	109.4	3	(VII-1)	180, ±55	$E_\sigma = -1.3$, $E_\psi \sim -1.7$	d
$Si(CH_3)_2$	235	115.4	3	(VII-1)	180, ±55	$E_\sigma = -0.4$, $E_\omega = \infty, E_\psi \sim -3.3$	d
S	206	106	2	(VII-8)	±90	$E_\omega = -1.1$	e
Se	234	104	2	(VII-8)	±90	$E_\omega = -1.7$	e

[a]The program listed in Appendix C can be used for calculation of C_∞ for all of the polymers listed in this table.
[b]Abe, A.; Jernigan, R. L.; Flory, P. J. *J. Am. Chem. Soc.* **1966**, *88*, 631. They also describe a five-state model with $\phi = 180°$, ±100°, ±65°. The model with five states splits each of the g states in order to take detailed account of the positions of local minima in the conformational energy surface in the vicinity of $g^\pm g^\mp$.
[c]Bates, T. W.; Stockmayer, W. H. *J. Chem Phys.* **1966**, *45*, 2321; *Macromolecules* **1968**, *1*, 12. The four-state model splits t into $t^\pm$. A recent six-state model also splits the g states (Smith, G. D.; Jaffe, R. L.; Yoon, D. Y. *Macromolecules*, in press). The three-state and four-state models can both account for the dipole moments of H–$(CF_2)_n$–H.
[d]Welsh, W. J.; DeBolt, L.; Mark, J. E. *Macromolecules* **1986**, *19*, 2978, using results of conformational energy calculations by Damewood, J. R., Jr.; West, R. *Macromolecules* **1985**, *18*, 159.
[e]Semlyen, J. A. *Trans. Faraday Soc.* **1967**, *63*, 743.

Incorporation of Fluctuations. Other variations on the model incorporate the influence of fluctuations in ϕ within each rotational isomeric state,[8,9] and also fluctuations in θ,[9] in the approximation that these fluctuations are independent. These models substitute $\langle \cos \theta \rangle$ and $\langle \sin \theta \rangle$ for each $\cos \theta$ and $\sin \theta$ in the transformation matrix, where the average is over the small range of θ accessible with thermal fluctuations. A similar substitution of $\langle \cos \phi_\eta \rangle$ and $\langle \sin \phi_\eta \rangle$ for $\cos \phi_\eta$ and $\sin \phi_\eta$ is made in each of the three rotational isomeric states, where the average is only over that range of ϕ associated with state η. Incorporation of these effects into the customary model for unperturbed polyethylene reduces the calculated value of C_∞ by ~6% (for fluctuations in ϕ) and ~1% (for fluctuations in θ). Larger reductions in C_∞ from these fluctuations are expected in other

[8]Cook, R.; Moon, M. *Macromolecules* **1978**, *11*, 1054; **1980**, *13*, 1537.
[9]Mansfield, M. L. *Macromolecules* **1983**, *16*, 1863.

chains that strongly prefer conformations in which long sequences of bonds occupy a particular rotational isomeric state.[10]

Incorporation of Third-Order Interactions. The highly repulsive *third-order interaction* in the $g^+g^-g^+$ and $g^-g^+g^-$ states can be included in Z by an expansion of the statistical weight matrix. It is not done in practice, because the first- and second-order interactions alone weight these conformations by $\sigma^3\omega^2$, which makes them very improbable when realistic values of ω and σ are assigned. Incorporation of the third-order interaction will merely further suppress conformations that are already strongly suppressed by the first- and second-order interactions. Nevertheless, it is useful at this stage to show how the third-order interaction, with statistical weight ω', could be incorporated, because the same methodology is used in applications to some other chains, where incorporation of higher order interactions is necessary for an adequate description of the unperturbed ensemble.

If an internal $\mathbf{U}_i$ containing statistical weights for first- and second-order interactions is represented as $\mathbf{V}_i\mathbf{D}_i$, Eq. (VII-3), the statistical weight matrix containing third-order interactions as well is generated as $\mathbf{W}_i(\mathbf{V}_i\mathbf{D}_i)^D$, where $(\mathbf{V}_i\mathbf{D}_i)^D$ has the elements of the $\nu_{i-1} \times \nu_i$ representation arranged in *reading order* along the main diagonal of a diagonal matrix, and $\mathbf{W}_i$ is a $(\nu_{i-2}\nu_{i-1}) \times (\nu_{i-1}\nu_i)$ matrix that contains the statistical weights for the third-order interactions. When $\mathbf{V}_i\mathbf{D}_i$ is given by Eq. (VII-3), the diagonal matrix denoted by $(\mathbf{V}_i\mathbf{D}_i)^D$ is

$$(\mathbf{V}_i\mathbf{D}_i)^D = \mathrm{diag}\,(\tau, \sigma, \sigma, 1, \sigma\psi, \sigma\omega, 1, \sigma\omega, \sigma\psi)_i \qquad \text{(VII-4)}$$

The strongly repulsive third-order interaction in the $g^+g^-g^+$ and $g^-g^+g^-$ states is incorporated with $\mathbf{W}_i$ constructed as

$\downarrow$ States of bonds $i - 1$ and i

	tt	tg^+	tg^-	g^+t	g^+g^+	g^+g^-	g^-t	g^-g^+	g^-g^-
tt	1	1	1	0	0	0	0	0	0
tg^+	0	0	0	1	1	1	0	0	0
tg^-	0	0	0	0	0	0	1	1	1
g^+t	1	1	1	0	0	0	0	0	0
g^+g^+	0	0	0	1	1	1	0	0	0
g^+g^-	0	0	0	0	0	0	1	ω'	1
g^-t	1	1	1	0	0	0	0	0	0
g^-g^+	0	0	0	1	1	ω'	0	0	0
g^-g^-	0	0	0	0	0	0	1	1	1

where the row label $\mathbf{W}_i =$ applies at g^+g^+, and $\uparrow$ States of bonds $i - 2$ and $i - 1$

$$\text{(VII-5)}$$

---

[10]These fluctuations can become the most important degrees of freedom in certain "rigid-rod" polymers, such as poly(benzobisoxazole), as described in Chapter IX.

Indexing of rows and columns is the same in $\mathbf{W}_i$ and in the statistical weight matrix, of dimensions $(\nu_{i-2}\nu_{i-1})\times(\nu_{i-1}\nu_i)$, formulated by this process. The rows are indexed by the states at bonds $i-2$ and $i-1$, and the columns are indexed by the states at bond $i-1$ and i. The same order is used for the indexing of the rows and columns of $\mathbf{W}_i$. The state of bond $i-1$ appears in the index for both the rows and the columns. The $\nu_{i-2}(\nu_{i-1}-1)\nu_{i-1}\nu_i$ null elements in $\mathbf{W}_i$ reject any contribution to Z from the nonsense sequences in which the state of bond $i-1$ specified by the row is negated by the state of the same bond, as specified by the column. The statistical weight matrix with third-order interactions is

$$\mathbf{U}_i = \begin{bmatrix} \tau & \sigma & \sigma & 0 & 0 & 0 & 0 & 0 & 0 \\ 0 & 0 & 0 & 1 & \sigma\psi & \sigma\omega & 0 & 0 & 0 \\ 0 & 0 & 0 & 0 & 0 & 0 & 1 & \sigma\omega & \sigma\psi \\ \tau & \sigma & \sigma & 0 & 0 & 0 & 0 & 0 & 0 \\ 0 & 0 & 0 & 1 & \sigma\psi & \sigma\omega & 0 & 0 & 0 \\ 0 & 0 & 0 & 0 & 0 & 0 & 1 & \sigma\omega\omega' & \sigma\psi \\ \tau & \sigma & \sigma & 0 & 0 & 0 & 0 & 0 & 0 \\ 0 & 0 & 0 & 1 & \sigma\psi & \sigma\omega\omega' & 0 & 0 & 0 \\ 0 & 0 & 0 & 0 & 0 & 0 & 1 & \sigma\omega & \sigma\psi \end{bmatrix}_i \qquad \text{(VII-6)}$$

At 300 K, the values of C_∞ computed with $E_{\omega'} = 0$ and $E_{\omega'} = \infty$ differ by only 0.038%.

B. Polytetrafluoroethylene

Another important chain sometimes treated with a 3×3 statistical weight matrix is *poly(tetrafluoroethylene)*, in which each hydrogen atom of polyethylene is replaced with a fluorine atom. Since fluorine is larger than hydrogen, the first- and second-order interactions that were repulsive in polyethylene are even more repulsive in poly(tetrafluoroethylene), as summarized in Table VII-1. The t state is more highly populated in this model for poly(tetrafluoroethylene) than in the model of polyethylene.

There is another model for polytetrafluoroethylene that more accurately reflects the details of the conformational energy surface in the vicinity of ϕ_{i-1}, $\phi_i = 180°$, $180°$.[11] In this more refined four-state model, the single t state of the three-state model is split into two states, denoted $t^\pm$:

[11]Bates, T. W.; Stockmayer, W. H. *J. Chem. Phys.* **1966**, *45*, 2321; *Macromolecules* **1968**, *1*, 12.

$$
\mathbf{U}_i \;=\; \begin{array}{c} \\ t^+ \\ g^+ \\ g^- \\ t^- \end{array}
\begin{array}{cccc} t^+ & g^+ & g^- & t^- \\ \left[\begin{array}{cccc} 1 & \sigma & 0 & \omega \\ 1 & \sigma & 0 & 0 \\ 0 & 0 & \sigma & 1 \\ \omega & 0 & \sigma & 1 \end{array}\right] \end{array}
\qquad\text{(VII-7)}
$$

The physical origin of the splitting arises from the fact that fluorine has a larger van der Waals radius than does hydrogen. As a consequence the shortest $F \cdots F$ distance between fluorines bonded to carbon atoms i and $i+2$ is *less* than the sum of their van der Waals radii if the two intervening torsional angles both have values of $180°$. In contrast, the shortest $H \cdots H$ distance in the comparable arrangement of polyethylene is *larger* than the sum of the van der Waals radii of two hydrogen atoms. The repulsion of the fluorine atoms when ϕ_{i-1}, $\phi_i = 180°$, $180°$ is relieved by small torsions *in the same direction* at both bonds. The preferred torsion angles for polytetrafluoroethylene are $\pm 165°$ for $t^\pm$ and $\pm 60°$ for $g^\pm$. The second-order interactions in the $g^\pm t^\mp$, $t^\pm g^\mp$, and $g^\pm g^\mp$ states are strongly repulsive, causing statistical weights of zero for these conformations. The second-order interactions also penalize the $t^\pm t^\mp$ sequence relative to the $t^\pm t^\pm$ sequence, and ω is used as the statistical weight for this interaction in Eq. (VII-7). The solid-state conformation is not a planar zigzag, as in polyethylene, but instead is a helix of low pitch produced by propagation of t^+ or t^-.[12] A recent model also splits the *gauche* states, thereby producing a description of polytetrafluoroethylene in terms of six rotational isomers.[13] This model provides improved agreement with the results of a measurement by light scattering of the characteristic ratio of polytetrafluoroethylene.[14]

The simpler three-state model is an adequate approximation for those conformation-dependent properties, such as $\langle \mu^2 \rangle_0$ for $H(CF_2)_n H$, $4 \le n \le 10$, that are insensitive to the splitting of the t state.[15]

C. Chains with Backbones of Silicon

Substitution of silicon for carbon in polyethylene is accompanied by an increase of ~50% in the lengths of the bonds in the main chain. For this reason, the first- and second-order interactions that were strong and repulsive in polyethylene become weaker, and even slightly attractive, in *polysilane*. Polysilane is predicted to have a smaller C_∞ than polyethylene, because of the closer approximation to free rotation in the silicon-containing polymer.[16] The parameters in Table VII-1 yields $C_\infty = 1.4$ at 300 K.

[12] Bunn, C. W.; Howells, E. R. *Nature (London)* **1954**, *174*, 549. *Macromolecules* **1968**, *1*, 12.
[13] Smith, G. D.; Jaffe, R. L.; Yoon, D. Y. *Macromolecules*, in press.
[14] Chu, B.; Wu, C.; Buck, W. *Macromolecules* **1989**, *22*, 831.
[15] Abe, A.; Furuya, H.; Toriumi, H. *Macromolecules* **1984**, *17*, 684.
[16] Welsh, W. J.; DeBolt, L.; Mark, J. E. *Macromolecules* **1986**, *19*, 2978.

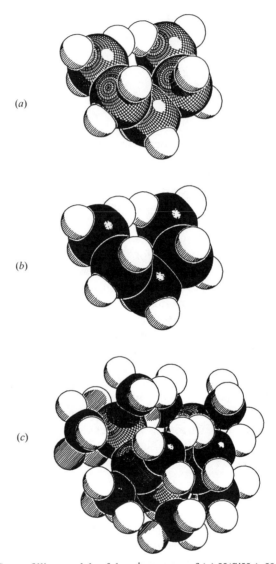

(a)

(b)

(c)

Figure VII-1. Space-filling models of the g^+g^- states of (a) $H(SiH_2)_5H$, (b) $H(CH_2)_5H$, and (c) $CH_3[Si(CH_3)_2]_5CH_3$.

In *poly(dimethylsilylene)*, where all hydrogen atoms of polysilane have been replaced with methyl groups, the increased bulk of the substituents produces an increase in θ_{SiSiSi} and also makes the $g^{\pm}g^{\mp}$ states untenable.[17] Figure VII-1 depicts the g^+g^- states of $H(SiH_2)_5H$, $H(CH_2)_5H$, and $CH_3[Si(CH_3)_2]_5$-CH_3, showing the differences in the severity of the repulsive interaction be-

[17]Damewood, J. R., Jr.; West, R. *Macromolecules* **1985**, *18*, 159.

tween the terminal groups. Complete suppression of $g^{\pm}g^{\mp}$ in poly(dimethyl-silylene), and a larger bond angle, results in the prediction of a C_{∞} that is nearly a factor of 10 larger than the value for polysilane.[16]

D. Polymeric Sulfur and Selenium

Two very simple chains, the linear homopolymers of *sulfur* and *selenium*, have been treated with a simpler model that has *two* symmetric, interdependent rotational states. There is no weighting factor for first-order interactions because they are equivalent in the two symmetric rotational isomeric states. The rather long bond lengths and absence of any pendant "side-chain" atoms cause the second-order interactions to be weak. The weak second-order interaction actually provides a small *preference* for rotational isomeric states of opposite sign at consecutive bonds, because atoms $i-3$ and $i+1$ are separated by a distance that slightly exceeds the sum of the van der Waals radii (and hence provide a weak attractive interaction) when ϕ_{i-1} and ϕ_i have opposite signs. The statistical weight matrix can be written as

$$\mathbf{U} = \begin{bmatrix} 1 & \omega \\ \omega & 1 \end{bmatrix} \tag{VII-8}$$

which yields a very simple expression for Z after condensation of the 2×2 form of $\mathbf{U}$ to a scalar.

$$Z = 2(1 + \omega)^{n-3} \tag{VII-9}$$

The polymers of sulfur and selenium have a high equilibrium flexibility, due to the absence of first-order interactions and strong second-order interactions.[18] The models in Table VII-1 predict C_{∞} very close to 1.

3. POLY(A–B)

Poly(A–B) denotes chains with sequences written as –A—B—A—B—A—B– and so on. Polymers in which two types of chain atoms occupy alternating positions in the main chain have the same bond length, l_{AB}, at all bonds, and the first-order interaction always involves A with B. However, the bond angles and second-order interactions alternate as one proceeds bond by bond along the chain. The bond angles are θ_{ABA} and θ_{BAB}, and the second-order interactions either involve A with A, or B with B. In general, the number of distinguishable $\mathbf{D}_i$ is at least as large as the number of distinguishable l_i, and it may be larger.

[18]Mark, J. E.; Curro, J. G. *J. Chem. Phys.* **1984**, *80*, 5262.

TABLE VII-2. Number of Distinguishable l_i, D_i, θ_i, and V_i

Repeat Unit	l_i	D_i	θ_i	V_i
A	1	1	1	1
AB	1	1	2	2
AAB	2	2	2	2
AAAB or ABAC	2	2	3	3
AAAAB or AABAB	2	3	3	3
AABAC	3	5	5	5
A_aB, $a > 4$	2	3	3	4
AABAAC	3	3	4	4

Also, the number of distinguishable V_i is at least as large as the number of distinguishable θ_i, and it may be larger. These assertions are illustrated in Table VII-2.

When the two bond angles assume different values, the all-*trans* conformation will no longer describe a fully extended chain, but instead describes a *cyclic polygon* that closes in n_{ring} bonds, given by

$$n_{\text{ring}} = \frac{2\pi}{|\theta_{\text{ABA}} - \theta_{\text{BAB}}|} \qquad \text{(VII-10)}$$

The nonequivalence of bond angles frequently produces a decrease in C_∞ for flexible chains, compared to the result that would have been obtained if every θ_i had the value of the average of all θ_i for the real bond angles in the chain.

Substituents larger than a hydrogen atom will be allowed in this section, but only if those substituents can be approximated as being rigidly attached to the backbone in a symmetric fashion, that is, the substituents do not have internal conformational degrees of freedom that must be considered. We will include polyisobutylene here, but we will defer consideration of polypropylene until the next chapter, because in that case the attachment of the methyl group does not produce a symmetric rotation potential energy function. Several poly(A–B) for which RIS models are available are listed in Table VII-3.

A. Statistical Weight Matrices, Z, and Probabilities

Statistical Weight Matrices. For purposes of construction of the statistical weight matrices, it is convenient to view the chain as bonds of type A—B that alternate with those of type B—A. The chain atoms involved in the first-order interactions are those at the ends of the sequences B–A—B–A and A–B—A–B, respectively. Thus the first-order interactions are identical in the two types of bonds, because one sequence is converted into the other by a change in the direction for indexing the bonds in the chain. All U_i will contain the same σ, and they will have rotational isomeric states with the same set of values of ϕ.

TABLE VII-3. Selected RIS Models for Poly(A–B) with Symmetric Rotation Potential Energy Functions[a]

A	B	l (pm)	θ_{ABA}, θ_{BAB} (deg)	U (Eq.)	ϕ (deg)	Nonzero Energies (kJ mol^{-1})	Ref.
CH$_2$	O	142	112, 112	(VII-11), (VII-12)	180, ±65	$E_\sigma = -5.8$, $E_{\omega_{AA}} = \infty$, $E_{\omega_{BB}} = 6.3$	b
CH$_2$	CF$_2$	155	113, 116	(VII-17), (VII-18)	±175, ±65, ±55	$E_\alpha = -2.51$, $E_{\alpha'} = 10.65$, $E_\gamma = -6.69$, $E_{\gamma'} = 3.81$, $E_\delta = 2.78$	c
CH$_2$	CCl$_2$	157	114, 121	(VII-17), (VII-18)	±165, ±90, ±52	$E_\alpha = -2.09$, $E_{\alpha'} = 7.11$, $E_\gamma = -2.18$, $E_{\gamma'} = 3.56$, $E_\delta = 1.04$	d
N	PCl$_2$	152	118, 130	(VII-19), (VII-20)	180, 0 ±50,	$E_\sigma = -5.9$, $E_{\sigma'} = -6.7$, $E_\omega = -4.2$	e
O	PO$_2$	162	102, 130	(VII-11), (VII-12)	180, ±60	$E_{\psi_{AA}} = E_{\psi_{BB}} = \infty$, $E_{\omega_{AA}} = E_{\omega_{BB}} = \infty$	f
CH$_2$	C(CH$_3$)$_2$	153	110, 124	(VII-21), (VII-22)	±155, ±60	$E_\xi = 12$	g
CH$_2$	Si(CH$_3$)$_2$	187	109.5, 121	(VII-11), (VII-12)	180, ±80	$E_{\omega_{BB}} = \infty$	h
O	Si(CH$_3$)$_2$	163	110, 143	(VII-11), (VII-12)	180, ±60	$E_\sigma = 1$, $E_{\omega_{AA}} = 2$, $E_{\omega_{BB}} = 3$, $E_{\psi_{AA}} = 2$, $E_{\psi_{BB}} = 2$	i

[a]The program listed in Appendix C can be used for calculation of C_∞ for all of the polymers listed in this table.
[b]Abe, A.; Mark, J. E. *J. Am. Chem. Soc.* **1976**, *98*, 6468.
[c]Carballeira, L.; Pereiras, A. J.; Rios, M. A. *Macromolecules* **1990**, *23*, 1309.
[d]Boyd, R. H.; Kesner, L. *J. Polym. Sci., Polym. Phys. Ed.* **1981**, *19*, 393.
[e]Saiz, E. *J. Polym. Sci., Part B: Polym. Phys.* **1987**, *25*, 1565.
[f]Semlyen, J. A.; Flory, P. J. *Trans. Faraday Soc.* **1966**, *62*, 2622.
[g]DeBolt, L. C.; Suter, U. W. *Macromolecules* **1987**, *20*, 1424.
[h]Sundararajan, P. R. *Comput. Polym. Sci.* **1991**, *1*, 18.
[i]Bahar, I.; Zúñiga, I.; Dodge, R.; Mattice, W. L. *Macromolecules* **1991**, *24*, 2993.

The second-order interactions are different for alternate bonds, as shown in Fig. VII-2. They involve two atoms of A when bond i is A—B, and two atoms of B when bond i is B—A.

$$U_{AB} = V_{AB}D = \begin{bmatrix} \tau_{AA} & \sigma & \sigma \\ 1 & \sigma\psi_{AA} & \sigma\omega_{AA} \\ 1 & \sigma\omega_{AA} & \sigma\psi_{AA} \end{bmatrix} \qquad \text{(VII-11)}$$

$$U_{BA} = V_{BA}D = \begin{bmatrix} \tau_{BB} & \sigma & \sigma \\ 1 & \sigma\psi_{BB} & \sigma\omega_{BB} \\ 1 & \sigma\omega_{BB} & \sigma\psi_{BB} \end{bmatrix} \qquad \text{(VII-12)}$$

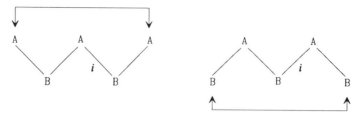

Figure VII-2. Chain atoms involved in the second-order interactions in U_{AB} (left) and U_{BA} (right) in poly(A–B).

The bond angles in the transformation matrices may also differ; T_i will contain θ_{ABA} when bond i is A—B, and it will contain θ_{BAB} when bond i is B—A.

Conformational Partition Function. The conformational partition function for a chain of n bonds is written as

$$Z = U_1 U_s (U_{AB} U_{BA})^{(n-4+n_I)/2} U_e U_n \qquad \text{(VII-13)}$$

where U_s and U_e are assigned as U_{BA}, U_{AB}, or I_3, depending on whether the ends are terminated by A or B. In the superscript in Eq. (VII-13), n_I denotes the number of U_s and U_e that are I_3. At large n, the dominant contribution to Z comes from $(U_{AB} U_{BA})^{(n-4+n_I)/2}$. The product $U_{AB} U_{BA}$ can be generated in an identical fashion from the statistical weight matrices defined in Eqs. (VII-11) and (VII-12), or from alternative symmetric matrices defined as

$$U_{AB} = V_{AB} = \begin{bmatrix} \tau_{AA} & 1 & 1 \\ 1 & \psi_{AA} & \omega_{AA} \\ 1 & \omega_{AA} & \psi_{AA} \end{bmatrix} \qquad \text{(VII-14)}$$

$$U_{BA} = D V_{BA} D = \begin{bmatrix} \tau_{BB} & \sigma & \sigma \\ \sigma & \sigma^2 \psi_{BB} & \sigma^2 \omega_{BB} \\ \sigma & \sigma^2 \omega_{BB} & \sigma^2 \psi_{BB} \end{bmatrix} \qquad \text{(VII-15)}$$

This sort of manipulation, in which all of the first-order interactions for two consecutive bonds are assigned to a single statistical weight matrix, often occurs in the literature on poly(A–B). Both sets of statistical weight matrices can multiply in the sequence $U_{AB} U_{BA}$ to give the product $V_{AB} D V_{BA} D$, where no subscript is required on D because the first-order interactions are the same at both bonds. This string of four matrices is grouped as $(V_{AB} D)(V_{BA} D)$ in Eqs. (VII-11) and (VII-12), but it is grouped as $(V_{AB})(D V_{BA} D)$ in Eqs. (VII-14) and (VII-15).

Probabilities. For bonds that are sufficiently far from both ends so that they are not influenced by end effects, $p_{\eta;i}$ is independent of i. For the pair probabilities, however, $p_{\xi\eta;i}$ need not be the same as $p_{\xi\eta;i-1}$, nor need $q_{\xi\eta;i}$ be the same as $q_{\xi\eta;i-1}$. The pair probabilities may depend on whether the two bonds are centered on A, or centered on B. For example, if $\omega_{AA} > 0$ and $\omega_{BB} = 0$, we will have $p_{g^+g^-;i}$ that alternate between being larger than zero, and zero, as i increases.

B. No Substituents Larger than Hydrogen–Polyoxymethylene

The simplest common poly(A–B) is *polyoxymethylene.* This chain is very different from polyethylene. It has a strong preference for g placements (due to the first-order interaction, which favors g over t) in long sequences of the same sign (due to the second-order interactions, which discriminate strongly against $g^{\pm}g^{\mp}$ relative to $g^{\pm}g^{\pm}$). The strong attraction in the g states is an example of the "anomeric effect," which many force fields have difficulty in estimating correctly.[19] (The typical error is underestimating the real preference for the g states in condensed media.) The second-order interaction of two methylene groups is more repulsive than the equivalent interaction of the smaller oxygen atoms, and hence $E_{CH_2,CH_2} > E_{O,O}$. The former interaction is more repulsive than the analogous second-order interaction in polyethylene because the C—O bonds are shorter than the C—C bonds, and therefore the two interacting methylene groups are more closely spaced in the g^+g^- conformation of polyoxymethylene than in the equivalent conformation of polyethylene, as shown in Fig. VII-3. It is sufficiently repulsive so that $g^{\pm}g^{\mp}$ states at O—CH$_2$—O are completely suppressed. Therefore E_{CH_2,CH_2}, which is large but finite, can be taken to be infinite. Equivalently stated, ω_{CH_2,CH_2} in polyoxymethylene is so small that it can be taken to be zero.

The large value of σ, coupled with $0 = \omega_{CH_2,CH_2} < \omega_{O,O} < 1$, causes the preferred conformation to be helices with sequences of *gauche* placements of the same sign. Right- and left-handed helices are equally probable. An example of each helix is depicted in Fig. VII-4. The strong preference for helical sequences in polyoxymethylene causes it to have a higher C_{∞} (~10) than polyethylene.[20] The melting of the helices[21] causes the temperature coefficient of the unperturbed dimensions to be large and negative.[20]

C. Head-to-Tail Poly(vinylidene halides)

Examples of poly(A–B) with nonarticulated substituents larger than hydrogen are provided by the *poly(vinylidene halides)* with exclusively *head-to-tail*

[19] Abe, A. *J. Am. Chem. Soc.* **1976**, *98*, 6477.
[20] Kokle, V.; Billmeyer, F. W., Jr. *J. Polym. Sci.,* **1965**, *3*, 47. Stockmayer, W. H.; Chan, L.-L. *J. Polym. Sci., Part A-2,* **1966**, *4*, 437.
[21] Curro, J. G.; Schweizer, K. S.; Adolf, D.; Mark, J. E. *Macromolecules* **1986**, *19*, 1739.

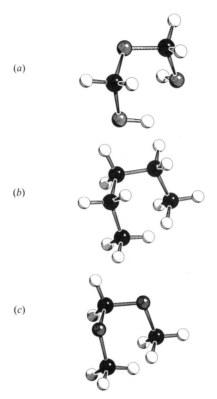

Figure VII-3. Repulsive interactions of terminal groups in the g^+g^- conformations of (a) HO–CH$_2$–O–CH$_2$–OH, (b) n-pentane, and (c) CH$_3$–O–CH$_2$–O–CH$_3$.

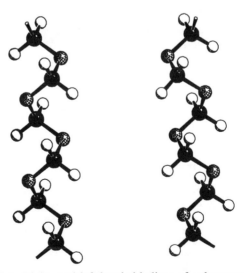

Figure VII-4. Right- and left-handed helices of polyoxymethylene.

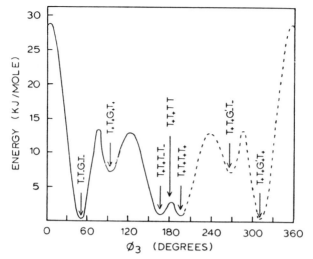

Figure VII-5. Least-energy profile for the variation in energy as a function of an internal torsion angle in a model for poly(vinylidene chloride). From Boyd, R. H.; Kesner, L. *J. Polym. Sci., Polym. Phys. Ed.* **1981**, *19*, 393. Copyright © 1981 John Wiley & Sons, Inc. Reprinted with permission from John Wiley & Sons, Inc.

addition. Head-to-tail is the favored mode of addition, with the largest departure from this pattern occurring in *poly(vinylidene fluoride)*, where the halogen is the smallest and the difference in sizes of CH_2 and CX_2 is least. A typical poly(vinylidene fluoride) might contain ~5% head-to-head, tail-to-tail placements.[22] The unperturbed dimensions do not appear to be sensitive to head-to-head, tail-to-tail placements when they are present at such low frequency, on the basis of calculations with an approximate model for the chain statistics.[23]

The poly(vinylidene halides) have been treated with a *six-state model* in which each of the t, g^+, and g^- states is divided into two closely spaced states.[24,25]

$$ t \rightarrow \left\{ \begin{array}{c} t_+ \\ t_- \end{array} \right\} \qquad g^+ \rightarrow \left\{ \begin{array}{c} g_+^+ \\ g_-^+ \end{array} \right\} \qquad g^- \rightarrow \left\{ \begin{array}{c} g_+^- \\ g_-^- \end{array} \right\} \qquad \text{(VII-16)} $$

This six-state model is an adaptation of one described originally for polyisobutylene.[26] The energy as a function of torsion angle has the form depicted in Fig. VII-5. The symmetric statistical weight matrices at CH_2—CX_2

[22]Wilson, C. H.; Santee, E. R. *J. Polym. Sci., Part C* **1965**, *8*, 97.
[23]Tonelli, A. E. *Macromolecules* **1976**, *9*, 547.
[24]Boyd, R. H.; Kesner, L. *J. Polym. Sci., Polym. Phys. Ed.* **1981**, *19*, 393.
[25]Carballeira, L.; Pereiras, A. J.; Rios, M. A. *Macromolecules* **1990**, *23*, 1309.
[26]Boyd, R. H.; Breitling, S. M. *Macromolecules* **1972**, *5*, 279.

and CX_2—CH_2 in the six-state model are written as

$$
\mathbf{U}_{CH_2CX_2} = \mathbf{D}\mathbf{V}_{CH_2CX_2}\mathbf{D} =
\begin{array}{c}
\\
t_+ \\ t_- \\ g_+^+ \\ g_-^+ \\ g_+^- \\ g_-^-
\end{array}
\begin{array}{c}
\begin{array}{cccccc} t_+ & t_- & g_+^+ & g_-^+ & g_+^- & g_-^- \end{array} \\
\left[
\begin{array}{cccccc}
1 & 0 & \alpha' & 0 & \alpha & 0 \\
0 & 1 & 0 & \alpha & 0 & \alpha' \\
\alpha' & 0 & \alpha'^2 & 0 & \alpha'\alpha & 0 \\
0 & \alpha & 0 & \alpha^2 & 0 & \alpha\alpha' \\
\alpha & 0 & \alpha\alpha' & 0 & \alpha^2 & 0 \\
0 & \alpha' & 0 & \alpha'\alpha & 0 & \alpha'^2
\end{array}
\right]
\end{array}
$$

$$\text{(VII-17)}$$

$$
\mathbf{U}_{CX_2CH_2} = \mathbf{V}_{CX_2CH_2} =
\begin{array}{c}
\\
t_+ \\ t_- \\ g_+^+ \\ g_-^+ \\ g_+^- \\ g_-^-
\end{array}
\begin{array}{c}
\begin{array}{cccccc} t_+ & t_- & g_+^+ & g_-^+ & g_+^- & g_-^- \end{array} \\
\left[
\begin{array}{cccccc}
1 & 1 & 1 & \delta & 1 & 1 \\
1 & 1 & 1 & 1 & \delta & 1 \\
1 & 1 & \gamma' & \gamma & 0 & 0 \\
\delta & 1 & \gamma & 0 & 0 & 0 \\
1 & \delta & 0 & 0 & 0 & \gamma \\
1 & 1 & 0 & 0 & \gamma & \gamma'
\end{array}
\right]
\end{array}
\qquad \text{(VII-18)}
$$

respectively. Each bond in a g state engenders a first-order interaction with statistical weight α or α'. The former statistical weight is used for g_-^+ and g_+^-, and the latter for g_+^+ and g_-^-. The separation of the first-order interaction into α and α' arises because the torsion angles for g_-^+ and g_+^- are closer to $\phi = 180°$ than are the torsion angles for g_+^+ and g_-^-, and hence they experience different first-order interactions. The first-order interactions for both bonds between successive CX_2 groups, CX_2—CH_2—CX_2 are written in the statistical weight matrix in Eq. (VII-17).

The pattern of the null elements in Eq. (VII-17) prohibits the sequences $t_{\pm}t_{\mp}$ and $g_{\pm}^{\pm}g_{\pm}^{\mp}$ at the two bonds flanking CH_2. Successive $g^{\pm}g^{\mp}$ placements at the two bonds flanking CH_2 are completely suppressed in Eq. (VII-18), with the bulk of the two CX_2 groups that participate in the "pentane effect." There are also second-order interactions in the $g^{\pm}g^{\pm}$ states at these two bonds. The statistical weight is denoted by γ if both states have subscripts of opposite sign. When the subscripts are of the same sign, the statistical weight is γ' if the superscripts also have that sign, and zero otherwise. A second-order interaction denoted by δ occurs when these two bonds have $t_{\pm}$ and $g_{\mp}^{\pm}$ states, with the states occuring in either sequence. Figure VII-6 depicts the conformations of the fragments that incur the second-order interactions, the statistical weights of which are γ, γ', and δ. With the parameters listed in Table VII-3, the calculations with the six-state model yield C_{∞} of 5.5 and 9.3 for poly(vinylidene fluoride) and poly(vinylidene chloride), respectively, at 300 K. Experimental results are 5.6

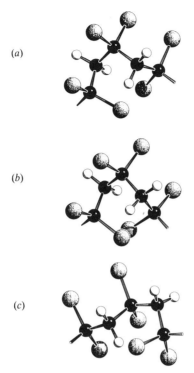

Figure VII-6. Conformations of CX_2—CH_2—CX_2—CH_2—CX_2 weighted by (*a*) γ, (*b*) γ', and (*c*) δ.

for poly(vinylidene fluoride) at 190°C[27] and 8 for poly(vinylidene chloride) at 25°C.[28] The chains become more extended as the bulk of the halogen atom increases.

D. Phosphorus in the Backbone

Poly(dichlorophosphazene). The conformations of the polyphosphazenes, with repeating sequence N—PR_1R_2, are poorly understood. Experimentally, the values of C_∞ can vary over an enormous range, depending on the selection of R_1 and R_2. A simple member of this series, *poly(dichlorophosphazene)*, R_1 = R_2 = Cl, has been treated with a *four-state model*, t, $g^\pm$, and c. The symmetric statistical weight matrices for the N—P and P—N bonds are

[27]Welch, G. J. *Polymer* **1974**, *15*, 429.
[28]Matsuo, K.; Stockmayer, W. H. *Macromolecules* **1975**, *8*, 660.

$$
\mathbf{U} = \begin{array}{c} \\ t \\ g^+ \\ c \\ g^- \end{array} \begin{array}{cccc} t & g^+ & c & g^- \\ \left[\begin{array}{cccc} 1 & \sigma & \sigma & \sigma \\ \sigma & \sigma' & \sigma' & \sigma' \\ \sigma & \sigma' & 0 & \sigma' \\ \sigma & \sigma' & \sigma' & \sigma' \end{array}\right] \end{array} \qquad \text{(VII-19)}
$$

$$
\mathbf{U} = \begin{array}{c} \\ t \\ g^+ \\ c \\ g^- \end{array} \begin{array}{cccc} t & g^+ & c & g^- \\ \left[\begin{array}{cccc} 1 & \omega & \omega & \omega \\ \omega & 0 & 0 & 0 \\ \omega & 0 & 0 & 0 \\ \omega & 0 & 0 & 0 \end{array}\right] \end{array} \qquad \text{(VII-20)}
$$

respectively. Successive c states are suppressed by the null values of u_{33} in Eq. (VII-19) and (VII-20). The 3×3 array of zeros in Eq. (VII-20) requires at least one of the bonds to each phosphorus atom must adopt a t state, and ω weights the tt combination at these two bonds differently from the other allowed conformations. The σ and σ' (which have nearly identical values) weight tt at the two bonds to a nitrogen atom differently from the other allowed states. With the assignments in Table VII-3, $C_\infty = 13.8$ at $25°C$, and $\partial \ln C_\infty / \partial T$ is -3×10^{-3} K^{-1}. The computed dimensions are quite sensitive to the value of E_σ.[29]

Polyphosphate. The bonds in *polyphosphate* do not have discrete rotational isomers, but the RIS formalism is a useful mathematical device for treating the continuous range of torsion angles. Although this polymer is structurally an example of poly(A–B), the severity of the second-order interactions whenever two consecutive bonds occupy g states, independent of their signs, or of whether they meet at O or PO_2, causes the statistical weight matrices for O—PO_2 and PO_2—O to become identical. The transformation matrices at the two bonds are nonequivalent because of the pronounced difference in the bond angles. The values of C_∞ determined from measurements in aqueous solution are near 7 at $25°C$.[30] The model in Table VII-3 gives $C_\infty = 7.39$.

E. Two Methyl Substituents on Alternate Chain Atoms

Now we turn to chains in which alternate chain atoms bear two methyl groups. Rotation about the bond between the methyl group and the atom in the main chain will be ignored in the determination of the number of rotational isomers; that is, the methyl groups are treated as spherical "united atoms." However,

[29] Saiz, E. *J. Polym. Sci., Part B: Polym. Phys.* **1987**, *25*, 1565.
[30] Strauss, U. P.; Windeman, P. L. *J. Am. Chem. Soc.* **1958**, *80*, 2366. *J. Phys. Chem.* **1962**, *66*, 2235.

the torsion angle at the C—CH_3 bond (denoted χ_1) is frequently treated as a variable in the conformational energy calculations used to determine values of the energies associated with the statistical weights in the expression for Z.

Polyisobutylene. The prototype is *polyisobutylene.* Historically, the treatment of this chain has presented a challenge because of the presence of mandatory, strongly repulsive second-order interactions. The origin of these interactions becomes apparent on recalling the conformation that produces the "pentane effect" in polyethylene. This repulsive interaction occurs whenever two successive internal bonds adopt g placements of opposite sign. Now consider the two bonds in polyisobutylene that meet at a CH_2 group. Rotations of 120° about these bonds in the small-molecule analog 2,2,4,4-tetramethylpentane, $(CH_3)_3C$—CH_2—$C(CH_3)_3$, merely interchanges the positions of the methyl groups. In polyisobutylene, the equivalent rotations interchange two methyl groups and one methylene group. If the two bonds adopt conformations that are tt with respect to the backbone, they will place two distinct pairs of methyl groups in positions that engender the "pentane effect." If the bonds were confined to the same three rotational isomers as in polyethylene, namely, t, $g^{\pm}$, pentane-effect interactions would be mandatory. Changes in the rotational isomers would merely change the identity of the groups participating in these interactions, with the possible combinations being methyl–methyl, methyl–methylene, and methylene–methylene.

The strength of the repulsive interactions can be reduced if the bond angles and torsion angles are displaced somewhat from the ones expected in polyethylene. For example, the crowding of the methyl groups in the all-*trans* conformation is alleviated by opening of θ_{C-CH_2-C}. The repulsion is alleviated further if the torsion angles at both C—C bonds to a methylene group are displaced by $\Delta\phi$ from 180°, $\Delta\phi \sim 20°$, with both displacements being of the same chirality.[31] The reduction in the repulsive steric interaction is depicted in Fig. VII-7. The t state of polyethylene is split into two $t_{\pm}$ states in polyisobutylene. A similar splitting occurs in the g states, which implies the use of a six-state model for polyisobutylene.[26,32] However, very good results for the dimensions can be obtained with a simpler, four-state model that does not split the g states. The $t_{\pm}$ region has twice the population of either of the $g^{\pm}$ regions, and hence the splitting of the latter can be ignored without a strong influence on the results. The symmetric statistical weight matrices for the CH_2—$C(CH_3)_2$ and $C(CH_3)_2$—CH_2 bonds are

[31] Allegra, G.; Benedetti, E.; Pedone, C. *Macromolecules* **1970**, *3*, 727.
[32] Suter, U. W.; Saiz, E.; Flory, P. J. *Macromolecules* **1983**, *16*, 1317. Vacatello, M.; Yoon, D. Y. *Macromolecules* **1992**, *25*, 2502.

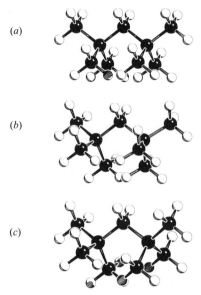

Figure VII-7. Conformation of 2,2,4,4-tetramethylpentane when the internal torsion angles are (a) 180°, 180°, (b) −160°, 160°, and (c) −160°, −160°.

$$
\mathbf{U} = \begin{array}{c} \\ t_+ \\ t_- \\ g^+ \\ g^- \end{array}
\begin{array}{cccc} t_+ & t_- & g^+ & g^- \end{array}
\left[\begin{array}{cccc}
1 & 0 & 1 & 0 \\
0 & 1 & 0 & 1 \\
1 & 0 & 1 & 0 \\
0 & 1 & 0 & 1
\end{array} \right] \tag{VII-21}
$$

$$
\mathbf{U} = \begin{array}{c} \\ t_+ \\ t_- \\ g^+ \\ g^- \end{array}
\begin{array}{cccc} t_+ & t_- & g^+ & g^- \end{array}
\left[\begin{array}{cccc}
0 & 0 & 1 & \xi \\
0 & 0 & \xi & 1 \\
1 & \xi & 0 & 0 \\
\xi & 1 & 0 & 0
\end{array} \right] \tag{VII-22}
$$

respectively, in the four-state model. The pattern of the null elements in Eq. (VII-21) enforces the maintenance of chirality through the $C(CH_3)_2$—CH_2—$C(CH_3)_2$ bonds. The null elements in Eq. (VII-22) require one bond be $t_\pm$, and the other $g^\pm$, in CH_2—$C(CH_3)_2$—CH_2, and $E_\xi > 1$ produces a preference for the maintenance of chirality at these two bonds. With the parameters in Table VII-3, this model yields $C_\infty = 6.6$ at 300 K, in agreement with experiment.[33]

[33]Fox, T. G., Jr.; Flory, P. J. *J. Am. Chem. Soc.* **1951**, *73*, 1909.

Poly(dimethylsilmethylene). *Poly*(*dimethylsilmethylene*), also known as *poly-*(*dimethylsilaethylene*), replaces alternate carbon atoms in polyisobutylene with silicon atoms. Every C—C bond in polyisobutylene is replaced by an Si—C bond in poly(dimethylsilmethylene). This replacement has important consequences for the conformation of the chain because the Si—C bond is ~34 pm longer than the C—C bond.[34] The lengthening of the bonds increases the separation of the groups that interacted repulsively in the pentane-effect conformations of polyisobutylene, thereby greatly reducing the magnitude of the repulsions in these conformations. Consequently it is no longer necessary to consider splitting of the *t* state, and Z can be formulated with a model that uses three rotational isomeric states. The only important short-range interaction is the interaction of the $Si(CH_3)_2$ groups when they participate in the pentane effect. This interaction can be taken to be infinitely repulsive, thereby producing an assignment of $\omega_{BB} = 0$. The model in Table VII-3 yields a calculated result for C_∞ of 5.5. The unperturbed dimensions are independent of temperature in this model, because none of the u_{ij} depends on temperature.

Poly(dimethylsiloxane). Replacement of the methylene group of poly(dimethylsilmethylene) by an oxygen atom yields *poly(dimethylsiloxane)*, a polymer well-known for its flexibility. In view of its importance, rotational isomeric state models for poly(dimethylsiloxane) date back three decades.[35] More recent structural information about the siloxane backbone, coupled with access to algorithms and computational power that permit incorporation of the information into the construction of the rotational isomeric analysis, have provided more accurate treatments. The unusually low force constant for bending at θ_{SiOSi}, and the unusually low intrinsic torsional potential energy function at the Si—O bond, have important effects on the conformation, whether evaluated by conformational energy analysis[36] or by molecular dynamics trajectories.[37] Figure V-3 depicted the probability density at 300 K for both bond pairs in poly(dimethylsiloxane). Both profiles exhibit continuous regions of high probability, in marked contrast to the discrete rotational isomers seen in the equivalent representation for polyethylene (Fig. IV-6). When the probabilities are represented as a function of a single torsion angle, as in Fig. V-4, they vary over a range only slightly greater than 2.

A description in terms of the RIS model is useful for poly(dimethylsiloxane) as a mathematical device in which a continuous distribution is represented by discrete values. The three-state model described in Table VII-3 was constructed initially from the probability profiles such as those depicted in Figs. V-3 and

[34]Bowen, H. J. M.; Sutton, L. E. *Tables of Interatomic Distances and Configuration in Molecules and Ions*, The Chemical Society, London, **1958**; Sutton, L. E. *Interatomic Distances, Supplement* The Chemical Society, London, **1965.**

[35]Flory, P. J.; Crescenzi, V.; Mark, J. E. *J. Am. Chem. Soc.* **1964**, *86*, 146.

[36]Grigoras, S. *Polym. Prepr.* (*Am. Chem. Soc., Div. Polym. Chem.*) **1991**, *32(2)*, 720.

[37]Bahar, I.; Zúñiga, I.; Dodge, R.; Mattice, W. L. *Macromolecules* **1991**, *24*, 2986.

V-4, with t and $g^{\pm}$ defined to provide nonoverlapping $120°$ ranges for ϕ, and with adoption of discrete values for the bond angles. (A more elaborate model using nine nonoverlapping $40°$ ranges for ϕ is also available.[38]) The energetic parameters derived initially from these profiles were then optimized in order to provide the best agreement with measured values for C_{∞} (6.3 in methylethylketone,[39] calculated 6.3) and D_{∞} (0.33 from experiment,[40] calculated 0.25), and qualitative agreement with measured macrocyclization equilibrium constants.[41] This model does not correctly predict the temperature coefficient of C_{∞}, perhaps because conformational degrees of freedom important for this property are not adequately represented by the limited set of discrete values used in the model.

4. POLY(A—A—B)

Polymers with the repeating sequence A—A—B have two different bond lengths, l_{AA} and l_{AB}, two different bond angles, θ_{AAB} and θ_{ABA}, two different types of first-order interactions, A $\cdots$ A and B $\cdots$ B, and two different types of second-order interactions, A $\cdots$ A and A $\cdots$ B, as summarized in Table VII-2. The first- and second-order statistical weights denoted by σ and ω occur as

$$\mathbf{U}_{AA} = \begin{bmatrix} 1 & \sigma_{BB} & \sigma_{BB} \\ 1 & \sigma_{BB} & \sigma_{BB}\omega_{AB} \\ 1 & \sigma_{BB}\omega_{AB} & \sigma_{BB} \end{bmatrix} \qquad \text{(VII-23)}$$

$$\mathbf{U}_{AB} = \begin{bmatrix} 1 & \sigma_{AA} & \sigma_{AA} \\ 1 & \sigma_{AA} & \sigma_{AA}\omega_{AB} \\ 1 & \sigma_{AA}\omega_{AB} & \sigma_{AA} \end{bmatrix} \qquad \text{(VII-24)}$$

$$\mathbf{U}_{BA} = \begin{bmatrix} 1 & \sigma_{AA} & \sigma_{AA} \\ 1 & \sigma_{AA} & \sigma_{AA}\omega_{AA} \\ 1 & \sigma_{AA}\omega_{AA} & \sigma_{AA} \end{bmatrix} \qquad \text{(VII-25)}$$

For bonds sufficiently far removed from either end, the $p_{\eta;i}$ for bonds involving B are identical, but they may differ from the $p_{\eta;i}$ at the A—A bonds. The pair probabilities, $p_{\xi\eta;i}$ and $q_{\xi\eta;i}$, are different for A—B, B—A, and A—A.

[38] Neuburger, N. A.; Bahar, I.; Mattice, W. L. *Macromolecules* **1992**, *25*, 2447.
[39] Crescenzi, V.; Flory, P. J. *J. Am. Chem. Soc.* **1964**, *86*, 141.
[40] Sutton, C.; Mark, J. E. *J. Am. Chem. Soc.* **1971**, *54*, 5011.
[41] Beevers, M. S.; Semlyen, J. A. *Polymer* **1972**, *13*, 385.

TABLE VII-4. Selected Models for Poly(A–A–B) with Symmetric Rotation Potential Energy Functions, Using Eqs. (VII-23), (VII-24), and (VII-25)[a]

A	B	l_{AA}, l_{AB} (pm)	θ_{AAB}, θ_{ABA} (deg)	ϕ_{AA}, ϕ_{AB} (deg)	Nonzero Energies (kJ mol^{-1})	Ref.
CH_2	O	153, 143	111.5, 111.5	180, ±68; 180, ±80	$E_{\sigma_{BB}} = -2.2, E_{\sigma_{AA}} = 3.2,$ $E_{\omega_{AA}} = \infty, E_{\omega_{AB}} = 2.2$	b
CH_2	S	153, 181.5	114, 100	180, ±70; 180, ±70	$E_{\sigma_{BB}} = 1.7, E_{\sigma_{AA}} = -0.4,$ $E_{\omega_{AA}} = 1.7, E_{\omega_{AB}} = 4.6$	c

[a]The program listed in Appendix C can be used for calculation of C_∞ for all of the polymers listed in this table.
[b]Abe, A.; Tasaki, K.; Mark, J. E. *Polym. J.* **1985**, *17*, 883.
[c]Abe, A. *Macromolecules* **1980**, *13*, 546.

A. Polyoxyethylene

The most famous example of poly(A—A—B) is *polyoxyethylene* [also known as *poly(ethylene oxide)* or *poly(ethylene glycol)*]. This polymer, like poly(dimethylsiloxane), has been the subject of RIS treatments for three decades.[42,43] It presents an interesting challenge because experiment unequivocally shows *g* states are preferred to *t* states at the C—C bond in 1,2-dimethoxyethane[44] and in polyoxyethylene,[44,45] but conformational energy calculations over the years have badly estimated the size of the preference, if indeed they even predicted *g* was favored over *t*.[43] At issue here is the size of the statistical weight denoted by σ_{BB} in U_{AA}. Since the *ttt* conformation of dimethoxyethane has no dipole moment, in contrast with the $tg^{\pm}t$ conformations, the population of the $g^{\pm}$ states at the C—C bond may be lower *in vacuo* than in a condensed phase, thereby affecting the type of calculation that must be done if the results are to be pertinent to polyoxyethylene.[46] A recent refinement of Z for polyoxyethylene is presented in Table VII-4. It yields $C_\infty = 5.1$ at 25°C, in good agreement with the experimental result.[47] Recently another model, which includes third-order interactions, has been proposed.[48]

B. Polythiaethylene

The sulfur analog of polyoxyethylene can be described by the same three statistical weight matrices, although with different values of the parameters, as shown

[42]Mark, J. E.; Flory, P. J. *J. Am. Chem. Soc.* **1965**, *87*, 1415.
[43]Abe, A.; Mark, J. E. *J. Am. Chem. Soc.* **1976**, *98*, 6468.
[44]Abe, A.; Inomata, K. *J. Mol. Struct.* **1991**, *245*, 399. Inomata, K.; Abe, A. *J. Phys. Chem.* **1992**, *96*, 7934.
[45]Tasaki, K.; Abe, A. *Polym. J.* **1985**, *17*, 641.
[46]Müller-Plathe, F.; van Gunsteren, W. F., submitted for publication.
[47]Beech, D. R.; Booth, C. *J. Polym. Sci. Part A-2* **1969**, *7*, 575.
[48]Smith, G. D.; Yoon, D. Y.; Jaffe, R. L. *Macromolecules*, **1993**, *26*, 5213.

in Table VII-4. The interaction energies tend to be weaker in *polythiaethylene* because the C—S bond is nearly 40 pm longer than the C—O bond, thereby increasing the distances between interacting groups. The value of C_∞ at 25°C is calculated to be 4.3 for polythioethylene, which is ~20% smaller than the result for polyoxyethylene.

5. BACKBONES WITH MORE COMPLICATED REPEATING PATTERNS

Conformational partition functions have been reported for numerous other linear macromolecules with pairwise interdependent bonds, symmetric threefold rotation potential energy functions, and the groups in the chain being selected from CH_2, $C(CH_3)_2$, O, or S. Citations are collected in Table VII-5. Two polymers from this table deserve special mention.

A. Poly(trimethylene oxide)

Poly(trimethylene oxide) has two different types of first-order interactions, between the terminal groups in CH_2—CH_2—CH_2—O and CH_2—CH_2—O—CH_2. In the statistical weight matrices presented below, they are represented by σ_{OC} and σ_{CC}, with the subscripts identifying the interacting atoms. Starting at an O—C bond and proceeding bond by bond along the chain, the statistical weight matrices are

$$\mathbf{U}_{OC} = \begin{bmatrix} 1 & \sigma_{CC} & \sigma_{CC} \\ 1 & \sigma_{CC} & \sigma_{CC}\omega'_{CC} \\ 1 & \sigma_{CC}\omega'_{CC} & \sigma_{CC} \end{bmatrix} \tag{VII-26}$$

$$\mathbf{U}_{CC,a} = \begin{bmatrix} 1 & \sigma_{OC} & \sigma_{OC} \\ 1 & \sigma_{OC} & \sigma_{OC}\omega_{CC} \\ 1 & \sigma_{OC}\omega_{CC} & \sigma_{OC} \end{bmatrix} \tag{VII-27}$$

$$\mathbf{U}_{CC,b} = \begin{bmatrix} 1 & \sigma_{OC} & \sigma_{OC} \\ 1 & \sigma_{OC} & \sigma_{OC}\omega_{OO} \\ 1 & \sigma_{OC}\omega_{OO} & \sigma_{OC} \end{bmatrix} \tag{VII-28}$$

$$\mathbf{U}_{CO} = \begin{bmatrix} 1 & \sigma_{CC} & \sigma_{CC} \\ 1 & \sigma_{CC} & \sigma_{CC}\omega_{CC} \\ 1 & \sigma_{CC}\omega_{CC} & \sigma_{CC} \end{bmatrix} \tag{VII-29}$$

TABLE VII-5. More Chains with Symmetric Rotation Potential Energy Functions

Bonds in Repeating Unit	Repeating Sequence	Ref.
3	CH_2—$C(CH_3)_2$—O	a
4	$(CH_2)_3$—O	b
4	$(CH_2)_3$—S	c
4	CH_2—$C(CH_3)_2$—CH_2—O	d
4	CH_2—$C(CH_3)_2$—CH_2—S	e
5	$(CH_2)_4$—O	b
5	$(CH_2)_2$—O—CH_2—O	f
6	$(CH_2)_5$—S	g
6	$(CH_2)_2$—O—$(CH_2)_2$—S	h
7	$(CH_2)_4$—O—CH_2—O	i
8	$(CH_2)_5$—O—CH_2—O	j
8	$(CH_2)_5$—S—CH_2—S	k
9	$(CH_2)_2$—O—$(CH_2)_2$—O—$(CH_2)_2$—S	l
9	$(CH_2)_5$—S—$(CH_2)_2$—S	m
9	$(CH_2)_6$—O—CH_2—O	j
10	$(CH_2)_5$—S—CH_2—O—CH_2—S	n
10	$(CH_2)_2$—O—CH_2—O—$(CH_2)_2$—S—CH_2—S	o
11	$(CH_2)_{10}$—O	p

[a] Kato, K.; Araki, K.; Abe, A. *Polym. J.* **1981**, *13*, 1055.
[b] Abe, A.; Mark, J. E. *J. Am. Chem. Soc.* **1976**, *98*, 6468.
[c] Guzmán, J.; Riande, E.; Welsh, W. J.; Mark, J. E. *Makromol. Chem.* **1982**, *183*, 2573.
[d] Saiz, E.; Riande, E.; Guzmán, J.; de Abajo, J. *J. Chem. Phys.* **1980**, *73*, 958.
[e] Riande, E.; Guzmán, J.; Saiz, E.; de Abajo, J. *Macromolecules* **1981**, *14*, 608.
[f] Riande, E.; Mark, J. E. *Macromolecules* **1978**, *11*, 956.
[g] Riande, E.; Guzmán, J. *Macromolecules* **1979**, *12*, 952.
[h] Riande, E.; Guzmán, J.; Welsh, W. J.; Mark, J. E. *Makromol. Chem.* **1982**, *183*, 2555.
[i] Riande, E.; Mark, J. E. *J. Polym. Sci., Polym. Phys. Ed.* **1979**, *17*, 2013.
[j] Riande, E.; Mark, J. E. *Polymer* **1979**, *20*, 1188.
[k] Welsh, W. J.; Mark, J. E.; Guzmán, J.; Riande, E. *Makromol. Chem.* **1982**, *183*, 2565.
[l] Riande, E.; Guzmán, J. *Macromolecules* **1979**, *12*, 1117.
[m] Riande, E.; Guzmán, J. *Macromolecules* **1981**, *14*, 1234.
[n] Riande, E.; Guzmán, J. *Macromolecules* **1986**, *19*, 2956.
[o] Riande, E.; Guzmán, J. *Macromolecules* **1981**, *14*, 1511.
[p] Riande, E. *Makromol. Chem.* **1977**, *178*, 2001.

There are three types of second-order interactions, between the terminal groups in CH_2—CH_2—CH_2—O—CH_2, CH_2—CH_2—O—CH_2—CH_2, and O—CH_2—CH_2—CH_2—O. The statistical weights are represented by ω_{CC}, ω'_{CC}, and ω_{OO} in the equations above. Equations (VII-27) and (VII-28) contain the same first-order statistical weight, σ_{OC}, and Eqs. (VII-26) and (VII-29) both contain another first-order statistical weight, σ_{CC}. Equations (VII-27) and (VII-29) contain the same second-order interaction, but each of the other two

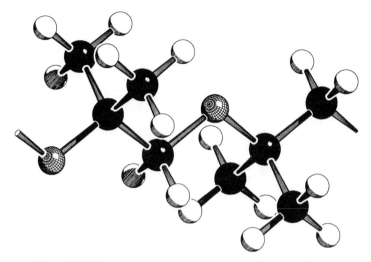

Figure VII-8. Fragment of poly(oxy-1,1-dimethylethylene).

statistical weight matrices has a second-order interaction unique to that matrix. No two of the statistical weight matrices are identical.[49]

The two second-order interactions that involve methylene groups are so strongly repulsive that both ω values can be set equal to zero.[42] The measured value of C_∞ is 3.94 ± 0.17,[50] significantly smaller than the results obtained with polyoxymethylene[20] and polyoxyethylene.[47] Calculation yields $C_\infty = 4.3$.[43]

B. Poly(oxy-1,1-dimethylethylene)

Poly(oxy-1,1-dimethylethylene) deserves special comment because of simplifications in Z that arise from the strong interactions present. A fragment of the chain is depicted in Fig. VII-8. Consider the repeating sequence CH_2—$C(CH_3)_2$—O—, located in the interior of a long chain, and sufficiently far from either end so that end effects can be ignored. The severe interactions between the bulky $C(CH_3)_2$ groups completely suppress the two g states at the intervening O—CH_2 bond.[51] The statistical weight matrix for this bond will have only one column, because the t state is the only state allowed. The bonds on either side of the O—CH_2 have access to the t and $g^{\pm}$ states, and hence the statistical weight matrix for O—CH_2 must be a column of three elements:

[49] But there is a perfectly equivalent formulation that uses only three (rather than four) distinct statistical weight matrices (see Problem VII-15).

[50] Takahashi, Y.; Mark, J. E. *Macromolecules* **1976**, *98*, 3756.

[51] Kato, K.; Araki, K.; Abe, A. *Polym. J.* **1981**, *13*, 1055.

$$\mathbf{U}_{O-CH_2} = \begin{bmatrix} 1 \\ 1 \\ 1 \end{bmatrix} \qquad \text{(VII-30)}$$

The statistical weight matrix for the following CH_2—$C(CH_3)_2$ bond must contain three elements arranged in a row, because the preceding bond had only one state:

$$\mathbf{U}_{CH_2-C(CH_3)_2} = \begin{bmatrix} 1 & \sigma & \sigma \end{bmatrix} \qquad \text{(VII-31)}$$

The $C(CH_3)_2$—O bond also has access to three rotational isomeric states. Since the preceding bond also can occupy three states, the statistical weight matrix is of dimensions 3×3. The only important statistical weight is the second-order interaction in the $g^{\pm}g^{\mp}$ states:

$$\mathbf{U}_{C(CH_3)_2-O} = \begin{bmatrix} 1 & 1 & 1 \\ 1 & 1 & \omega' \\ 1 & \omega' & 1 \end{bmatrix} \qquad \text{(VII-32)}$$

These three matrices, evaluated in the order $\mathbf{U}_{CH_2-C(CH_3)_2}\mathbf{U}_{C(CH_3)_2-O}\mathbf{U}_{O-CH_2}$, yield a *scalar.*

$$\begin{bmatrix} 1 & \sigma & \sigma \end{bmatrix} \begin{bmatrix} 1 & 1 & 1 \\ 1 & 1 & \omega' \\ 1 & \omega' & 1 \end{bmatrix} \begin{bmatrix} 1 \\ 1 \\ 1 \end{bmatrix} = 3 + 4\sigma + 2\sigma\omega' \qquad \text{(VII-33)}$$

Denoting this scalar by u^3, the conformational partition function for a very long chain with x such internal units contains the factor $(u^3)^x$, which is the xth power of a scalar, and hence very easily computed. The complete expression for Z will be obtained as $Z = \mathbf{S}(u^3)^x\mathbf{E}$, where $\mathbf{S}$ and $\mathbf{E}$ are whatever products of statistical weight matrices might be required by the groups at either end of the long chain. The important point is the local rigidity at the O—CH_2 bond, imposed in this chain by severe steric interactions that make the g states untenable, greatly simplifies the computation of Z. This feature will be exploited in Chapter IX, when we consider chains that contain other bonds for which only a single rotational isomer exists, usually as a consequence of incorporation of the bond into a ring, or the presence of double bond character.

The calculated value of C_∞ for poly(oxy-1,1-dimethylethylene) is 6.1.[52]

[52] Abe, A.; Ande, I.; Kato, K.; Uematsu, I. *Polym. J.* **1981**, *13*, 1069.

C. End-Labeled *n*-Alkanes

Another permutation is the design of conformational partition functions for *end-labeled n-alkanes*, $X—(CH_2)_{n-1}—X$, $X \neq CH_3$. The relative orientation of the two end groups may be important, as when the dipole moment of the end-labeled *n*-alkane is dominated by the contributions from the group dipole associated with the two $X—CH_2$ bonds. These end-labeled *n*-alkanes require new first-order interactions for rotation about the first and last $CH_2—CH_2$ bonds in the chain, $X–CH_2—CH_2-CH_2$. New second-order interactions are required for the first and last pair of $CH_2—CH_2$ bonds, $X–CH_2—CH_2—CH_2–CH_2$. The first $CH_2—CH_2$ in the chain will require a second-order interaction if X is articulated, as in the case where X is $O—CH_3$. Conformational partition functions have been obtained for *n*-alkanes end-labeled with bromine,[53] iodine,[54] or methoxy.[55] The conformational partition function has also been formulated for polytetrafluoroethylene end-labeled with hydrogen or bromine.[15]

PROBLEMS

Appendix B contains answers for Problems VII-1 through VII-7, VII-9, VII-11, and VII-13 through VII-16.

VII-1. What local conformations of a simple chain with identical bonds and a symmetric threefold rotation potential energy function are favored when

    (a) $\sigma \ll 1$?

    (b) $1 \ll \sigma, \omega \ll 1 \ll \psi$?

    (c) $1 \ll \sigma, \psi \ll 1 \ll \omega$?

VII-2. Is p_t more sensitive to temperature in polyethylene, polysilane, or poly(dimethylsilmethylene)?

VII-3. Using the models described in Table VII-1, determine which differences for the following pairs of chains are primarily responsible for their differences in C_∞ at 300 K.[56]

    (a) Polyethylene and polytetrafluoroethylene: θ, ϕ_g, or E_σ?

    (b) Polysilane and polydimethylsilylene: θ, E_σ, E_ψ, or E_ω?

    (c) Polymers of sulfur and selenium: θ or E_ω?

VII-4. In the four-state model for polytetrafluoroethylene, how important for C_∞ is the differential weighting of $t^\pm t^\pm$ versus $t^\pm t^\mp$?[56]

[53] Khanarian, G.; Tonelli, A. E. *J. Chem. Phys.* **1981**, *75*, 5031.

[54] Abe, A.; Tasaki, K. *Macromolecules* **1986**, *19*, 2647.

[55] Inomata, K.; Phataralaoha, N.; Abe, A. *Comput. Polym. Sci.* **1991**, *1*, 126.

[56] The program listed in Appendix C can be used for calculation of C_∞ for the polymers in this problem.

VII-5. How important for C_∞ at 300 K is the nonidentity of the bond angles in poly(A–B)? Examine the effect on polyphosphate, polyisobutylene, polydimethylsiloxane, and polydimethylsilmethylene of the replacement of both θ_{ABA} and θ_{BAB} by their average.[56]

VII-6. When the nonidentity of bond angles is important, is it also important that they not be interchanged? Examine the effect on C_∞ at 300 K for polyphosphate, polyisobutylene, polydimethylsiloxane, and polydimethylsilmethylene of the interchange of θ_{ABA} and θ_{BAB}.[56]

VII-7. How large must a second-order interaction energy become before it can be considered to be infinite? Evaluate C_∞ for polyethylene, polyoxymethylene, polyphosphate, and polydimethylsilmethylene, using zero and nonzero values of ω. How much larger than zero must ω be in order to affect C_∞ by 2%? In which direction is the effect on C_∞? What is the energy calculated as $E_\omega = -RT \ln \omega$?[56]

VII-8. The RIS model for the sulfur analog of polyoxymethylene is unknown. Using the models presented here for polyoxymethylene, polyoxyethylene, and polythiaethylene, construct a plausible model for polythiamethylene. Does the model predict polyoxymethylene, or its sulfur analog, should have the higher C_∞? Which properties of the model for polythiamethylene are most responsible for this prediction? How confident are you of the accuracy of the estimation of the critical properties of the model?[56]

VII-9. How many distinct σ and ω values occur in Z for the polymers with the repeating sequences listed below? Which of the ω values can be approximated as zero?[56]
(a) $CH_2-CH_2-CH_2-CH_2-O$
(b) $CH_2-CH_2-O-CH_2-O$
(c) $CH_2-CH_2-O-CH_2-CH_2-S$

VII-10. Critically evaluate the hypothesis that the real short-range interactions in a polymer always produce an increase in the unperturbed dimensions. Is it always true that $\partial C_\infty/\partial w_i > 0$ if $w_i > 1$, and $\partial C_\infty/\partial w_i < 0$ if $w_i < 1$, where w_i is one of the statistical weights in the RIS model of a chain? What are the implications for the sign of the temperature coefficient of the unperturbed dimensions, $\partial \ln C_\infty/\partial T$?[56]

VII-11. Construct an RIS formulation for Z for a polymer with the repeating sequence $-CH_2CH_2C(CH_3)_2-$. How is your model related to those described in this chapter for polyethylene, polyisobutylene, and poly(oxy-1,1-dimethylethylene)?

VII-12. Polyethers of the sequence $-(CH_2)_mO-$ have values of C_∞ that pass through a minimum when $m = 3$.[43] How can this result be rationalized?

VII-13. What are the condensed forms of the statistical weight matrices for polyisobutylene?

VII-14. The limiting components of the end-to-end vector can be written as

$$\lim_{n \to \infty} \langle \mathbf{r} \rangle_0 = \begin{bmatrix} \langle r_x \rangle \\ \langle r_y \rangle \\ \langle r_z \rangle \end{bmatrix}_0 \qquad \text{(VII-34)}$$

From the RIS models described in this chapter, are there any chains for which the largest component of $\lim_{n \to \infty} \langle \mathbf{r} \rangle_0$ is

(a) $\langle r_x \rangle_0$ **(b)** $\langle r_y \rangle_0$? **(c)** $\langle r_z \rangle_0$?

VII-15. What set of *symmetric* statistical weight matrices can be used for poly(trimethylene oxide)?

VII-16. Why does the limit for $\langle \mathbf{r} \rangle_0$ as $n \to \infty$ for polyoxyethylene depend strongly on whether a C—C, C—O, or O—C bond is indexed as bond 1?

VIII VINYL AND VINYLIDENE POLYMERS WITH SIMPLE SIDE CHAINS

The formulation of the conformational partition function becomes more complicated when substituents are present that produce $E(\phi) \neq E(-\phi)$. The rotation potential energy function is no longer symmetric, and the *stereochemical composition* of the chain must be specified before Z can be written. In this chapter we focus attention first on chains bearing substituents that consist of a single atom, as in the poly(vinyl halides). Then we consider vinyl polymers with polyatomic side chains that, with the combination of symmetry considerations and conformational constraints, may be approximated as a rigid body affixed to the backbone in a single orientation. The polymers considered under this description will include polypropylene, polystyrene and several of its derivatives, poly(methyl acrylate) and poly(vinyl acetate). We defer until Chapter XI the consideration of *articulated* side chains, by which we denote side chains with distinguishable conformations that interact differently with the conformations of the main chain. Vinyl polymers with articulated side chains, as in poly(1-butene), will be considered after the development of the treatment appropriate for articulated branches.

The number of examples of poly(A) with an asymmetric, nonarticulated side chain is much smaller than the number of poly(A–B), the latter of which includes several important vinyl polymers. Poly(A) is represented by polyfluoromethylene, where the repeating unit is –CFH–. A rotational isomeric state (RIS) model with $\nu = 3$ has been proposed for this polymer,[1] but in view of the much greater interest in the vinyl polymers, we will proceed directly to that class.

1. GENERAL DESCRIPTION OF VINYL POLYMERS

Figure VIII-1 depicts three fragments of a poly(vinyl bromide) chain, as a simple example of a vinyl polymer. Fragments in Figs. VIII-1*a* and VIII-1*b* are found in *isotactic* chains. These two fragments are distinguishable if one end of the chain is different from the other. If the ends are indistinguishable, as in methyl-terminated poly(vinyl bromide), the fragments describe the same

---

[1]Tonelli, A. E. *Macromolecules* **1980**, *13*, 734.

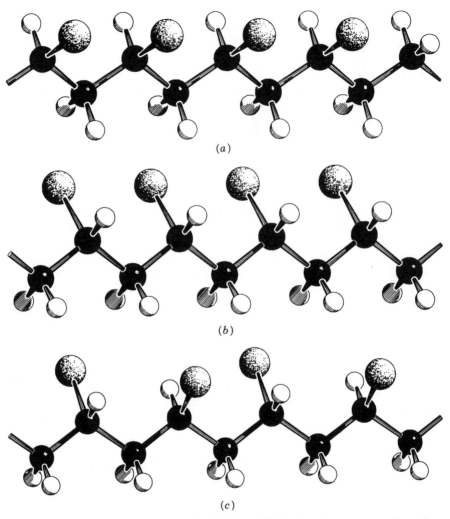

Figure VIII-1. Fragments from a poly(vinyl bromide) in the planar *trans* conformation: chains composed of (*a*) *meso* diads with pseudoasymmetric centers in the *d* configuration, (*b*) *meso* diads with pseudoasymmetric centers in the *l* configuration, and (*c*) *racemo* diads with pseudoasymmetric centers that alternate between *d* and *l* configurations.

molecule as viewed from different sides. Indexing the bonds in the chain from 1 through *n* imposes a conceptual distinction (although the distinction may be artificial) on the two ends. Therefore one might anticipate formulations for Z that would employ different sets of statistical weight matrices for the chains depicted in Figs. VIII-1*a* and VIII-1*b*. These two sets of statistical weight matrices must be formulated so that the completion of the serial matrix multiplication required for Z produces identical values for the conformational partition function for both chains, because the distinction between the ends is artificial. The formulation that retains the artificial distinction between the chain ends considers the *pseudoasymmetric centers* in Fig. VIII-1*a* to have one configuration, and the pseudoasymmetric centers in Fig. VIII-1*b* to have a different configuration. An alternative treatment views both chains as being composed exclusively of *meso* diads. Two successive pseudoasymmetric centers with the same stereochemistry constitute a *meso* diad. A formulation of Z based on *meso* diads utilizes a different set of statistical weight matrices from either of the sets employed for chains based on pseudoasymmetric centers. All three sets yield the same numeric value of Z.

The discussion of vinyl polymers is facilitated by introduction of terms that refer specifically to the pseudoasymmetric centers in Fig. VIII-1*a* or Fig. VIII-1*b*. We will make that distinction with the aid of the local coordinate system already defined for bond i in Chapter VI. Let bond i be a C^α—C bond, where C^α denotes the carbon atom in the main chain that bears the nonhydrogen substituent, and C is the carbon atom in the following methylene group. The nonhydrogen substituent bonded to C^α must have a nonzero component along the z axis of the local coordinate system of bond i. We *define* C^α as having the *d* configuration if the z component of the nonhydrogen substituent is positive, and as having the *l* configuration if the z component is negative.[2]

According to this definition, all pseudoasymmetric centers are in the *d* configuration in Fig. VIII-1*a*, and they are all in the *l* configuration in Fig. VIII-1*b*. The chain of Fig. VIII-1*c* has pseudoasymmetric centers that alternate between *d* and *l* configurations. When two successive pseudoasymmetric centers have different stereochemistry, whether ordered as *dl* or *ld*, they constitute a *racemo* diad. The expression for Z for this chain can be formulated from two different sets of statistical weight matrices, the one of which is based on *d* and *l* pseudoasymmetric centers, the other based on *racemo* diads. Both sets of statistical weight matrices yield exactly the same result for Z. That result is different from the one obtained for the chain constructed from *meso* diads.

A. Chains Constructed from *meso* Diads

Consideration begins with the chain depicted in Fig. VIII-1*a* and the treatment based on pseudoasymmetric centers in the *d* configuration.[2,3] Two types of sta-

---

[2]Our definition is the reverse of the one adopted by Flory et al.[3] This reversal is intentional. We use $\phi = 0°$ in the *cis* state, but Flory and coworkers placed the zero for ϕ in the *trans* state. Our g^+ (obtained by rotation of ~ +60° from our reference, *cis*) is Flory's g^- (obtained by rotation of ~ −120° from his reference, *trans*). If we adopted Flory's definitions for *d* and *l* pseudoasymmetric centers, our statistical weight matrices for chains with *d* pseudoasymmetric centers would be the same as his statistical weight matrices for chains with *l* pseudoasymmetric centers, and vice versa. By switching the definitions of *d* and *l* pseudoasymmetric centers (in conjunction with our switch in the location of the 0 for ϕ), the statistical weight matrices we introduce in this chapter will appear identical with the ones defined by Flory and coworkers.

[3]Flory, P. J.; Mark, J. E.; Abe, A. *J. Am. Chem. Soc.* **1966**, *88*, 639.

tistical weight matrices are required. They are denoted by U_d and U_{dd} for the C^α—C and C—C^α bonds. The first-order interactions that must be included in U_d are easily determined with the aid of Fig. VIII-2. This figure depicts three views along an axis drawn from C^α to the following C. It shows the disposition of the atoms bonded to both carbon atoms in each of the three rotational isomeric states. The rotational isomeric states are denoted t, g^+, and g^-, with the assignment based on the disposition of the pendant carbon atoms. Note that R = X and R′ = H.

First-order interactions between these pendant carbon atoms occur in the two g states. If X were equivalent to a hydrogen atom, no further first-order interactions would be considered. For the polymers presently under consideration, however, X is certainly larger than a hydrogen atom. It might also be larger than a methylene group, as would be the case in poly(vinyl bromide). Therefore U_d must incorporate the first-order interaction of X with the groups centered on the carbon atom at the other end of the fragment. These first-order interactions occur in the t and g^- states. When the two types of first-order interactions are combined, no two rotational isomeric states have the same combination. One set of first-order interactions can arbitrarily be assigned a statistical weight of 1, with the other two sets assigned different statistical weights. The scheme often

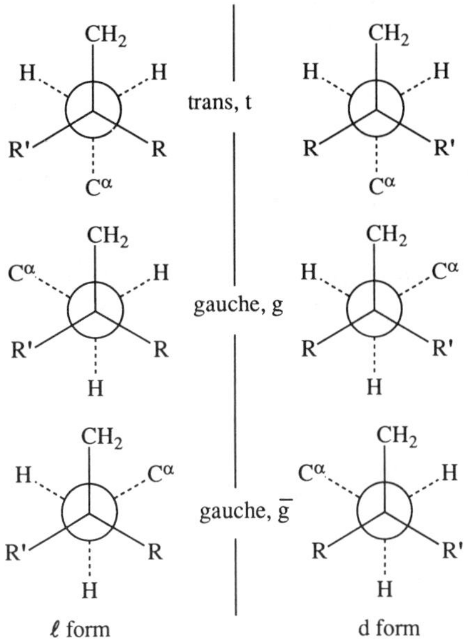

Figure VIII-2. Newman diagrams for the view from the carbon atom in C^α to the following C. Reprinted with permission from Flory, P. J.; Sundararajan, P. R.; DeBolt, L. C. *J. Am. Chem. Soc.* **1974**, *96*, 5015. Copyright 1974 American Chemical Society. The symbols *d* and *l* have been interchanged so that the figure corresponds to the convention used in the text. In the development of the statistical weight matrices we assume R′ = H and R = X ≠ H.

adopted[3] is represented by the diagonal matrix

$$\mathbf{D}_d = \text{diag}(\eta, 1, \tau) \tag{VIII-1}$$

where the elements along the main diagonal are the statistical weights assigned to the combinations of first-order interactions in the t, g^+, and g^- states, with the first-order interaction energy being assigned as zero in the g^+ state.[4] The same set of first-order interactions must occur in the statistical weight matrix for the C—C$^\alpha$ bond, as is apparent from Fig. VIII-2. However, there must be a change in the order of the appearance of the second and third statistical weights:

$$\mathbf{D}_{d,CC^\alpha} = \text{diag}(\eta, \tau, 1) = \mathbf{Q}\mathbf{D}_d\mathbf{Q} \tag{VIII-2}$$

$$\mathbf{Q} = \begin{bmatrix} 1 & 0 & 0 \\ 0 & 0 & 1 \\ 0 & 1 & 0 \end{bmatrix} \tag{VIII-3}$$

The matrix $\mathbf{Q}$ will be used repeatedly in the formulation of Z for vinyl polymers. It has the convenient property that $\mathbf{Q}\mathbf{Q} = \mathbf{I}_3$.

The second-order interactions that appear in $\mathbf{U}_d$ are those between the groups centered on the terminal carbon atoms in the fragment C$^\alpha$—C–C$^\alpha$–C—C$^\alpha$. In a first approximation, they should be nearly equivalent to the second-order interactions in the linear polyethylene chain. When the consideration is restricted to those second-order interactions that place interacting groups in a conformation equivalent to the "pentane effect" in polyethylene, matrix $\mathbf{V}_d$ is[3]

$$\mathbf{V}_d = \begin{bmatrix} 1 & 1 & 1 \\ 1 & 1 & \omega \\ 1 & \omega & 1 \end{bmatrix} \tag{VIII-4}$$

and $\mathbf{U}_d$ is obtained as the product $\mathbf{V}_d\mathbf{D}_d$.

$$\mathbf{U}_d = \mathbf{V}_d\mathbf{D}_d = \begin{bmatrix} \eta & 1 & \tau \\ \eta & 1 & \tau\omega \\ \eta & \omega & \tau \end{bmatrix} \tag{VIII-5}$$

[4]The use of τ in $\mathbf{D}_d$ is different from its use in the general form for $\mathbf{U}$ for a chain with pairwise interdependent bonds subject to a symmetric threefold rotation potential energy function. In $\mathbf{D}_d$, τ is a statistical weight for the *first-order* interaction in the g^- state at a bond subject to an *asymmetric* threefold rotation potential energy function, but in Eq. (VII-1) it denotes the statistical weight for a *second-order* interaction in the tt state. Since both usages for τ are widespread in the literature, we will retain the duplicate usage here. Subscripts 1 or 2 (for first- or second-order) will be appended in those rare cases where the meaning of τ is not immediately obvious from the context.

Additional second-order interactions must be incorporated in $\mathbf{U}_{dd}$. They arise from the second-order interaction between the groups centered on C and X, as $C \cdots C$, $C \cdots X$, and $X \cdots X$, when they are the terminal atoms in the fragments $C-C^{\alpha}-C-C^{\alpha}-C$, $X-C^{\alpha}-C-C^{\alpha}-C$, $C-C^{\alpha}-C-C^{\alpha}-X$, and $X-C^{\alpha}-C-C^{\alpha}-X$, in the conformations that generate the pentane effect for these pairs of atoms. The statistical weight for the first type of second-order interaction, between C and C, is assigned to the 2,3 and 3,2 elements of $\mathbf{V}_{dd}$, as in polyethylene. Second-order interactions between two X atoms occur in the tt state, as is evident in Fig. VIII-1a, and in the $g^{-}g^{+}$ state. Four other states produce a second-order interaction between C at one end and X at the other end of the fragment. The statistical weights for the three types of pentane effect second-order interactions are denoted by ω, ω_{XX}, and ω_X, respectively. The matrix of second-order interactions is

$$\mathbf{V}_{dd} = \begin{bmatrix} \omega_{XX} & \omega_X & 1 \\ 1 & \omega_X & \omega \\ \omega_X & \omega\omega_{XX} & \omega_X \end{bmatrix} \qquad \text{(VIII-6)}$$

Figure VIII-3 depicts the nine conformations of $meso$-2,4-dibromopentane, arranged in a 3×3 array that corresponds to the arrangement of the elements in $\mathbf{V}_{dd}$. $\mathbf{U}_{dd}$ is obtained as $\mathbf{V}_{dd}\mathbf{D}_d$:

$$\mathbf{U}_{dd} = \mathbf{V}_{dd}\mathbf{D}_d = \begin{bmatrix} \eta\omega_{XX} & \tau\omega_X & 1 \\ \eta & \tau\omega_X & \omega \\ \eta\omega_X & \tau\omega\omega_{XX} & \omega_X \end{bmatrix} \qquad \text{(VIII-7)}$$

The stereochemistry of all of the pseudoasymmetric centers is inverted if the bonds in the isotactic chain are indexed so that the picture changes from Fig. VIII-1a to Fig. VIII-1b. Consequently $\mathbf{U}_l$ can be obtained from $\mathbf{U}_d$, and $\mathbf{U}_{ll}$ can be obtained from $\mathbf{U}_{dd}$, by exchange of columns 2 and 3, followed (or preceded) by exchange of rows 2 and 3. The net effect is exchange of elements according to the scheme

$$u_{12} \rightleftharpoons u_{13} \qquad u_{21} \rightleftharpoons u_{31} \qquad u_{22} \rightleftharpoons u_{33} \qquad u_{23} \rightleftharpoons u_{32} \qquad \text{(VIII-8)}$$

which is accomplished by premultiplication and postmultiplication with $\mathbf{Q}$.

$$\mathbf{U}_l = \mathbf{Q}\mathbf{U}_d\mathbf{Q} = \begin{bmatrix} \eta & \tau & 1 \\ \eta & \tau & \omega \\ \eta & \tau\omega & 1 \end{bmatrix} \qquad \text{(VIII-9)}$$

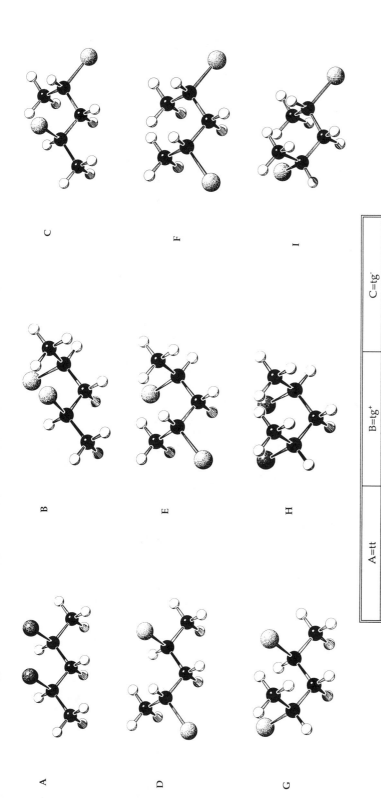

Figure VIII-3. The nine conformations of *meso*-2,4-dibromopentane, arranged in a 3 × 3 array that corresponds to the arrangement of the elements in $\mathbf{V}_{dd}$.

A=tt	B=tg⁺	C=tg⁻
D=g⁺t	E=g⁺g⁺	F=g⁺g⁻
G=g⁻t	H=g⁻g⁺	I=g⁻g⁻

179

$$\mathbf{U}_{ll} = \mathbf{QU}_{dd}\mathbf{Q} = \begin{bmatrix} \eta\omega_{XX} & 1 & \tau\omega_X \\ \eta\omega_X & \omega_X & \tau\omega\omega_{XX} \\ \eta & \omega & \tau\omega_X \end{bmatrix} \tag{VIII-10}$$

The conformational partition function for a long isotactic chain with methyl end groups can be generated identically by either Eq. (VIII-11) or (VIII-12):[5]

$$Z = \begin{bmatrix} 1 & 0 & 0 \end{bmatrix} (\mathbf{U}_d\mathbf{U}_{dd})^{(n-2)/2} \begin{bmatrix} 1 \\ 1 \\ 1 \end{bmatrix} \tag{VIII-11}$$

$$Z = \begin{bmatrix} 1 & 0 & 0 \end{bmatrix} (\mathbf{U}_l\mathbf{U}_{ll})^{(n-2)/2} \begin{bmatrix} 1 \\ 1 \\ 1 \end{bmatrix} \tag{VIII-12}$$

A third expression, Eq. (VIII-13), will give exactly the same result.

$$Z = \begin{bmatrix} 1 & 0 & 0 \end{bmatrix} (\mathbf{U}_p\mathbf{U}_m)^{(n-2)/2} \begin{bmatrix} 1 \\ 1 \\ 1 \end{bmatrix} \tag{VIII-13}$$

Here $\mathbf{U}_p$ and $\mathbf{U}_m$ are statistical weight matrices constructed using a different definition of the g states that facilitates treatment of the conformational statistics in terms of *meso* diads.[6] The rotational isomeric states are now denoted by t, g, and $\bar{g}$. There is no change in the definition of the state denoted by t. The *gauche* state that has τ in its statistical weight is denoted by $\bar{g}$, and the remaining *gauche* state is denoted simply by g. Rows and columns are indexed in the order t, g, $\bar{g}$.

---

[5] Substitution of $\mathbf{U}_l = \mathbf{QU}_d\mathbf{Q}$ and $\mathbf{U}_{ll} = \mathbf{QU}_{dd}\mathbf{Q}$ into Eq. (VIII-12) yields

$$Z = \begin{bmatrix} 1 & 0 & 0 \end{bmatrix} (\mathbf{QU}_d\mathbf{QQU}_{dd}\mathbf{Q})^{(n-2)/2} \begin{bmatrix} 1 \\ 1 \\ 1 \end{bmatrix}$$

Equation (VIII-11) is recovered with $\begin{bmatrix} 1 & 0 & 0 \end{bmatrix}\mathbf{Q} = \begin{bmatrix} 1 & 0 & 0 \end{bmatrix}$, $\mathbf{QQ} = \mathbf{I}_3$, and

$$\mathbf{Q}\begin{bmatrix} 1 \\ 1 \\ 1 \end{bmatrix} = \begin{bmatrix} 1 \\ 1 \\ 1 \end{bmatrix}$$

[6] Flory, P. J.; Sundararajan, P. R.; DeBolt, L. C. *J. Am. Chem. Soc.* **1974**, *96*, 5015. We use $\mathbf{U}_p$ for their $\mathbf{U}'$, in order to avoid any confusion of the matrix in Eq. (VIII-14) with the $\mathbf{U}'$ defined in Chapter V.

$$
\mathbf{U}_p =
\begin{array}{c}
\\ t \\ g \\ \bar{g}
\end{array}
\begin{array}{c}
\begin{array}{ccc} t & g & \bar{g} \end{array} \\
\left[
\begin{array}{ccc}
\eta & 1 & \tau \\
\eta & \omega & \tau \\
\eta & 1 & \tau\omega
\end{array}
\right]
\end{array}
\qquad \text{(VIII-14)}
$$

$$
\mathbf{U}_m =
\begin{array}{c}
\\ t \\ g \\ \bar{g}
\end{array}
\begin{array}{c}
\begin{array}{ccc} t & g & \bar{g} \end{array} \\
\left[
\begin{array}{ccc}
\eta\omega_{XX} & 1 & \tau\omega_X \\
\eta & \omega & \tau\omega_X \\
\eta\omega_X & \omega_X & \tau\omega\omega_{XX}
\end{array}
\right]
\end{array}
\qquad \text{(VIII-15)}
$$

The switch from rotational isomeric states g^+ and g^- to states denoted by g and $\bar{g}$ is achieved by an interchange of two rows, or two columns, in the statistical weight matrix. The interchange of the rows (or columns) is accomplished by $\mathbf{Q}$. The matrix $\mathbf{U}_p$ can be obtained from $\mathbf{U}_d$ by exchange of the second and third rows, or from $\mathbf{U}_l$ by exchange of the second and third columns:

$$
\mathbf{U}_p = \mathbf{Q}\mathbf{U}_d = \mathbf{U}_l\mathbf{Q}
\qquad \text{(VIII-16)}
$$

and $\mathbf{U}_m$ is obtained from $\mathbf{U}_{dd}$ by exchange of the second and third columns, and from $\mathbf{U}_{ll}$ by exchange of the second and third rows:

$$
\mathbf{U}_m = \mathbf{U}_{dd}\mathbf{Q} = \mathbf{Q}\mathbf{U}_{ll}
\qquad \text{(VIII-17)}
$$

Equation (VIII-12) is recovered from Eq. (VIII-13) with $\mathbf{U}_p\mathbf{U}_m = \mathbf{U}_l\mathbf{Q}\mathbf{Q}\mathbf{U}_{ll} = \mathbf{U}_l\mathbf{U}_{ll}$.

Whenever statistical weight matrices indexed by t, g, and $\bar{g}$ are combined with $\mathbf{G}_\xi$ for the construction of $\mathscr{G}$, special care must be paid to the value of ϕ_ξ.[7] If a $\mathbf{U}_p$ or $\mathbf{U}_m$ was obtained by *postmultiplication* with $\mathbf{Q}$, as in $\mathbf{U}_l\mathbf{Q}$ and $\mathbf{U}_{dd}\mathbf{Q}$, *the order of the torsion angles associated with the last two columns has been switched*. When bond i is in state ξ, the torsion angle is ϕ_ξ, and the statistical weights are found in the column of $\mathbf{U}_i$ that bears index ξ. Switching the columns in $\mathbf{U}_i$ requires that ϕ_ξ must be conceptually carried along with the column that contains the statistical weights for state ξ.

It is a matter of individual preference whether the extra complication in the definition of the signs of the torsion angles does, or does not, compensate for the smaller set of $\mathbf{U}$ required when one uses the *meso,racemo* notation rather than the notation based on d,l pseudoasymmetric centers. When properly employed, both procedures yield exactly the same result. There may be less opportunity for error in the calculation of C_∞ if the formulation is based on d,l pseudoasymmetric centers, because of the simpler form for assignment of ϕ_ξ in each $\mathbf{G}_\xi$. The primary advantage to the formulation based on diads will

[7] Flory, P. J.; Fujiwara, Y. *Macromolecules* **1970**, *3*, 280.

become apparent when chains with different stereochemical compositions are considered.

The values of the statistical weights will depend on the assignment of X. Statistical weight matrices identical with those used for the internal bonds in polyethylene, Eq. (IV-19), can be recovered with the assignments $\eta = 1/\sigma$, $\tau = \omega_X = \omega_{XX} = 1$, followed by renormalization by multiplication of all elements in the matrix by σ. In most applications, only the relative size of the elements in $\mathbf{U}$ is important, making the renormalization unnecessary. If the size of the statistical weights is determined entirely by repulsive steric interactions, an increase in the bulk of atom X might be accompanied by a decrease in the sizes of η, τ, ω_X, and ω_{XX}. This simplistic interpretation focused solely on repulsive steric interactions may be misleading. Electrostatic contributions can become important if the C—X bond has substantial polar character.

B. Chains Constructed from *racemo* Diads

A fragment of a syndiotactic chain is depicted in Fig VIII-1c. If the statistical weight matrices are to be formulated using the concept of d and l pseudoasymmetric centers, $\mathbf{U}_d$ and $\mathbf{U}_l$ can be used for alternate C^α—C bonds. These two matrices are specified by Eq. (VIII-5) and (VIII-9), respectively. New statistical weight matrices must be formulated for the intervening C—C^α bonds. These matrices will be of two types, depending on whether the sequence of the two pseudoasymmetric centers is dl or ld. Implementation of the same analysis as used for the development of $\mathbf{U}_{dd}$ in Eq. (VIII-7) ultimately leads to[6]

$$\mathbf{U}_{dl} = \begin{bmatrix} \eta & \omega_X & \tau\omega_{XX} \\ \eta\omega_X & 1 & \tau\omega \\ \eta\omega_{XX} & \omega & \tau\omega_X^2 \end{bmatrix} \qquad (\text{VIII-18})$$

$$\mathbf{U}_{ld} = \mathbf{Q}\mathbf{U}_{dl}\mathbf{Q} = \begin{bmatrix} \eta & \tau\omega_{XX} & \omega_X \\ \eta\omega_{XX} & \tau\omega_X^2 & \omega \\ \eta\omega_X & \tau\omega & 1 \end{bmatrix} \qquad (\text{VIII-19})$$

Figure VIII-4 depicts the nine conformations of *racemic*-2,4-dibromopentane, arranged in a 3×3 array that corresponds to the 3×3 array of elements in $\mathbf{U}_{dl}$. The matrices $\mathbf{U}_{dl}$ and $\mathbf{U}_{ld}$ cannot be converted into $\mathbf{U}_{dd}$ and $\mathbf{U}_{ll}$ by any rearrangement of their elements. The elements that contain τ are $\tau\omega$, $\tau\omega_{XX}$, and $\tau\omega_X^2$ in $\mathbf{U}_{dl}$ and $\mathbf{U}_{ld}$, but they are $\tau\omega_X$ (which occurs twice) and $\tau\omega\omega_{XX}$ in $\mathbf{U}_{dd}$ and $\mathbf{U}_{ll}$.

A completely syndiotactic chain with methyl end groups will have a conformational partition function given by

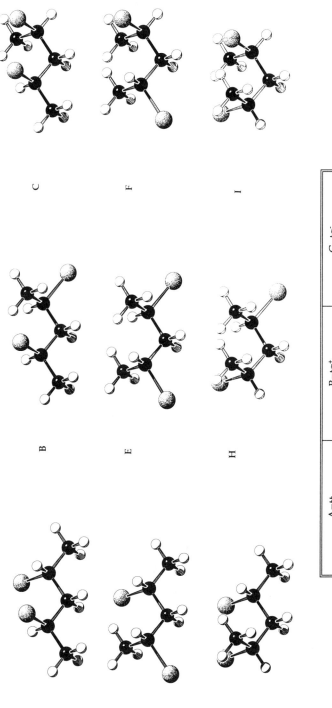

A=tt	B=tg⁺	C=tg⁻
D=g⁺t	E=g⁺g⁺	F=g⁺g⁻
G=g⁻t	H=g⁻g⁺	I=g⁻g⁻

Figure VIII-4. The nine conformations of *racemic*-2,4-dibromopentane, arranged in a 3 × 3 array that corresponds to the arrangement of the elements in $\mathbf{V}_{dl}$.

$$Z = \begin{bmatrix} 1 & 0 & 0 \end{bmatrix} (U_d U_{dl} U_l U_{ld})^{(n-2)/4} \begin{bmatrix} 1 \\ 1 \\ 1 \end{bmatrix} \qquad \text{(VIII-20)}$$

or by

$$Z = \begin{bmatrix} 1 & 0 & 0 \end{bmatrix} (U_d U_{dl} U_l U_{ld})^{(n-4)/4} U_d U_{dl} \begin{bmatrix} 1 \\ 1 \\ 1 \end{bmatrix} \qquad \text{(VIII-21)}$$

depending on whether the number of C^α atoms is odd or even, respectively. Four different statistical weight matrices are required for the computation of Z via Eq. (VIII-20) or (VIII-21). The statistical weights (η, τ, ω, ω_X, ω_{XX}) that occur in these matrices are exactly the same as those found in the matrices required for the treatment of the isotactic chain. However, they occur in different locations in the statistical weight matrices. In general, the value of Z specified by Eq. (VIII-20) or (VIII-21) for a syndiotactic chain with a specified number of bonds will differ from the value of Z obtained from Eq. (VIII-11), (VIII-12), or (VIII-13) for the isotactic chain with the same number of bonds.[8]

A simpler formulation, requiring only two different statistical weight matrices, is accessible with the use of the *meso,racemo* notation and rotational isomers defined as t, g, and $\bar{g}$. The statistical weight matrix in Eq. (VIII-14) is used for all C^α—C bonds. The statistical weight matrix for all C—C^α bonds, denoted U_r, is[6]

$$U_r = U_{dl} = QU_{ld}Q = \begin{bmatrix} \eta & \omega_X & \tau\omega_{XX} \\ \eta\omega_X & 1 & \tau\omega \\ \eta\omega_{XX} & \omega & \tau\omega_X^2 \end{bmatrix} \qquad \text{(VIII-22)}$$

With this statistical weight matrix and U_p, Z for the chains described by Eq. (VIII-20) and (VIII-21) can be more simply written as

$$Z = \begin{bmatrix} 1 & 0 & 0 \end{bmatrix} (U_p U_r)^{(n-2)/2} \begin{bmatrix} 1 \\ 1 \\ 1 \end{bmatrix} \qquad \text{(VIII-23)}$$

Equation (VIII-23) provides the same numeric result as Eq. (VIII-20) for syn-

---

[8]This observation implies that different concentrations of isotactic and syndiotactic chains might be obtained if a mechanism were available for equilibration of the stereochemistry at each C^α in the chain. The ratio of the concentrations could be predicted from the expressions for Z. Further development of this observation will be presented in Chapter XV, in the consideration of epimerization of vinyl polymers to stereochemical equilibrium.

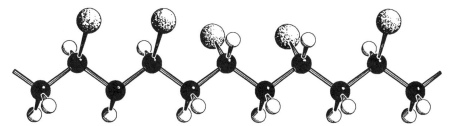

Figure VIII-5. Fragment from a poly(vinyl bromide) chain with intermediate stereo-chemical composition.

TABLE VIII-1. Relationships between the U Values for Vinyl Polymers

Relationship	Equations for the U Values
$U_p = QU_d = U_lQ$	(VIII-14), (VIII-5), (VIII-9)
$U_m = U_{dd}Q = QU_{ll}$	(VIII-15), (VIII-7), (VIII-10)
$U_r = U_{dl} = QU_{ld}Q$	(VIII-22), (VIII-18), (VIII-19)

diotactic chains with an odd number of C^α,[9] and the same result as Eq. (VIII-21) for syndiotactic chains with an even number of C^α. Exactly the same set of statistical weights is used in both formulations. However, only two distinct matrices (U_p and U_r) are required if the formulation is based on diads and t, g, and $\bar{g}$ states, while four distinct matrices (U_d, U_{dl}, U_l, U_{ld}) are required if the formulation uses dl pseudoasymmetric centers and t, g^+, and g^- states. The formalism based on the *meso,racemo* notation is more elegant, because it uses fewer types of matrices. Neither formulation is more correct than the other, because both formulations produce the same result. When $\mathscr{G}$ is constructed from U_r and the G_ξ, one must remember that postmultiplication by Q, as in $QU_{ld}Q$, switches the torsion angles associated with the second and third columns.

The relationships between the statistical weight matrices defined thus far for vinyl polymers are summarized in Table VIII-1.

C. Chains with Intermediate Stereochemical Composition

The advantages of the formulation based on the *meso,racemo* notation over the formulation based on d and l pseudoasymmetric centers is most apparent in the treatment of chains of intermediate stereochemical compositions. Consider, for instance, the fragment depicted in Fig. VIII-5. The five pseudoasymmetric centers, in order from left to right, are *ddlld*. The contribution to Z from this fragment can be written as $U_dU_{dd}U_dU_{dl}U_lU_{ll}U_lU_{ld}$. This expression uti-

[9]The product $U_pU_rU_pU_r$ in Eq. (VIII-23) is also $(QU_d)(U_{dl})(U_lQ)(QU_{ld}Q)$, which recovers Eq. (VIII-20).

lizes six distinct statistical weight matrices. When described in terms of diads, the stereochemical composition of the same fragment is *mrmr*, where *m* and *r* denote *meso* and *racemo* diads, respectively. The contribution made by the fragment to Z is $\mathbf{U}_p\mathbf{U}_m\mathbf{U}_p\mathbf{U}_r\mathbf{U}_p\mathbf{U}_m\mathbf{U}_p\mathbf{U}_r$. There are only three distinct matrices in this product. Of course, both formulations lead to the same value for Z. The sum of the elements in the first row of $\mathbf{U}_d\mathbf{U}_{dd}\mathbf{U}_d\mathbf{U}_{dl}\mathbf{U}_l\mathbf{U}_{ll}\mathbf{U}_l\mathbf{U}_{ld}$ is identical with the sum of the elements in the first row of $\mathbf{U}_p\mathbf{U}_m\mathbf{U}_p\mathbf{U}_r\mathbf{U}_p\mathbf{U}_m\mathbf{U}_p\mathbf{U}_r$.

Whether expressed in terms of *d* and *l* pseudoasymmetric centers or in terms of *meso* and *racemo* diads, the value of Z obtained from the matrix formulation is for a chain of specified stereochemical sequence. If the chain contains both *meso* and *racemo* diads, the value of Z will depend on the sequence in which the diads appear. It is unlikely that all chains in a sample of an atactic vinyl polymer will have the same sequence of diads. Consequently the sample must contain chains with different values of Z. The conformational properties for such a sample are best evaluated by an appropriate averaging over a suitably large set of representative chains. For example, if the probability of a *meso* diad is denoted by p_m, and the stereochemical sequence is Bernoullian, the stereochemical sequence of a representative chain with x diads can be assigned with a string of x random numbers, each of which lies in the range 0–1. The ith diad is assigned as *meso* if the ith random number is less than p_m; otherwise it is assigned as *racemo*. The value of Z, as well as other conformation-dependent physical properties of interest, can be calculated for this chain when its stereochemical sequence is known. The process is repeated for other chains specified by new strings of x random numbers, but the same p_m, until the averages of the desired conformation dependent properties settle down to a constant value, within the limit of accuracy required by the application. Any other known distribution of configurations can, of course, be analogously implemented.

Frequently in the literature $\mathbf{U}_p$, $\mathbf{U}_m$, and $\mathbf{U}_r$ are written with a different (but completely equivalent) arrangement of the statistical weights for the first-order interactions. Recognizing that these matrices will appear in Z as the products $\mathbf{U}_p\mathbf{U}_m$ and $\mathbf{U}_p\mathbf{U}_r$, any factor that appears in all elements in column i of $\mathbf{U}_p$ could be placed instead in all elements in row i for $\mathbf{U}_m$ and $\mathbf{U}_r$, without affecting the products $\mathbf{U}_p\mathbf{U}_m$ and $\mathbf{U}_p\mathbf{U}_r$. When this change is made, Eqs. (VIII-14), (VIII-15), and (VIII-22) become

$$\mathbf{U}_p = \begin{bmatrix} 1 & 1 & 1 \\ 1 & \omega & 1 \\ 1 & 1 & \omega \end{bmatrix} \tag{VIII-24}$$

$$\mathbf{U}_m = \begin{bmatrix} \eta^2\omega_{XX} & \eta & \eta\tau\omega_X \\ \eta & \omega & \tau\omega_X \\ \eta\tau\omega_X & \tau\omega_X & \tau^2\omega\,\omega_{XX} \end{bmatrix} \tag{VIII-25}$$

$$\mathbf{U}_r = \begin{bmatrix} \eta^2 & \eta\omega_X & \eta\tau\omega_{XX} \\ \eta\omega_X & 1 & \tau\omega \\ \eta\tau\omega_{XX} & \tau\omega & \tau^2\omega_X^2 \end{bmatrix} \qquad \text{(VIII-26)}$$

These matrices suggest tg (or equivalently gt) will often be the preferred conformation in *meso* diads. The 1,2 and 2,1 elements in $\mathbf{U}_m$ are the only ones that do not incur at least one of the ω values. Since the ω values usually weight strongly repulsive interactions, the tg and gt sequences are often preferred in isotactic vinyl polymers. The preferred conformations for *racemo* diads will often be tt or gg, which avoid the ω values in $\mathbf{U}_r$. The balance between the nonequivalent tt and gg conformations of a syndiotactic polymer will be strongly affected by η. The gg sequence at C^α—C—C^α may occur interspersed within strings of tt states.

2. APPLICATIONS TO POLY(VINYL HALIDES)

The poly(vinyl halides) are examples of vinyl polymers in which X is a single atom. Rotational isomeric state models used for these polymers are summarized in Table VIII-2.

Poly(vinyl fluoride) has an unusually small value for $E_{\omega X}$, which is probably due to the small size of the fluorine atom, but $E_{\omega XX}$ is large, presumably because it arises in conformations where two C—F dipoles are parallel, and hence inter-

TABLE VIII-2. Models for Poly(Vinyl Halides)[a]

X	$\theta_{C-C^\alpha-C}, \theta_{C^\alpha-C-C^\alpha}$	E_η	E_τ	E_ω	$E_{\omega X}$	$E_{\omega XX}$	Ref.
F	113, 114	−2.5	2.1	7.1	1.3	6.7	b
Cl	112, 112	−3.6	2.1	8.4	6.7	8.4	c
Cl	112, 112	−1.3	0.4	10.5	6.7	9.6	d
Br	113, 116	−1.3	0.4	10.9	8.4	14	b
Br	112, 114	−0.6	2.1	10.5	8.4	15	e

[a]With $l_{CC} = 153$ pm, $\{\phi\} = \{180°, \pm60°\}$, θ in deg, energies in kJ mol^{-1}. Appendix C contains a program that can be used to calculate C_∞ for these chains when $p_m = 0$ or 1.
[b]Carballeira, L.; Pereiras, A. J.; Ríos, M. A. *Macromolecules* **1989**, *22*, 2668. They include $E_{\omega p}$, as described by Boyd and Kesner for poly(vinyl chloride), with values of 0.4 and 0.8 kJ mol^{-1} for poly(vinyl fluoride) and poly(vinyl bromide), respectively.
[c]Flory, P. J.; Williams, A. D. *J. Am. Chem. Soc.* **1969**, *91*, 3118.
[d]Boyd, R. H.; Kesner, L. *J. Polym. Sci., Polym. Phys. Ed.* **1981**, *19*, 375. They include another second-order interaction, of electrostatic origin and with $E_{\omega p} = 2.1$ kJ mol^{-1}, in the 2,2 elements of $\mathbf{U}_m$, $\mathbf{U}_r$, and $\mathbf{U}_{dl}$, the 2,3 element of $\mathbf{U}_{ll}$, the 3,2 element of $\mathbf{U}_{dd}$, and the 3,3 element of $\mathbf{U}_{ld}$.
[e]Saiz, E.; Riande, E.; Delgado, M. P.; Barrales-Rienda, J. M. *Macromolecules* **1982**, *15*, 1152.

act repulsively.[10] The model for poly(vinyl fluoride) predicts $C_\infty \sim 5.0$ at ambient temperature for p_m near 0.5. No experimental test of this prediction seems to be available.

Both poly(vinyl fluoride) and *poly(vinyl chloride)* have $E_\eta < 0$, in part because the halogen atom, being smaller than a methylene group, can more easily participate in a *syn* interaction with a methylene group. [But arguments based on size alone would not explain why E_η is also negative in poly(vinyl bromide).] The value of the energy for this first-order interaction is one of the most important differences between the models proposed for poly(vinyl chloride) by Flory and Williams,[11] and by Boyd and Kesner.[12] The two models predict similar unperturbed dimensions ($C_\infty \sim 11$–12) at ambient temperature for polymers with stereochemical compositions typical of this polymer, $p_m \sim 0.43$, in reasonable agreement with experiment. Experimental results in a Θ solvent that does not promote aggregation of poly(vinyl chloride) yield C_∞ of 9.6–10.9 for the atactic polymer.[13] The molar Kerr constant is more easily rationalized with $E_\eta \sim -1.7$ kJ mol^{-1} than with a more negative value.[14] The value of C_∞ for the completely syndiotactic poly(vinyl chloride) is large if $E_\eta = -3.6$ kJ mol^{-1} because the chain strongly prefers long sequences of t placements, as suggested by the appearance of η^2, and absence of τ and the ω values, in the 1,1 element of U_r, Eq. (VIII-26).

Smaller unperturbed dimensions ($C_\infty \sim 6$–7 at ambient temperature for $p_m \sim 0.5$) are calculated[15] and observed[16] for atactic *poly(vinyl bromide)* than for poly(vinyl chloride), in response to the smaller value of η and greater differences in the bond angles.

3. POLYATOMIC SIDE CHAINS

When the X in poly(vinyl X) contains more than one atom, the contribution to Z from a specified unit may depend on the torsion angle, denoted by χ, adopted at the C^α—X bond. For present purposes, polymers with polyatomic X can be divided into two categories. The simpler class is the one where only a single value of χ needs to be considered, either because X is rigidly attached to the backbone, or because the symmetry of X causes identical interactions in all rotational isomeric states at the C^α—X bond. Examples are provided by the –CH$_3$ group of polypropylene, and the –C$_6$H$_5$ group of polystyrene. This simpler class of vinyl polymers with polyatomic side chains is considered here.

[10]Carballeira, L.; Pereiras, A. J.; Ríos, M. A. *Macromolecules* **1989**, *22*, 2668.

[11]Flory, P. J.; Williams, A. D. *J. Am. Chem. Soc.* **1969**, *91*, 3118.

[12]Boyd, R. H.; Kesner, L. *J. Polym. Sci., Polym. Phys. Ed.* **1981**, *19*, 375.

[13]Sato, M.; Koshiishi, Y.; Asahine, M. *J. Polym. Sci., Part B* **1963**, *1*, 223. Ludovice, P. J. doctoral thesis, Massachusetts Institute of Technology (MIT), January 1989.

[14]Khanarian, G.; Schilling, F. C.; Cais, R. E.; Tonelli, A. E. *Macromolecules* **1983**, *16*, 287.

[15]Saiz, E.; Riande, E.; Delgado, M. P.; Barrales-Rienda, J. M. *Macromolecules* **1982**, *15*, 1152.

[16]Ciferri, A.; Lauretti, M. *Ann. Chim. (Roma)* **1958**, *48*, 198.

TABLE VIII-3. Vinyl Polymers with Polyatomic Side Chains with Indistinguishable States at the C^α—X Bond^a

X	$\theta_{C-C^\alpha-C}$, $\theta_{C^\alpha-C-C^\alpha}$	$\{\phi\}$	E_η	E_ω	$E_{\omega X}$	$E_{\omega XX}$	Ref.
CH_3	112, 112	5-state^b	0.4	11^b			b
C_6H_5	112, 114	170, 70	−1.7	8.4	8.4	9.2	c
C_6H_5	112, 114	170, 70	−3.7	6.7	13	∞	d
p-C_6H_4-Y	112, 114	170, 70	−1.7	8.4	8.4	9.2	e
N-Carbazole	112, 114-117	160–180, 70–80	−2.5 ± 0.4	8	8	13 ± 1	f
2-Pyridine	112, 114	180, 60	−0.4	2.8	6.9	3.8	g
N-Pyrrolidone	112, 114	180, 60	0.4	5.0	7.1	9.2	h
$COOCH_3$	112, 114	170, 70	−1.3 ± 0.4	6.7	5.9 ± 0.4	6.3 ± 1.3	i
$OOCCH_3$	112, 114	170, 70	−1.7 ± 0.4	6.7	6.7	6.3	j

^{a}With l_{CC} = 153 pm, angles in deg, energies in kJ mol^{-1}. Appendix C contains a program that can be used to calculate C_∞ for these chains when p_m = 0 or 1.

^{b}Suter, U. W.; Flory, P. J. *Macromolecules* **1975**, *8*, 765. The model has five states, described in conjunction with Eqs. (VIII-27)–(VIII-29), and requires E_τ = 3.3 kJ mol^{-1}. Statistical weights τ and ω^* have preexponential factors of 0.4 and 0.9, respectively, and E_{ω^*} is tabulated in the column headed E_ω.

^{c}Yoon, D. Y.; Sundararajan, P. R.; Flory, P. J. *Macromolecules* **1975**, *8*, 776. Preexponential factors are 0.8 for η, 1.3 for ω and ω_X.

^{d}Rapold, R.; Suter, U. W. *Macromol. Theory Simul.* **1994**, *3*, 1. Preexponential factors are 0.5, 1.1, and 1.7 for η, ω, and ω_X, respectively. The ω in U_p was set equal to 0.

^{e}Used for poly(p-halostyrenes) with Y = F, Cl, Br, and for poly(p-nitrostyrene). See text for references.

^{f}Abe, A.; Kobayashi, H.; Kawamura, T.; Date, M.; Uryu, T.; Matsuzaki, K. *Macromolecules* **1988**, *21*, 3414, using θ and ϕ from Sundararajan, P. R. *Macromolecules* **1980**, *13*, 512.

^{g}Tonelli, A. E. *Macromolecules* **1985**, *18*, 2579. Preexponential factors are 1.0, 1.4, 1.6, and 0.9 for η, ω, ω_X, and ω_{XX}, respectively. Explicit incorporation of the absence of a twofold symmetry axis at χ is not important for the computation of C_∞ for this polymer.

^{h}Tonelli, A. E. *Polymer* **1982**, *23*, 676. Preexponential factors are 0.73, 0.69, 1.5, and 1.3 for η, ω, ω_X, and ω_{XX}, respectively. Explicit incorporation of the absence of a twofold symmetry axis at χ is not important for the computation of C_∞ for this polymer.

^{i}Yoon, D. Y.; Suter, U. W.; Sundararajan, P. R.; Flory, P. J. *Macromolecules* **1975**, *8*, 784. Preexponential factors are 1.0, 1.3, 1.4, and 1.2 (each ±0.1) for η, ω, ω_X, and ω_{XX}, respectively.

^{j}Riande, E.; Saiz, E.; Mark, J. E. *J. Polym. Sci., Polym. Phys. Ed.* **1984**, *22*, 863. Preexponential factors are 1.3, 1.4, and 1.2 for ω, ω_X, and ω_{XX}, respectively.

The more complicated class, where there are distinguishable states at the C^α—X bond that present different types of interactions with the main chain, will be considered in Chapter XI.

Table VIII-3 summarizes RIS models for several vinyl polymers with polyatomic side chains represented in reasonable approximation with a single value of χ. Several of these chains have been treated successfully with Z formulated from **U** presented previously in this chapter, but others require different **U**.

A. Polypropylene

Two rotational isomeric state models for *polypropylene* are available. The more accurate of the two models, which employs five rotational isomeric states, is based on careful analysis of the conformational energies of a fully atomistic

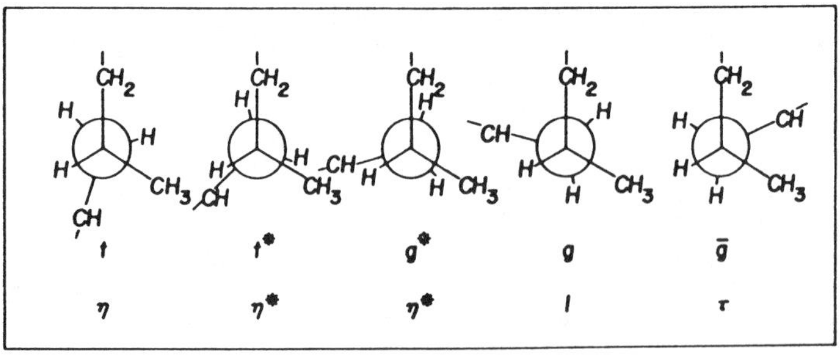

Figure VIII-6. Newman projections along $CH(CH_3)$—CH_2 in polypropylene, depicting the five rotational isomeric states. Reprinted with permission from Suter, U. W.; Flory, P. J. *Macromolecules* **1975**, *8*, 765. Copyright 1975 American Chemical Society.

description of the fragment $-CH_2CH(CH_3)CH_2CH(CH_3)CH_2-$.[17] Accurate representation of the conformational energy surface requires the use of five rotational isomers, with ϕ of 165, 130, 110, 75, and $-65°$, as depicted in Fig. VIII-6. The five states are denoted t, t^*, g^*, g, and $\bar{g}$, respectively. In their simplest form, the statistical weight matrices are

$$\mathbf{U} = \begin{bmatrix} 1 & 1 & 1 & 1 & 1 \\ 1 & 1 & 1 & 1 & 1 \\ 1 & 1 & 0 & 0 & 1 \\ 1 & 1 & 0 & 0 & 1 \\ 1 & 1 & 1 & 1 & 0 \end{bmatrix} \tag{VIII-27}$$

$$\mathbf{U}_m = \begin{bmatrix} 0 & \eta\omega^* & 0 & \eta & 0 \\ \eta\omega^* & 0 & 0 & 0 & \tau\omega^* \\ 0 & 0 & 0 & \omega^* & \tau\omega^* \\ \eta & 0 & \omega^* & 0 & 0 \\ 0 & \tau\omega^* & \tau\omega^* & 0 & 0 \end{bmatrix} \tag{VIII-28}$$

$$\mathbf{U}_r = \begin{bmatrix} \eta^2 & 0 & \eta\omega^* & 0 & 0 \\ 0 & 0 & 0 & \omega^* & \tau\omega^* \\ \eta\omega^* & 0 & 0 & 0 & \tau\omega^* \\ 0 & \omega^* & 0 & 1 & 0 \\ 0 & \tau\omega^* & \tau\omega^* & 0 & 0 \end{bmatrix} \tag{VIII-29}$$

The ω^* in $\mathbf{U}_m$ and $\mathbf{U}_r$ is a compact notation for the product of the statistical weights for a first- and second-order interaction that always occur together in

[17] Suter, U. W.; Flory, P. J. *Macromolecules* **1975**, *8*, 765.

the matrices. The five-state model predicts C_∞ at 140°C that range from ~12 at $p_m = 0$ to ~4 at $p_m = 1$, with a monotonic change in C_∞ at intermediate p_m. The dependence of C_∞ on p_m is stronger in the syndiotactic region than in the isotactic region. The computed values of C_∞ for atactic chains are in good agreement with experimental results of 5.9 at 37.6°C[18] and 5.4 at 153°C.[19] The results computed for the isotactic chain are close to values of 4.7–6.4 measured at 125–183°C.[20] Agreement is less satisfactory for syndiotactic chains, with the calculations producing C_∞ much larger than the results from experiment for materials described as syndiotactic.[21]

The five-state RIS model does not reduce exactly to a three-state model. However, the proper characteristic ratio for isotactic and atactic chains of polypropylene can also be obtained with appropriately parameterized nine-state[22] and three-state models.[23] Three-state models have often been used for calculations of copolymers of propylene with other monomers subject to a threefold rotation potential energy function, usually with energies in the range $E_\eta = 0$, $0 < E_\tau \le 2$, $E_\omega = E_{\omega X} = E_{\omega XX} = 8$ kJ mol^{-1}. Analysis of an alternating copolymer of ethylene and propylene suggests the uncertainty in E_τ should be resolved in favor of $E_\tau = 2$ kJ mol^{-1}.[24]

B. Rings: Polystyrene and Its Derivatives

A side chain consisting of a ring attached directly to the main chain can sometimes be treated as being rigidly attached to the backbone, insofar as the formulation of Z is concerned. This approximation is justified if the ring is either confined to a specific conformation, or if it has multiple conformations that each present similar interactions to the main chain.

Polystyrene. The C^α—X bond is a twofold symmetry axis in *polystyrene*. Repulsive steric interactions between the *ortho* CH groups in the aromatic ring and the CH_2 groups in the backbone produce a marked preference for a torsion angle close to zero at the C^α—X bond.[25,26] A segment of an isotactic polystyrene chain with χ in this range is depicted in Fig. VIII-7. In the approximation that each $\chi = 0°$, the phenyl group can be viewed as an extended

---

[18]Zhongde, X.; Mays, J.; Xuexin, C.; Hadjichristidis, N.; Schilling, F. C.; Bair, H. E.; Pearson, D. S.; Fetters, L. J. *Macromolecules* **1985**, *18*, 2560.

[19]Kinsinger, J. B.; Hughes, R. E. *J. Phys. Chem.* **1963**, *67*, 1922.

[20]Kinsinger, J. B.; Hughes, R. E. *J. Phys. Chem.* **1959**, *63*, 2002. Nakajima, A.; Saiyo, A. *J. Polym. Sci., Part A-2* **1968**, *6*, 735. Heatley, F.; Salovey, R.; Bovey, F. A. *Macromolecules* **1969**, *2*, 619.

[21]Inagaki, H.; Miyamoto, T.; Ohta, S. *J. Phys. Chem.* **1966**, *70*, 3420.

[22]Boyd, R. H.; Breitling, S. M. *Macromolecules* **1972**, *5*, 729.

[23]Biskup, U.; Cantow, H. J. *Macromolecules* **1972**, *5*, 546; *Makromol. Chem.* **1973**, *168*, 315.

[24]Mathur, S. C.; Mattice, W. L. *Makromol. Chem.* **1988**, *189*, 2893.

[25]Williams, A. D.; Flory, P. J. *J. Am. Chem. Soc.* **1969**, *91*, 3111.

[26]The atoms that define χ in polystyrene are H–C^α–C^{ar}–C^{ar}, where C^{ar} denotes an aromatic carbon atom.

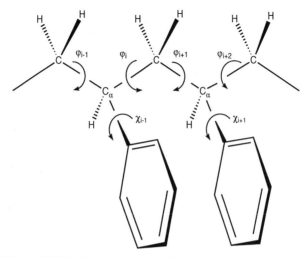

Figure VIII-7. Segment from an isotactic polystyrene chain.

atom rigidly attached to the backbone. The polystyrene chain is then susceptible to a treatment that is a minor modification of the one described above for the poly(vinyl halides), where X is a single atom. The minor modification is actually a simplification.

When X is as bulky as the C_6H_5 group, the chain experiences large repulsive energies in the conformations weighted by $\tau\omega$, $\tau\omega_X$, and $\tau\omega_{XX}$. They are sufficiently repulsive so that any element of a statistical weight matrix that contains τ and one of the ω values can be replaced by zero in excellent approximation.[25] If the three products of statistical weights denoted by $\tau\omega$, $\tau\omega_X$, and $\tau\omega_{XX}$ all approach zero, the $\bar{g}$ states are suppressed. Thus, in the limit where each of these three products approaches zero, Z can be formulated from 2×2 statistical weight matrices that retain the t and g states, and reject the third row and third column of the matrices in Eqs. (VIII-24), (VIII-25), and (VIII-26).[27]

$$\mathbf{U}_p = \begin{bmatrix} 1 & 1 \\ 1 & \omega \end{bmatrix} \qquad \text{(VIII-30)}$$

$$\mathbf{U}_m = \begin{bmatrix} \eta^2\omega_{XX} & \eta \\ \eta & \omega \end{bmatrix} \qquad \text{(VIII-31)}$$

$$\mathbf{U}_r = \begin{bmatrix} \eta^2 & \eta\omega_X \\ \eta\omega_X & 1 \end{bmatrix} \qquad \text{(VIII-32)}$$

[27]Yoon, D. Y.; Sundararajan, P. R.; Flory, P. J. *Macromolecules* **1975**, *8*, 776.

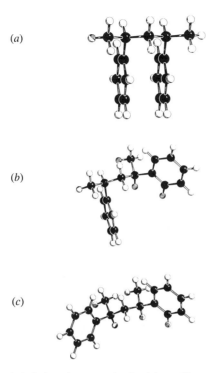

Figure VIII-8. *Meso*-2,4-diphenylpentane in the (*a*) *tt*, (*b*) *tg ≡ gt*, and (*c*) *gg* conformations.

The statistical weight τ no longer occurs in the conformational partition function when this approximation is allowed.

The two phenyl rings interact strongly with one another in the *tt* conformation of the *meso* diad, as shown in Fig. VIII-8. When a realistic charge distribution is included in the calculation, this interaction is so strongly repulsive that there is no accessible minimum in the *tt* region of the conformational energy surface of the *meso* diad.[28] For the atactic chain, experimental values of C_∞ of 9.8–10.7 were measured in a dozen Θ solvents covering a temperature range of 8.5–75.0°C.[29] The RIS models give good agreement with the results for atactic chains. The model of Rapold and Suter predicts much higher C_∞ for the isotactic chain because the *tt* states are suppressed at a *meso* diad. The C_∞ for material described as isotactic has been reported as 17.3,[30] which is much lower than the prediction from the model. Since the model finds $\partial C_\infty/\partial p_m$ is large and positive for p_m near 1, the comparison of calculation with experiment requires an accurate characterization of p_m.

[28] Rapold, R.; Suter, U. W. *Macromol. Theory Simul.* **1994**, *3*, 1.
[29] Mays, J. W.; Hadjichristidis, N.; Fetters, L. J. *Macromolecules* **1985**, *18*, 2231.
[30] Utiyama, H. *J. Phys. Chem.* **1965**, *69*, 4138.

Poly(p-halostyrene). Figure VIII-8 shows that a small substituent in the *para* position will produce a minor perturbation on the interaction of the side chain with atoms in the backbone. Substituents on neighboring aromatic rings may interact with one another in those conformations weighted by ω_{XX}. Nevertheless, one anticipates the RIS scheme for polystyrene should be adaptable to poly(*p*-halostyrenes). This supposition is supported by measurements of the unperturbed dimensions of poly(*p*-chlorostyrene), which yield values of C_∞ in the range 10.6 ± 0.6. These values are indistinguishable from those obtained with polystyrene.[31] Rotational isomeric state schemes derived originally for polystyrene have been used for poly(*p*-fluorostyrene),[32] poly(*p*-chlorostyrene),[33] poly(*p*-bromostyrene),[34] and poly(*p*-nitrostyrene).[35]

Placement of a bulkier group in the *para* position may necessitate a revision of the model, or at least a revision of the values assigned to the parameters that occur in the model. For example, poly(*p-tert*-butylstyrene) has a value of C_∞ that is ~20% larger than the one reported for polystyrene at the same temperature.[36]

Poly(N-vinyl carbazole). *Poly(N-vinyl carbazole)*, like polystyrene, has a symmetric twofold rotational potential energy function for χ. The plane of the carbazole ring system prefers those conformations that bisect $\theta_{CC^\alpha C}$, as depicted in Fig. VIII-9. The RIS model for this polymer utilizes statistical weight matrices of the same form as those used for polystyrene, but with slightly different assignments for the statistical weights that enter those matrices. Analysis of the nmr of small molecule analogs leads to the assignments presented in Table VIII-3. The severe interaction of the two bulky aromatic units in the *tt* state of the *meso* diad causes $\theta_{C^\alpha CC^\alpha}$ to open up from 114° to 117° in this conformation, and also causes the preferred torsion angle in the *t* and *g* states to depend on the state of the neighboring bond and the stereochemistry of the diad.[37] This information was incorporated in the generator matrices used by Abe et al.[38] The matrix formulation permits uses of different values of $\theta_{C^\alpha CC^\alpha}$ for different conformations and diads, and also permits the value of ϕ for a particular state at bond *i* to depend on the stereochemistry of the diad and the state of the other bond in the diad. The parameters adopted by Abe et al. provide

[31] Kurata, M.; Stockmayer, W. H. *Fortschr. Hochpolymer. Forsch.* **1963**, *3*, 196. Kuwahara, N.; Ogino, K.; Kasai, A.; Ueno, S.; Kaneko, M. *J. Polym. Sci., Part A* **1965**, *3*, 985. Mohite, R. B.; Gundiah, S.; Kapur, S. L. *Makromol. Chem.* **1968**, *116*, 280. Noguchi, Y.; Aoki, A.; Tanaka, G.; Yamakawa, H. *J. Chem. Phys.* **1970**, *52*, 2651.

[32] Lin, Y.-Y.; Mark, J. E.; Ackerman, J. L. *ACS Symp. Ser.* **1982**, *193*, 279.

[33] Mark, J. E. *J. Chem. Phys.* **1972**, *56*, 458. Saiz, E.; Mark, J. E.; Flory, P. J. *Macromolecules* **1977**, *10*, 967. Bahar, I.; Baysal, B. M.; Erman, B. *Macromolecules* **1986**, *19*, 1703.

[34] Khanarian, G.; Cais, R. E.; Kometani, J. M.; Tonelli, A. E. *Macromolecules* **1982**, *15*, 866.

[35] Khanarian, G. *J. Chem. Phys.* **1982**, *77*, 2684.

[36] Mays, J. W.; Nan, S.; Whitfield, D. *Macromolecules* **1991**, *24*, 315.

[37] Sundararajan, P. R. *Macromolecules* **1980**, *13*, 512.

[38] Abe, A.; Kobayashi, H.; Kawamura, T.; Date, M.; Uryu, T.; Matsuzaki, K. *Macromolecules* **1988**, *21*, 3414.

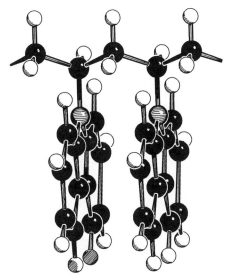

Figure VIII-9. The *tt* conformation of a *meso* diad in poly(*N*-vinyl carbazole).

reasonable agreement with measured values for C_∞, which are in the range 15–19 for samples having p_m of ~0.25–0.57.[39]

Poly(2-vinyl pyridine), Poly(N-vinyl pyrrolidone). A few vinyl polymers with planar cyclic substituents have been treated using statistical weight matrices of the same form as those employed for polystyrene, even though their torsion at χ does not have a symmetric twofold torsion potential energy function. Examples are *poly(2-vinyl pyridine)*,[40] where the nitrogen atom breaks the twofold symmetry present in polystyrene, and *poly(N-vinyl pyrrolidone)*,[41] where, in addition, the ring is not aromatic (see Fig. VIII-10). These two polymers have E_η close to zero. The energies for the second-order interactions are larger for poly(*N*-vinyl pyrrolidone) than for poly(2-vinyl pyridine).

C. Other Relatively Inflexible, Approximately Planar Side Chains

The ester group, which prefers a planar conformation,[42] is frequently encountered as a side chain in vinyl polymers. Small-molecule models for the attach-

[39]Kuwahara, N.; Higashida, S.; Nakata, M.; Kaneko, M. *J. Polym. Sci., Polym. Phys. Ed.* **1969**, *7*, 285. Sitaramaiah, G.; Jacobs, D. *Polymer* **1970**, *11*, 165. Leon, L. M.; Katime, I.; Rodriguez, M. *Eur. Poly. J.* **1979**, *15*, 29. Urizar, M.; Rodriguez, M.; Leon, L. M. *Makromol. Chem.* **1984**, *185*, 76.
[40]Tonelli, A. E. *Macromolecules* **1985**, *18*, 2579.
[41]Tonelli, A. E. *Polymer* **1982**, *23*, 676.
[42]Pauling, L.; Sherman, J. *J. Chem. Phys.* **1933**, *1*, 606. Curl, R. F. *J. Chem. Phys.* **1959**, *30*, 1529.

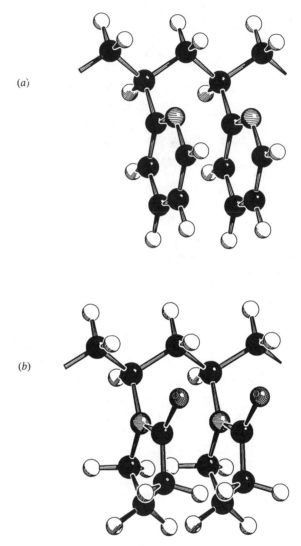

(a)

(b)

Figure VIII-10. The preferred conformations of the side chains in (a) poly(2-vinyl pyridine) and (b) poly(N-vinyl pyrrolidone)

ment of the ester in *poly(methyl acrylate)* (see Fig. VIII-11) showing the preferred conformation at χ will produce a *cis* conformation at H–C$^\alpha$—C=O, where the bond with the long dash denotes the attachment of the ester to the main chain.[43] The conformation produced by a rotation about this bond of 180° to the *trans* conformation increases the energy by only a few kilojoules per

[43]Dirikov, S.; Stokr, J.; Schneider, B. *Collect. Czech. Chem. Commun.* **1971**, *36*, 3028.

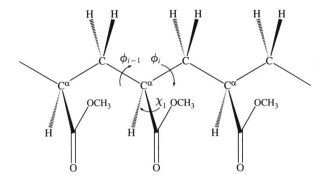

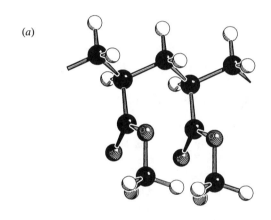

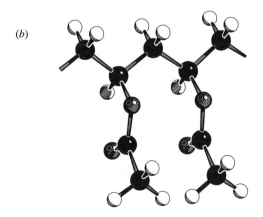

VIII-11. The preferred conformations of the side chains in (*a*) poly(methyl acrylate) and (*b*) poly(vinyl acetate).

mole. These two conformations will dominate the population at χ, and they have been viewed as equivalent, in first approximation, insofar as the formulation of Z for the main chain is concerned. The conformational partition function has been formulated using statistical weight matrices of the same form as those employed for polystyrene.[44] Explicit treatment of the two different orientations of the ester groups, corresponding to two different rotational isomers for χ, separated by rotations of 180°, has been achieved by increasing the dimensions of the statistical weight matrices from 2×2 to 4×4.[45] The chain obtained by the reversal of the orientation of the attachment of the ester group to the main chain, as in *poly(vinyl acetate)*, can also be treated using statistical weight matrices of the same form.[46]

4. VINYLIDENE POLYMERS

Here we consider polymers of $H_2C{=}CXY$ with $X \neq Y$, where neither X nor Y is hydrogen, and both X and Y can be approximated as being nonarticulated. These polymers represent a special challenge in the formulation of Z because, as in the case of polyisobutylene, repulsive second-order interactions must be considered.

Let the definition of the pseudoasymmetric center be the one adopted at the beginning of this chapter, using X as the defining group. The statistical weight matrices denoted by U_d, U_l, and U_p take the same form as those introduced at the beginning of this chapter. Additional second-order interactions, all involving Y, must be incorporated in U_{dd}, U_{ll}, U_{dl}, U_{ld}, U_m, and U_r. Every element in these matrices contains the product of two ω values:

$$U_{dd} = \begin{bmatrix} \eta\,\omega_{XX}\omega_{YY} & \tau\omega_X\omega_{XY} & \omega_{XY}\omega_Y \\ \eta\,\omega_{XY}\omega_Y & \tau\omega_X\omega_Y & \omega\,\omega_{YY} \\ \eta\,\omega_X\omega_{XY} & \tau\omega\,\omega_{XX} & \omega_X\omega_Y \end{bmatrix} \tag{VIII-33}$$

$$U_{dl} = \begin{bmatrix} \eta\,\omega_{XY}^2 & \omega_X\omega_{YY} & \tau\omega_{XX}\omega_Y \\ \eta\,\omega_X\omega_{YY} & \omega_Y^2 & \tau\omega\,\omega_{XY} \\ \eta\,\omega_{XX}\omega_Y & \omega\,\omega_{XY} & \tau\omega_X^2 \end{bmatrix} \tag{VIII-34}$$

The relationships in the first column of Table VIII-1 generate U_{ll}, U_{ld}, U_m, and U_r from U_{dd} and U_{dl}.

[44]Yoon, D. Y.; Suter, U. W.; Sundararajan, P. R.; Flory, P. J. *Macromolecules* **1975**, *8*, 784.
[45]Ojalvo, E. A.; Saiz, E.; Masegosa, R. M.; Hernández-Fuentes, I. *Macromolecules* **1979**, *12*, 865.
[46]Riande, E.; Saiz, E.; Mark, J. E. *J. Polym. Sci., Polym. Phys. Ed.* **1984**, *22*, 863.

A. Poly(α-methylstyrene)

Poly(α-methylstyrene), where $X = C_6H_5$ and $Y = CH_3$, has been treated using the t and g states employed for polystyrene; the $\bar{g}$ state is of such high energy that it can be ignored.[47] The chain has C—C bond lengths of 153 pm and $\theta_{CC^\alpha C}$ of 109.5°, virtually identical with the values used for polystyrene. The increased severity of the repulsive second-order interactions, however, produces an opening of $\theta_{C^\alpha CC^\alpha}$ to 122°, which is substantially larger than the value of 114° employed for polystyrene. Even after this bond angle has been opened, the repulsions produced by the second-order interactions cannot be completely alleviated. The result is a splitting of each conformational energy minimum found for polystyrene into two minima, separated by ~20°, in poly(α-methylstyrene). Nevertheless, a first approximation permits treatment of the chain using only two states per bond, the splitting of each conformational energy minimum being ignored; U_p is still given by Eq. (VIII-30). Since Y is CH_3, the statistical weight denoted by ω_{XY} can be replaced by ω_X, and ω_{YY} and ω_Y can be replaced with ω:

$$\mathbf{U}_m = \begin{bmatrix} \eta^2 \omega \, \omega_{XX} & \eta \, \omega \, \omega_X \\ \eta \, \omega \, \omega_X & \omega^2 \end{bmatrix} \tag{VIII-35}$$

$$\mathbf{U}_r = \begin{bmatrix} \eta^2 \omega_X^2 & \eta \, \omega \, \omega_X \\ \eta \, \omega \, \omega_X & \omega^2 \end{bmatrix} \tag{VIII-36}$$

An appropriate normalization shows the number of parameters requiring definition to be two instead of the four suggested by these two equations. With $\alpha \equiv \omega_X / \eta \, \omega_{XX}$ and $\beta \equiv \omega_X^2 / \omega \, \omega_{XX}$, the matrices become

$$\mathbf{U}_m = \begin{bmatrix} 1 & \alpha \\ \alpha & \alpha^2/\beta \end{bmatrix} \tag{VIII-37}$$

$$\mathbf{U}_r = \begin{bmatrix} \beta & \alpha \\ \alpha & \alpha^2/\beta \end{bmatrix} \tag{VIII-38}$$

The measured[48] C_∞ of ~11–12 for several chains with p_m in the range 0.05–0.5

[47] Sundararajan, P. R. *Macromolecules* **1977**, *10*, 623.
[48] Cowie, J. M. G.; Bywater, S.; Worsfold, D. J. *Polymer* **1967**, *8*, 105. Cowie, J. M. G.; Bywater, S. *J. Polym. Sci., Part A-2* **1968**, *6*, 499. Noda, I.; Mizutani, K.; Kato, T.; Fujimoto, T.; Nagasawa, M. *Macromolecules* **1970**, *3*, 787.

are reproduced fairly well using $E_\alpha = 2.1$ kJ mol^{-1}, $E_\beta = -4.2$ kJ mol^{-1}, with preexponential factors of 0.6 and 0.9 for α and β, respectively.

B. Poly(methyl methacrylate)

Poly(methyl methacrylate) is the example with X = COOCH$_3$ and Y = CH$_3$. In view of the importance of this polymer and its availability over a wide range of stereochemical compositions, RIS models date back two decades.[49] The severity of the mandatory repulsive second-order interactions opens $\theta_{C^\alpha CC^\alpha}$ to $\sim124°$.[50,51] Even so, the conformational energy surfaces show splitting of the customary three rotational isomeric states, suggesting application of a six-state model if the conformational energy surface is to be accurately represented. On the other hand, a much simpler approximation to Z would be accessible if the splitting were not incorporated explicitly into the model. Conformational partition functions using both approaches were developed independently and simultaneously.[50,51]

The simpler three-state $(t, g, \bar{g})$ model requires[51]

$$\mathbf{U}_p = \begin{bmatrix} 1 & 1 & 1 \\ 1 & 0 & \psi \\ 1 & \psi & 0 \end{bmatrix} \tag{VIII-39}$$

The expressions for $\mathbf{U}_m$ and $\mathbf{U}_r$ contain five distinct statistical weights $(\eta, \tau, \omega, \omega_X, \omega_{XX})$, but an appropriate renormalization reduces the number of independent parameters to three, defined as $\alpha \equiv \omega_X/\eta\omega_{XX}$, $\beta \equiv \omega_X^2/\omega\omega_{XX}$, $\rho \equiv \tau/\eta$.

$$\mathbf{U}_m = \begin{bmatrix} 1 & \alpha & \beta\rho \\ \alpha & \alpha^2/\beta & \alpha\rho \\ \beta\rho & \alpha\rho & \rho^2 \end{bmatrix} \tag{VIII-40}$$

$$\mathbf{U}_r = \begin{bmatrix} \beta & \alpha & \rho \\ \alpha & \alpha^2/\beta & \alpha\rho \\ \rho & \alpha\rho & \beta\rho^2 \end{bmatrix} \tag{VIII-41}$$

[49]Sundararajan, P. R.; Flory, P. J. *J. Am. Chem. Soc.* **1974**, *96*, 5025.
[50]Vacatello, M.; Flory, P. J. *Macromolecules* **1986**, *19*, 405.
[51]Sundararajan, P. R. *Macromolecules* **1986**, *19*, 415.

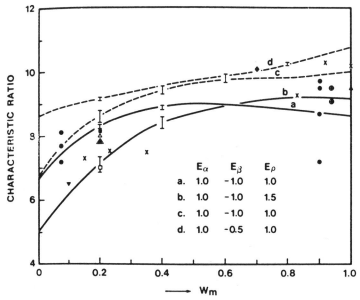

Figure VIII-12. Fits (lines) to the measured characteristic ratios (points) of poly(methyl methacrylate), using the three-state model with the indicated assignments (in kcal mol^{-1}) of $E_\alpha, E_\beta,$ and E_ρ. Reprinted with permission from Sundararajan, P. R. *Macromolecules* **1986**, *19*, 415. Copyright 1986 American Chemical Society.

Fitting to the measured[52] C_∞ over a broad range of p_m suggests $E_\alpha \sim 4$, $E_\beta \sim -4$, $E_\rho \sim 4$, each in kJ mol^{-1}, as shown in Fig. VIII-12. The shapes of the conformational energy surfaces suggest preexponential factors of 0.85, 0.77, and 0.87 for α, β, and ρ, respectively.

Retention of the splitting of the t, g, $\bar{g}$ states produces a six-state model for poly(methyl methacrylate). Extensive analysis of the conformational energies of isotactic and syndiotactic fragments of this polymer produced a Z formulated from symmetric 6×6 matrices that, after appropriate normalization, contain 18 statistical weights:[50]

$$
\mathbf{U}_p =
\begin{array}{c}
 \\
t_- \\
t_+ \\
g_- \\
g_+ \\
\bar{g}_- \\
\bar{g}_+
\end{array}
\begin{array}{cccccc}
t_- & t_+ & g_- & g_+ & \bar{g}_- & \bar{g}_+
\end{array}
\left[
\begin{array}{cccccc}
u_{t_-t_-} & 1 & u_{t_-g_-} & u_{g_-g_+} & u_{t_-\bar{g}_-} & u_{t_-\bar{g}_+} \\
 & u_{t_+t_+} & u_{t_+g_-} & u_{t_+g_+} & u_{t_+\bar{g}_-} & u_{t_+\bar{g}_+} \\
 & & 0 & 0 & u_{g_-\bar{g}_-} & u_{g_-\bar{g}_+} \\
 & & & 0 & 0 & u_{g_+\bar{g}_+} \\
 & & & & 0 & 0 \\
 & & & & & 0
\end{array}
\right]
\qquad \text{(VIII-42)}
$$

[52]Work of several investigators, summarized by Jenkins, R.; Porter, R. S. *Adv. Polym. Sci.* **1980**, *36*, 1; *Polymer* **1982**, *23*, 105.

$$
\mathbf{U}_m =
\begin{array}{c}
\\
t_- \\
t_+ \\
g_- \\
g_+ \\
\overline{g}_- \\
\overline{g}_+
\end{array}
\begin{array}{cc}
\begin{array}{cccccc}
t_- & t_+ & g_- & g_+ & \overline{g}_- & \overline{g}_+
\end{array} \\
\left[
\begin{array}{cccccc}
0 & 1 & 0 & \alpha\beta & 0 & \alpha\overline{\alpha} \\
 & 0 & \alpha\beta & 0 & \alpha\overline{\alpha} & 0 \\
 & & 0 & \beta^2 & 0 & \overline{\alpha}\beta \\
 & & & 0 & \overline{\alpha}\beta & 0 \\
 & & & & 0 & \rho^* \\
 & & & & & 0
\end{array}
\right]
\end{array}
\qquad \text{(VIII-43)}
$$

$$
\mathbf{U}_m =
\begin{array}{c}
\\
t_- \\
t_+ \\
g_- \\
g_+ \\
\overline{g}_- \\
\overline{g}_+
\end{array}
\begin{array}{cc}
\begin{array}{cccccc}
t_- & t_+ & g_- & g_+ & \overline{g}_- & \overline{g}_+
\end{array} \\
\left[
\begin{array}{cccccc}
\alpha^2 & 0 & \alpha\beta & 0 & \overline{\rho} & 0 \\
 & \alpha^2 & 0 & \alpha\beta & 0 & \overline{\rho} \\
 & & \beta^2 & 0 & \overline{\alpha}\beta & 0 \\
 & & & \beta^2 & 0 & \overline{\alpha}\beta \\
 & & & & \overline{\alpha}^2 & 0 \\
 & & & & & \overline{\alpha}^2
\end{array}
\right]
\end{array}
\qquad \text{(VIII-44)}
$$

where the missing elements are defined by the fact that each matrix is symmetric. The α and β in Eqs. (VIII-43) and (VIII-44) are different from the statistical weights represented by the same symbols in the three-state model, because they are defined in terms of the second-order statistical weights of the six-state model. All of the nonzero elements in $\mathbf{U}_p$ were evaluated numerically from the conformational energy surfaces.

The increase in the attention to the actual conformational energy surface, and the increase in the number of parameters in the model, lead to the expectation that the six-state model should perform at least as well as the simpler three-state model. It does indeed provide an adequate representation of measured values[53] for C_∞, as shown in Fig. VIII-13, although it is not immediately obvious that the quality of the fit is significantly better than can be obtained with the simpler three-state model shown in Fig. VIII-12. Distinctions between the two models become more apparent when the property of interest arises from a more local aspect of the conformation of the chain, as in the scattering of neutrons and X-rays at intermediate angles.[54] The six-state model provides a better representation of this data than does the three-state model, as expected.

[53] Chinai, S. N.; Valles, R. J. *J. Polym. Sci.* **1959**, *39*, 363. Sakurada, I.; Nakajima, A.; Yoshizaka, O.; Nakamae, K. *Kolloid Z.* **1962**, *186*, 41. Fox, T. G. *Polymer* **1962**, *3*, 111. Krause, S.; Cohn-Ginsberg, E. *J. Phys. Chem.* **1963**, *67*, 1479. Shulz, G. V.; Wunderlich, W.; Kirste, R. *Makromol. Chem.* **1964**, *75*, 22. Vadusevan, P.; Santappa, M. *J. Polym. Sci., Part A-2* **1971**, *9*, 483. Jenkins, R.; Porter, R. S. *Polymer* **1982**, *23*, 105.
[54] Vacatello, M.; Yoon, D. Y.; Flory, P. J. *Macromolecules* **1990**, *23*, 1993.

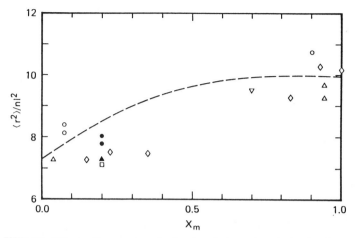

Figure VIII-13. Fits to the measured characteristic ratios (points) of poly(methyl methacrylate), using the six-state model. Reprinted with permission from Vacatello, M.; Flory, P. J. *Macromolecules* **1986**, *19*, 405. Copyright 1986 American Chemical Society.

5. SILICON-CONTAINING ANALOGS

Rotational isomeric state models have been developed for several chains with the repeating sequence W—SiXY, where W might be oxygen, carbon, or another silicon atom. Conformation-dependent properties of such chains are often treated with Z formulated in a manner similar to that described above for vinyl and vinylidene polymers.

Polysilapropylene replaces alternate carbon atoms in the polypropylene backbone with silicon atoms, such that every C—C bond becomes an Si—C bond. The repeat unit is CH_2—$SiHCH_3$. The increase in bond length (to 187 pm for Si—C) is expected to reduce the severity of short-range interactions that are repulsive in polypropylene. The conformational partition function for polysilapropylene has been formulated using a three-state (t, g, $\bar{g}$) model, using statistical weight matrices of the form found in Eq. (VIII-25) and (VIII-26), along with[55]

$$\mathbf{U}_p = \begin{bmatrix} 1 & 1 & 1 \\ 1 & 0 & 1 \\ 1 & 1 & 0 \end{bmatrix} \qquad \text{(VIII-45)}$$

The bond angle was 109.5° for θ_{CSiC}, and θ_{SiCSi} was in the range 115–119°.

[55] Sundararajan, P. R. *Macromolecules* **1990**, *23*, 3179.

Torsion angles were displaced a few degrees from those expected for perfectly staggered rotational isomers. The energies associated with the five statistical weights, η, τ, ω, ω_X, and ω_{XX} are consistent with weak short-range interactions. Only E_τ was estimated to be outside the range ±0.4 kJ mol^{-1}, with E_τ estimated at -1.3 kJ mol^{-1} from the calculations. Computed C_∞ were ~4 for all p_m, thereby suggesting a chain with a much smaller C_∞ than polypropylene. There does not appear to be any experimental data pertinent to the C_∞ for polysilapropylene.

Polysilastyrene has the repeating sequence $-SiH_2-SiHC_6H_5-$, where every carbon atom in the main chain of polystyrene has been replaced with a silicon atom. The bond lengths in the main chain are 234 pm, and all θ_{SiSiSi} are very close to tetrahedral.[56] Two three-state (t, g, $\bar{g}$) models have been described.[56,57] The absence of experimental data for C_∞ makes discrimination between the models difficult. Both models predict that the syndiotactic chain has a much larger C_∞ than the isotactic chain, with the model of Welsh et al. always predicting the larger dimensions. According to the model of Sundararajan, which employs Eq. (VIII-25), (VIII-26), and (VIII-45), the first-order interactions are repulsive, with $E_\eta = 0.6$ and $E_\tau = 12$ kJ mol^{-1}, while the second-order interactions are weakly attractive ($E_\omega = -1.0$, $E_{\omega X} = -0.6$, and $E_{\omega XX} = -0.6$ kJ mol^{-1}). In contrast, the second-order interactions are strongly repulsive in polystyrene, with the shorter bond lengths in the carbon-based polymer.

Attachment of a phenyl and methyl group to every backbone Si atom, instead of to alternate Si atoms as in polysilastyrene, yields *poly(methylphenylsilylene)* with repeating sequence $-SiCH_3C_6H_5-$. The repulsive short-range interactions in this polymer cause an opening of the bond angle by ~5–7° compared to its value in polysilastyrene. Large values of C_∞ for atactic polymers are predicted with two-state (t, $\bar{g}$, with a large, but not exclusive, preference for t)[58] and "three-state" (t, g^+, g^- initially, but only t survives with the statistical weights used)[56] models, in qualitative agreement with the very large extension observed experimentally.[59]

Poly(methylphenylsiloxane) presents issues of tacticity similar to those seen in the vinylidene polymers. The repeating sequence can be derived from poly(α-methylstyrene) by appropriate replacement of the carbon atoms in the backbone with silicon and oxygen atoms. A RIS model for this chain was described by Mark and Ko.[60] This RIS model is based on a conformational energy analysis using a fixed θ_{SiOSi} of 143°. Using the *meso,racemo* notation, $\mathbf{U}_p$ in the three-state model was assembled as

[56]Welsh, W. J.; Damewood, J. R., Jr.; West, R. C. *Macromolecules* **1989**, *22*, 2947.

[57]Sundararajan, P. R. *Macromolecules* **1991**, *24*, 1420.

[58]Sundararajan, P. R. *Macromolecules* **1988**, *21*, 1256.

[59]Cotts, P. M.; Miller, R. D.; Trefonas, P. T.; West, R.; Fickes, G. N. *Macromolecules* **1987**, *20*, 1046.

[60]Mark, J. E.; Ko, J. H. *J. Polym. Sci., Polym. Phys. Ed.* **1975**, *13*, 2221.

$$\mathbf{U}_p = \begin{bmatrix} 1 & \sigma & \sigma \\ 1 & 0 & \sigma \\ 1 & \sigma & 0 \end{bmatrix} \qquad \text{(VIII-46)}$$

where σ is the first-order interaction of Si with O, and all other short-range interactions [except for the highly repulsive Si$\cdots$Si interaction in the $g^{\pm}g^{\mp}$ states, represented by the two null elements in Eq. (VIII-46)] are reserved for inclusion in $\mathbf{U}_m$ and $\mathbf{U}_r$:

$$\mathbf{U}_m = \begin{bmatrix} \omega_{pp}\omega_{mm} & \sigma\omega_{mp} & \sigma\delta\omega_{mp} \\ \omega_{mp} & \sigma\omega_{oo}\omega_{mm} & \sigma\delta \\ \delta\omega_{mp} & \sigma\delta & \sigma\omega_{oo}\omega_{pp} \end{bmatrix} \qquad \text{(VIII-47)}$$

$$\mathbf{U}_r = \begin{bmatrix} \omega_{mp}^2 & \sigma\delta\omega_{mm} & \sigma\omega_{pp} \\ \delta\omega_{mm} & \sigma & \sigma\omega_{oo}\omega_{mp} \\ \omega_{pp} & \sigma\omega_{oo}\omega_{mp} & \sigma\delta^2 \end{bmatrix} \qquad \text{(VIII-48)}$$

The interacting groups in the ω-type interactions retained in these matrices are denoted by "m" for methyl, "o" for oxygen, and "p" for phenyl, and δ is a comparable interaction of a phenyl group with an oxygen atom. Only σ and ω_{oo} were estimated to be smaller than one, with $E_\sigma = 1.5$ and $E_{\omega_{oo}} = 6.7$ kJ mol^{-1}. The energies of the remaining second-order interactions retained in the matrices were estimated to be in the range -10.5 to -2.1 kJ mol^{-1}, with the most attractive second-order interaction being the one between two phenyl groups. Conceivably these estimates might change if the unusual flexibility of θ_{SiOSi} were incorporated into the analysis, as in the study of poly(dimethylsiloxane).[61] With the parameters of Mark and Ko, C_∞ at 30°C is predicted to be a nearly linear function of p_m, ranging from ~8 for the syndiotactic chain to ~2 for the isotactic chain.

PROBLEMS

Appendix B contains answers for Problems VIII-1, VIII-3, VIII-4, VIII-7, and VIII-9 through VIII-13.

VIII-1. Do the *meso* and *racemic* isomers of 2,4-dichloropentane have the same Z?

VIII-2. Convince yourself exactly the same Z is obtained for isotactic chains considered as a string of d pseudoasymmetric centers, a string of l pseudoasymmetric centers, and a string of *meso* dyads.

[61] Bahar, I.; Zúñiga, I.; Dodge, R.; Mattice, W. L. *Macromolecules* **1991**, *24*, 2993.

VIII-3. Determine the preferred conformations for isotactic and syndiotactic poly(vinyl fluoride), poly(vinyl chloride), and poly(vinyl bromide). Which of the short-range interactions play the most important roles in establishing these preferences?

VIII-4. Electronic excitation of benzene enhances its tendency to form a face-to-face complex with a ground-state benzene, such that the two rings are separated by only ~350 pm. (This complex is called an *excimer*, Section II.2.E.) If the phenyl group in unit x of a polystyrene is electronically excited, and all other phenyl groups are in their electronic ground state, which $\mathbf{U}$ should be modified in order to incorporate this effect? Which elements in these $\mathbf{U}$ need to be changed? Does the answer depend on the stereochemical sequence in the vicinity of unit x?

VIII-5. Polyoxypropylene and polythiopropylene present the issue of an asymmetric rotation potential energy function, but with a three-bond repeat, rather than a two bond-repeat, in the main chain. The repeating units are $-O-CH(CH_3)-CH_2-$ and $-S-CH(CH_3)-CH_2-$, respectively. Construct the forms of the statistical weight matrices required for generation of Z for a three-state model for these polymers, and identify the first- and second-order interactions incorporated in the model. Compare your expressions with those described in Abe, A.; Hirano, T.; Tsuruta, T. *Macromolecules* **1979**, *12*, 1092 and Abe, A. *Macromolecules* **1980**, *13*, 541.

VIII-6. The repeating sequence is $-CH_2-CH_2-CHCH_3-CH_2-O-$ in poly(3-methyltetrahydrofuran). Construct Z using a three-state RIS model. Compare your formulation with the one described in Saiz, E.; Tarazona, M. P.; Riande, E.; Guzmán, J. *J. Polym. Sci., Polym. Phys. Ed.* **1984**, *22*, 2165.

VIII-7. Polystyrene, poly(2-vinyl pyridine), and poly(*N*-vinyl carbazole) each contains an aromatic side chain. Which polymer has the largest C_∞ at 300 K? Does the answer depend on p_m?

VIII-8. Estimate the size of the product of τ and the ω values at which the two-state (t, g) model for a vinyl polymer becomes a good substitute for the three-state $(t, g, \bar{g})$ model.

VIII-9. Describe a set of short-range interactions that will cause C_∞ to be strongly dependent on p_m for highly syndiotactic polymers, $0 \leq p_m < 0.05$. Provide an explanation, in terms of the local conformations favored by the chain.

VIII-10. Describe a set of short-range interactions that will cause C_∞ to be dependent on p_m for highly isotactic polymers, $0.95 < p_m \leq 1$. Provide an explanation, in terms of the local conformations favored by the chain.

VIII-11. Describe a set of short-range interactions that will cause C_∞ to be nearly independent of p_m. Provide an explanation, in terms of the local conformations favored by the chain.

VIII-12. Describe a set of short-range interactions that will cause C_∞ for a vinyl polymer to pass through a minimum as p_m increases from 0 to 1. Provide an explanation, in terms of the local conformations favored by the chain.

VIII-13. How many nonzero components are there in $\langle \mathbf{r} \rangle_0$ for methyl-terminated isotactic and syndiotactic polystyrene? Which component is the largest?

VIII-14. Methyl-terminated polypropylene is a special case of a vinyl polymer, because the end groups and side chains are identical. What are the implications for the statistical weight matrices for the bonds near the ends of the chain? Compare your answer with the treatment in Suter, U. W.; Flory, P. J. *Macromolecules* **1975**, *8*, 765.

IX CHAINS WITH RIGID UNITS AND VIRTUAL BONDS

With one exception, all chains considered in Chapters VI, VII, and VIII were composed of bonds having access to two or more rotational isomers, such that $\nu_i > 1$ at every internal bond. The most common value of ν_i was 3, but examples were also presented where ν_i was 2, 4, 5, 6, and 9. An important exception was poly(oxy-1,1-dimethylethylene), which was considered in Section VII.5. The severe steric interactions of the bulky $C(CH_3)_2$ groups in this polymer cause the g states to be of very high energy at the bond represented by the long dash in $C(CH_3)_2–O—CH_2–C(CH_3)_2$. The rotational isomeric state (RIS) model for poly(oxy-1,1-dimethylethylene) retains only the t state at this bond, although $\nu_i = 3$ at the other two bonds in the repeating unit.[1] Adoption of this scheme permits the formulation of Z by a repeating pattern of a row of three elements, a 3×3 matrix, and a column of three elements, the serial product of which is a scalar, as shown in Eq. (VII-33). This chapter describes additional formulations of Z that take advantage of simplifications that result from the presence of rigid units within the chain.

Several different types of structural features may cause a chain to contain bonds for which $\nu_i = 1$ is a useful approximation. In the case of poly(oxy-1,1-dimethylethylene), repulsive steric interactions between substituents nearby in the chain suppress all but one of the potential rotational isomeric states at the $O—CH_2$ bond. The repulsive part of the potential energy function describing nonbonded interactions restricts the rotation about this bond. Electronic effects localized within a bond may also produce $\nu_i = 1$, as in double bonds, and in bonds with partial double-bond character. The form of the intrinsic torsion potential energy function is responsible for the restriction on rotation about these bonds. Incorporation of bonds in the chain into small rings, either aromatic or alicyclic, will produce severe constraints on the torsion angles. The necessity for maintenance of ring closure restricts the torsion angles. All of these different phenomena produce similar simplification in the structure of Z.

Often the simplification in Z is accompanied by a redefinition of the chain, making use of the concept of *virtual bonds*.[2] When l_i is a bond vector, it points from atom $i-1$ to atom i, where these two atoms are connected by a chemical bond of length l_i, typically in the range 140–240 pm. The very limited magni-

[1] Kato, K.; Araki, K.; Abe, A. *Polym. J.* **1981**, *13*, 1055.
[2] Benoit, H. *J. Polym. Sci.* **1948**, *3*, 376.

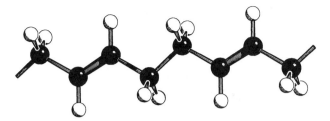

Figure IX-1. A short segment of poly(1,4-*trans*-butadiene).

tude of the stretching of real bonds at the temperatures of interest usually allows assignment of a constant value to l_i. When $\mathbf{l}_i$ is a virtual bond vector, the two atoms at its ends are connected by several real bonds, but they act as a rigid segment about which torsion can occur, as a result of constraints imposed on the torsion about one or more of the real bonds. The lengths of virtual bonds are usually larger than 300 pm. As we shall see, it is sometimes convenient to define virtual bonds such that there are no atoms at their ends.

1. DOUBLE BONDS

Polymers of 1,3-butadiene and isoprene are simple, and important, examples of chains in which the treatment of the conformational statistics is simplified by the population of only a single rotational isomer at the C=C bond.

A. Poly(1,4-*trans*-butadiene)

The covalent structure of a fragment of *poly(1,4-trans-butadiene)* is depicted in Fig. IX-1. The chain contains three different types of bonds, CH=CH, CH—CH$_2$ (and its inverse), and CH$_2$—CH$_2$. Only a single conformation, t, is accessible at CH=CH. The other bonds have access to three rotational isomeric states. These states are the customary t, g^+, g^- at the CH$_2$—CH$_2$ bond, as in polyethylene. The minima in the torsion potential-energy function are displaced by 60° at the CH—CH$_2$ bonds, giving rise to states denoted by c, s^+, s^- (for *cis,skew$^+$,skew$^-$*), with torsion angles at 0° and near ±120°.[3] Since the interdependence of the bonds is confined to the three single bonds between the double bonds, and does not extend across the double bond, an internal repeating unit will contribute a scalar factor, z_{trans}, to Z. This factor can be generated as the serial product of a row of three elements, two 3×3 matrices, and a column of three elements.[4] The column is the statistical weight matrix for the CH=CH bond, because that bond has only one state, t.

[3]Lide, D. R., Jr.; Mann, D. E. *J. Chem. Phys.* **1957**, *27*, 868. Lide, D. R., Jr. *Annu. Rev. Phys. Chem.* **1964**, *15*, 225.
[4]Suter, U. W.; Höcker, H. *Makromol. Chem.* **1988**, *189*, 1603.

Z **with First- and Second-Order Interactions Only.** The formulation conveniently starts with the CH—CH_2 bond immediately following a double bond. The statistical weight matrix for this bond is a row because the preceding bond, CH=CH, has access to only one rotational isomeric state. The row has three elements, one each for the c, s^+, s^- states at CH—CH_2. The symmetry of the rotation potential energy function demands that both s states be weighted equally, but c may have a different weight. Assigning the s states a statistical weight of 1, the weight for c is denoted by ρ, with E_ρ = 1.3 kJ mol^{-1}:[4,5]

$$\mathbf{U}_{CHCH_2} = t \begin{array}{ccc} c & s^+ & s^- \\ \left[\rho 1 1 \right] \end{array}$$

(IX-1)

The following CH_2—CH_2 bond requires a 3×3 statistical weight matrix:

$$\mathbf{U}_{CH_2CH_2} = \begin{array}{c} c \\ s^+ \\ s^- \end{array} \begin{array}{c} \begin{array}{ccc} t & g^+ & g^- \end{array} \\ \left[\begin{array}{ccc} 1 & \beta & \beta \\ 1 & 1 & 1 \\ 1 & 1 & 1 \end{array} \right] \end{array}$$

(IX-2)

The formulation described by Abe and Flory[5] included a first-order interaction in the two g states, but the statistical weight for this interaction is sufficiently close to 1 to justify writing the matrix with that value so assigned.[4] The statistical weight denoted by β appears in $\mathbf{U}_{CH_2CH_2}$ as a second-order interaction specific to the sequences $cg^\pm$. It actually has a more complex origin. It weights a variety of interactions between nonbonded carbon and hydrogen atoms separated by distances of 212–248 pm in the sequences $cg^\pm c$, $cg^\pm s^\pm$ (and $s^\pm g^\pm c$), and $cg^\pm s^\mp$ (and $s^\mp g^\pm c$), as well as interactions between nonbonded hydrogen atoms separated by 160–194 pm in the $cg^\pm s^\pm$ (and $s^\pm g^\pm c$) and $cg^\pm s^\mp$ (and $s^\mp g^\pm c$) sequences.[5] These conformations are depicted in Fig. IX-2. The sum of the interactions contributes about the same energy in each of these sequences. They can all be represented by the same statistical weight, β, with $E_\beta \sim 8$ kJ mol^{-1}.[4,5] The appearance of β in the last two elements of the first row in $\mathbf{U}_{CH_2CH_2}$ says we know β will be required if the first two bonds in a triplet are $cg^\pm$, no matter what state might be adopted by the third bond in the sequence.

The following CH_2—CH bond also requires a 3×3 statistical weight matrix:

$$\mathbf{U}_{CH_2CH} = \begin{array}{c} t \\ g^+ \\ g^- \end{array} \begin{array}{c} \begin{array}{ccc} c & s^+ & s^- \end{array} \\ \left[\begin{array}{ccc} \rho & 1 & 1 \\ \rho\beta & 1 & 1 \\ \rho\beta & 1 & 1 \end{array} \right] \end{array}$$

(IX-3)

[5] Abe, Y.; Flory, P. J. *Macromolecules* **1971**, *4*, 219.

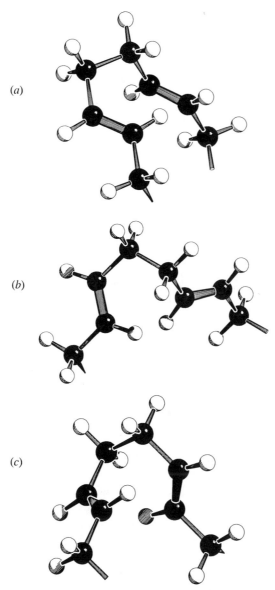

Figure IX-2. The conformations of poly(1,4-*trans*-butadiene) that produce higher-order interactions weighted by β. The bonds represented by long dashes in CH_2–CH=CH—CH_2—CH_2—CH=CH–CH_2 are in the states (*a*) cg^+s^-, (*b*) cg^+s^+, or (*c*) cg^+c.

This matrix weights the c state with ρ, as did the row in $\mathbf{U}_{CHCH_2}$. The appearance of β guarantees it will weight all triplets where the last two bonds are $g^{\pm}c$, independent of the state of the first bond in the triplet. Therefore the appearance of β in $\mathbf{U}_{CH_2CH_2}$ and $\mathbf{U}_{CH_2CH}$ promises the use of this weight whenever the three-bond sequence is $cg^{\pm}x$ or $xg^{\pm}c$, where x may be c, s^+, or s^-. Momentarily we will deal with the situation where $x = c$. Then we will get β twice, once from $\mathbf{U}_{CH_2CH_2}$ and again from $\mathbf{U}_{CH_2CH}$. Passing over this observation for the moment, we can write for the double bond

$$
\mathbf{U}_{CH=CH} = \begin{array}{c} \\ c \\ s^+ \\ s^- \end{array} \begin{array}{c} t \\ \begin{bmatrix} 1 \\ 1 \\ 1 \end{bmatrix} \end{array}
\qquad \text{(IX-4)}
$$

and from Eq. (IX-1)–(IX-4) we obtain the scalar, z_{trans}, as

$$
z_{trans} \equiv \mathbf{U}_{CHCH_2}\mathbf{U}_{CH_2CH_2}\mathbf{U}_{CH_2CH}\mathbf{U}_{CH=CH} = (2 + \rho)^2 + 2(2 + \rho\beta)^2 \quad \text{(IX-5)}
$$

The anticipated simplification of the expression for Z has been achieved. For a very long poly(1,4-*trans*-butadiene) chain, the contribution from all of the chain except the end groups is contained in z_{trans}^x, where x denotes the number of internal 1,4-*trans*-butadiene units. There is a corresponding simplification in the computation of $\langle r^2 \rangle_0$. Following Eq. (IX-5), we can define

$$
\mathscr{G}_{trans} \equiv \mathscr{G}_{CHCH_2}\mathscr{G}_{CH_2CH_2}\mathscr{G}_{CH_2CH}\mathscr{G}_{CH=CH}
\qquad \text{(IX-6)}
$$

where every matrix in $\mathscr{G}_{trans}$ contains five times as many rows, and five times as many columns, as its counterpart $\mathbf{U}$ in z_{trans}, because each $\mathbf{U}$ has been expanded by application of the 5×5 matrix $\mathbf{G}_\eta$. The bulk of the effort in the evaluation of $\langle r^2 \rangle_0$ for a very long poly(1,4-*trans*-butadiene) chain, with x internal units unaffected by chain ends, is achieved by computation of $\mathscr{G}_{trans}^x$, the dimensions of which are only 5×5.

Special Weighting for a Specific Three-Bond Sequence. In Eq. (IX-5), each triplet that has a $g^{\pm}$ placement in the middle with at least one flanking c placement (and hence at least one factor of ρ) must include β in its statistical weight. But the appearance of β in the factor $(2 + \rho\beta)^2$ means some of the triplets will be weighted by β^2. Those triplets are the ones also weighted by ρ^2, that is, the triplets with the sequence $cg^{\pm}c$. How important is the weighting with β^2, if the energetics actually suggest β instead? The difference is rather unimportant, because with $E_\beta \sim 8$ kJ mol^{-1}, we are concerned here with the details of the weighting of a rather improbable conformation.[4] Nevertheless, it is instructive to investigate how Z should be formulated if the weight for $cg^{\pm}c$

is to contain a factor of β instead of a factor of β^2. Here the objective is to modify the weight for a specific triplet, $cg^{\pm}c$, while leaving the weights for all other sequences unchanged. The modification requires a matrix formulation that allows assignment of specific elements to specific combinations of rotational isomeric states that span three successive bonds. This ability is required for the rigorous introduction of third-order interactions. The approach is analogous to the one adopted in Chapter VII for the incorporation of the third-order interaction in the $g^{\pm}g^{\mp}g^{\pm}$ states into Z for polyethylene, but differs in that the statistical weight matrix for polyethylene was square, but in the present case the statistical weight matrices will be rectangular.

The changes required are achieved by increases in the dimensions for $U_{CH_2CH_2}$ and U_{CH_2CH}.[5] The number of rows in the statistical weight matrix for the CH_2—CH_2 bond that follows CH—CH_2 must remain at three, because it is defined by the number of columns in U_{CHCH_2}. However, in order to accumulate and retain information that will be employed later for examination of specific three-bond sequences, the number of columns must be increased from three to nine. The increase permits assignment of a column to every combination of rotational isomeric states at both bonds in the sequence CH—CH_2—CH_2:

$$U_{CH_2CH_2} = \begin{array}{c} \\ c \\ s^+ \\ s^- \end{array} \begin{array}{ccccccccc} ct & s^+t & s^-t & cg^+ & s^+g^+ & s^-g^+ & cg^- & s^+g^- & s^-g^- \\ \left[\begin{array}{ccccccccc} 1 & 0 & 0 & 1 & 0 & 0 & 1 & 0 & 0 \\ 0 & 1 & 0 & 0 & 1 & 0 & 0 & 1 & 0 \\ 0 & 0 & 1 & 0 & 0 & 1 & 0 & 0 & 1 \end{array} \right] \end{array}$$

$$(IX\text{-}7)$$

Here the first three columns are for the CH_2—CH_2 bond in the t state, the next three columns have it in the g^+ state, and the last three columns have it in the g^- state. With each set of three columns, the preceding CH—CH_2 bond can be c, s^+, or s^-. Exactly two-thirds of the elements in this matrix are null, because the matrix must reject contributions to Z from elements in which the state of the preceding CH—CH_2 bond (as defined by the index for the row) contradicts the state for that very same bond (as defined by the first of the two indices for each column). The nonzero elements are those expected for the first-order statistical weights at a CH_2—CH_2 bond in poly(1,4-*trans*-butadiene).[6] Introduction of β is deferred until we can examine every three-bond sequence, one at a time.

The third-order interactions are incorporated in the statistical weight matrix for the next bond in the sequence. This matrix must have nine rows, indexed in the same manner as the columns for $U_{CH_2CH_2}$. There are three columns, indexed by the c, s^+, s^- at the CH_2—CH bond:[7]

---

[6]Therefore $U_{CH_2CH_2}$ can be written more concisely as $[1 \ 1 \ 1] \otimes I_3$.

[7]In Eq. (IX-8), the c, s^+, s^- in the indexing of the *rows* refer to the CH—CH_2 bond *preceding* the CH_2—CH_2 bond, but the c, s^+, s^- indexing the *columns* refer to the CH_2—CH bond *following* the CH_2—CH_2 bond.

$$
\mathbf{U}_{CH_2CH} =
\begin{array}{c}
 \\
ct \\
s^+t \\
s^-t \\
cg^+ \\
s^+g^+ \\
s^-g^+ \\
cg^- \\
s^+g^- \\
s^-g^-
\end{array}
\begin{array}{ccc}
c & s^+ & s^- \\
\left[\begin{array}{ccc}
\rho & 1 & 1 \\
\rho & 1 & 1 \\
\rho & 1 & 1 \\
\rho\beta & \beta & \beta \\
\rho\beta & 1 & 1 \\
\rho\beta & 1 & 1 \\
\rho\beta & \beta & \beta \\
\rho\beta & 1 & 1 \\
\rho\beta & 1 & 1
\end{array}\right]
\end{array}
\qquad \text{(IX-8)}
$$

The ρ in every element in the first column in Eq. (IX-8) is the same statistical weight that appeared in the first element of the row in Eq. (IX-1). It weights the c states relative to the s states at a CH—CH_2 (or CH_2—CH) bond. Utilization of the matrices in Eqs. (IX-1), (IX-7), (IX-8), and (IX-4) in the expression for z_{trans} yields

$$
z_{trans} \equiv \mathbf{U}_{CHCH_2}\mathbf{U}_{CH_2CH_2}\mathbf{U}_{CH_2CH}\mathbf{U}_{CH=CH} = (2 + \rho)^2 + 2(4 + 4\rho\beta + \rho^2\beta)
$$

$$
\text{(IX-9)}
$$

where every conformation weighted by ρ also bears a factor of β, but no sequence is weighted by β^2. The $cg^{\pm}c$ sequences are now weighted by $\rho^2\beta$, whereas in Eq. (IX-5) they were weighted by $\rho^2\beta^2$. The computed C_∞ is 6.2 at 50°C,[4] which compares well with an experimental result of 5.8 ± 0.2.[8] However, nearly the same result is obtained with the simpler formulation that uses no statistical weight matrix larger than 3×3. With physically realistic assignments for ρ and β, there is little difference in the expressions for z_{trans} in Eq. (IX-5) and Eq. (IX-9).

The procedure described here for introduction of longer-range interactions can be generalized to the incorporation of fourth-, fifth-, and higher-order interactions, when the need arises. The statistical weight matrices are expanded, and the rows and columns are indexed by states at sequences of bonds, rather than states at one bond only. The expansion in the dimensions of the matrices is accompanied by the introduction of many null elements, as illustrated in Eq. (IX-7). Efficient evaluation of the products of the expanded matrices utilizes algorithms that take advantage of the fact that the large matrices are sparse. These algorithms do not waste time evaluating those portions of the matrix products that involve use of null elements. Efficient computational algorithms include information that identifies the position and composition of all the nonzero elements in every **U**. With this strategy, computations can be performed

[8]Mark, J. E. *J. Am. Chem. Soc.* **1967**, 89, 6829.

with U that would have dimensions as large as $12{,}402 \times 12{,}402$, if they were to be written out in full.[9] In a few systems, including one described in the next chapter, the higher-order interactions can play a dominant role, in contrast to the comparatively minor role of the exact details of the third-order interaction in poly(1,4-*trans*-butadiene).

B. Poly(1,4-*cis*-butadiene)

When the double bond in polybutadiene is in the *cis* conformation, there are very severe steric repulsions if either of the flanking CH—CH$_2$ bonds also adopts a c state. For this reason the c state is suppressed at the CH—CH$_2$ bonds in *poly(1,4-cis-butadiene)*, but the two s states survive. The energy barrier between the two s states is lower than was the case in poly(1,4-*trans*-butadiene), and hence a state is assigned at t.[10] This state, however, makes a minor contribution to Z. The statistical weight matrix for a CH—CH$_2$ bond following a *cis* double bond becomes[5]

$$\mathbf{U}_{\text{CHCH}_2} = c \begin{array}{ccc} t & s^+ & s^- \\ \left[\zeta \right. & 1 & \left. 1 \right] \end{array} \tag{IX-10}$$

with $E_\zeta = 2.9 \text{ kJ mol}^{-1}$ and a preexponential factor of 0.3.[4] Confinement of both CH—CH$_2$ bonds flanking the CH$_2$—CH$_2$ bond to torsion angles within $\pm 60°$ of 180° eliminates the higher-order interactions that were present in $\mathbf{U}_{\text{CH}_2\text{CH}_2}$ for poly(1,4-*trans*-butadiene). The statistical weight matrix for this bond takes the simpler form[4]

$$\mathbf{U}_{\text{CH}_2\text{CH}_2} = \begin{bmatrix} 1 & 1 & 1 \\ 1 & 1 & 1 \\ 1 & 1 & 1 \end{bmatrix} \tag{IX-11}$$

It follows that the next matrix in the serial product is

$$\mathbf{U}_{\text{CH}_2\text{CH}} = \begin{array}{c} t \\ g^+ \\ g^- \end{array} \begin{array}{ccc} t & s^+ & s^- \\ \begin{bmatrix} \zeta & 1 & 1 \\ \zeta & 1 & 1 \\ \zeta & 1 & 1 \end{bmatrix} \end{array} \tag{IX-12}$$

and the serial product of the statistical weight matrices for the four bonds starting with CH—CH$_2$ and ending with the *cis* double bond is

[9]Mattice, W. L. *Biopolymers* **1985**, *24*, 2231.
[10]Mark, J. E. *J. Am. Chem. Soc.* **1966**, *88*, 4354.

$$z_{cis} \equiv \mathbf{U}_{CHCH_2}\mathbf{U}_{CH_2CH_2}\mathbf{U}_{CH_2CH}\mathbf{U}_{CH=CH} = 3(2 + \zeta)^2 \qquad \text{(IX-13)}$$

This model yields C_∞ of 5.08 at 50°C,[5] in good agreement with experimental results in the range 4.9 ± 0.2.[11] Poly(1,4-*cis*-butadiene) has a smaller C_∞ than poly(1,4-*trans*-butadiene).

The expressions for Z for the poly(1,4-butadiene) have been generalized to linear chains with double bonds separated by larger numbers of single bonds, that is, to chains with repeating units of $(CH_2)_yCH=CH$, $y > 2$.[4]

C. Poly(1,4-*trans*-isoprene)

The conformational partition function for *poly(1,4-trans-isoprene)* requires modification of two of the four statistical weight matrices used for poly(1,4-*trans*-butadiene), and a shift in the positions of the RIS at one of the bonds.[12]

If the direction for indexing the bonds in the chain is chosen so that the methyl group is bonded to the first atom in the double bond, as depicted in Fig. IX-3, the c state at the following CH—CH$_2$ bond is suppressed for the same reason as in poly(1,4-*cis*-butadiene). The preferred states at this bond will be $s^\pm$, with a minor contribution from torsion angles near 180°. Abe and Flory adopted the statistical weight matrix in Eq. (IX-10) for $\mathbf{U}_{CHCH_2}$.[12] Changes occur in $\mathbf{U}_{CH_2C(CH_3)}$, which now incorporates a second-order interaction with statistical weight ω. In the approximation of using first- and second-order interactions only, it can be written as

$$\mathbf{U}_{CH_2C(CH_3)} = \begin{bmatrix} \rho & 1 & 1 \\ \rho\beta & \omega & 1 \\ \rho\beta & 1 & \omega \end{bmatrix} \qquad \text{(IX-14)}$$

with $E_\rho = -1.3$ kJ mol^{-1} and $E_\omega = 5$ kJ mol^{-1}.[12] These matrices yield

$$z_{trans} = (2 + \zeta)[\rho + 2(2 + \rho\beta + \omega)] \qquad \text{(IX-15)}$$

The computed C_∞ at 50°C is ~7,[12] in reasonable agreement with experimental results of 6.6[13] or 7.4.[8,14] The change in sign of E_ρ is primarily responsible

[11] Moraglio, G. *Eur. Poly. J.* **1965**, *1*, 103. Abe, M.; Fujita, H. *J. Phys. Chem.* **1965**, *69*, 3263.
[12] Abe, Y.; Flory, P. J. *Macromolecules* **1971**, *4*, 230.
[13] Poddubnyi, Ya.; Erenburg, E. G.; Eryomina, M. A. *Vysokomol. Soedin., Ser. A* **1968**, *10*, 1381.
[14] Wagner, H. L.; Flory, P. J. *J. Am. Chem. Soc.* **1952**, *74*, 195.

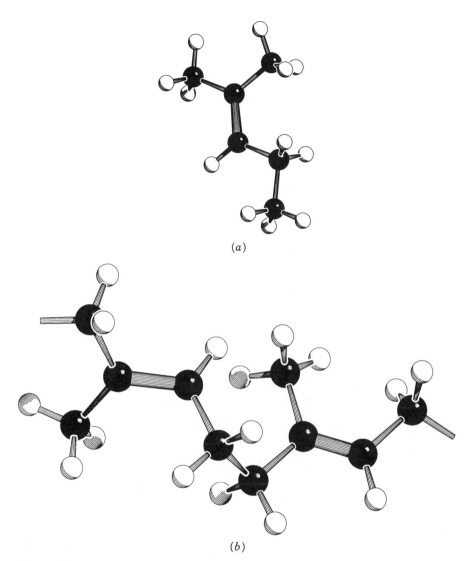

Figure IX-3. Conformations of fragments of poly(1,4-*trans*-isoprene) that generate (*a*) the first-order interaction denoted by ζ and (*b*) the second-order interaction denoted by ω.

for poly(1,4-*trans*-isoprene) having a larger C_∞ than poly(1,4-*trans*-butadiene). The introduction of ω has a smaller effect on the unperturbed dimensions.

D. Poly(1,4-*cis*-isoprene)

For *poly(1,4-cis-isoprene)*, U_{CHCH_2} and $U_{CH_2CH_2}$ are given by Eq. (IX-10) and (IX-11), respectively.[12] Neither the *t* nor *c* state is tenable at the CH_2—$C(CH_3)$

bond, and hence the third matrix in z_{cis} takes the simple form

$$
\mathbf{U}_{CH_2C(CH_3)} = \begin{array}{c} \\ t \\ g^+ \\ g^- \end{array} \begin{array}{cc} s^+ & s^- \\ \begin{bmatrix} 1 & 1 \\ \omega & 1 \\ 1 & \omega \end{bmatrix} \end{array}
\tag{IX-16}
$$

The statistical weight matrix for the double bond is simply

$$
\mathbf{U}_{C(CH_3)=CH} = \begin{bmatrix} 1 \\ 1 \end{bmatrix}
\tag{IX-17}
$$

which yields

$$
z_{cis} = 2(2 + \zeta)(2 + \omega)
\tag{IX-18}
$$

Calculation gives $C_\infty = 4.55$,[12] in reasonable agreement with experimental results of 4.7.[8,10]

2. PARTIAL DOUBLE BONDS

The restriction on rotation at double bonds, described in the preceding section in conjunction with the CH=CH bonds in polybutadiene and polyisoprene, can also occur at bonds with *partial* double-bond character. Frequently these bonds are written with the single dash that implies a single bond, as in the representation of the ester bond in RO—(C'=O)R. This notation seems to imply the double-bond character is confined to the bond between the carbonyl carbon, denoted by C', and the carbonyl oxygen. This interpretation is erroneous. An appreciable portion of the π-electron density is distributed over the C'—O bond, as shown by the fact that this bond is 10 pm shorter than the C—O bond in typical ethers.[15] The partial double-bond character at the C'—O bond causes a pronounced preference for a planar conformation of the five atoms, C–(C'=O)—O–C, in the ester unit. Furthermore, the planar *trans* conformation is strongly preferred to the planar *cis* conformation.[16] In the approximation that all ester groups are confined to the planar *trans* conformation, a polyester with the repeating sequence $-C^\alpha R_1 R_2-(C'=O)-O-$ can be represented by a string of *virtual bonds* connecting successive main-chain carbon atoms denoted by C^α, as depicted in Fig.

[15] Bowen, H. J. M.; Sutton, L. E. *Tables of Interatomic Distances and Configurations in Molecules and Ions*, The Chemical Society, London, **1958**.
[16] O'Gorman, J. M.; Shand, W., Jr.; Schomaker, V. *J. Am. Chem. Soc.* **1950**, *72*, 4222. Curl, R. F., Jr. *J. Chem. Phys.* **1959**, *30*, 1529. Miyazawa, T. *Bull. Chem. Soc. Japan* **1961**, *34*, 691.

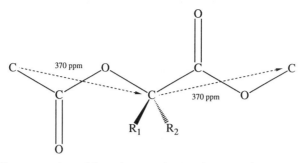

Figure IX-4. Representation of the polyester of HO–C$^\alpha$R$_1$R$_2$–C$'$OOH by virtual bonds.

IX-4. With standard bond lengths and bond angles, the virtual bonds have a length of 370 pm.[17]

For a similar reason, partial double-bond character is found at the amide bond. If an amide is derived from a primary amine, the six atoms in the C$^\alpha$–(C$'$=O)–NH–C$^\alpha$ unit have a pronounced preference for the planar conformation that produces a *trans* conformation for the C$^\alpha$–C$'$–N–C$^\alpha$ sequence.[18] The simplest such polyamide, polyglycine, can be represented by a string of virtual bonds with a length of 380 pm when bond lengths and bond angles are assigned standard values, as depicted in Fig. IX-5.[19] The energetic discrimination between *trans* and *cis* conformations at the amide bond is blurred if the amide is derived from a secondary amine. In this case there may be significant populations of both *cis* and *trans* conformations at the amide bond.

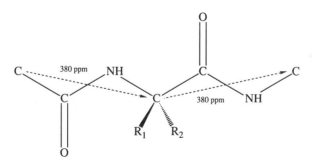

Figure IX-5. Representation of the polyamide of H$_2$N–CR$_1$R$_2$–C$'$OOH by virtual bonds.

[17] Brant, D. A.; Tonelli, A. E.; Flory, P. J. *Macromolecules* **1969**, *2*, 228.
[18] Kurland, R. J.; Wilson, E. B. *J. Chem. Phys.* **1957**, *27*, 585. LaPlanche, L. A.; Rogers, M. T. *J. Am. Chem. Soc.* **1964**, *86*, 337.
[19] Brant, D. A.; Flory, P. J. *J. Am. Chem. Soc.* **1965**, *87*, 2791.

The simplifications in the RIS treatment that arise from the use of virtual bonds are well illustrated by the simple homopolyesters and polyamides for which the repeating sequences are $-C^{\alpha}R_1R_2-O-C'O-$ or $-C^{\alpha}R_1R_2-NH-C'O-$, as shown in the sections that follow.

A. Esters

The poly(lactic acids) introduce the concepts and notation that is also adopted for the poly(α-amino acids), which are addressed in the following section.

Poly(L-lactic acid). The simplest polyester for which a detailed RIS model is available is *poly*(L-*lactic acid*), which has the structure depicted in Fig. IX-4 with $R_1 = H$ and $R_2 = CH_3$.[17] The conventional notation denotes the torsion angles at the $O-C^{\alpha}$ and $C^{\alpha}-C'$ bonds in unit i as ϕ_i and ψ_i, respectively. These two rotations are *interdependent*. As can be seen from Fig. IX-4, the orientation of virtual bond $i+1$ with respect to virtual bond i is determined by ϕ_i, ψ_i when the bond lengths and bond angles are constant, and the ester unit is in the *trans* conformation.

The spacing of consecutive lactyl units by the *trans* ester unit causes the ϕ_i, ψ_i pair to be independent of the ϕ_{i-1}, ψ_{i-1} pair, in good approximation. The rotations that determine the orientation of virtual bond $i+1$ with respect to virtual bond i are *independent* of the rotations that determine the orientation of virtual bond i with respect to virtual bond $i-1$. When viewed as a chain composed of virtual bonds, poly(L-lactic acid) can be treated as a chain of *independent* bonds, with all the mathematical simplifications that attend the independence of bonds.

In order to take full advantage of the simplifications that follow from a chain with independent bonds, we need the proper formulation for the matrix that transforms a vector from its representation in the local coordinate system of virtual bond $i+1$ into its representation in the local coordinate system of virtual bond i. This transformation matrix must allow for the facts that the angle between two successive virtual bonds is variable, as can be seen by imagining a rotation of $180°$ about either ϕ_i or ψ_i in Fig. IX-4. The angle between these two virtual bonds is a maximum when $\phi_i = \psi_i = 180°$, and it is a minimum when $\phi_i = \psi_i = 0°$. The range is from $\theta^{\alpha} + \eta + \xi$ to $\theta^{\alpha} - \eta - \xi$, where θ^{α} is the angle between the real $O-C^{\alpha}$ and $C^{\alpha}-C'$ bonds, η is the angle between virtual bond $i+1$ and the $C^{\alpha}-C'$ bond, and ξ is the angle between virtual bond i and the $O-C^{\alpha}$ bond. Since neither η nor ξ is zero, neither of the two rotations ϕ_i nor ψ_i occurs about a real bond parallel with a virtual bond. The desired transformation matrix can be formulated by using three successive transformations, each of which has a well-defined counterpart of the "bond angle" and torsion angle used in the transformation matrix defined in Eq. (VI-5):

$$\mathbf{T}_i = \mathbf{T}_3 \mathbf{T}_{2,\phi_i} \mathbf{T}_{1,\psi_i} \qquad \text{(IX-19)}$$

Starting from the extreme right of Eq. (IX-19), $\mathbf{T}_{1,\psi_i}$ transforms a vector from its representation in the local coordinate system for virtual bond $i + 1$ into the coordinate system for real bond C^α—C' in lactyl unit i. The "bond angle" input to this transformation depends on the angle between virtual bond $i + 1$ and the C^α—C' bond. This angle, denoted by η is a constant if the ester unit is confined to the planar *trans* conformation, with constant bond lengths and bond angles. With standard geometry, $\eta = 18.9°$.[17] The torsion angle in $\mathbf{T}_{1,\psi_i}$ is ψ_i. These two angles enter the expression in Eq. (VI-5) as though the bond angle and torsion angle were $\eta + \pi$ and $\psi + \pi$, respectively.

$$\mathbf{T}_{1,\psi_i} = \begin{bmatrix} \cos \eta & -\sin \eta & 0 \\ -\sin \eta \cos \psi_i & -\cos \eta \cos \psi_i & \sin \psi_i \\ -\sin \eta \sin \psi_i & -\cos \eta \sin \psi_i & -\cos \psi_i \end{bmatrix} \qquad \text{(IX-20)}$$

The next transformation, $\mathbf{T}_{2,\phi_i}$, takes the vector from its expression in the coordinate system for the C^α—C' bond into its representation in the coordinate system for the O—C^α bond. The bond angle, denoted by θ^α, is $\theta_{OC^\alpha C'}$, and the torsion angle is ϕ_i. From Eq. (VI-5), we obtain

$$\mathbf{T}_{2,\phi_i} = \begin{bmatrix} -\cos \theta^\alpha & \sin \theta^\alpha & 0 \\ -\sin \theta^\alpha \cos \phi_i & -\cos \theta^\alpha \cos \phi_i & -\sin \phi_i \\ -\sin \theta^\alpha \sin \phi_i & -\cos \theta^\alpha \sin \phi_i & \cos \phi_i \end{bmatrix} \qquad \text{(IX-21)}$$

The final transformation is from the coordinate system of real bond O—C^α to the coordinate system of virtual bond i. This transformation has a "bond angle" that depends on the geometry of the ester unit, in perfect analogy to η. This angle, denoted by ξ, is $19.9°$ for an ester unit of standard geometry.[17] It enters Eq. (VI-5) as though the bond angle were $\pi - \xi$. There is no change in the z coordinate.

$$\mathbf{T}_3 = \begin{bmatrix} \cos \xi & \sin \xi & 0 \\ -\sin \xi & \cos \xi & 0 \\ 0 & 0 & 1 \end{bmatrix} \qquad \text{(IX-22)}$$

It is the average of the transformation matrix that plays the crucial role in the evaluation of $\langle \mathbf{r} \rangle_0$ via Eq. (VI-39) and (VI-40) and $\langle r^2 \rangle_0$ via Eq. (VI-56) and (VI-57) [or, when applicable, Eq. (VI-64)]. Since η, θ^α, and ξ are constants determined by the bond lengths and bond angles and the confinement of the ester unit to the planar *trans* conformation, the two variables that determine $\mathbf{T}_i$ are ϕ_i and ψ_i. The *averaged transformation matrix* we seek is hence

$$\langle \mathbf{T}_i \rangle = \mathbf{T}_3 \langle \mathbf{T}_{2,\phi_i} \mathbf{T}_{1,\psi_i} \rangle \qquad \text{(IX-23)}$$

where $\mathbf{T}_3$ can be written outside the angle brackets because it depends on neither ϕ_i nor ψ_i. The information required for constructing the average is usually presented as a *conformational energy map*, depicting the conformational energy for a suitable fragment as a function of ϕ_i and ψ_i. For example, the conformational energy surface computed for $C^\alpha C'OO—C^\alpha HCH_3—C'OOC^\alpha$ by Brant et al.[17] is depicted in Fig. IX-6. Their representation, based on calculations for 36^2 conformations generated with $10°$ intervals for ϕ_i and ψ_i, identifies the conformation of lowest energy (denoted by $\times$ in Fig. IX-6) and draws contours at energies of 1, 2, 3, 4, and 5 kcal mol^{-1} above the minimum. Regions with energies higher than 5 kcal mol^{-1} above the minimum can be ignored because they have negligible population at ordinary temperatures. The averaging required for $\langle \mathbf{T}_i \rangle$ is achieved by a discrete summation using specified intervals for $\Delta\phi_i$ and $\Delta\psi_i$:

$$\langle \mathbf{T}_{2,\phi_i}\mathbf{T}_{1,\psi_i} \rangle = \frac{\sum_{j=1}^{2\pi/\Delta\phi}\sum_{l=1}^{2\pi/\Delta\psi} \mathbf{T}_{2,j\Delta\phi}\mathbf{T}_{1,l\Delta\psi} \exp\left[-E(\phi_i, \psi_i)/kT\right]}{\sum_{j=1}^{2\pi/\Delta\phi}\sum_{l=1}^{2\pi/\Delta\psi} \exp\left[-E(\phi_i, \psi_i)/kT\right]} \tag{IX-24}$$

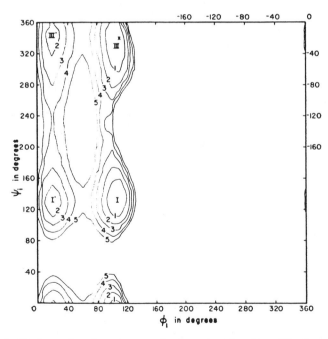

Figure IX-6. Conformational energy map for the L-lactyl residue. The figure was constructed using *trans* states as the zeros for ϕ_i and ψ_i. Energies for the contour lines are in kcal mol^{-1}. Reprinted with permission from Brant, D. A.; Tonelli, A. E.; Flory, P. J. *Macromolecules* **1969**, *2*, 228. Copyright 1969 American Chemical Society.

Using the conformational energy surface depicted in Fig. IX-6, and $\Delta\phi = \Delta\psi = 30°$, the averaging procedure leads ultimately to[17]

$$\langle \mathbf{T}_i \rangle_{\text{L-Lac}} = \begin{bmatrix} 0.332 & 0.167 & 0.017 \\ -0.273 & 0.085 & 0.257 \\ 0.778 & -0.355 & 0.082 \end{bmatrix}_i \qquad \text{(IX-25)}$$

The characteristic ratio for this chain is defined as

$$C \equiv \frac{\langle r^2 \rangle_0}{n_p l_p^2} \qquad \text{(IX-26)}$$

where l_p is the length of the virtual bond [370 pm for poly(L-lactic acid)], and n_p is the number of virtual bonds in the chain. For chains comprised of virtual bonds, this formulation of the denominator of the defining expression for C often replaces nl^2, which is the sum of the squares of the lengths of the *chemical* bonds in the main chain. When this $\langle \mathbf{T} \rangle$ is inserted in Eq. (VI-64), it yields $C_\infty = 2.02$,[17] in reasonable agreement with the experimental result of 2.13.[20]

The treatment of polyesters with larger numbers of methylene groups between the ester groups, specifically, repeating sequences of $(CH_2)_y COO$ with $y > 1$, is similar to the treatment of the polyisoprenes, in that Z can be formulated as a product of scalar factors, z, which span $y + 2$ bonds.[21] Polyesters containing rings will be addressed later in this chapter.

Poly(D-lactic acid). Replacement of every conformation in the ensemble by its mirror image cannot affect the value of C_∞. That change, however, converts every residue of L-lactic acid into a residue of D-lactic acid, thereby changing both the conformational energy map and $\langle \mathbf{T}_i \rangle$. The change in the conformational energy map is obtained by reflection through the point $\phi_i, \psi_i = 0°, 0°$ (or, equivalently, through the point $\phi_i, \psi_i = 180°, 180°$). The consequences for $\langle \mathbf{T} \rangle$ are the change in the sign of every element in the third column, followed by the change in sign of every element in the third row:

$$\langle \mathbf{T}_i \rangle_{\text{D-Lac}} = \text{diag}\,(1, 1, -1)\,\langle \mathbf{T}_i \rangle_{\text{L-Lac}}\,\text{diag}\,(1, 1, -1) \qquad \text{(IX-27)}$$

The value of $\langle r^2 \rangle_0$, and hence C_∞, is unaffected by this transformation in achiral media:

$$\langle r^2 \rangle_{0,\text{D-Lac}} = \langle r^2 \rangle_{0,\text{L-Lac}} \qquad \text{(IX-28)}$$

[20] Tonelli, A. E.; Flory, P. J. *Macromolecules* **1969**, *2*, 225.

[21] Flory, P. J.; Williams, A. D. *Polym. Sci., Part A-2*, **1967**, *5*, 399.

The sign of the third element in $\langle \mathbf{r} \rangle_0$ is altered by the change in stereochemistry. The average end-to-end vectors for the two polymers point in different directions from the plane established by the first two virtual bonds:

$$\langle \mathbf{r} \rangle_{0,\text{D-Lac}} = \text{diag}\,(1, 1, -1)\,\langle \mathbf{r} \rangle_{0,\text{L-Lac}} \qquad \text{(IX-29)}$$

B. Amides

The poly(α-amino acids) are analogous in some respects to poly(L-lactic acid), as is apparent by comparison of Figs. IX-4 and IX-5. Both chains are conveniently represented by virtual bonds that reach from one C^α to the next. With standard geometry, the virtual bond is slightly longer in the polyamides (380 pm) than in the polyesters (370 pm). The presence of the amide hydrogen in the poly(α-amino acids) opens the possibility of the intramolecular formation of hydrogen bonds, which was not an issue with the polyesters. These hydrogen bonds play a minor role in the unperturbed random coils addressed in this section, but they play a very important role in the ordered structures formed by poly(amino acids). Their influence on the intramolecular formation of helices and antiparallel sheets by homopolypeptides will be addressed in the next chapter. We will omit entirely the important subject of the ability of certain copolypeptides of well-defined sequence to fold into specific globular structures. That topic is of enormous importance in biology, but its treatment requires different methods than those that are the focus of this book.[22]

The usual approach to the unperturbed dimensions of poly(amino acids) is similar to the one described above for poly(L-lactic acid). When applied to poly(amino acids), ϕ_i and ψ_i denote the torsions at the N—C^α and C^α—C' bonds, respectively. The symbol ω is used for the torsion at the C'—N bond, but frequently that degree of freedom is constrained at $\omega = 180°$, because of the strong preference of most amides for the planar *trans* conformation. A conformational energy map, depicting the conformational energy as a function of ϕ_i and ψ_i, is used for evaluation of $\langle \mathbf{T}_i \rangle$ via Eqs. (IX-20)–(IX-24), with the only important changes (other than the conformational energy surface itself) residing in the values of η and ξ. The standard values for bond angles and bond lengths in the amide unit yield $\eta = 22.2°$ and $\xi = 13.2°$ for simple poly(amino acids).[19]

The details of the conformational energy map depend on the force field used in the computation. With the great interest in the theoretical study of the folding of globular proteins, several different force fields have been developed for

---

[22]The strength of RIS theory lies in providing a connection between an atom-based description of the short-range interactions in a polymer and the conformation-dependent properties of an unperturbed ensemble in which there are important contributions by an enormous number of different conformations. The statistical weight for the most probable conformation typically contributes much less than 1% of the value of Z. In contrast, the ordered structures formed by globular proteins require methods that seek out a single conformation with a statistical weight that may contribute well over 90% of the value of Z for a copolypeptide of definite sequence, under conditions where long-range interactions play a decisive role.

this class of macromolecules. They differ, sometimes in important ways, in the conformational energy surfaces predicted. Rotational isomeric state theory provides a method of determining whether the differences in the appearance of the conformational energy surfaces are important insofar as the prediction of conformation-dependent physical properties of unperturbed chains is concerned. When experimental measurements for those conformation-dependent properties are available, RIS theory may provide a means for discrimination between the various force fields. This use of RIS theory, and potential limitations on its use, are illustrated in the examples that follow.

Polyglycine. *Polyglycine* is the simplest poly(α-amino acid), and the most important one that does not contain an asymmetric center. The repeating sequence is $-NH-C^{\alpha}H_2-C'O-$. The absence of an asymmetric center implies that the operation used to convert a conformational energy surface for an L-residue into the surface for a D-residue cannot alter the conformational energy surface of the glycyl residue. One consequence of this implication is that the conformational energy surface must be unaffected by inversion through ϕ_i, $\psi_i = 0°, 0°$ or $180°, 180°$. That feature is apparent in the conformational energy surface calculated for $CC'ONH-C^{\alpha}H_2-C'ONHC$ by Brant et al.[23] and depicted in Fig. IX-7. Another consequence, which follows from application of Eq. (IX-27) to the glycyl residue, is that four of the elements in $\langle \mathbf{T}_i \rangle$—namely, the 1,3, 2,3, 3,1, and 3,2 elements—must be null:

$$\langle \mathbf{T}_i \rangle_{Gly} = diag\,(1, 1, -1)\,\langle \mathbf{T}_i \rangle_{Gly}\,diag\,(1, 1, -1) \qquad \text{(IX-30)}$$

That feature is apparent in the averaged transformation matrix extracted from the conformational energy surface depicted in Fig. IX-7.

$$\langle \mathbf{T}_i \rangle_{Gly} = \begin{bmatrix} 0.36 & -0.077 & 0 \\ -0.092 & -0.37 & 0 \\ 0 & 0 & -0.12 \end{bmatrix}_i \qquad \text{(IX-31)}$$

This averaged transformation matrix, when used in Eq. (VI-64), yields $C_{\infty} = 2.15$.[23]

No experimental result is available for C_{∞} for polyglycine, because of the extremely low solubility of the polymer. However, even if an experiment were to produce $C_{\infty} \sim 2$, it would not be a sensitive test of the validity of the details of the conformational energy surface for the glycyl residue. This statement follows from the fact that C_{∞} for polyglycine is responding primarily to two features of the map: (1) broad regions of low energy occur in all four quadrants, and (2) the surface is unaffected by inversion through the origin. A perfectly flat conformational energy surface, corresponding to free rotation about the $N-C^{\alpha}$

[23] Brant, D. A.; Miller, W. G.; Flory, P. J. *J. Mol. Biol.* **1967**, *23*, 47.

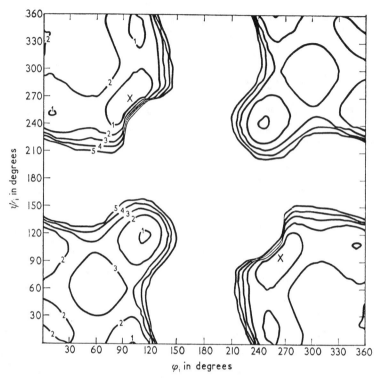

Figure IX-7. Conformational energy map for the glycyl residue. The axes were labeled using *trans*, rather than *cis*, placements as the zeros for ϕ_i and ψ_i. Energies for the contour lines are in kcal mol^{-1}. From Brant, D. A.; Miller, W. G.: Flory, P. J. *J. Mol. Biol.* **1967**, *23*, 47. Copyright © John Wiley & Sons 1967. Reprinted by permission of John Wiley & Sons, Inc.

and C^α—C' bonds, yields a value of C_∞ of 1.93, which is only 11% smaller than the result predicted from Fig. IX-7.

Poly(L-alanine) and Poly(D-alanine). The predicted values of C_∞ can become more sensitive to the details of the calculation, and hence to the force field employed, when the symmetry of the rotations is abolished by the chiral attachment of a side chain at C^α. The simplest example is provided by *poly*(L-*alanine*), where the side chain is a methyl group. The absence of the inversion center is readily apparent in the conformational energy surface reported by Brant et al.,[23] and depicted in Fig. IX-8. This conformational energy surface yields

$$\langle \mathbf{T}_i \rangle_{\text{L-Ala}} = \begin{bmatrix} 0.51 & 0.20 & 0.59 \\ -0.046 & -0.61 & 0.21 \\ 0.65 & -0.23 & -0.30 \end{bmatrix}_i \qquad \text{(IX-32)}$$

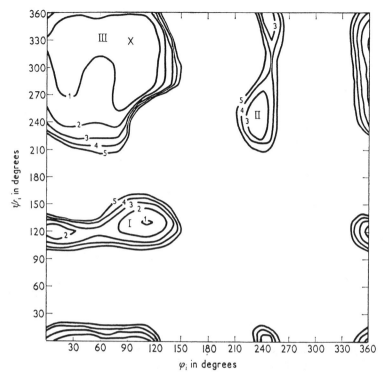

Figure IX-8. Conformational energy map for the L-alanyl residue. The axes were labeled using *trans*, rather than *cis*, placements as the zeros for ϕ_i and ψ_i. Energies for the contour lines are in kcal mol^{-1}. From Brant, D. A.; Miller, W. G.; Flory, P. J. *J. Mol. Biol.* **1967**, *23*, 47. Copyright © John Wiley & Sons 1967. Reprinted by permission of John Wiley & Sons, Inc.

with no null elements. The prediction for C_∞ is 9.44, which is larger by a factor of 4.3 than the result predicted for polyglycine from Fig. IX-7. No experimental result is available for C_∞ for poly(L-alanine) itself, because it is not soluble as a random coil in solvents in which the experiments can easily be performed. The predicted C_∞ seems reasonable because measured values for several other poly(α-amino acids) with –CH$_2$R side chains yield results in the range 9.8 ± 1.2.[24] These experiments were not obtained under Θ conditions, but were instead performed in good solvents. The measured dimensions were corrected for the expansion produced by excluded volume, using the second virial coefficient. Experiments cannot be performed in solvents of poor quality, approximating Θ solvents, because the polymers either are insoluble, or con-

[24]Brant, D. A.; Flory, P. J. *J. Am. Chem. Soc.* **1965**, *87*, 2788. Mattice, W. L.; Lo, J.-T. *Macromolecules* **1972**, *5*, 734. The side chains are –CH$_2$COOCH$_2$C$_6$H$_5$, –(CH$_2$)$_2$COOCH$_2$C$_6$H$_5$, –(CH$_2$)$_2$CO$_2^-$, –(CH$_2$)$_4$NH$_3^+$, and –(CH$_2$)$_2$CONH(CH$_2$)$_2$OH.

vert from the random coil to an ordered structure. Two of the order $\rightleftharpoons$ disorder transitions will be discussed in the next chapter.

The prediction of $C_\infty = 9.44$ for poly(L-alanine) *is* sensitive to the details of the conformational energy surface. If the surface were flat, meaning free rotation about N—C^α and C^α—C', C_∞ would be 1.93, as was the case for polyglycine in the approximation of free rotation. (Both polymers have the same bond lengths and bond angles in the main chain, and hence they would have exactly the same dimensions if rotation were free.) A variety of other conformational energy surfaces for the same or very similar structure, but computed with other force fields, predict values of C_∞ that vary over an order of magnitude, from ~2 to ~20. The comparison of these predictions with the expectation from experiment of $C_\infty = 9.8 \pm 1.2$ has been used as a basis for deciding which force fields most accurately represent the conformational energy surface accessible to a simple unperturbed poly(α-amino acid).[25] The predicted dimensions are strongly influenced by the relative weights for the large low-energy region in the upper left quadrant of the conformational energy map and the smaller low-energy region in the lower left quadrant. Electrostatic interaction of the strong dipoles in the two amide units favors the upper left over the lower left region, leading to larger values of C_∞.[23] Some force fields assign an attractive energy to an incipient bent hydrogen bond in the low-energy region in the lower left quadrant, stabilizing that region and leading to lower values of C_∞.[25] The manner in which these polar interactions appear in a force field can have important implications for C_∞ for homopoly(amino acids) with –CH_2R side chains in the L configuration.

Poly(L-proline). The computed values of C_∞ become even more sensitive to the details of the calculation of the conformational energy surface in the case of *poly*(L-*proline*) (Fig. IX-9). The side chain in this polymer completes a pyrrolidine ring, with the N—C^α bond being incorporated into the ring.[26] The requirement of ring closure produces severe restraints on the permissible range for ϕ_i, with the allowed region being very close to $\phi_i = -60°$. Some computations treat ϕ_i as fixed, allowing representation of the conformational energy surface as energy versus ψ_i. In other computations, a small range is

[25]Tonelli, A. E. *Macromolecules* **1971**, *4*, 618. Roterman, I. K.; Lambert, M. H.; Gibson, K. D.; Scheraga, H. A. *J. Biomol. Struct. Dyn.* **1989**, *7*, 421. Gibson, K. D.; Scheraga, H. A. *J. Biomol. Struct. Dyn.* **1991**, *8*, 1109.

[26]The absence of a hydrogen atom at the amide nitrogen atom blurs the energetic distinction between the *cis* and *trans* conformations of the C'—N bond. Both conformations can be observed in poly(L-proline). Severe steric interactions between neighboring residues cause the transition from one rotational isomer at the C'—N bonds to the other rotational isomer to be highly cooperative. Chains prefer conformations where the C'—N bonds are either all-*trans* or all-*cis*. The latter form is obtained only in very poor solvents, and hence there are experimental measurements for the unperturbed dimensions only for the form where all C'—N bonds are *trans*. For that reason the discussion will focus on poly(L-proline) with the *trans* conformation at the C'—N bonds.

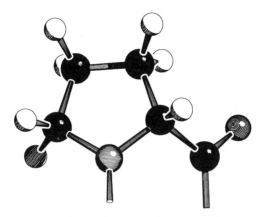

Figure IX-9. Segment of poly(L-proline).

allowed for ϕ_i, which requires an allowance for different puckering of the pyrrolidine ring.

In a qualitative sense, the conformational energy surfaces can be described by imaging a slice through the conformational energy map for the L-alanyl residue at the value of ϕ_i demanded by ring closure. All computations find that the minimum energy is obtained near the upper extreme of this slice, and the minimum is more constrained by repulsive interactions between successive pyrrolidine rings than would be expected from the vertical slice through the L-alanyl conformational energy map. The details in the magnitude of the additional confinement in this region, and in whether the region in the lower left quadrant can or cannot make a very minor contribution to Z, are responsible for the differences in the predictions for C_∞.

The conformational energy surfaces that have been published imply C_∞ with a very broad range, at least ~5–200.[27] Experiment yields $C_\infty = 19 \pm 1$ in several organic solvents, and a slightly lower value, $C_\infty \sim 14$, in water.[28] Poly(γ-hydroxy-L-proline) has a C_∞ about 15% larger than C_∞ of poly(L-proline).[29,30] These values are higher than the results obtained for typical flexible polymers, and verify the anticipated tendency for stiffness due to ring closure and the steric interaction of neighboring rings in poly(L-proline). Nevertheless, a much higher value would have been predicted from many of the conformational energy sur-

[27]Mandelkern, L.; Mattice, W. L. *Conformation of Biological Molecules and Polymers*, Israel Academy of Sciences and Humanities, Jerusalem, **1973**, p 121.

[28]Mattice, W. L.; Mandelkern, L. *J. Am. Chem. Soc.* **1971**, *93*, 1769.

[29]Clark, D. S.; Mattice, W. L. *Macromolecules* **1977**, *10*, 369.

[30]The hydroxyl group is placed in a position on the pyrrolidine ring where it does not interact directly with neighboring hydroxyprolyl units, but it may still influence the dimensions slightly by modification in the puckering of the ring. Changes in the conformation of the pyrrolidine ring modify the interactions between successive residues, and hence produce small, but potentially important, alterations in the conformational energy surface.

faces. It is difficult to account for values of C_∞ this low for poly(L-proline) without invoking fluctuations of ϕ_i within the small range that is consistent with closure of the pyrrolidine ring.[31]

For inherently stiff polymers, minor degrees of freedom, legitimately ignored in typical flexible polymers, must be allowed to exert their influence on the computed $\langle T_i \rangle$, and hence on the dimensions. Degrees of freedom inconsequential in the prediction of C_∞ for a very flexible polymer, such as polyglycine, may become of crucial importance in the consideration of a stiffer polymer, such as poly(L-proline).

Nylons. Insertion of additional methylene groups into the backbone of polyglycine produces the family of *nylons* that are polymers of $NH_2(CH_2)_x$-COOH, $x > 1$. The statistical weight matrix for polyethylene can be used for $x - 3$ of these bonds.[21] For example, *poly(6-aminocaproamide)* has a conformational partition function that contains a scalar given by[32]

$$z = \begin{bmatrix} 1 & 1 & 1 \end{bmatrix} U_a U_{PE} U_{PE} U_a U_a \begin{bmatrix} 1 \\ 1 \\ 1 \end{bmatrix} \qquad \text{(IX-33)}$$

where U_{PE} denotes the statistical weight matrix used for polyethylene, Eq. (IV-19), and

$$U_a = \begin{bmatrix} 1 & 1 & 1 \\ 1 & 1 & \omega_a \\ 1 & \omega_a & 1 \end{bmatrix} \qquad \text{(IX-34)}$$

In Eq. (IX-33), the row vector is the statistical weight matrix for the equally weighted t, g^+, and g^- states at the N—C bond, and the column vector is the statistical weight matrix for the C′—N bond. Bonds near the amide, and using U_a, are subject to first-order interactions that do not discriminate between the three states, and a second-order interaction with $E_{\omega_a} \sim 5.5$ kJ mol^{-1}.

C. The Elements of $\langle T \rangle^i$

Thus far Section IX.2 has described several chains for which a model based on independent virtual bonds has proved useful. The examples discussed were poly(L-lactic acid), polyglycine, poly(L-alanine), and poly(L-proline). An equivalent treatment has been used successfully for several polysaccharides, which will be discussed later in this chapter. For such chains, the average of the end-

[31] Mattice, W. L.; Nishikawa, K.; Ooi, T. *Macromolecules* **1973**, *6*, 443. Ooi, T.; Clark, D. S.; Mattice, W. L. *Macromolecules* **1974**, *7*, 337.
[32] Mutter, M.; Suter, U. W.; Flory, P. J. *J. Am. Chem. Soc.* **1976**, *98*, 5745.

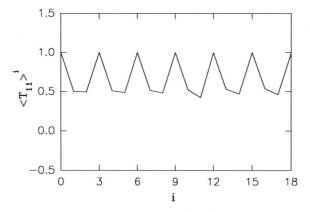

Figure IX-10. Behavior of $T^i_{h,11}$ when the chain has the structure of poly(L-proline) Form II ($\phi = -76.3°$, $\psi = 145.1°$).

to-end vector can be written [see Eq. (VI-39)] as

$$\langle \mathbf{r} \rangle_0 = \sum_{i=0}^{n_p-1} \langle \mathbf{T} \rangle^i \mathbf{l}_p \qquad (IX\text{-}35)$$

A typical term, $\langle \mathbf{T} \rangle^i \mathbf{l}_p$, is a vector that contains the components contributed by virtual bond $i + 1$ to $\langle \mathbf{r} \rangle_0$. The size of its three components depends in a trivial fashion on the length of $\mathbf{l}_p$, and in a potentially more interesting fashion on the elements in the first column of $\langle \mathbf{T} \rangle^i$. The influence of l_p can be eliminated by replacement of $\mathbf{l}_p$ with a unit vector pointing along the x axis of the local coordinate system for virtual bond $i + 1$. The average components of this unit vector along the x, y, and z axes of the local coordinate system for bond 1 are $\langle \mathbf{T} \rangle^i_{11}$, $\langle \mathbf{T} \rangle^i_{21}$, and $\langle \mathbf{T} \rangle^i_{31}$, respectively. More generally, $\langle \mathbf{T} \rangle^i_{jk}$ is the average projection of a unit vector along the kth axis (x axis, $k = 1$; y axis, $k = 2$; z axis, $k = 3$) of local coordinate system $i + 1$ onto the jth axis (x axis, $j = 1$; y axis, $j = 2$; z axis, $j = 3$) for the local coordinate system of bond 1.

The behavior of $\langle \mathbf{T} \rangle^i_{jk}$, as a function of i, is of interest in poly(α-amino acids) and polysaccharides that have access to regular helical structures. Temporarily setting aside the disordered conformation of the chain, and focusing completely on the helical structures, we can recognize two distinct types of behavior for $T^i_{h,jk}$.[33]

If the helix contains an integer number of virtual bonds per turn, n_t, the $T^i_{h,jk}$ will repeat every n_t units. This behavior is illustrated in Fig. IX-10 for the

---

[33]The "h" subscript calls attention to the fact that there is only a single conformation for the chain, that conformation having the ϕ_i and ψ_i appropriate for that helix. Angle brackets are unnecessary, because there is no averaging over conformation when only one conformation is present.

structure known as "poly(L-proline) Form II," which is a left-handed helix with three residues per turn.[34] The repeating pattern is easily observed. Some other helices, such as the well-known α helix,[35] do not have an integral number of residues per turn. Here $T^i_{h,jk}$ exhibits oscillations as i increases, but a precisely repeating pattern may not be evident until several turns have been considered. For both types of helices, $T^i_{h,jk}$ will not decay to zero as i increases, because the perpetuation of a single conformation causes the end-to-end distance to be proportional to the number of virtual bonds, and $\mathbf{r}$ increases without limit.

Now we return to the unperturbed chain, where ϕ_i and ψ_i are distributed over the conformational energy surface according to $\exp(-E_{\phi_i,\psi_i}/kT)$. As the number of bonds increases, $\langle r^2 \rangle_0$ becomes proportional to n_p, $\langle \mathbf{r} \rangle_0$ approaches a limit, and every element in $\langle \mathbf{T} \rangle^{n_p-1}$ becomes null. It is often of interest to know whether this unperturbed chain contains any evidence of the types of helices that are present when the same chain is ordered. One means of searching for evidence of these helices is to see whether the $\langle \mathbf{T} \rangle^i_{jk}$ exhibit an oscillation, with a period expected for the helix, that is superimposed on their decay to zero. An example of this behavior for unperturbed poly(L-proline) is depicted in Fig. IX-11. A slight oscillation is apparent, with a period of about three virtual bonds, as expected from the pattern in Fig. IX-10. When oscillations are detected, they can often be described by a model based on a damped harmonic oscillator.[36] The conformational energy surface for poly(L-alanine), Fig. IX-8, specifies $\langle \mathbf{T} \rangle^i_{jk}$ that show no evidence of the α helix in their decay, as shown in Fig. IX-12.[37]

The same information can be obtained for chains with interdependent bonds, but that interdependence must be included in the computation. Instead of $\langle \mathbf{T} \rangle^i l_{i+1}/l_{i+1}$, we now must seek $\langle \mathbf{T}_1 \mathbf{T}_2 \cdots \mathbf{T}_i \rangle l_{i+1}/l_{i+1}$. For example, if the chain is composed of identical bonds with a symmetric threefold rotation potential energy function and pairwise interdependencies, the average projection of a unit vector along the x axis of bond $i+1$ onto the x axis of bond 1 is the 1,1 element of

$$\frac{1}{Zl_{i+1}}[\mathbf{T}_1 \quad 0 \quad 0] \begin{bmatrix} \tau\mathbf{T}_t & \sigma\mathbf{T}_{g^+} & \sigma\mathbf{T}_{g^-} \\ \mathbf{T}_t & \sigma\psi\mathbf{T}_{g^+} & \sigma\omega\mathbf{T}_{g^-} \\ \mathbf{T}_t & \sigma\omega\mathbf{T}_{g^+} & \sigma\psi\mathbf{T}_{g^-} \end{bmatrix}^{i-1} \begin{bmatrix} \tau 1 & \sigma 1 & \sigma 1 \\ 1 & \sigma\psi 1 & \sigma\omega 1 \\ 1 & \sigma\omega 1 & \sigma\psi 1 \end{bmatrix}_{i+1} \mathbf{U}^{n-i-2}\mathbf{U}_n$$

$$(IX\text{-}36)$$

[34]Cowan, P. M.; McGavin, S. *Nature (London)* **1955**, *176*, 501. Sasisekharan, V. *Acta Crystallogr.* **1959**, *12*, 897.

[35]Pauling, L.; Corey, R. B.; Branson, H. R. *Proc. Natl. Acad. Sci., USA* **1951**, *37*, 205.

[36]Darsey, J. A.; Mattice, W. L. *J. Math. Chem.* **1990**, *4*, 383.

[37]Erie, D.; Darsey, J. A.; Mattice, W. L. *Macromolecules* **1983**, *16*, 910.

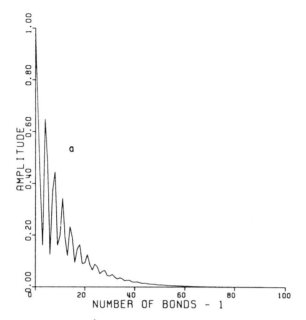

Figure IX-11. Behavior of $\langle \mathbf{T} \rangle^i_{h,11}$ for poly(L-proline), using a conformational energy surface that predicts a C_∞ of 12.4. Reprinted with permission from Darsey, J. A.; Mattice, W. L. *Macromolecules* **1982**, *15*, 1626. Copyright 1982 American Chemical Society.

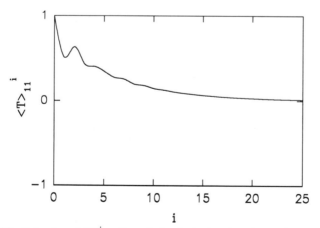

Figure IX-12. Behavior of $\langle \mathbf{T} \rangle^i_{h,11}$ for poly(L-alanine), using the conformational energy surface in Fig. IX-8.

3. AROMATIC RINGS

Aromatic rings, when incorporated into the main chain, provide an excellent opportunity for use of virtual bonds. Often they appear in conjunction with rigid groups discussed earlier in this chapter, as in the polyesters derived from aromatic dicarboxylic acids.

A. Polyesters with Flexible Spacers

The conformation-dependent physical properties of *polyesters* derived from an aromatic dicarboxylic acid and an aliphatic glycol have been studied extensively using RIS theory. A summary of some of the formulations of Z for polymers in this class is presented in Table IX-1. In order to illustrate the issues involved, we consider here the three polyesters that each have diethylene glycol as the flexible spacer, with the rigid unit being phthalate,[38] isophthalate,[39] or terephthalate.[40] The repeating units are depicted in Fig. IX-13.

The preferred conformation of the model compound methylbenzoate places all of the nonhydrogen atoms in a plane. The *trans* state is strongly preferred over the *cis* state at the ester bond in aliphatic esters,[16] and this preference is enhanced in methyl benzoate by repulsion between the CH_3 and *ortho* CH in the *cis* state.[41] These considerations justify the adoption of a scheme that maintains the ester unit in the planar *trans* conformation. The molecule strongly prefers the two rotational isomeric states at the C^{ar}—C' bond, with torsion angles of $0°$ and $180°$, that produce coplanarity of the carbonyl group and the aromatic ring, thereby maximizing the favorable interaction of the π electrons in the ring and in the ester. These two conformations must be equally weighted in methylbenzoate.

A new issue raised on the elaboration of the model compound to dimethylterephthalate is whether the rotations at the two C^{ar}—C' bonds are independent or interdependent. The two distinguishable types of conformations are depicted in Fig. IX-14. They can be denoted as *trans* or *cis*, depending on the orientation of the two ester units. Each ester unit remains in the *trans* state in both types of conformations of dimethylterephthalate. Arguments based on steric effects alone must lead to the conclusion that the bonds are independent, because the *para* substituents are separated by such large distances that their direct steric interaction is negligible. The possibility of interaction via the dipoles in the ester groups, or by delocalization of π electrons that affects the torsion potential energy functions, can be assessed by calculations or by experiments with the small model compounds. The conformational partition function for dimethyl-

[38]Riande, E.; de la Campa, J. G.; Schlereth, D. D.; de Abajo, J.; Guzmán, J. *Macromolecules* **1987**, *20*, 1641.

[39]Riande, E.; Guzmán, J.; de Abajo, J. *Makromol. Chem.* **1984**, *185*, 1943.

[40]Riande, E. *J. Polym. Sci., Polym. Phys. Ed.* **1977**, *15*, 1397. Riande, E.; Guzmán, J.; Tarazona, M. P.; Saiz, E. *J. Polym. Sci., Polym. Phys. Ed.* **1984**, *22*, 917.

[41]Williams, A. D.; Flory, P. J. *J. Polym. Sci., Part A-2* **1967**, *5*, 417.

TABLE IX-1. RIS Models for Polyesters

Rigid Unit	Flexible Spacer	Ref.
Terephthalate	CH_2CH_2	a
Terephthalate	$CH_2CH_2CH_2$	a
Terephthalate	$CH_2CH_2CH_2CH_2$	a
Terephthalate	$CH_2CH_2CH_2CH_2CH_2$	a
Terephthalate	$CH_2CH_2OCH_2CH_2$	b
Terephthalate	$CH_2CH_2SCH_2CH_2$	c
Terephthalate	$CH(CH_3)CH_2OCH(CH_3)CH_2$	d
Terephthalate	$CH_2CH_2CH_2CH_2CH_2CH_2$	a
Terephthalate	$CH_2CH_2CH_2OCH_2CH_2CH_2$	e
Terephthalate	$CH_2CH_2OCH_2CH_2OCH_2CH_2$	f
Terephthalate	Cyclohexanediols	g
Isophthalate	$CH_2CH_2OCH_2CH_2$	h
Isophthalate	$(CH_2)_m$ and $(CH_2CH_2O)_mCH_2CH_2$	i
Phthalate	$CH_2CH_2OCH_2CH_2$	j
Phthalate	$(CH_2)_m$ and $(CH_2CH_2O)_mCH_2CH_2$	i
2,6-Naphthoate	CH_2CH_2	k

[a]Menduti, F.; Rodrigo, M. M.; Tarazona, M. P.; Saiz, E. *Macromolecules* **1990**, *23*, 1139.

[b]Riande, E. *J. Polym. Sci., Polym. Phys. Ed.* **1977**, *15*, 1397. Riande, E.; Guzmán, J.; Tarazona, M. P.; Saiz, E. *J. Polym. Sci., Polym. Phys. Ed.* **1984**, *22*, 917.

[c]Riande, E.; Guzmán, J.; San Román, J. *J. Chem. Phys.* **1980**, *72*, 5263.

[d]Diaz-Calleja, R.; Riande, E.; Guzmán, J. *Macromolecules* **1989**, *22*, 3654.

[e]González, C. C.; Riande, E.; Bello, A.; Pereña, J. M. *Macromolecules* **1988**, *21*, 3230.

[f]Riande, E.; Guzmán, J. *J. Polym. Sci., Polym. Phys. Ed.* **1985**, *23*, 1235.

[g]Menduti, F.; Mattice, W. L. *Polymer* **1992**, *33*, 4180.

[h]Riande, E.; Guzmán, J.; de Abajo, J. *Makromol. Chem.* **1984**, *185*, 1943.

[i]Menduti, F.; Patel, B.; Waldeck, D. H.; Mattice, W. L. *Polymer* **1989**, *30*, 1680.

[j]Riande, E.; de la Campa, J. G.; Schlereth, D. D.; de Abajo, J.; Guzmán, J. *Macromolecules* **1987**, *20*, 1641.

[k]Menduti, F.; Saiz, E.; Mattice, W. L. *Polymer* **1992**, *33*, 4908. Gallego, J.; Menduti, F.; Saiz, E.; Mattice, W. L. *Polymer*, **1993**, *34*, 2475.

terephthalate might be written as

$$Z = 2(1 + \gamma) \tag{IX-37}$$

where γ is the weight of *cis*-dimethylterephthalate relative to *trans*-dimethylterephthalate, and the factor of two arises because each of the C^{ar}—C' bonds has two states. Conformational analysis suggests that the *trans* conformation is favored very slightly over the *cis* conformation.[42] However, the difference in energy is so small (~ 0.2 kJ mol^{-1}) that most RIS treatments employ the approx-

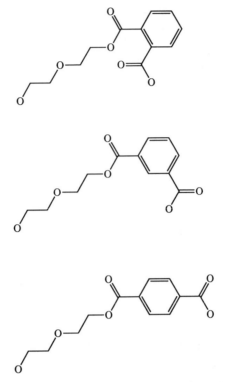

Figure IX-13. Repeating units of polyesters with diethylene glycol spacers and rigid units of phthalate, isophthalate, or terephthalate.

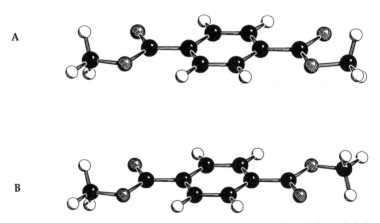

Figure IX-14. Two distinguishable planar conformations of dimethylterephthalate with *trans* ester units.

TABLE IX-2. First- and Second-Order Interaction Energies (kJ mol^{-1}) for Bonds with Symmetric Threefold Rotation Potential Energy Functions in Poly(diethylene glycol terephthalate),^{a} Poly(diethylene glycol isophthalate),^{b} and Poly(diethylene glycol phthalate)^{c,d}

Order	Fragment	Terephthalate	Isophthalate	Phthalate
First	C'-O—CH_2-CH_2	1.8	2.3	1.7
First	O-CH_2—CH_2-O	-3.3	-3.3	-3.3
First	CH_2-CH_2—O-CH_2	3.9	5.0	3.8
Second	C'-O—CH_2—CH_2-O	5.9	5.9	8.0
Second	O-CH_2—CH_2—O-CH_2	1.5	2.5	2.5
Second	CH_2-CH_2—O—CH_2-CH_2	∞	∞	∞

^{a}Riande, E.; Guzmán, J.; Tarazona, M. P.; Saiz, E. *J. Polym. Sci., Polym. Phys. Ed.* **1984**, *22*, 917. $E_\gamma = 0$.

^{b}Riande, E.; Guzmán, J.; de Abajo, J. *Makromol. Chem.* **1984**, *185*, 1943. $E_\gamma = 0$.

^{c}Riande, E.; de la Campa, J. G.; Schlereth, D. D.; de Abajo, J.; Guzmán, J. *Macromolecules* **1987**, *20*, 1641. $E_\gamma \sim 2$ kJ mol^{-1}.

^{d}The program in Appendix C can be used to calculate C_∞ for all of the polymers in this table.

imation $\gamma = 1$. The two distinguishable conformations of the terephthalate unit are taken to be of equal weights.

The RIS model is conveniently formulated using a virtual bond, of length 574 pm, that extends from one C' to the other in the unit C'—C_6H_4—C'. Starting with this virtual bond, an internal repeating unit makes a contribution to Z that can be formulated as the scalar specified by

$$z_u = 2[1 \quad \gamma]\begin{bmatrix} 1 \\ 1 \end{bmatrix}[1 \quad \sigma_\kappa \quad \sigma_\kappa]\mathbf{U}_{CC}\mathbf{U}_{CO}\mathbf{U}_{OC}\mathbf{U}_{CC}\mathbf{U}_{CO}\begin{bmatrix} 1 \\ 1 \\ 1 \end{bmatrix} \quad \text{(IX-38)}$$

The O—CH_2 and CH_2—CH_2 bonds are subject to interdependent symmetric threefold rotations. The energies corresponding to first- and second-order statistical weights in the statistical weight matrices for these bonds are listed in Table IX-2. As the bonds in the flexible unit become more remote from the rigid spacer, their statistical weight matrices approach those expected for the equivalent bond in polyoxyethylene.

A similar approach has been employed for the polymer in which isophthalate replaces terephthalate.[38] Here also the two C'—C^{ar} bonds independently occupy equally weighted rotational isomeric states with torsion angles of 0° and 180°. The virtual bond between the two C' atoms is shorter than in terephthalate, and the angle between the virtual bond and the C'—O bond is smaller, but otherwise the parameters are quite similar in the two polymers.

[42] Saiz, E.; Hummel, J. P.; Flory, P. J.; Plavšić, M. *J. Phys. Chem.* **1981**, *85*, 3211.

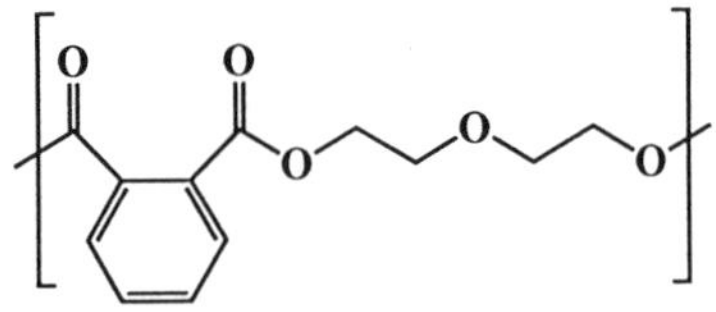

Figure IX-15. Fragment of poly(diethylene glycol phthalate).

Phthalate presents a different situation because there is severe steric repulsion between the two ester groups when they lie in the plane of the aromatic ring.[38] Conformational energy calculations exhibit maxima when the torsion angle is $0°$ or $180°$, and there are broad minima centered at $\pm 90°$. A two-state model is still a useful approximation for the $C'\!-\!C^{ar}$ bond, but the torsion angles are $\pm 90°$, which orients the ester group perpendicular to the plane of the aromatic ring. The ester dipoles are oriented differently in the two distinct conformations of dimethylphthalate. The conformation in which these two dipoles are nearly parallel is weighted by γ with respect to the conformation in which they are nearly antiparallel. The measured dipole moments for the polymer imply that E_γ is positive and in the range 1.5–2.5 kJ mol^{-1}. The contribution to the conformational partition function from the fragment depicted in Fig. IX-15 can be written as

$$z_u = [1 \quad 1]\begin{bmatrix} 1 & \gamma \\ \gamma & 1 \end{bmatrix}\begin{bmatrix} 1 \\ 1 \end{bmatrix}[1 \quad \sigma_\kappa \quad \sigma_\kappa]\mathbf{U_{CC}U_{CO}U_{OC}U_{CC}U_{CO}}\begin{bmatrix} 1 \\ 1 \\ 1 \end{bmatrix} \quad \text{(IX-39)}$$

using virtual bonds that extend from the C' atoms to the center of mass of the six carbon atoms in the aromatic ring.

Extension to aromatic esters that involve larger rings makes use of the same concepts presented in conjunction with phthalate, isophthalate, and terephthalate. Polymers in which the aromatic unit is derived from 2,6-naphthalene dicarboxylic acid have been treated with two independent rotational isomeric states at the $C'\!-\!C^{ar}$ bond, with torsion angles of $0°$ and $180°$.[43] If the carbonyl groups are attached to the naphthalene at the α position, the interaction of the ester group with the CH from the neighboring ring will perturb the conformations at the $C'\!-\!C^{ar}$ bond.[44]

---

[43] Mendicuti, F.; Patel, B.; Mattice, W. L. *Polymer* **1990**, *31*, 453. Mendicuti, F.; Saiz, E.; Zúñiga, I.; Patel, B.; Mattice, W. L. *Polymer* **1992**, *33*, 2031. Mendicuti, F.; Saiz, E.; Mattice, W. L. *Polymer* **1992**, *33*, 4908.
[44] Mendicuti, F.; Patel, B.; Mattice, W. L. *Polymer* **1990**, *31*, 1877.

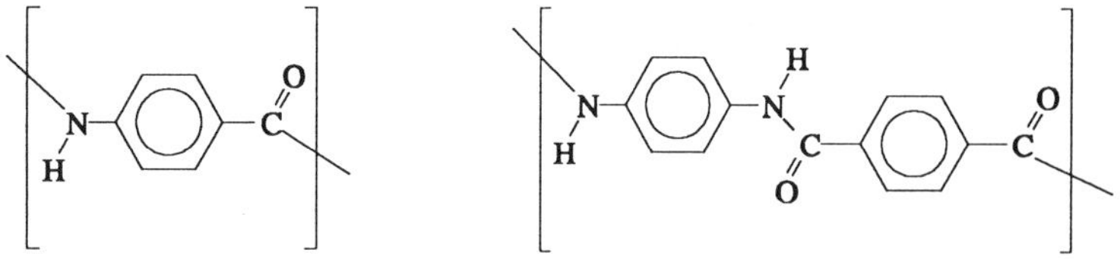

Figure IX-16. Repeating sequences of two *p*-phenylene polyamides.

Flexible polyesters derived from an aliphatic dicarboxylic acid and a partially aromatic glycol have been examined, although they are less common than the polyesters derived from aromatic dicarboxylic acids. Examples are the polyesters from aliphatic dicarboxylic acids and 2,2-bis[4-(2-hydroxyethoxy)phenyl]propane.[45]

B. Aromatic Polyesters and Polyamides without Flexible Spacers

Repeating units from two of the polymers of interest in this section are depicted in Fig. IX-16. Both molecules have *p*-phenylene units connected by amide units. They differ in whether the repeating unit has a single aromatic ring, as in $-NHC_6H_4CO-$, or two aromatic rings, as in $-NHC_6H_4NH-COC_6H_4CO-$. On the basis of the previous discussion of groups containing bonds with partial double-bond character, the C–CONH–C units (and the C–COO–C units in the corresponding polyesters) might be taken to be rigid. With rings that are also treated as being rigid, the only internal degrees of freedom are at the bonds that join the amide (or ester) to the rings. For chains with a repeat unit containing a single aromatic ring, virtual bonds were assigned from one C' to the next C'.[46] These virtual bonds are of lengths 650 and 644 pm in the amide and ester, respectively. In chains with repeating units composed of two aromatic units, the virtual bonds were drawn from C' in one amide (or ester) to N (or O) in the next amide (or ester). These virtual bonds are of lengths 649 and 651 pm in the amide, and both virtual bonds are 637 pm in the ester.

If the amide and ester links were constructed such that $\theta_{C^{ar}C'N} = \theta_{C'NC^{ar}}$ and $\theta_{C^{ar}C'O} = \theta_{C'OC^{ar}}$, torsion about the bond between the ring and the amide (ester) would produce no change in the component of **r** along this bond. The real structures display no such equalities in bond angles; the differences are 10° in the amides and 7.4° in the esters. The consequences of this nonidentity of bond angles can be assessed from the manner in which the orientation of successive virtual bonds is controlled by torsions about the bonds that link the amide (or ester) to the ring. Specifically, $\langle r \rangle_0$ is evaluated from Eq. (VI-40), using independent virtual bonds.

[45] Guzmán, J.; Riande, E.; Salvador, R.; de Abajo, J. *Macromolecules* **1991**, *24*, 5357.
[46] Erman, B.; Flory, P. J.; Hummel, J. P. *Macromolecules* **1980**, *13*, 484.

In the absence of electronic delocalization between the amide (or ester) and the aromatic ring, the torsion angles at these bonds would have minima at $\pm 90°$, as a result of repulsive steric interactions of the amide (or ester) with the *ortho*-hydrogen atoms in the conformation with a torsion angle of $0°$. However, the electronic delocalization stabilizes the conformations in which amide (or ester) and ring are coplanar.[47] The net effect from the steric repulsion and electronic delocalization is preferred torsion angles at $\pm 30°$ at C^{ar}—C' in the amide, and at $0°$ (or $180°$) in the ester. For the N—C^{ar} bond the minima are near $\pm 30°$, and for the O—C^{ar} bond they are near $\pm 60°$. The symmetry of the rotation potential energy function for each bond requires $\langle \sin \phi \rangle = 0$ in each $\langle T \rangle$. The averaged transformation matrices are of the form

$$\langle T \rangle = \begin{bmatrix} \alpha & \beta & 0 \\ -\beta c & \alpha c & 0 \\ 0 & 0 & c \end{bmatrix} \tag{IX-40}$$

where α and β are fixed by the bond angles in the chain, and c allows for a nonzero value for $E(\phi) - E(\pi - \phi)$ [$c = 0$ if $E(\phi) = E(\pi - \phi)$]. The value of c might be nonzero if there is a significant interaction of the dipoles in the polar amide (or ester) groups at either end of the phenylene unit.

When $c = 0$, the estimates for the x component of $\langle r \rangle_0$ were 41 and 74 nm for the polyamide and polyester with a single aromatic ring in the repeat unit.[46] The larger value for the polyester arises from the fact that the differences in the bond angles (see above) are smaller in the ester than in the amide, causing a smaller deviation from a parallel alignment of the bond vectors joining the ester to the rings. The estimates are likely to be upper limits, because they do not take account of fluctuations in the bond angles or in the torsion angle at the amide bond and ester bond. These degrees of freedom can be ignored in more flexible chains, such as polyglycine and poly(L-lactic acid). However, in very stiff chains these additional sources of limited flexibility may produce a significant reduction in the dimensions (estimated at 20–40% for these polyamides and polyesters).[46]

This approach can be extended to the RIS treatment of the unperturbed dimensions of polyesters consisting of more complicated units, such as *m*-phenylene, 2,6-disubstituted naphthylene, and 4,4′-disubstituted biphenylene.[48]

C. Polycarbonates

A repeating unit from poly(oxycarbonyloxy-1,4-phenyleneisopropylidene-1,4-phenylene), more commonly known as the *polycarbonate of bisphenol A*, is

[47]Hummel, J. P.; Flory, P. J. *Macromolecules*, **1980**, *13*, 479.
[48]Rutledge, G. C. *Macromolecules*, **1992**, *25*, 3984.

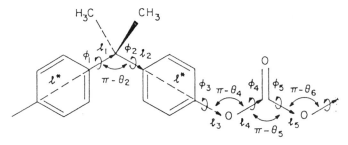

Figure IX-17. Repeat unit of the polycarbonate of bisphenol A. Reprinted with permission from Hutnik, M.; Argon, A. S.; Suter, U. W. *Macromolecules* **1991**, *24*, 5956. Copyright 1991 American Chemical Society.

depicted in Fig. IX-17. A rotational isomeric state model for this polycarbonate can be constructed from information on diphenyl carbonate and 2,2-diphenyl-propane. The conformation analysis of these two model compounds, by molecular mechanics[49,50] and by quantum mechanics methods,[51] dates back over two decades.

The RIS model must incorporate four equally weighted RIS at the $C-C^{ar}$ and $O-C^{ar}$ bonds, with torsional angles located ±45° from ideal *trans* and *cis* placements, and also unequally weighted, interdependent *trans* and *cis* states at the $O-C'$ bonds.[52] The two $C-C^{ar}$ bonds are interdependent, but the $O-C^{ar}$ bonds are independent. The contribution of the internal repeat unit to Z can be written as the product of four or six statistical weight matrices, depending on how the virtual bonds are assigned. One assignment employs two virtual bonds of length 571 pm that span $C-C_6H_4-O$, and two real bonds, of length 133 pm, for $O-C'$.[49] Another assignment splits each of the virtual bonds that span $C-C_6H_4-O$ into a virtual bond and a real bond (either $C-C_6H_4$ and $C-O$, or C-C and C_6H_4-O).[52] The former model has the virtue of simplicity, but the latter model is more easily adapted to use with other polycarbonates and related polymers. Both models yield the same result for $\langle r^2 \rangle_0$ when parameterized equivalently.

Statistical weight matrices are summarized in Table IX-3. The interdepen-

[49]Williams, A. D.; Flory, P. J. *J. Polym. Sci., Part A-2* **1968**, *6*, 1945.

[50]Tonelli, A. E. *Macromolecules* **1972**, *5*, 558. Erman, B.; Marvin, D. C.; Irvine, P. A.; Flory, P. J. *Macromolecules* **1982**, *15*, 664. Erman, B.; Wu, D.; Irvine, P. A.; Marvin, D. C.; Flory, P. J. *Macromolecules* **1982**, *15*, 670. Tekely, P.; Turska, E. *Polymer* **1983**, *24*, 667. Sundararajan, P. R. *Can. J. Chem.* **1985**, *63*, 103; Perez, S.; Scaringe, R. P. *Macromolecules*, **1987**, *20*, 68. Henrichs, P. M.; Luss, H. R.; Scaringe, R. P. *Macromolecules*, **1989**, *22*, 2731.

[51]Bendler, J. T. *Ann. N. Y. Acad. Sci.* **1981**, *371*, 299. Clark, D. T.; Munro, H. S. *Polym. Degrad. Stab.* **1982**, *4*, 83, *5*, 23. Bicerano, J.; Clark, H. A. *Macromolecules* **1988**, *21*, 585, 597. Laskowski, B. C.; Yoon, D. Y.; McLean, D.; Jaffe, R. L. *Macromolecules* **1988**, *21*, 1629. Fried, J. R.; Letton, A.; Welsh, W. J. *Polymer* **1990**, *31*, 1032. Sung, Y. J.; Chen, C. L.; Su, A. C. *Macromolecules* **1990**, *23*, 1941.

[52]Hutnik, M.; Argon, A. S.; Suter, U. W. *Macromolecules* **1991**, *24*, 5956.

TABLE IX-3. Rotational Isomeric State Model for the Polycarbonate of Bisphenol A [$\gamma =\exp{(-900/T)^a}$][b]

Bond	Atoms	l_i (pm)	θ_i^c (deg)	ϕ_i (deg)	U_i
1	C—C	154	109.5	±45, ±135	$\begin{bmatrix} 1 & 1 & 1 & 1 \\ 1 & 1 & 1 & 1 \\ 1 & 1 & 1 & 1 \\ 1 & 1 & 1 & 1 \end{bmatrix}$
2	C—C$_6$H$_4^d$	430	180	±45, ±135	$\begin{bmatrix} 1 & 0 & 1 & 0 \\ 0 & 1 & 0 & 1 \\ 1 & 0 & 1 & 0 \\ 0 & 1 & 0 & 1 \end{bmatrix}$
3	C—O	141	124	±45, ±135	$\begin{bmatrix} 1 & 1 & 1 & 1 \\ 1 & 1 & 1 & 1 \\ 1 & 1 & 1 & 1 \\ 1 & 1 & 1 & 1 \end{bmatrix}$
4	O—C	133	109.5	180, 0	$\begin{bmatrix} 1 & \gamma \\ 1 & \gamma \\ 1 & \gamma \\ 1 & \gamma \end{bmatrix}$
5	C—O	133	124	180, 0	$\begin{bmatrix} 1 & \gamma \\ 1 & 0 \end{bmatrix}$
6	O—C$_6$H$_4^d$	417	180	±45, ±135	$\begin{bmatrix} 1 & 1 & 1 & 1 \\ 1 & 1 & 1 & 1 \end{bmatrix}$

[a]Hutnik, M.; Argon, A. S.; Suter, U. W. *Macromolecules* **1991**, *24*, 5956.
[b]The program in Appendix C can be used to calculate C_∞ for this polymer.
[c]Angle between bonds i and $i + 1$.
[d]Virtual bond extending through the aromatic ring.

dence of the two C—C bonds is introduced via the null elements in U_2. These null elements prohibit conformations of the type $\phi_1, \phi_2 = 45°, 135°$, which produce an unacceptable repulsive interaction between the *ortho* hydrogen atoms on the two rings. The nonzero elements in the same matrix allow conformations of the type $\phi_1, \phi_2 = 45°, 45°$. The null element in U_5 rejects the sterically unacceptable *cis,cis* conformation of the carbonate. The first-order statistical weight for the *cis* conformation at these bonds is denoted by γ. At ambient temperature $\gamma \approx 0.05$ due to the strong preference for *trans* placements at the O—C bonds in the carbonate. This model yields $C_\infty = 8.2$, where the repeat unit is taken to consist of six bonds with the lengths given in the third column of Table IX-3. Experimental results based on light scattering,[53,54] intrinsic viscosity,[53] and neutron scattering[55] are in the range 8 ± 2.

[53]Berry, G. C. *J. Chem. Phys.* **1967**, *46*, 1338.
[54]de Chirico, A. *Chim. Ind.* **1960**, *42*, 248.
[55]Ballard, D. G. H.; Burgess, A. N.; Cheshire, P.; Janke, E. W.; Nevin, A.; Schelten, J. *Polymer* **1981**, *22*, 1353. Gawrisch, W.; Brereton, M. G.; Fischer, E. W. *Polym. Bull.* **1981**, *4*, 687.

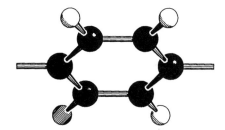

Figure IX-18. A segment from poly(p-phenylene).

Rotational isomeric state models have been described for many structural modifications of the polycarbonate of bisphenol A. The unperturbed dimensions are not sensitive to substitution of one or both of the methyl groups by hydrogen, phenyl, or cyclohexyl, or to substitution of $C(CH_3)_2$ by $C=CCl_2$, because the torsions remain symmetric, with little change in the relative energies of the preferred regions.[56] Models have also been developed for poly(thiocarbonates),[57] some of which have a methyl replaced by chlorophenyl or dichlorophenyl.[58]

D. Highly Aromatic Backbones

For typical flexible polymers, the only internal degrees of freedom retained in the RIS model are the torsion angles. Certain of the torsion angles may be held constant if local rigidity permits the assignment of virtual bonds. This section addresses a different class of polymers, in which additional degrees of freedom play an important role. The prototype is a chain in which aromatic units are bonded together such that all of the bonds between successive rings are collinear when bond angles are assigned their usual values. A segment of poly(p-phenylene) is depicted in Fig. IX-18. When the bonds between the rings are collinear, and the rings are undistorted, changes in the torsion angles of the bonds between the rings have no effect on the end-to-end distance. The value of r^2 is proportional to n^2, and C_n increases without limit as n increases.

The important variable internal degrees of freedom for these stiff polymers include those that are usually taken be to fixed in flexible polymers. Fluctuations in certain bond angles may now play an important role. In the case of poly(p-phenylene), for example, fluctuation in $\theta_{C-C=C}$, where C—C is the bond between two rings, breaks the collinear arrangement of the inter-ring

[56]Sundararajan, P. R. *Macromolecules*, **1989**, *22*, 2149.
[57]Saiz, E.; Fabre, M. J.; Gargallo, L.; Radić, D.; Hernández-Fuentes, I. *Macromolecules* **1989**, *22*, 3660.
[58]Saiz, E.; Abradelo, C.; Mogín, J.; Tagle, L. H.; Hernández-Fuentas, I. *Macromolecules* **1991**, *24*, 5594.

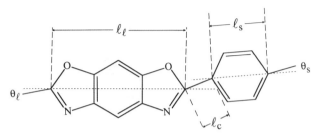

Figure IX-19. Definition of the virtual bonds used for poly(*cis*-benzobisoxazole).

bonds and allows torsion about these bonds to affect the end-to-end distance. Selected bond angles may become critically important degrees of freedom in intrinsically stiff chains.

One of the more famous stiff chains is *poly(cis-benzobisoxazole)*. A fragment of this chain is depicted in Fig. IX-19. The bonds between the rings are very nearly collinear when all bond angles have their preferred values. The persistence length was calculated using the bonds and virtual bonds defined in Figure IX-19.[59] Virtual bonds denoted as l_l and l_s span the large and small ring systems, respectively. The rings are connected by C—C bonds, with length l_c. The lengths are 648.5, 286.1, and 141.0 pm for l_l, l_s, and l_c, respectively.

The important degrees of freedom are θ and ϕ at each bond and virtual bond, with ϕ coming into play primarily because of the fluctuations in the bond angles. The simplest form for each $\langle \mathbf{T}_i \rangle$ is obtained by adoption of four assumptions: (1) each bond and virtual bond is independent; (2) fluctuations in θ_i and ϕ_i are uncorrelated; (3) the symmetry of each torsion potential energy function produces $E(\phi_i) = E(-\phi_i)$, and hence $\langle \sin \phi_i \rangle = 0$; and (4) the differences in $E(\phi_i)$ and $E(\pi - \phi_i)$ are either zero, as for rotation about l_s, or very small (on the order of 0.1 kJ mol^{-1} in the low-energy regions for rotation about l_l), and hence $\langle \cos \phi_i \rangle$ can also be taken to be zero. With these assumption each $\langle \mathbf{T}_i \rangle$ is of the form

$$\langle \mathbf{T}_i \rangle = \begin{bmatrix} -\langle \cos \theta_i \rangle & \langle \sin \theta_i \rangle & 0 \\ 0 & 0 & 0 \\ 0 & 0 & 0 \end{bmatrix} \qquad \text{(IX-41)}$$

This simplication of the form of each $\langle \mathbf{T}_i \rangle$ also reduces the number of different average transformation matrices to only two, because there are only two types of θ_i, those between l_l and l_c, and those between l_s and l_c.

With this formulation, $\langle \mathbf{r} \rangle_0$ can be computed with Eq. (VI-40), using the two distinct $\langle \mathbf{T}_i \rangle$ along with $\mathbf{l}_l$, $\mathbf{l}_s$, and $\mathbf{l}_c$. The crucial role is played by the fluctuations

[59]Zhang, R.; Mattice, W. L. *Macromolecules* **1992**, *25*, 4937.

in the bond angles (degrees of freedom usually taken to be constant in the more flexible chains), as is apparent from Eq. (IX-41). Numeric results at 300 K yield[59]

$$\langle \cos \theta_l \rangle = 0.9948, \langle \sin \theta_l \rangle = 0.0888,$$
$$\langle \cos \theta_s \rangle = 0.9958, \langle \sin \theta_s \rangle = 0.0794 \qquad \text{(IX-42)}$$

which lead to the calculation of a persistence length of ~64.8 nm. Measurements in solution[60] suggest the experimental value is somewhat larger than 50 nm.[61]

At 0 K the predicted persistence length would be nearly infinite, because the values of θ_l and θ_s are very nearly zero at the minimum in the conformational energy surface. The thermally induced fluctuations in θ_l and θ_s are responsible for the prediction of a finite (although large) persistence length at 300 K, and these fluctuations appear in the top row of the $\langle \mathbf{T}_i \rangle$ used in the computation. In the case of this very stiff molecule, the formalism of RIS theory is used to incorporate the influence of an important fluctuation in the angle between a C—C bond and a virtual bond.

This method has been applied to a few other stiff chains [poly(*trans*-benzobisoxazole), poly(*trans*-benzobisthiazole), poly(*cis*-benzobisthiazole)] in which real C—C bonds alternate with virtual bonds that span rings.[59] The method is also applicable to stiff chains in which the link between the aromatic rings is more complicated than the simple C—C bond of poly(*cis*-benzobisoxazole). It has been employed for several aromatic polyesters, such as poly(*p*-hydroxybenzoate),[48,62,63] poly(ethylene terephthalate),[62] poly(4-hydroxybicyclo[2.2.2]-octane-1-carboxylate),[62] poly(*p*-phenylene terephthalate),[64] and poly(*p*-phenylene isophthalate).[64]

4. ALIPHATIC RINGS

The incorporation of aliphatic rings into the backbone of a chain often presents the opportunity for the adoption of virtual bonds for simplification of the expression for Z. For several reasons, the design of the virtual bond scheme is frequently more complex for aliphatic rings than for aromatic rings. The stereochemistry of the attachment of the chain to the ring may become more complicated in aliphatic rings, as in the distinction between the α-1,4' and β-1,4' links between glucopyranoside units in amylose and cellulose, respectively. Fur-

[60]Wong, C.-P.; Ohnuma, H.; Berry G. C. *J. Polym. Sci., Polym. Symp.* **1978**, *65*, 173.
[61]Aharoni, S. M. *Macromolecules* **1983**, *16*, 1722. Kumar, S. in *International Encyclopedia of Composites*, Lee, S. M., ed., VCH, New York, **1990**, Vol. 4, p. 51.
[62]Jung, B.; Schürmann, B. L. *Makromol. Chem., Rapid Commun.* **1989**, *10*, 419.
[63]Jung, B.; Schürmann, B. L. *Macromolecules*, **1989**, *22*, 477.
[64]Depner, M.; Schürmann, B. L. *Polymer* **1992**, *33*, 398.

thermore, while the atoms in an aromatic ring may lie in a common plane, an aliphatic ring may be puckered sufficiently so that assignment of a common plane would produce important errors in the conformation-dependent physical properties extracted from the RIS treatment. Finally, the aliphatic ring may have access to distinguishable puckered conformations that contribute in different ways to the conformation-dependent physical properties of interest.

Nature provides several important classes of polymers that incorporate aliphatic rings in the backbone. The best known examples are the polysaccharides, which provide a rich variety of cyclic structures in their monomer units, and the nucleic acids, which incorporate either ribose or deoxyribose into a chain that also contains phosphate units. The condensed tannins incorporate into the backbone a fused-ring system in which one of the rings is aromatic, but the other ring is aliphatic. Several synthetic polymers have incorporated derivatives of cyclohexane into the backbone of the polymer. Representatives of each of these types of polymers will be described in turn.

A. Polysaccharides

The polysaccharides incorporate five- or six-membered rings into the backbone. These rings are usually connected by bridging oxygen atoms. The two polysaccharides that have received the most attention, cellulose and amylose, are both polymers of D-glucose in which the sugar forms a six-membered ring, known as glucopyranose.

Amylose. The repeating unit in *amylose* is depicted in Fig. IX-20. The α-D-glucose units prefer the chair conformation of the ring. They are linked by α-1,4' linkages, where 1 and 4' denote the carbon atoms in the two rings, and α denotes the stereochemistry of the attachment at carbon atom 1. In the approximation that bond lengths and bond angles are constant, and the ring maintains a single conformation, the amylose chain can be treated as a string of virtual bonds that link successive bridging oxygen atoms. With standard geometry for the glucose unit, the length of the virtual bond is 420 pm.[65] The only internal degrees of freedom for the maltose unit depicted in Fig. IX-20 are the torsions about the two bonds to the bridging oxygen atom. These two torsions are denoted by ϕ and ψ, in analogy with the description of the polypeptide chain. The torsion at the C_1—O bond is denoted by ϕ, and the torsion at the O—C_4' bond is denoted by ψ. The atoms used in defining these torsions are H_1–C_1—O_1–C_4' for ϕ, and C_1–O_1—C_4'–H_4 for ψ.[66]

[65]Rao, V. S. R.; Yathindra, N.; Sundararajan, P. R. *Biopolymers* **1969**, *8*, 325.
[66]Brant, D. A.; Goebel, K. D. *Macromolecules* **1975**, *8*, 522.

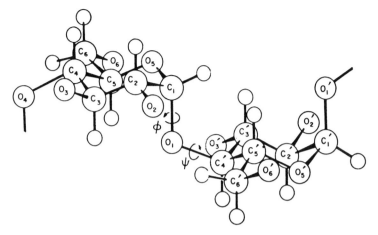

Figure IX-20. Two consecutive glucose units (or, equivalently, one maltose unit) from amylose. Reprinted with permission from Goebel, C. V.; Dimpfl, W. L.; Brant, D. A. *Macromolecules* **1970**, *3*, 644. Copyright 1970 American Chemical Society.

Conformational energy surfaces for the maltose unit have been computed by several groups.[65,67,68] The calculations produce a single region of low conformational energy in the vicinity of $\phi, \psi = 0°, 0°$. The details of the conformational energy surface in this region of ϕ, ψ space are sensitive to minor changes in the description of the structure of maltose. The value assigned to the bond angle centered on the bridging oxygen atom is especially important.[66] The conformational energy surfaces provide the basis for computation of an average transformation matrix for the transformation from the coordinate system of virtual bond $i + 1$ into the coordinate system of virtual bond i. The values of C_∞ at 25°, computed from the averaged transformation matrices via Eq. (VI-64), decrease fivefold (from 10 to 2) when the bond angle at the bridging oxygen atom changes from 113° to 118°.[66] Experimental results reported for unsubstituted amylose,[69] sodium carboxymethylamylose,[66] and diethylaminoethylamylose hydrochloride[66] yield C_∞ at 25° of 4–5, 4.7 ± 0.5, and 6.7 ± 0.3, respectively. The former two results are consistent with calculations that assign 114.75° to the bond angle at the bridging oxygen atom, and the last result is reproduced using a bond angle of 114°.[66] The averaged transformation matrices are

[67]Brant, D. A.; Dimpfl, W. L. *Macromolecules* **1970**, *3*, 655.
[68]Whittington, S. G.; Glover, R. W. *Macromolecules* **1972**, *5*, 55.
[69]Banks, W.; Greenwood, C. T. *Carbohyd. Res.* **1968**, *7*, 349, 414. Jordan, R. C.; Brant, D. A. *Macromolecules* **1980**, *13*, 491.

$$\langle \mathbf{T} \rangle_{\theta=114.75°} = \begin{bmatrix} 0.607 & 0.647 & -0.392 \\ -0.774 & 0.509 & -0.222 \\ 0.077 & 0.452 & 0.809 \end{bmatrix} \qquad \text{(IX-43)}$$

$$\langle \mathbf{T} \rangle_{\theta=114.00°} = \begin{bmatrix} 0.619 & 0.630 & -0.409 \\ -0.767 & 0.517 & -0.249 \\ 0.075 & 0.480 & 0.802 \end{bmatrix} \qquad \text{(IX-44)}$$

The model for sodium carboxymethylamylose yields an asymptotic limit for the averaged end-to-end vector [via Eq. (VI-44)] of

$$\langle \mathbf{r} \rangle_0 = \begin{bmatrix} 1.36 \\ -1.16 \\ -2.18 \end{bmatrix} \text{nm} \qquad \text{(IX-45)}$$

expressed in the coordinate system of the first virtual bond.[70]

Cellulose. *Cellulose* differs from amylose in the configuration at C_1. The repeating units are β-D-glucose in cellulose and α-D-glucose in amylose. The disaccharide repeating unit from cellulose is depicted in Fig. IX-21. With standard geometry, the virtual bonds connecting successive bridging oxygen atoms are of length 545 pm, which is 125 pm longer than the virtual bonds used for amylose. The dominant region of low energy in the conformational energy surface occurs near $\phi, \psi = 0°, 0°$, and the details in this region are very sensitive to the value assigned to the bond angle at the bridging oxygen atom, as was also the case with amylose. The computed mean square unperturbed dimensions are much larger for cellulose than for amylose. The values of C_∞ vary from ~80 to ~40 as the bond angle opens from $112°$ to $119°$.[71]

Experimental results for C_∞ for derivatives of cellulose are variable, generally falling in the range 30–60 (~30 for hydroxyethylcellulose in water,[72] ~40 for cellulose trinitrate in acetone,[73,74] ~50 for cellulose tricarbanilate in dioxane/methanol,[75] and ~60 for cellulose trinitrate in ethyl acetate[73]). The incorporation of a small fraction of β-D-glucose units puckered into other conformations has an important effect on the calculated values of C_∞, with a few

[70] Jordan, R. C.; Brant, D. A.; Cesáro, A. *Biopolymers* **1978**, *17*, 2617.

[71] Yathindra, N.; Rao, V. S. R. *Biopolymers* **1970**, *9*, 783.

[72] Brown, W. *Ark. Kem.* **1961**, *18*, 227.

[73] Holtzer, A. M.; Benoit, H.; Doty, P. *J. Phys. Chem.* **1954**, *58*, 624.

[74] Flory, P. J.; Spurr, O. K.; Carpenter, D. K. *J. Polym. Sci.* **1958**, *27*, 231. Huque, M. M.; Goring, D. A.; Mason, S. G. *Can. J. Chem.* **1958**, *36*, 952.

[75] Burchard, W. *Br. Polym. J.* **1971**, *3*, 214.

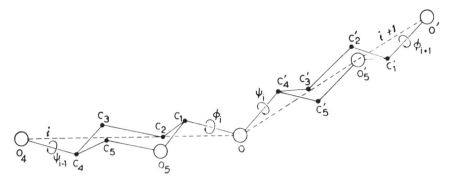

Figure IX-21. Two consecutive β-D-glucose units from cellulose (cellobiose unit). From Yathindra, N.; Rao, V. S. R. *Biopolymers* **1970**, *9*, 783. Copyright © 1970 John Wiley & Sons, Inc. Reprinted by permission of John Wiley & Sons, Inc.

percent of rings in the boat or skew conformation producing a drastic reduction in the unperturbed dimensions.[76]

Other Polysaccharides. The results described above for amylose and cellulose are summarized along with calculated C_∞ for several other polysaccharides in Table IX-4. The variability in the predictions for several of the polymers can be attributed to the extreme sensitivity of the computed C_∞ to minor changes in the values selected for those internal degrees of freedom that are held constant. Especially important in this regard are the conformation adopted for the ring and the bond angle at the bridging oxygen atom. Even with this variations, several trends are apparent. All of the calculations agree that C_∞ is larger for cellulose than for amylose. With 1,3 links, the α configuration produces higher C_∞ than the β configuration. Very low C_∞ are obtained if the chain is comprised of 1,6 links.

B. Nucleic Acids

The presence of a furanose ring in the backbone of the *polynucleotides* suggests that a virtual bond scheme might be useful in the description of the conformations of this important class of polymers. The task is more complicated than in the simple polysaccharides described in the previous section because the rings in the polynucleotides are connected by phosphate units. The underivatized furanose ring of D-ribose or D-2′-deoxyribose can assume different puckered conformations, but it becomes stiffer on the attachment of substituents, as occurs in the polynucleotides.[77] When incorporated into the chain of typical nucleic acid, the stiffness is somewhat greater in RNA(ribonucleic acid) than in DNA

[76]Brant, D. A.; Goebel, K. D. *Macromolecules* **1972**, *5*, 536.
[77]Olson, W. K.; Sussman, J. L. *J. Am. Chem. Soc.* **1982**, *104*, 270.

TABLE IX-4. Selected RIS Models for Homopolysaccharides[a]

Repeat Unit	C_∞	$\langle T \rangle$ Reported?	Conformational Energy Surface Reported?	Ref
1,4-α-D-Glucose	~ 6.9	-	+	b
	4.7–7.5	+	+	c
	10	-	-	d
	4.5	-	-	e
	5.0	-	-	f
	5.1–6.5	-	-	g
1,4-β-D-Glucose	35–80	-	+	h
	154	-	-	d
	93	-	-	e
	30–60	-	-	i
	100	-	-	f
1,3-α-D-Glucose	37	-	-	d
	32	-	+	f
1,3-β-D-Glucose	6.1	-	-	d
	3.1	-	+	f
1,6-α-D-Glucose	1.48	-	-	f
1,6-β-D-Glucose	1.66	-	-	f
1,3-α-D-Galactose	32	-	-	d
1,3-β-D-Galactose	3.7	-	-	d
1,4-α-D-Galactose	235	-	-	d
	193	-	-	f
1,4-β-D-Galactose	27	-	-	d
	42	-	-	f
1,4-α-D-Mannose	2	-	-	d
	0.50	-	-	e
1,4-β-D-Mannose	96	-	-	d
	67	-	-	e
1,3-α-D-Mannose	14	-	-	d
1,3-β-D-Mannose	2.7	-	-	d
1,3-α-D-Xylose	36	-	-	d
1,3-β-D-Xylose	5.7	-	-	d
1,4-α-D-Xylose	1	-	-	d
1,4-β-D-Xylose	31	-	-	d
1,3-α-L-Arabinose	3.3	-	-	d

[a]Most of the references describe the method used for the calculation, but do not show the conformational energy surfaces or $\langle T \rangle$. Temperatures are near 25° C.
[b]Rao, V. S. R.; Yathindra, N.; Sundararajan, P. R. *Biopolymers* **1969**, *8*, 325.
[c]Brant, D. A.; Dimpfl, W. L. *Macromolecules* **1970**, *3*, 655.
[d]Whittington, S. G. *Macromolecules* **1971**, *4*, 569.
[e]Whittington, S. G.; Glover, R. M. *Macromolecules* **1972**, *5*, 55.
[f]Burton, B. A.; Brant, D. A. *Biopolymers* **1983**, *22*, 1769.
[g]Buliga, G. S.; Brant, D. A. *Int. J. Biol. Macromol.* **1987**, *9*, 77.
[h]Yathindra, N.; Rao, V. S. R. *Biopolymers* **1970**, *9*, 783.
[i]Brant, D. A.; Goebel, K. D. *Macromolecules* **1972**, *5*, 536.

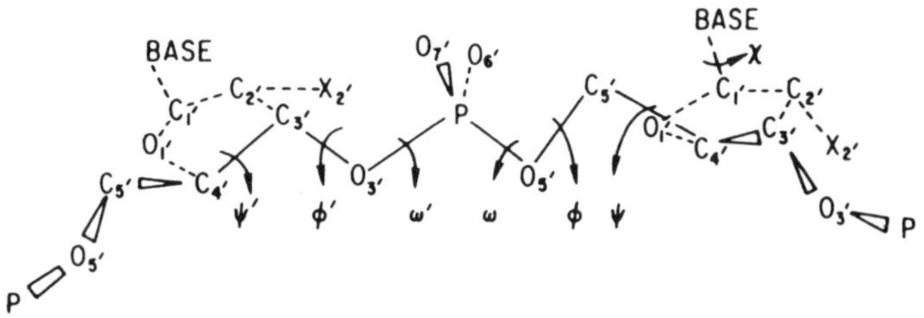

Figure IX-22. Depiction of a segment of a polynucleotide chain, with definitions of the torsion angles. Reprinted with permission from Olson, W. K. *Macromolecules* **1980**, *13*, 721. Copyright 1980 American Chemical Society.

(deoxyribonucleic acid). In both chains, the stiffness is sufficient so that virtual bonds were adopted even in the earliest descriptions of the conformations of the chains.[78]

A segment of a polynucleotide is depicted in Fig. IX-22. Each monomer unit contributes six bonds to the main chain. The torsions about these bonds are denoted by ψ', ϕ', ω', ω, ϕ, and ψ, as shown in the figure. The rings act as spacers that effectively cause the rotations about the five skeletal bonds preceding a furanose ring to be independent of the rotations about the five skeletal bonds that follow that ring. For this reason, the conformations of successive units can be treated as being independent. With this approximation, the conformational partition function for a long chain (apart from end effects) can be written as

$$Z = z^{n/6} \tag{IX-46}$$

$$z = U_{\phi'} U_{\omega'} U_{\omega} U_{\phi} U_{\psi} \tag{IX-47}$$

where z is a scalar. It is the conformational partition function for the sequences of five bonds whose torsion angles are denoted by ϕ', ω', ω, ϕ, and ψ.

The description of the conformations is simplified by the adoption of a virtual bond scheme. The repeating unit has been described using one[79] or two[77,80] virtual bonds. A model can be constructed with two virtual bonds so that it will permit the incorporation of the effects of base stacking. This model is based on the strong preference for *trans* states at the C—O bonds in the main chain in most nucleic acid structures. Each unit in the chain can be represented by two virtual bonds, one from C_4' to P, the next from P to the following C_4', as depicted in Fig. IX-23. The lengths of these virtual bonds are in the range 370–400 pm,

[78] Olson, W. K.; Flory, P. J. *Biopolymers* **1972**, *11*, 1.
[79] Olson, W. K. *Macromolecules* **1975**, *8*, 272.
[80] Olson, W. K. *Macromolecules*, **1980**, *13*, 721.

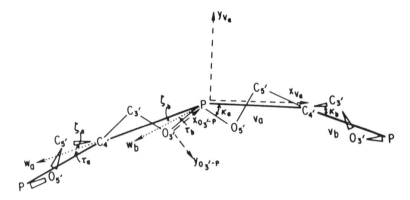

Figure IX-23. Section of a polynucleotide with two virtual bonds per unit. Each virtual bond connects a C_4' and a P. Reprinted with permission from Olson, W. K. *Macromolecules* **1980**, *13*, 721. Copyright 1980 American Chemical Society.

depending on the precise value used for ϕ and ϕ'. The orientation of a virtual bond with respect to its predecessor depends on two torsions, either ψ and ψ' flanking a C_4', or ω and ω' flanking a P. The mean square unperturbed end-to-end distance for a single-chain polynucleotide is calculated in the two-virtual bond scheme as

$$\langle r^2 \rangle_0 = [1 \quad 2l_b\langle \mathbf{T}_b \rangle \quad l_b^2](\langle \mathbf{G}_a \rangle \langle \mathbf{G}_b \rangle)^{x-1} \begin{bmatrix} l_a^2 \\ l_a \\ 1 \end{bmatrix} \tag{IX-48}$$

where l_a and l_b are the two virtual bond vectors. The calculated results for C_∞ decrease with increasing temperature, and are sensitive to the conformational distribution at ψ', ψ.[79] They can reproduce the experimental results for poly(rA) with $x = 1740$ nucleotides, which show that $\langle r^2 \rangle / nl^2$ decreases from slightly above 40 at 10° to about 15 at 60°.[81] (Here l^2 denotes the average of l^2 for the six bonds contributed by each unit to the main chain.)

The virtual bond model can be applied also to double-stranded DNA. In these applications, the virtual bonds are not defined by atoms in the deoxyribose phosphate backbone, but are defined instead using the base pairs.[82,83,84] If the bases were fixed at their positions in the well-known regular B-DNA helix, virtual bonds of length 340 pm, positioned normal to the plane of the base pair, would describe the structure. Flexibility is introduced by assigning each

[81] Eisenberg, H.; Felsenfeld, G. *J. Mol. Biol.* **1967**, *30*, 17.
[82] Maroun, R. C.; Olson, W. K. *Biopolymers* **1988**, *27*, 561, 585.
[83] Olson, W. K.; Marky, N. L.; Jernigan, R. L.; Zhurkin, V. B. *J. Mol. Biol.* **1993**, *232*, 530.
[84] Marky, N. L.; Olson, W. K. *Biopolymers* **1994**, *34*, 109.

virtual bond as

$$\mathbf{l}_i = \begin{bmatrix} l_i^0 \\ 0 \\ 0 \end{bmatrix}_i + \begin{bmatrix} \Delta l_x \\ \Delta l_y \\ \Delta l_z \end{bmatrix}_i \qquad\qquad (IX\text{-}49)$$

where the first vector on the right-hand side is the form of the virtual bond in absence of fluctuations, and the fluctuations along the three axes of the local coordinate system are incorporated by the second vector.[84] If $\Delta l_y = \Delta l_z = 0$, the fluctuation is a stretching along the normal to the plane of the base pair. The fluctuations are expected to be dependent on the sequence of the base pairs in the DNA. Fluctuations of the sizes that can be expected have large effects on $\langle r^2 \rangle_0$ and on the asymmetry of the distribution of the base pairs.[83]

C. Condensed Tannins

Condensed tannins provide an example of a chain that contains a fused-ring system with only one of the two rings being aromatic. These polymers are widespread in the plant kingdom. A trimer composed of (+)-catechin or (−)-epicatechin with the most common type of linkage between the monomeric units is depicted in Fig. IX-24. Bold lines highlight the chemical bonds that are considered to constitute the chain for the purposes of defining $\mathbf{r}$ and r^2. The attachment of the interflavan bond (the bond between the rings) at the heterocyclic ring can be via either α or β stereochemistry. This attachment is part of the configuration of the chain, and can be taken to be a constant insofar as

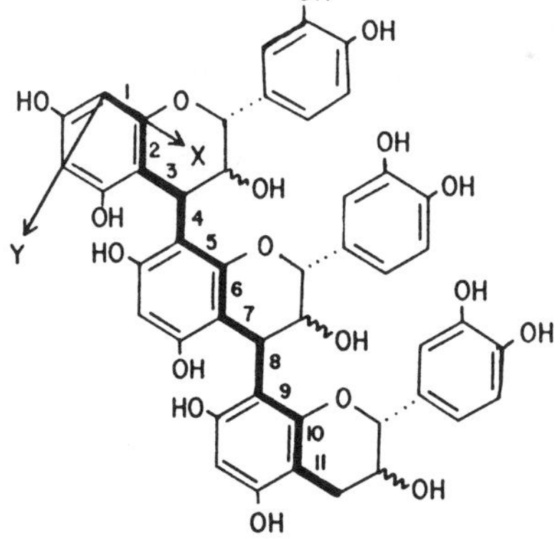

Figure IX-24. A trimeric procyanidin with $4 \rightarrow 8$ interflavan bonds. Reprinted with permission from Viswanadhan, V. N.; Bergmann, W. R.; Mattice, W. L. *Macromolecules* **1987**, *20*, 1539. Copyright 1987 American Chemical Society.

the conformational averaging is concerned. If the heterocyclic ring is treated as rigid, the chain can be represented as a sequence of virtual bonds.

The most important degrees of freedom are the torsions at the interflavan bonds (bonds 4 and 8 in Fig. IX-24) that connect consecutive monomer units. Severe steric interactions between successive monomer units produced two well-defined, nonequivalent rotational isomers at this bond. The unperturbed dimensions are strongly affected by the relative weights for these two rotational isomers, with C_∞ being as low as 8.9 if the two rotational isomers are populated equally.[85] The chain becomes more extended as the populations of the rotational isomers become nonequivalent, approaching very large values when one rotational isomer significantly dominates the other.

D. Other Synthetic Polymers

Rotational isomeric state treatments have been developed for several synthetic polymers that contain cyclohexane units connected by flexible spacers. Examples include alicyclic polyformals,[86] polythioformals,[87] polyesters,[88] and polyethers.[89] Usually the configuration of the chain through the cyclohexane unit has been *trans*-1,4,[86,87] but *cis*-1,4[88] and *trans*-1,2[89] units have also been used.

PROBLEMS

Appendix B contains answers for Problems IX-1 through IX-3, IX-5, IX-9, IX-11, and IX-12.

   IX-1. Suggest a form for Z for a chain in which 1,4-*cis*-butadiene units alternate with 1,4-*trans*-butadiene units.

   IX-2. Find $\langle \mathbf{T} \rangle$ for polyglycine in the approximation that both ϕ and ψ are subject to free rotation. What value of C_∞ is specified by this $\langle \mathbf{T} \rangle$?

   IX-3. Find $\langle \mathbf{T} \rangle$ for poly(L-proline) with *trans* peptide units, in the (rather poor) approximation that ring closure fixes ϕ at $\sim -60°$, and ψ is subject to free rotation. Does this approximation yield C_∞ compatable with experiment?

   IX-4. Set up the matrices required for calculation of $\langle \mathbf{T} \rangle$ for poly(L-proline)

[85] Viswanadhan, V. N.; Bergmann, W. R.; Mattice, W. L. *Macromolecules* **1987**, *20*, 1539.
[86] Riande, E.; Guzmán, J.; Saiz, E. *Polymer* **1981**, *22*, 465. Riande, E.; Guzmán, J.; Saiz, E.; Tarazona, M. P. *J. Polym. Sci., Polym. Phys. Ed.* **1985**, *23*, 1031.
[87] de la Peña, J.; Riande, E.; Guzmán, J. *Macromolecules* **1985**, *18*, 2739.
[88] Riande, E.; Guzmán, J.; de la Campa, J. G.; de Abajo, J. *Macromolecules* **1985**, *18*, 1583.
[89] Vega, S.; Riande, E.; Guzmán, J. *Macromolecules* **1990**, *23*, 3573.

with *cis* peptide units. How do they differ from the expressions used for the same polymer with *trans* peptide units?

IX-5. For poly(amino acids) with standard peptide units, what is the conversion between C_∞ defined in terms of virtual bonds, and C_∞ defined in terms of the lengths of the real chemical bonds in the main chain?

IX-6. For polyamides with repeating sequence $NH(CH_2)_yCO$, how large must y be in order for the polyamide and a polyethylene of the same (large) molecular weight to have values of $\langle r^2 \rangle_0$ that differ by no more than 5%?

IX-7. How might one assign virtual bonds in the polyester obtained from terephthalic acid and 1,4-*trans*-cyclohexanediol? How many statistical weight matrices would be required for each repeating unit?

IX-8. From Eq. (IX-40), find a closed form for $\langle \mathbf{r} \rangle_0 / l$, in the limit $n \to \infty$, in terms of α, β, and c. Here l is the length of the virtual bond. [For the answer, see Eq. (13) in Erman et al.[46]] How does a nonzero value of c affect the result?

IX-9. Find $\langle \mathbf{T} \rangle$, in terms of E/kT, for a poly(α-amino acid) in which

$E = \infty$ if $0° < \phi < 180°$

$E = 0$ if $-180° < \phi < 0°$ and $0° < \psi < 180°$

$E = E$ if $-180° < \phi < 0°$ and $-180° < \psi < 0°$

Then find C_∞ as a function of E/kT.

IX-10. Suggest a formulation for Z for poly(β-alanine), which has the repeating sequence $-NH-CH_2-CH_2-CO-$. In what ways must the formulation be different from the one described in the text for poly(L-alanine)?

IX-11. What aspects of the covalent structure and short-range interactions cause C_∞ for poly(L-alanine) to be ~4 times as large as C_∞ for poly(L-lactic acid)?

IX-12. Consider an achiral amino acid residue with bulky side chains that allow only two conformations, one with ϕ_i, ψ_i and the other with $-\phi_i, -\psi_i$, both conformations being equally weighted. Construct a contour map showing C_∞ as a function of ϕ_i, ψ_i. How sensitive is C_∞ to the location of ϕ_i, ψ_i? Why is C_∞ a strong function of ϕ_i, ψ_i in some portions of the map, but only weakly dependent on ϕ_i, ψ_i in other portions?

X Intramolecular Formation of Helices and Sheets

The description of the formation of intramolecular helices and antiparallel sheets by homopolymers is susceptible to treatment by the same matrix formalism used in rotational isomeric state (RIS) theory.[1] Helices are periodic structures obtained by propagation of a repeating pattern of torsion angles along a constitutionally regular chain. The type of helix is specified by the repeating pattern of torsion angles. Once the type of helix has been identified, the structure of the helix is completely specified by the number of monomer units it contains. In this sense the helix is a one-dimensional ordered structure, because it is characterized by its length (or the number of units in it).

An intramolecular sheet contains two or more strands that interact with one another. Two consecutive strands can be classified as antiparallel or parallel, depending on whether they run in opposite directions, or in the same direction. Matrix methods are much more easily adapted to the description of antiparallel sheets than to parallel sheets. The antiparallel sheet is a two-dimensional order structure. The number of units in an antiparallel sheet does not completely specify its structure, because a large sheet might be comprised of a few long strands, or of many short strands.

Helices are usually stabilized by interactions of relatively short range that occur between unit i and unit $i+j$, where j is a small integer. The conformational partition function must be formulated in a manner that takes account of this stabilizing interaction for the helix $\rightleftharpoons$ coil transition[2] in a satisfactory way. Sometimes a satisfactory treatment can be achieved using a matrix of smaller dimensions than might be expected. For example, in the case where the helix is the well-known α helix,[3] the important stabilizing interaction is a hydrogen bond the formation of which depends on appropriate assignments for six consecutive degrees of freedom. However, this sixth-order interaction can often be

---

[1] Poland, D.; Scheraga, H. A. *Theory of Helix-Coil Transitions in Biopolymers*, Academic Press, New York, **1970**.

[2] The term "transition" is used here in a much vaguer sense than is common in physics and physical chemistry. The transitions described here do not make sudden changes similar to those observed in phase transitions. The models involved are special cases of one-dimensional Ising models and, hence, not capable of reproducing first-order transitions. Rather, "transition" means here a remarkable large change over a narrow range of conditions. The extensive literature on this "transition" uses the word "coil" to denote the disordered state.

[3] Pauling, L.; Corey, R. B.; Branson, H. R. *Proc. Natl. Acad. Sci., USA* **1951**, *37*, 205.

treated in adequate approximation in homopoly(amino acids) with a statistical weight matrix of dimensions 2×2, as will be shown in this chapter.

The important stabilizing interactions in the antiparallel sheets differ from the stabilizing interaction in the helix in that the pertinent range for j is larger (often much larger) and variable. If, for example, one folds a segment of n units into an antiparallel sheet with two strands, each of $n/2$ units, the interacting pairs of units in the two strands are characterized by larger and larger values of j as one moves from the site of the fold to the free ends of the two strands. The formulation of the conformational partition function must therefore take satisfactory account for the sheet ⇌ coil transition of an important interaction of variable, and often long, range. In this chapter, we will deal first with the easier problem (the helix ⇌ coil transition), and then proceed to the sheet ⇌ coil transition.

The helix ⇌ coil and sheet ⇌ coil transitions are often produced experimentally by changing the quality of the solvent. This change modifies the values of the statistical weights for propagation of the ordered structure. The formulation is frequently applied under conditions where the solvent is not a Θ solvent for the polymer. Strictly speaking, the results obtained from the formulation of these order ⇌ disorder transitions apply to the hypothetical state where the influence of interactions of longer range than those specifically included in the formulation of Z is nil, either because the system is actually in a Θ state, or because the data obtained in a non-Θ state have been corrected for the influence of the long-range interactions.

1. THE INTRAMOLECULAR HELIX ⇌ COIL TRANSITION

One of the more famous order ⇌ disorder transitions in a chain molecule in dilute solution is the *helix* ⇌ *coil* transition in poly(L-α-amino acids) with $-CH_2R$ side chains.[1,4] The conformational partition function relevant for this transition is easily combined with a virtual bond model for the unperturbed random coil for study of the behavior of the unperturbed dimensions during this conformational transition. The techniques required for the accounting of interactions of higher than second-order, which were introduced in conjunction with ω' in polyethylene in Chapter VII, are greatly elaborated in this chapter. However, a suitable approximation for many purposes can be devised with a very small statistical weight matrix.

When every residue[5] in a poly(α-amino acid) chain adopts the same values for ϕ, ψ, the chain adopts the conformation of a helix. If these values are selected from a small area in the region of low conformational energy in the

---

[4]The poly(D-α-amino acids) form similar helices, but with opposite chirality. Polymers of the L-α-amino acids are more commonly studied because of their occurrence in proteins.
[5]The amino acid residue denotes NH—C$^\alpha$HR—CO, which contains the atoms contributed to the chain by a single amino acid.

lower left or upper right quadrant in Fig. IX-8, intramolecular hydrogen bonds are possible in the helix.[6] The handedness of the helix depends on which quadrant is selected, and the type of hydrogen bond formed depends on the details of the selection of ϕ, ψ within either quadrant. The most famous of these helices is the α helix.[3] It has 3.6 residues per turn, a translation per residue along the helix axis of 0.15 nm, a pitch of 0.54 nm, and intramolecular hydrogen bonds from the NH of residue i to the CO of residue $i - 4$. The atoms that form this hydrogen bond are depicted in Fig. X-1. They are nowhere near the proper position for formation of the hydrogen bond when the chain is fully extended or in the vast majority of the conformations adopted by the random coil. However, if the ϕ, ψ at residues $i - 3$, $i - 2$, and $i - 1$ are all assigned as either $-47°$, $-57°$, or as $47°$, $57°$, the conformation produces the hydrogen bond between the NH in residue i and the CO in residue $i - 4$.

Polyglycine has no preference between right- and left-handed helices,[7] but a preference is exercised by chiral amino acid residues. Nearly all of the L-amino acids that form the α helix strongly prefer the right-handed structure over the left-handed one. The right-handed α helix uses ϕ, $\psi = -47°$, $-57°$.

If residues $i - 3$, $i - 2$, and $i - 1$ are the only ones in the chain that have adopted the ϕ, ψ appropriate for the α helix, there will be only a single hydrogen bond in the helix. However, if residue i also adopts this ϕ, ψ, a second hydrogen bond can form, between the NH of residue $i + 1$ and the CO of residue $i - 3$. The chain must sacrifice conformational entropy at three successive residues in order to form the first hydrogen bond of an α helix, but a subsequent similar sacrifice at a neighboring residue will generate a second hydrogen bond. This fact suggests initiation of the helix should be a less probable event than propagation of an existing helix. Any successful formulation of Z for this transition must distinguish between initiation and propagation.[1] Several formulations for the conformational partition function were published in 1959–1961.[8] The description employed here adopts the widely used notation of Zimm and Bragg.

A. The Conformational Partition Function

The formulation of Z must take account of the hydrogen bond that is an important source of the stabilization of the α helix. The depiction in Fig. X-1 shows

---

[6]Poly(L-proline) is excluded from this discussion, because it cannot form intramolecular hydrogen bonds.

[7]Polyglycine does not form an α helix at all, because it has access to another type of helix that is more stable. This helix, the polyglycine Form II helix, has exactly three residues per turn. Interchain hydrogen bonds are possible in this structure only when side chains are absent, as they are in polyglycine. The polyglycine Form II helix can be either right- or left-handed, with both structures being of equal probability.

[8]Gibbs, J. H.; DiMarzio, E. A. *J. Chem. Phys.* **1959**, *30*, 271. Peller, L. *J. Phys. Chem.* **1959**, *63*, 1194. Zimm, B. H.; Bragg, J. K. *J. Chem. Phys.* **1959**, *31*, 526. Nagai, K. *J. Phys. Soc. Japan* **1960**, *15*, 407. Miyake, A.; Chûjô *J. Polym. Sci.* **1960**, *46*, 163. Lifson, S.; Roig, A. *J. Chem. Phys.* **1961**, *34*, 1963.

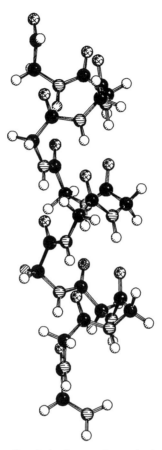

Figure X-1. Conformation of a chain that produces the hydrogen bond in an α helix.

this hydrogen bond is a sixth-order interaction, because it depends on the appropriate selection of torsion angles at three N—C^α bonds and at three C^α—C' bonds. A simplification is achieved by division of the conformational space accessible at each ϕ, ψ into two regions. The pair of ϕ, ψ is denoted by h if it assumes the values $-47°, -57°$, within a small tolerance. Thus h denotes a pair of ϕ, ψ that is capable of supporting the hydrogen bonding in a right-handed α helix, if that pattern is propagated over three successive amino acid residues. Any other combination of ϕ, ψ is denoted by c, for random coil. With this notation, the hydrogen bond can be treated with a formalism appropriate for a third-order interaction. Its formation requires that three successive residues must adopt the state h.

The formulation of **U** follows the approach anticipated for $\nu = 2$ (the two rotational isomeric states are c and h) when a third-order interaction is important. The dimensions of **U** would be $\nu \times \nu$ if only first- and second-order interactions were considered. Inclusion of third-order interactions increases the dimensions to $\nu^2 \times \nu^2$. The statistical weight matrix can be written as[9]

$$
\mathbf{U}_i =
\begin{array}{c}
\\ cc \\ ch \\ hc \\ hh
\end{array}
\begin{array}{cccc}
cc & ch & hc & hh \\
\begin{bmatrix} 1 & \hat{\sigma}^{1/2} & 0 & 0 \\ 0 & 0 & 0 & \hat{\sigma}^{1/2} \\ 1 & \hat{\sigma}^{1/2} & 0 & 0 \\ 0 & 0 & 1 & s \end{bmatrix}_i
\end{array}
\tag{X-1}
$$

where the rows are indexed by the states at residues $i - 2$ and $i - 1$, and the columns are indexed by the states at residues $i - 1$ and i. Statistical weights of zero in Eq. (X-1) reject contributions from the eight elements of $\mathbf{U}_i$ (two in each row) in which the row and column disagree on the state of residue $i - 1$. A zero is also used for the third element in the second row, which weights the *chc* sequence. The justification for this assignment is that the shortest sequence with one hydrogen bond of the type found in the α helix must contain three consecutive *hs*.

One might wonder whether we should also suppress contributions from the sequence *chhc*. To do so, the statistical weight matrix must be designed so that we can identify elements that correspond to specific quadruplets, rather than merely the triplets defined by each element in Eq. (X-1). That objective can be obtained with a matrix of dimensions 8×8.

$$
\mathbf{U}_i =
\begin{array}{c}
\\ ccc \\ cch \\ chc \\ chh \\ hcc \\ hch \\ hhc \\ hhh
\end{array}
\begin{array}{cccccccc}
ccc & cch & chc & chh & hcc & hch & hhc & hhh \\
\begin{bmatrix} 1 & \hat{\sigma}^{1/2} & 0 & 0 & 0 & 0 & 0 & 0 \\ 0 & 0 & 0 & \hat{\sigma}^{1/2} & 0 & 0 & 0 & 0 \\ 0 & 0 & 0 & 0 & 0 & 0 & 0 & 0 \\ 0 & 0 & 0 & 0 & 0 & 0 & 0 & s \\ 1 & \hat{\sigma}^{1/2} & 0 & 0 & 0 & 0 & 0 & 0 \\ 0 & 0 & 0 & \hat{\sigma}^{1/2} & 0 & 0 & 0 & 0 \\ 0 & 0 & 0 & 0 & 1 & \hat{\sigma}^{1/2} & 0 & 0 \\ 0 & 0 & 0 & 0 & 0 & 0 & 1 & s \end{bmatrix}_i
\end{array}
\tag{X-2}
$$

which can immediately be reduced to a 7×7 matrix because the third row and third column are null. The seventh element in the fourth row is the one that specifies the sequence *chhc*.

The statistical weights assigned in Eq. (X-2) cause each conformation of the chain, that is, each string of x letters selected from c and h, to have a statistical weight, w, given by

---

[9]The symbol σ was introduced by Zimm and Bragg for the end effect in an α helix, and it has been widely adopted in helix $\rightleftharpoons$ coil transition theory. It should not be confused with the statistical weight for a g state, which is also commonly denoted by σ in the literature. To avoid confusion, we employ the symbol $\hat{\sigma}$ here.

$$w = \hat{\sigma}^{n_{ch}} s^{n_h} \qquad (X\text{-}3)$$

where n_{ch} denotes the number of hs that are the first in a string of hs (and therefore n_{ch} is the number of distinct helices), and n_h denotes the number of hydrogen bonds in these helices. Each helix must contain at least one hydrogen bond (at least three hs). The statistical weight contributed by each distinct helix contains a term that depends on its number of hydrogen bonds, through the contribution of the factor containing s, and another term, $\hat{\sigma}$, that incorporates the end effects in each distinct helix, without regard to its length. Thus $\hat{\sigma}$ must incorporate the sacrifice in conformational entropy on initiation of the helix. The statistical weight for propagation of a helix is s; Z for a chain of x residues (with $x - 1$ virtual bonds connecting x successive C^{α} atoms) is generated as

$$Z = [1 \quad 0 \quad \cdots \quad 0]\mathbf{U}^{x-1}\begin{bmatrix} 1 \\ \vdots \\ 1 \end{bmatrix} \qquad (X\text{-}4)$$

The essence of this model can be retained with a much more compact matrix if $\hat{\sigma} \ll 1$, that is, if the penalty for initiation is severe. Then the only stable helices will be long ones, and we can generate an excellent approximation to Z with

$$\mathbf{U}_i = \begin{matrix} c \\ h \end{matrix}\begin{matrix} c \quad\ h \\ \begin{bmatrix} 1 & \hat{\sigma}s \\ 1 & s \end{bmatrix} \end{matrix} \qquad (X\text{-}5)$$

The 2×2 statistical weight matrix in Eq. (X-5) allows the sequences chc and $chhc$, and n_h in Eq. (X-3) becomes the number of hs in the chain, rather than the number of hydrogen bonds. If $\hat{\sigma} \ll 1$, the very short sequences of hs are strongly suppressed by $\hat{\sigma}$, and the expressions for Z derived from Eqs. (X-2) and (X-5) are nearly equivalent.

The 2×2 statistical weight matrix in Eq. (X-5) is of the same form as the reduced statistical weight matrix for a chain with pairwise interdependent bonds and a symmetric threefold rotation potential-energy function, an example of which is polyethylene, Eq. (IV-28). Both matrices can be written as

$$\mathbf{U}_i = \begin{bmatrix} 1 & AB \\ 1 & B \end{bmatrix} \qquad (X\text{-}6)$$

where $A = \hat{\sigma}$ and $B = s$ if we are discussing helix ⇌ coil transition theory, but $A = 2(1 + \omega)^{-1}$ and $B = \sigma(1 + \omega)$ for polyethylene. In both cases, one can utilize the procedures for matrix diagonalization (Section IV.6) with

$$\lambda_1, \lambda_2 = \frac{1 + B \pm \sqrt{(1+B)^2 - 4B(1-A)}}{2} \tag{X-7}$$

Thus the fraction of the residues in the helix is given by

$$p_h = \frac{1}{x} \frac{\partial \ln Z}{\partial \ln s} \tag{X-8}$$

$$\lim_{x \to \infty} p_h = \frac{\lambda_1 - 1}{\lambda_1 - \lambda_2} \tag{X-9}$$

and, with the fraction of residues that initiate new helices obtained as

$$p_{ch} = \frac{1}{x} \frac{\partial \ln Z}{\partial \ln \hat{\sigma}} \tag{X-10}$$

$$\lim_{x \to \infty} p_{ch} = \frac{(\lambda_1 - 1)(1 - \lambda_2)}{\lambda_1(\lambda_1 - \lambda_2)} \tag{X-11}$$

the average number of residues in a helix is

$$\nu_h = \frac{(\partial \ln Z)/(\partial \ln s)}{(\partial \ln Z)/(\partial \ln \hat{\sigma})} \tag{X-12}$$

$$\lim_{x \to \infty} \nu_h = \frac{\lambda_1}{1 - \lambda_2} \tag{X-13}$$

If $\hat{\sigma} \ll 1$, as expected, the value of ν_h specified by Eq. (X-13) will be large whenever s is large enough to produce a significant helix content in a long chain.

We can think of the statistical weight matrix in Eq. (X-6) as having application to a rather large number of problems involving the conformations of chain molecules, with the behavior predicted by the formalism depending on the region of A—B space occupied by a particular chain. Typical orders of magnitude are $A \approx 10^{-4}$ and $B \approx 1$ in applications to helix $\rightleftharpoons$ coil transition theory, but $A \approx 2$ and $B \approx 0.5$ for polyethylene. These polymers occupy quite different regions of A—B space, as shown in Fig. X-2. The physically significant sizes of A in particular are quite different in the two applications.

The helix $\rightleftharpoons$ coil transition is often induced by changes in solvent composition or temperature, which affect the stability of the hydrogen bond in the α helix, as well as side chain–side chain interactions that may affect the stability

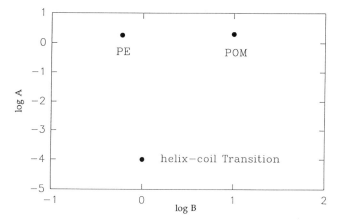

Figure X-2. Regions of A—B space pertinent for unperturbed polyethylene, a simple model for unperturbed polyoxymethylene, and for the helix ⇌ coil transition in poly(α-amino acids).

of the helix. It is instructive to examine the behavior of p_h for a long chain when $\hat{\sigma}$ is constant, and the change in conformation is produced by changes in s. According to Eq. (X-9), $p_h = 1/2$ when $s = 1$, independent of the assignment of $\hat{\sigma}$. The value of $\hat{\sigma}$ affects the cooperativity of the transitions, with the transition becoming closer to all-or-none behavior as $\hat{\sigma} \to 0$, as depicted in Fig. X-3.

If $\hat{\sigma}$ and s are held constant, p_h becomes a function of x at sufficiently small x. Recall that Eq. (X-13) predicts the helices tend to be long in very long

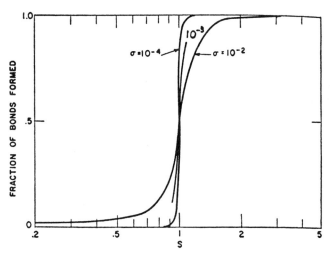

Figure X-3. Variation in p_h with s at various $\hat{\sigma}$, in the limit as $x \to \infty$. Reproduced with permission from Zimm, B. H.; Bragg, J. K. *J. Chem. Phys.* **1959**, *31*, 526. Copyright 1959 American Physical Society.

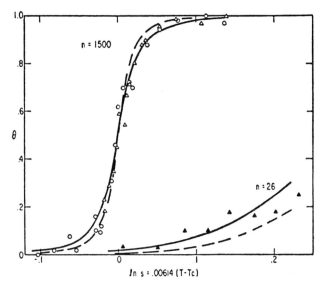

Figure X-4. Variation in p_h with s at $x = 26$ and $x = 1500$ when $\hat{\sigma} = 2 \times 10^{-4}$ (solid lines) and $\hat{\sigma} = 1 \times 10^{-4}$ (dashed lines). The points are experimental data for poly (γ-benzyl-L-glutamate) reported by (o) Doty, P.; Yang, J. T. *J. Am. Chem. Soc.* **1956**, *78*, 498 and ($\triangle$) Doty, P.; Iso, K. private communication cited by Zimm and Bragg. Reproduced with permission from Zimm, B. H.; Bragg, J. K. *J. Chem. Phys.* **1959**, *31*, 526. Copyright 1959 American Physical Society.

chains. Therefore we should anticipate that short chains, and especially chains for which x is smaller than the ν_h given by Eq. (X-13), will have a lower helix content than long chains. For these shorter chains, p_h can be evaluated from Eq. (X-8), but the complete expression for Z (rather than the approximation as $x \to \infty$) must be employed. This effect is demonstrated in Fig. X-4. For values of p_h close to unity, and $\lambda_1(1 - \lambda_2)^{-1} > x$, most chains will have the perfectly helical conformation.

Data that depict p_h as a function of x can be used to determine the best values of $\hat{\sigma}$ and s.[10] Manifestations of optical activity (optical rotatory dispersion,[11] circular dichroism[12]) are the most common experimental routes to p_h.

B. Mean-Square Unperturbed Dimensions

The computation of the mean-square unperturbed dimensions utilizes the expression for U_i in Eq. (X-5). The second column contains statistical weights

[10]There have been many such studies. See, for example, von Dreele, P. J.; Lotan, N.; Ananthanarayanan, V. S.; Andreatta, R. H.; Poland, D.; Scheraga, H. A. *Macromolecules* **1971**, *4*, 408.
[11]Greenfield, N.; Davidson, B.; Fasman, G. D. *Biochemistry* **1967**, *6*, 1630.
[12]Greenfield, N.; Fasman, G. D. *Biochemistry* **1969**, *8*, 4108.

for a specific state (the right-handed α helix), but the first column has the statistical weights for the random coil. The random coil is recovered in the limit as $s \rightarrow 0$. The unperturbed dimensions in this limit were discussed earlier (Section IX.2). There the treatment was based on a chain composed of independent virtual bonds. Incorporation of the α helix into the ensemble introduces an interdependence of the virtual bonds, because an h at residue i may represent either initiation or propagation, depending on the state at residue $i - 1$. The mean square end-to-end distance, unperturbed by long-range interactions, is obtained for a chains with $x > 2$ as

$$\langle r^2 \rangle_0 = Z^{-1} \mathscr{G}_1 \mathscr{G}^{x-3} \mathscr{G}_{x-1} \tag{X-14}$$

$$\mathscr{G}_i = \begin{bmatrix} \langle \mathbf{G}_c \rangle & \hat{\sigma} s \mathbf{G}_h \\ \langle \mathbf{G}_c \rangle & s \mathbf{G}_h \end{bmatrix}_i \quad 1 < i < x - 1 \tag{X-15}$$

$$\mathscr{G}_1 = [1 \quad 2\mathbf{l}_p^T \langle \mathbf{T}_c \rangle \quad l_p^2 \quad \hat{\sigma} s \quad 2\hat{\sigma} s \mathbf{l}_p^T \mathbf{T}_h \quad \hat{\sigma} s l_p^2] \tag{X-16}$$

$$\mathscr{G}_{x-1} = \begin{bmatrix} (1 + \hat{\sigma} s) l_p^2 \\ (1 + \hat{\sigma} s) \mathbf{l}_p \\ 1 + \hat{\sigma} s \\ (1 + s) l_p^2 \\ (1 + s) \mathbf{l}_p \\ (1 + s) \end{bmatrix}_{x-1} \tag{X-17}$$

Here the chain contains x residues (and x C^α atoms), and $x - 1$ virtual bonds connecting the C^α atoms. The end-to-end distance is defined by the positions of the C^α atoms in residues 1 and x. The generator matrices are the 5×5 matrices defined in Eq. (VI-21) (for $\mathbf{G}_h$), and in Eq. (VI-57) (for $\langle \mathbf{G}_c \rangle$). The average transformation matrix in $\langle \mathbf{G}_c \rangle$ is the $\langle \mathbf{T} \rangle$ for the appropriate amino acid residue, e.g., Eq. (IX-32) for poly(L-alanine). The transformation matrix in $\mathbf{G}_h$ is the expression for $\mathbf{T}_3 \mathbf{T}_{2,\phi} \mathbf{T}_{1,\psi}$ when ϕ and ψ are assigned the values found in the α helix. In both $\mathbf{G}_h$ and $\langle \mathbf{G}_c \rangle$, $\mathbf{l}$ is the virtual bond vector, and l is its length, 380 pm.

As $x \rightarrow \infty$ the helix must be the more extended structure, because r^2 for the helix is proportional to x^2, but $\langle r^2 \rangle_0$ for the random coil becomes proportional to x as $x \rightarrow \infty$. However, at small x the situation may be reversed, because the α helix has a small translation per residue, only 150 pm. If the averaged transformation matrix for the random coil is the one specified for poly(L-alanine) in Eq. (IX-32), r^2 for the helix will be smaller than $\langle r^2 \rangle_0$ for the random coil when x is smaller than ~100, as depicted in Fig. X-5.

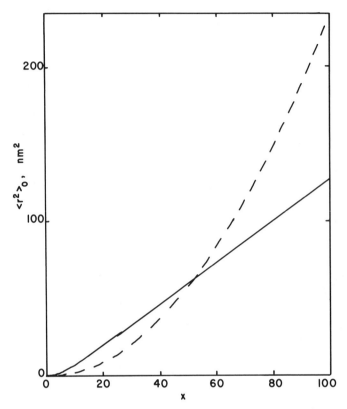

Figure X-5. $\langle r^2 \rangle_0$ for an unperturbed poly(L-alanine) (solid line) and r^2 for an α helix (dashed line), both as a function of x.

In the limit where $\hat{\sigma} \to 0$, a helix $\rightleftharpoons$ coil transition produced by changing s for a very long chain will approach all-or-none behavior. The value of C_∞ will approach a discontinuous change at $s = 1$. An actual first-order transition cannot be obtained with this model, however. If, however, $\hat{\sigma}$ is on the order of 10^{-4}, C_∞ experiences a monotonic increase as s (and hence p_h) increases. This increase is extremely sharp as $p_h \to 1$, because $r^2 \to \infty$ in this limit. In contrast, the increase is relatively gentle in the region of small p_h.[13]

At somewhat larger $\hat{\sigma}$, on the order of 10^{-3}, the mean square unperturbed dimensions may pass through a minimum as s increases. This minimum arises because shorter helices become more important as $\hat{\sigma}$ increases and, as noted above, α helices with fewer than ~100 residues may have r^2 smaller than $\langle r^2 \rangle_0$ for the random coil with the same x.[13,14]

[13] Miller, W. G.; Flory, P. J. *J. Mol. Biol.* **1966**, *15*, 298.
[14] Mattice, W. L. *Macromolecules* **1980**, *13*, 904.

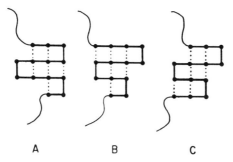

Figure X-6. Three illustrative antiparallel sheets with tight bends. From Mattice, W. L.; Scheraga, H. A. *Biopolymers* **1984**, *23*, 1701. Copyright © 1984 John Wiley & Sons, Inc. Reprinted by permission of John Wiley & Sons, Inc.

2. ANTIPARALLEL SHEET ⇌ COIL TRANSITION

A more extensive expansion of the dimensions of **U** permits the formulation of Z for the transition from a random coil to an intramolecular antiparallel sheet in which consecutive strands are connected by tight bends. The strands in the sheet need not all have the same number of residues. Examples of three small antiparallel sheets are depicted in Fig. X-6. Each of these three antiparallel sheets contains 13 residues, but they differ in shape. The potential differences in shape must be accommodated in the statistical weights. A matrix formulation has been described in which each sheet is weighted as[15,16]

$$\text{Weight} = \delta^{n_{st}-1}\hat{\tau}^{n_b} t^{n_{sh}} \tag{X-18}$$

where n_{sh} denotes the number of residues in the sheet, n_{st} is the number of strands (and hence $n_{st} - 1$ is the number of tight bends connecting successive strands), and n_b is the number of residues in the sheet that do not have a partner in a preceding strand. The term $t^{n_{sh}}$ is determined completely by the number of residues in the sheet, but $\delta^{n_{st}-1}\hat{\tau}^{n_b}$ depends on its size and shape. Thus t in the sheet ⇌ coil transition theory is analogous to s in the helix ⇌ coil transition theory. There is a weaker analogy between $\delta^{n_{st}-1}\hat{\tau}^{n_b}$ and $\hat{\sigma}$. Both are "end effects," but $\hat{\sigma}$ always occurs in exactly the same way for each helix, independent of its length, whereas the end effect for an antiparallel sheet depends on its size and shape. A very large **U** is required for the correct generation of $\delta^{n_{st}-1}\hat{\tau}^{n_b}$.

The procedure adopted for the computation of Z takes advantage of the fact that incorporation of longer-range interactions into **U** causes this matrix to

[15]Mattice, W. L.; Scheraga, H. A. *Biopolymers* **1984**, *23*, 1701.
[16]To avoid confusion with the earlier defined τ, we use $\hat{\tau}$ here for the *anti*-parallel sheet.

become sparse as well as large. Examination of the expressions in Eqs. (X-1), (X-2), and (X-5) for **U** for the helix $\rightleftharpoons$ coil transition shows that the fraction of nonzero elements decreases from 1 to 7/16 to 11/64 as the expression moves from evaluation of doublets to triplet to quadruplets. As a formulation of Z requires an accounting of longer and longer multiplets, a point will be reached at which it is no longer desirable (or even possible) to write out **U** directly. However, computations are still quite feasable if they can be based on summarizing statements that identify the location of every nonzero element in **U**. That procedure was adopted for the treatment of the sheet $\rightleftharpoons$ coil transition.[15]

The formulation envisions an accounting of all antiparallel sheets that can be formed with strands containing no more than I residues. The dimensions of **U** are $I(I+3)/2 \times I(I+3)/2$, which is of the modest size 5×5 if the longest strand contains only two residues, but becomes 230×230 when the longest strand contains 20 residues. The fraction of the elements that is nonzero becomes inversely proportion to I as I increases. The location and identity of every nonzero element is given by six simple summarizing statements that are easily incorporated in a computational algorithm.[17]

The correct calculation should use $I = x$, in order to include in the ensemble the conformation where all units in the chain form a single strand. For large x, this approach would require a very large matrix. Often the longest strands in the conformations that make the most important contributions to a sheet $\rightleftharpoons$ coil transition are only a small fraction of the chain length when x is large. For such transitions, a series of calculations performed with the same x and same set of statistical weights, but with increasing values of I, will approach a limit as I increases. Often it is useful to identify this limit with the behavior that might have been calculated using $I = x$, even when the limit is attained at $I \ll x$. For the usual cases where $\delta < 1$ and $\hat{\tau} < 1$, the largest I that should be retained in the calculation is approximated by[18]

$$I_{\max} = \left(x \frac{\ln \delta}{\ln \hat{\tau}} \right)^{1/2} \qquad \text{(X-19)}$$

Incorporation of the generator matrices into the calculation is formally similar to the approach adopted in Eq. (X-14) for the helix $\rightleftharpoons$ coil transition. The algorithm does not actually construct the very large, sparse $\mathscr{G}$, but instead is using the summarizing statements that denote the positions of the nonzero elements in **U**. Computations of $\langle r^2 \rangle_0$ based on this formulation of Z have been performed for I as large as 35.[19] If $\mathscr{G}_i$ for an internal unit were written out explicitly, it would be as large as 3325×3325. An I of this size means that the statistical weight matrix includes elements that specify the interaction of

[17]Mattice, W. L.; Tilstra, L. F. *Biopolymers* **1987**, *26*, 203.
[18]Mattice, W. L.; Scheraga, H. A. *Macromolecules* **1984**, *17*, 2690.
[19]Tilstra, L. F.; Mattice, W. L. *Biopolymers* **1988**, *27*, 805.

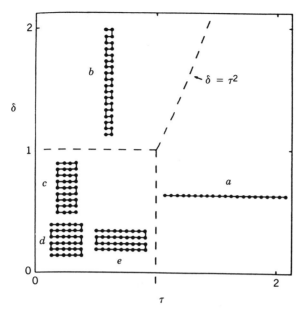

Figure X-7. Optimal intramolecular antiparallel sheets as $t \to \infty$. The symbol τ in this figure corresponds to $\hat{\tau}$ in the text (see footnote 16 on p. 267). Reprinted with permission from Mattice, W. L.; Scheraga, H. A. *Macromolecules* **1984**, *17*, 2690. Copyright 1984 American Chemical Society.

residues i and $i - 69$. Clearly one can incorporate interactions of much higher than second order into tractable formulations of Z.

The shape of the antiparallel sheets formed, and hence the behavior of $\langle r^2 \rangle_0$ during a sheet ⇌ coil transition brought about by increasing t, depends strongly on the values of δ and $\hat{\tau}$. The shapes of the preferred antiparallel sheets at the completion of the transition are depicted in Fig. X-7. The most interesting region of the diagram is at the lower left, since usually $\delta < 1$ and $\hat{\tau} < 1$. If $\delta < \hat{\tau}$, there is a preference for antiparallel sheets with a few long strands, but when $\hat{\tau} < \delta$ the preferred sheet has many short strands.

Antiparallel sheets with a large number of short strands are described as "cross-β" sheets because the predominant chain direction in the strands is perpendicular to the long axis of the sheet.[20] An elaboration of the formulation of Z that incorporates connections of successive strands by short loops as well as by tight bends is important for the treatment of the transition from the random coil to the cross-β sheet, because in this structure the ratio of the number of strands to the number of residues is large.[21] The transition from the random coil to an antiparallel "cross-β" sheet becomes more diffuse when consecutive strands can be connected by loops, as well as by tight bends.[21]

[20]Geddes, A. J.; Parker, K. D.; Atkins, E. D. T.; Beighton, E. *J. Mol. Biol.* **1968**, *32*, 343.
[21]Mattice, W. L.; Scheraga, H. A. *Biopolymers* **1985**, *24*, 565.

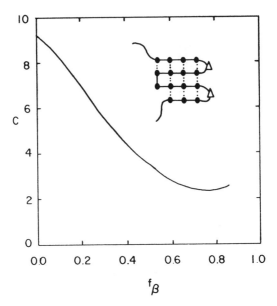

Figure X-8. Variation of C_x with the fraction of residues in the antiparallel sheets for $x = 300$, $\delta = 0.2$, and $\hat{\tau} = 0.4$. The transition was produced by changing t. The formulation of Z allowed for connection of consecutive strands by small loops as well as by tight bends. The structure drawn in the figure depicts a typical sheet when C_x is at its minimum. From Tilstra, L. F.; Mattice, W. L. *Biopolymers* **1988**, *27*, 805. Copyright © 1988 John Wiley & Sons, Inc. Reprinted by permission of John Wiley & Sons, Inc.

Often the sheet $\rightleftharpoons$ coil transition will produce a collapse in the chain, as depicted by the example in Fig. X-8.[19] The collapse seen here is more extensive than the comparatively minor reduction in $\langle r^2 \rangle_0$ produced during the helix $\rightleftharpoons$ coil transition when $\hat{\sigma}$ is at the upper end of its likely range. Pertinent experimental measurements of the mean square dimensions are extremely difficulty for the intramolecular sheet $\rightleftharpoons$ coil transition, because conditions which promote β sheet formation in a homopoly(α-amino acid) also promote aggregation.

The conformational partition function for the sheet $\rightleftharpoons$ coil transition can be made to closely resemble the conformational partition function for a helix $\rightleftharpoons$ coil transition by restricting consideration to only those antiparallel sheets that have a specified, and small, number of residues per strand. For example, if every strand in the sheet must contain two residues, the statistical weight matrix is[15]

$$\mathbf{U}_i = \begin{bmatrix} 1 & \hat{\tau}t & 0 & 0 & 0 \\ 0 & 0 & \hat{\tau}t & 0 & 0 \\ 1 & 0 & 0 & \delta t & 0 \\ 0 & 0 & 0 & 0 & t \\ 1 & 0 & 0 & \delta t & 0 \end{bmatrix}_i \qquad (\text{X-20})$$

which specifies statistical weights of $\hat{\tau}^2 \delta^{(x-2)/2} t^x$ for antiparallel sheets of x units.[22] When the helix $\rightleftharpoons$ coil transition is treated using the 2×2 statistical weight matrix, Eq. (X-5), a helix of x residues has a statistical weight of $\hat{\sigma} s^x$. These two weighting schemes are brought nearly into correspondence with the assignments

$$\hat{\sigma} = \frac{\hat{\tau}^2}{\delta} \tag{X-21}$$

$$s^2 = \delta t^2 \tag{X-22}$$

With this set of assignments, the only difference between the two formulations is that the helix (but not the sheet) can contain an odd number of residues.

3. HELIX–HELIX INTERACTION

In some systems, helix formation is promoted by aggregation of the chains. With homopolymers, the aggregation can be expected to produce particles with varying numbers of chains. The number of chains participating in the aggregate may be of a specific size for suitably constructed amino acid copolymers of carefully selected sequences, as, for example, in the case of the formation of dimers by tropomyosin.[23] A simple model for the influence of in-register dimerization on the helix content can be constructed by an elaboration of the statistical weight matrix in Eq. (X-5).[23]

Imagine two amino acid homopolymers, each with x units, which can participate in the types of dimers depicted in Fig. X-9. The statistical weight of a helical segment in one of the chains is unaffected by the dimerization, unless there is a helical segment at the corresponding position in the other chain. Whenever helices in the two chains are in contact, there is an interaction per residue with statistical weight $\hat{\omega}$ and energy $-RT \ln \hat{\omega}$. Let the chains be labeled A and B, with the helix $\rightleftharpoons$ coil transition of each of the isolated chains described by the statistical weight matrix in Eq. (X-5). Then the combined statistical weight matrix for residue i in chain A *and* residue i in chain B can be written as

$$
\mathbf{U}_{i,AB} =
\begin{array}{c}
\\
c_A c_B \\
c_A h_B \\
h_A c_B \\
h_A h_B
\end{array}
\begin{array}{cccc}
c_A c_B & c_A h_B & h_A c_B & h_A h_B
\end{array}
\left[
\begin{array}{cccc}
1 & \hat{\sigma}_B s_B & \hat{\sigma}_A s_A & \hat{\sigma}_A s_A \hat{\sigma}_B s_B \hat{\omega} \\
1 & s_B & \hat{\sigma}_A s_A & \hat{\sigma}_A s_A s_B \hat{\omega} \\
1 & \hat{\sigma}_B s_B & s_A & s_A \hat{\sigma}_B s_B \hat{\omega} \\
1 & s_B & s_A & s_A s_B \hat{\omega}
\end{array}
\right]_{i,AB}
\tag{X-23}
$$

---

[22] The stipulation that every strand must contain two residues requires that x must be even.
[23] Skolnick, J.; Holtzer, A. *Macromolecules* **1982**, *15*, 303.

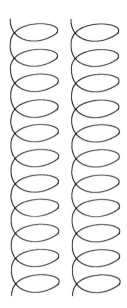

Figure X-9. A dimer in which there is interaction between helices whenever there are helical segments at equivalent positions in the two chains.

where the columns are indexed by the state of residue i in each chain, and the rows are indexed by the state of residue $i - 1$. If $\hat{\omega}$ were assigned the value of 1, this matrix would be just the direct product of the 2×2 statistical weight matrices for the ith residues in the two chains:

$$\mathbf{U}_{i,AB} = (\mathbf{U}_{i,A} \otimes \mathbf{U}_{i,B})\,\mathrm{diag}\,(1, 1, 1, \hat{\omega}_i) \qquad (\text{X-24})$$

The presence of the factor $\hat{\omega}$ in the last column allows a modification in the statistical weight whenever the ith residues in both chains are in a helical state. An attractive helix–helix interaction requires $\hat{\omega} > 1$. The absence of $\hat{\omega}$ in the other columns implies no interaction of residues i in the two chains if either, or both, is in the disordered state.

If $\hat{\sigma}$ is small, say on the order of 10^{-4}, and s is slightly smaller than 1, the isolated chain will have a very low helix content. However, when helices in the two chains are allowed to interact via Eq. (X-23), a value of $\hat{\omega}$ that is somewhat greater than 1 will be sufficient to produce a very large helix content in the dimer. When $\hat{\omega} > 1$, it has the approximate effect of increasing the effective size of s.

It is not obvious how one should combine Eq. (X-23) with the geometry of the chain in order to compute the mean square unperturbed radius of gyration for the dimer. Conformations of the dimer that have two interacting helices, as in Fig. X-10, require modifications of the transformation matrices for the interven-

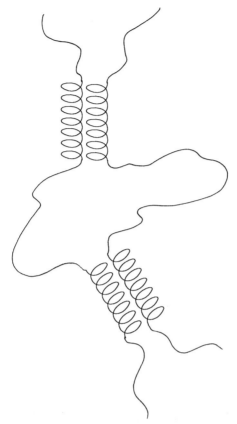

Figure X-10. A dimer in which helices interact in two different regions, causing the formation of loops between the interacting helices.

ing disordered segments in order to produce ring closure, and those modifications present formidable problems. In addition, the requirement of ring closure also suggests that the c states between interacting helices should be weighted differently from the c states that are between a free end and an interacting helix.

PROBLEMS

Appendix B contains answers for all problems in this chapter.

  X-1. What is **T** for the right-handed α helix? For the left-handed α helix?

  X-2. What is the most compact form (most reduced form) of the statistical weight matrix in Eqs. (X-1)?

X-3. Devise a statistical weight matrix for the helix $\rightleftharpoons$ coil transition when both right- and left-handed helices may be present, but with different energies for propagation. Assume that all units in each individual helix must be of the same chirality. If $\hat{\sigma} = 10^{-4}$, and s for the right-handed helix is always 10% larger than s for the left-handed helix, what is the upper limit for the fraction of residues in the left-handed helix, as $n_p \rightarrow \infty$?

X-4. Equations (X-2) and (X-5) describe nearly the same transition. If the chain contains a very large number of residues, and the transition is produced by changing s, for which values of $\hat{\sigma}$ is the difference between the two descriptions perceptible?

X-5. If two chains with $x = 500$, $\hat{\sigma} = 10^{-4}$, and $s = 0.9$ are allowed to interact, what value of $\hat{\omega}$ is required in order to bring the helix content to 0.9 in the dimer?

X-6. Compute **r** for a right-handed α helix with 500 units. Then compute $\langle \mathbf{r} \rangle_0$ for a poly(L-alanine) chain of 500 units, using $\hat{\sigma} = 10^{-4}$ and various values of s, so that the helix content ranges from 0 to nearly 1. Why isn't $\langle \mathbf{r} \rangle_0$ a linear function of the helix content?

X-7. The units in polyisocyanates have access to two conformations, denoted here as h^+ and h^-. Junctions of unlike chirality, either as h^+h^- or as h^-h^+, have much higher conformational energy than the pairs, h^+h^+ and h^-h^-, that propagate the same chirality. In the absence of a chiral side chain, h^+ and h^- must have the same probability. Suggest a formulation for Z for this chain. In what ways is the statistical weight matrix similar to, and different from, the 2×2 matrix used for the helix $\rightleftharpoons$ coil transition in a chiral poly(α amino acid)?

X-8. In the chiral polyisocyanate, poly[(R)-1-deuterio-n-hexyl isocyanate)], the states denoted by h^+ and h^- in Problem X-7 are no longer equally probable. Experiments indicate the following sizes for two energy differences:[24]

$$E_{h^+} - E_{h^-} = 4\,\mathrm{J\ mol^{-1}}$$
$$E_{h^+h^-} - E_{h^-h^-} = 16000\,\mathrm{J\ mol^{-1}}$$

Devise a matrix formulation for Z for this chain. In the limit as $n \rightarrow \infty$, what are the values of p_{h^-} and the average number of units in a sequence of h^- values? How can you rationalize these results?

X-9. The statistical weight matrix in Eq. (X-5) assigned a weight of $\hat{\sigma} s^{n_h}$ to a sequence of $n_h + 2$ consecutive residues in the states ch $\cdots$ hc.

[24]Lifson, S.; Andreola, C.; Peterson, N. C.; Green, M. M. *J. Am. Chem. Soc.* **1989**, *111*, 8850.

All of the end effect (the factor of $\hat{\sigma}$) is associated with the first h in the sequence of hs. Devise a statistical weight matrix that would apportion the end effect in equal manner between the first and last h in the sequence (each of these hs should have a factor of $\hat{\sigma}^{1/2}$ in its statistical weight). What is the most compact form of this matrix? Will the Z formulated from this matrix predict different properties than the Z formulated from Eq. (X-5)?

XI Stars, Grafts, and Articulated Side Chains

The conformational partition function for a chain molecule is formulated in rotational isomeric state (RIS) theory as an ordered string of statistical weight matrices. This string has a single well-defined origin, at bond 1, using the row vector denoted by U_1, and a single well-defined end, at bond n, using the column vector denoted by U_n. The classic application envisions a molecule that can be described as a string of n bond vectors, connected head-to-tail. For each internal bond i, the preceding and succeeding bonds, $i - 1$ and $i + 1$ respectively, are well defined.

This chapter describes methods for the construction of conformational partition functions for macromolecules that contain one or more branches with internal degrees of freedom that should not be suppressed in the formulation of Z. The prototype is a *trifunctional star*, consisting of three branches that emanate from a common atom. Application of random-flight statistics to trifunctional stars shows that $\langle s^2 \rangle_0$ is smaller than the result for the linear chain with the same number of bonds. The ratio of the mean-square unperturbed radii of gyration for the branched molecule and the linear chain with the same number of bonds is denoted by g:[1]

$$g \equiv \frac{\langle s^2 \rangle_{0,\text{branched}}}{\langle s^2 \rangle_{0,\text{linear}}} \tag{XI-1}$$

The size of the reduction in $\langle s^2 \rangle_0$ depends on the relationship between the numbers of bonds in the three branches. The most severe reduction in $\langle s^2 \rangle_0$ for a trifunctional star, $g = 7/9$, occurs when all three branches contain the same number of bonds. This result from random flight statistics shows that the branches must be correctly incorporated into Z for an accurate description of the conformation-dependent physical properties, such as $\langle s^2 \rangle_0$, in the RIS model. If the indexing of the bonds begins at the free end of a branch that contains n_1 bonds, an ambiguity needs to be resolved when the indexing reaches the branch point. Which bond succeeds the $C\text{—}C^\alpha$ bond, where C^α denotes the atom at the branch point? An error will be introduced if the question is resolved in favor of the first bond in either of the remaining branches. A correct solution must

[1]Zimm, B. H.; Stockmayer, W. H. *J. Chem. Phys.* **1949**, *17*, 1302.

consider both remaining branches simultaneously, without favoring one over the other.

The application to trifunctional stars can be generalized to two linear chains connected by a *crosslink*, as well as *grafts* with a few (or many) branches attached to the chain. Also included in this chapter are monomer units containing a *branch with internal degrees of freedom*, as in a polymer of 1-butene. The problem common to all of these molecules is the necessity for treatment of trifunctional branch points in the formulation of Z.

Excluded from this chapter is consideration of molecules in which the branching leads to closed rings. Description of such systems would require the formulation of a Z in which $\mathbf{r}$ for a subchain is constrained to be $\mathbf{0}$, by the requirement of ring closure. A general solution in closed form is not available for the formulation of Z for such molecules. Approximations suitable for the evaluation of macrocyclization equilibrium constants will be discussed in Chapter XIII.

1. TRI- AND TETRAFUNCTIONAL STARS

The class of macromolecules known as *stars* includes the simplest branched polymers. An f-functional star contains f branches that emanate from a common site. The linear chain is the special case where $f = 2$. The word "star" is reserved for those macromolecules where $f > 2$. The number of bonds in the star is denoted by n, and n_i denotes the number of bonds in the ith branch, such that $n = \sum_i n_i$. It is not necessary that each branch contain the same number of bonds. The basic concepts required for the formulation of Z can be demonstrated with the trifunctional star, where $f = 3$.[2]

A. U for Bonds Near a Trifunctional Branch Point

Consider a trifunctional polyethylene star, with the branches arbitrarily numbered 1, 2, and 3, as depicted in Fig. XI-1. The prototype is isobutane, where $n_i = 1$ for all i. This molecule is uninteresting for present purposes, because the conformation is fixed if constant l and θ are adopted. The longest linear chain through isobutane contains only two C—C bonds, and a chain must contain a minimum of three bonds before there is an internal bond at which rotational isomerism becomes possible. The further development here assumes that all of the n_i are much greater than one. No two of the n_i need be equal.

The formulation of Z incorporates all first- and second-order interactions that might occur in a simple chain with bonds subject to a symmetric threefold rotation potential energy function and nearest-neighbor interdependencies. The first- and second-order interactions were weighted by σ, ω, τ, and ψ in Eq. (VII-1). In the present application, this τ will be replaced by τ_2, to empha-

[2]Mattice, W. L. *Macromolecules* **1975**, *8*, 644.

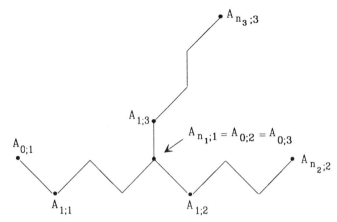

Figure XI-1. A trifunctional star with bonds indexed from 1 to n_i in branch i. Indexing begins at the free end for branch 1, and at the branch point for branches 2 and 3, and $A_{i;j}$ denotes the ith atom in branch j. The atom at the branch point can be denoted equivalently by $A_{n_1;1}, A_{0;2}$, or $A_{0;3}$.

size that it weights a second-order interaction. The symbol τ_1 will be reserved for a first-order interaction that is similar to the τ first defined, in Eq. (VIII-1), during the discussion of $\mathbf{D}_d$ for a vinyl polymer. The statistical weight matrix was written in Eq. (VII-3) as $\mathbf{VD}$, where $\mathbf{D}$ is a diagonal matrix that includes the first-order interactions and $\mathbf{V}$ is a matrix that includes the second-order interactions. That equation is reproduced here, with τ_2 replacing τ, because extensive use will be made of $\mathbf{D}$ and $\mathbf{V}$ in this section:

$$\mathbf{VD} \equiv \begin{bmatrix} \tau_2 & 1 & 1 \\ 1 & \psi & \omega \\ 1 & \omega & \psi \end{bmatrix} \text{diag} (1, \sigma, \sigma) \qquad \text{(XI-2)}$$

The formulation of Z begins at the free end of the branch that is arbitrarily designated as branch 1. The final result for Z will be independent of which branch was selected. When indexing for bonds in this branch commences at the free end, the contribution to Z from bonds 1 through $n_1 - 1$ in branch 1 is written as $\mathbf{U}_1\mathbf{U}_2 \cdots \mathbf{U}_{n_1-1}$, where $\mathbf{U}_1 = \begin{bmatrix} 1 & 0 & 0 \end{bmatrix}$, and $\mathbf{U}_2$ through $\mathbf{U}_{n_1-1}$ are given by Eq. (XI-2). Thus far, the presence of the branch has exerted no influence on the formulation of Z. Consideration of the branch begins with $\mathbf{U}_{n_1}$.

The Last Bond in Branch 1. The statistical weight matrix for bond n_1 in branch 1 must include the first- and second-order interactions of the initial atoms within branches 2 and 3 with atoms in branch 1. In order to define precisely the interactions involved, we introduce the notation $A_{i;j}$ for the ith atom in branch j, as illustrated in Fig. XI-1. The statistical weight matrix for the last bond in branch 1 must contain the first-order interactions of $A_{1;2}$ and $A_{1;3}$ with $A_{n_1-2;1}$, and the second-order interactions of the same two atoms ($A_{1;2}$ and $A_{1;3}$) with $A_{n_1-3;1}$, as shown in Fig. XI-2. The strategy used in the formulation of $\mathbf{U}_{n_1;1}$ has much in common with that presented for $\mathbf{U}_d$ and $\mathbf{U}_l$ in the discussion of the vinyl polymers at the beginning of Chapter VIII. There the t state for the bond preceding the branch point was well defined, because the vinyl polymer has a long main chain that is distinguishable by its length from the much shorter chains that might be drawn such that they end (or start) at a side group.

In order to unambiguously define the rotational isomeric states at the last bond of branch 1 of the trifunctional star, we arbitrarily shall use atoms $A_{n_1-2;1}, A_{n_1-1;1}, A_{n_1;1}$, and $A_{1;2}$. This arbitrary choice has the appearance of defining a main chain that consists of branches 1 and 2, with branch 3 playing the role of a side chain. The formulation that will be developed here yields exactly the same result for Z for all of the six possible arbitrary assignments of the branches as branches 1, 2, and 3. When the arbitrary assignment of branches 1 and 2 has been completed, the local coordinate system for the first bond in branch 2 is well defined. The branch point will be described as a d pseudoasymmetric center if $A_{1;3}$ has a positive z component in this local coordinate system, and as an l pseudoasymmetric center otherwise. This definition is the same one employed for vinyl polymers in Chapter VIII.

The first- and second-order interactions for $\mathbf{U}_{n_1;1}$ at a d pseudoasymmetric center are incorporated correctly as the product of two matrices, $\mathbf{V}_{n_1;1;d}\mathbf{D}_{n_1;1;d}$:

$$\mathbf{D}_{n_1;1;d} = \text{diag}\,(1, \sigma, \sigma)\,\text{diag}\,(\sigma, \sigma, 1) \qquad (\text{XI-3})$$

The first diagonal matrix on the right-hand side contains the first-order interactions of $A_{1;2}$ with $A_{n_1-2;1}$, and the second diagonal matrix has the first-order

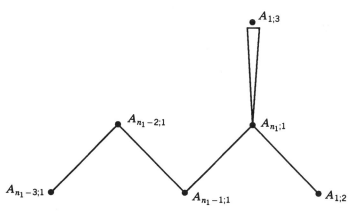

Figure XI-2. Atoms that participate in the first- and second-order interactions in $\mathbf{U}_{n_1}$.

interactions of $A_{1;3}$ with $A_{n_1-2;1}$. In both diagonal matrices, the columns are indexed in the order of t, g^+, g^-, with the state defined by $A_{n_1-2;1}, A_{n_1-1;1}, A_{n_1;1}$, and $A_{1;2}$, as required by the decision to let the state at the last bond of branch 1 be defined by the "main chain" consisting of branches 1 and 2. These two diagonal matrices contain the same elements, but written in a different order that corresponds to a cyclic permutation. The cyclic permutation must account for the fact that when the torsion angle at the last bond in branch 1 places $A_{1;2}$ in the positions of t, g^+, g^-, respectively, atom $A_{1;3}$ appears to be in positions g^+, g^-, t, respectively, as shown in Fig. XI-3. Alternatively, Eq. (XI-3) can be written as

$$\mathbf{D}_{n_1;1;d} = \mathbf{D}\mathbf{Q}_d^T\mathbf{D}\mathbf{Q}_d \qquad \text{(XI-4)}$$

where $\mathbf{D}$ is the diagonal matrix defined in Eq. (XI-2), and $\mathbf{Q}_d$ is the matrix that accomplishes the desired cyclic permutation for representation of the first-order interaction of $A_{1;3}$ with $A_{n_1-2;1}$:

$$\mathbf{Q}_d \equiv \begin{bmatrix} 0 & 0 & 1 \\ 1 & 0 & 0 \\ 0 & 1 & 0 \end{bmatrix} \qquad \text{(XI-5)}$$

The cyclic permutation accomplished with $\mathbf{Q}_d$ will be used extensively in this chapter. The result obtained from Eq. (XI-4) can be written as

$$\mathbf{D}_{n_1;1;d} = \text{diag}\,(1, \tau_1, 1) \qquad \text{(XI-6)}$$

where the statistical weight for g^+, which has simultaneous first-order interactions of $A_{n_1-2;1}$ with both $A_{1;2}$ and $A_{1;3}$, is assigned as $\sigma\tau_1$ instead of σ^2, and

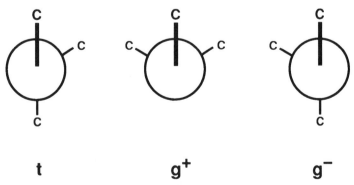

Figure XI-3. Projection along the last bond in branch 1, showing the positions of the atoms that participate in first-order interactions in the t, g^+, g^- states when the branch point is a d pseudoasymmetric center.

each element in $D_{n_1;1;d}$ has been divided by σ. For branched alkanes, the value of τ_1 is close to the value of σ.[3]

The second-order interactions that must appear in $U_{n_1;1;d}$ are conveniently written using V and the matrix that accomplishes the necessary cyclic permutation for the second-order interactions that involve $A_{1;3}$. Let $[V]_{jk}$ denote the element in row j, column k of a matrix V. Then the desired matrix of second-order interactions, denoted by $V_{n_1;1;d}$, has elements that are constructed as

$$[V_{n_1;1;d}]_{ij} = [V]_{ij}[VQ_d]_{ij} \tag{XI-7}$$

where the first matrix on the right-hand side contains the second-order interactions of $A_{1;2}$ with $A_{n_1-3;1}$, and VQ_d contains the second-order interactions of $A_{1;3}$ with $A_{n_1-3;1}$. The final result is

$$U_{n_1;1;d} = V_{n_1;1;d}D_{n-1;1;d} = \begin{bmatrix} \tau_2 & \tau_1 & \tau_2 \\ \psi & \tau_1\psi\omega & \omega \\ \omega & \tau_1\psi\omega & \psi \end{bmatrix} \tag{XI-8}$$

With a different arbitrary scheme for numbering the branches, the branch point might meet the definition of an l pseudoasymmetric center, instead of a d pseudoasymmetric center. Then the statistical weight matrix for the last bond in branch 1, $U_{n_1;1;l}$, is obtained from $U_{n_1;1;d}$ by the operation used to convert between statistical weight matrices for vinyl polymers considered as d or l pseudoasymmetric centers, Table VIII-1, which employs the matrix denoted by Q [see Eq. (VIII-3)].

$$U_{n_1;1;l} = QU_{n_1;1;d}Q \tag{XI-9}$$

The First Bonds in Branches 2 and 3: The Rectangular Matrix. The construction of $U_{n_1;1}$ was a straightforward implementation of the procedures adopted in Chapter VIII for the treatment of vinyl polymers. The situation becomes more complicated on consideration of the next statistical weight matrix in the serial product. The atoms that participate in the first- and second-order interactions are depicted in Fig. XI-4. Note that $A_{2;2}$ and $A_{2;3}$ can potentially participate in second-order interactions with each other and with $A_{n_1-2;1}$. The second-order interactions that involve $A_{n_1-2;1}$ depend on the state at the last bond in branch 1. This state indexes the columns of $U_{n_1;1}$, and will index the rows of the statistical weight matrix that follows $U_{n_1;1}$ in the expression for Z. The second-order

[3]Mathur, S. C.; Mattice, W. L. *Makromol. Chem.* **1988**, *189*, 2893.

interaction of $A_{2;2}$ with $A_{2;3}$, however, is independent of any states in branch 1, and depends instead on the states at the first bond in branch 2 and branch 3. This problem is handled by an expansion in the number of columns in the statistical weight matrix.

The rationale has much in common with the expansion in dimensions that was employed in Eq. (IX-7), in order to account for the third-order interaction denoted by β in poly(1,4-*trans*-butadiene). There it was necessary to formulate a statistical weight matrix where one could associate an element with the states at three consecutive bonds. Here also we need to associate an element with the states at three bonds, but the bonds are not consecutive. Instead they are the three bonds to the atom at the trifunctional branch point.

The expanded matrix is of dimensions 3×9. The rows are indexed by the states of the last bond in branch 1. The first set of three columns is for a t state at the first bond in branch 2, the next set of three columns is for a g^+ state at this bond, and the final set of three columns is for the g^- state. The state at the first bond in branch 2 is defined by $A_{n_1-1;1}, A_{n_1;1}, A_{1;2}$, and $A_{2;2}$. Within each set of three columns, the first, second, and third columns are also indexed by the t, g^+, and g^- states at the first bond in branch 3, in that order. The state at the first bond in branch 3 is defined by $A_{n_1-1;1}, A_{n_1;1}, A_{1;3}$, and $A_{2;3}$. In the case of a d pseudoasymmetric center, this procedure ultimately leads to

$$
\mathbf{U}_{b;d} = \begin{array}{c} \\ t \\ g^+ \\ g^- \end{array}
\begin{array}{c}
\begin{array}{ccccccccc} tt & tg^+ & tg^- & g^+t & g^+g^+ & g^+g^- & g^-t & g^-g^+ & g^-g^- \end{array} \\
\left[
\begin{array}{ccccccccc}
\tau_2\omega & \tau_1\tau_2\psi^2 & \tau_2\omega & 1 & \tau_1\psi & \tau_2\omega & \tau_1\psi & \tau_1^2\psi\omega & \tau_1\omega \\
\omega & \tau_1\psi\omega & \psi & \psi & \tau_1\psi\omega & \tau_2\psi^2 & \tau_1\psi\omega & \tau_1^2\omega^3 & \tau_1\psi\omega \\
\tau_2\omega & \tau_1\psi & 1 & \tau_2\omega & \tau_1\omega & \tau_2\omega & \tau_1\tau_2\psi^2 & \tau_1^2\psi\omega & \tau_1\psi
\end{array}
\right]
\end{array}
$$

(XI-10)

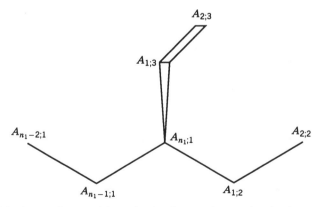

Figure XI-4. Atoms that participate in the first- and second-order interactions in the rectangular statistical weight matrix.

This matrix appears to have a complicated composition. It results from the large number of potential first- and second-order interactions in the vicinity of a trifunctional branch point.

In spite of its complicated appearance, $U_{b;d}$ can be formulated in a systematic manner, using only $D, I_3, Q_d, V, [1 \quad 1 \quad 1]$, and the expression

$$U_{b;d} = V_{b;d}D_{b;d} \tag{XI-11}$$

The matrix with the first-order interactions can be written as

$$D_{b;d} = (DQ_d DQ_d^T) \otimes (DQ_d^T DQ_d)$$
$$= \sigma^2 \text{ diag}(1, 1, \tau_1) \otimes \text{diag}(1, \tau_1, 1) \tag{XI-12}$$

where $\sigma \text{ diag}(1, 1, \tau_1)$ arises from the first-order interactions of $A_{2;2}$ with $A_{n_1-1;1}$ and $A_{1;3}$, and $\sigma \text{ diag}(1, \tau_1, 1)$ arises from the first-order interaction of $A_{2;3}$ with $A_{n_1-1;1}$ and $A_{1;2}$. The leading factor of σ^2, which is common to all elements in $D_{b;d}$, was deleted in Eq. (XI-10).

The matrix of the second-order interactions is conveniently considered in three parts:

$$[V_{b;d}]_{ij} = [V_{12}]_{ij}[V_{13;d}]_{ij}[V_{23;d}]_{ij} \tag{XI-13}$$

where $V_{hk;d}$ incorporates the second-order interactions of the pair of atoms that are two bonds removed from the branch point in branches h and k when the branch point constitutes a d pseudoasymmetric center. The atoms that contribute to each term in Eq. (XI-13) are identified in Fig. XI-4. The second-order interactions of $A_{2;2}$ and $A_{n_1-2;1}$ appear in V_{12} as

$$V_{12} = V \otimes [1 \quad 1 \quad 1] \tag{XI-14}$$

These second-order interactions are independent of the state of the first bond in branch 3, and are also independent of the stereochemistry of the attachment of branch 3 to the branch point. The second-order interaction of $A_{2;3}$ with $A_{n_1-2;1}$ is independent of the state of the first bond in branch 2. The statistical weights for this second-order interactions must appear in a 3×3 block that is repeated three times. A cyclic permutation on V is required because the definition of the state of the last bond of branch 1, and hence the indexing of the rows, uses $A_{1;2}$ instead of $A_{1;3}$:

$$V_{13;d} = [1 \quad 1 \quad 1] \otimes (Q_d^T V) \tag{XI-15}$$

The contribution represented by $V_{23;d}$ is independent of the state of the last bond in branch 1. Consequently the three rows of $V_{23;d}$ must be identical. Each block of three elements in a row corresponds to a row in an appropriate cyclic permutation of $\mathbf{V}$:

$$\mathbf{V}_{23;d} = \begin{bmatrix} 1 \\ 1 \\ 1 \end{bmatrix} \otimes (\mathbf{Q}_d^R [\mathbf{I}_3 \otimes (\mathbf{V}\mathbf{Q}_d)]) \qquad \text{(XI-16)}$$

where $\mathbf{Q}_d^R$ is a row vector with the nine elements of $\mathbf{Q}_d$ in reading order:

$$\mathbf{Q}_d^R \equiv [0 \quad 0 \quad 1 \quad 1 \quad 0 \quad 0 \quad 0 \quad 1 \quad 0] \qquad \text{(XI-17)}$$

The result of the operation in Eq. (XI-16) is

$$\mathbf{V}_{23;d} = \begin{bmatrix} \omega & \psi & 1 & 1 & 1 & \tau_2 & \psi & \omega & 1 \\ \omega & \psi & 1 & 1 & 1 & \tau_2 & \psi & \omega & 1 \\ \omega & \psi & 1 & 1 & 1 & \tau_2 & \psi & \omega & 1 \end{bmatrix} \qquad \text{(XI-18)}$$

Multiplication of corresponding elements in the three rectangular matrices, as required by Eq. (XI-13), yields

$$\mathbf{V}_{b;d} = \begin{bmatrix} \tau_2\omega & \tau_2\psi^2 & \tau_2\omega & 1 & \psi & \tau_2\omega & \psi & \psi\omega & \omega \\ \omega & \psi\omega & \psi & \psi & \psi\omega & \tau_2\psi^2 & \psi\omega & \omega^3 & \psi\omega \\ \tau_2\omega & \psi & 1 & \tau_2\omega & \omega & \tau_2\omega & \tau_2\psi^2 & \psi\omega & \psi \end{bmatrix} \qquad \text{(XI-19)}$$

The final expression for $\mathbf{U}_{b;d}$, Eq. (XI-10), is obtained systematically as $\mathbf{V}_{b;d}\mathbf{D}_{b;d}$.

Replacements for Eqs. (XI-11), (XI-12), (XI-13), (XI-15), (XI-16), and (XI-10) for the case where the arbitrary numbering of the branches yields a branch point that meets the definition of an l pseudoasymmetric center are, respectively,

$$\mathbf{U}_{b;l} = \mathbf{V}_{b;l}\mathbf{D}_{b;l} \qquad \text{(XI-20)}$$

$$\mathbf{D}_{b;l} = (\mathbf{Q} \otimes \mathbf{Q})\mathbf{D}_{b;d}(\mathbf{Q} \otimes \mathbf{Q}) \qquad \text{(XI-21)}$$

$$[\mathbf{V}_{b;l}]_{ij} = [\mathbf{V}_{12}]_{ij}[\mathbf{V}_{13;l}]_{ij}[\mathbf{V}_{23;l}]_{ij} \qquad \text{(XI-22)}$$

$$\mathbf{V}_{13;l} = \mathbf{Q}\mathbf{V}_{13;d}(\mathbf{I}_3 \otimes \mathbf{Q}) \qquad \text{(XI-23)}$$

$$\mathbf{V}_{23;l} = \mathbf{V}_{23;d}(\mathbf{Q} \otimes \mathbf{Q}) \qquad \text{(XI-24)}$$

$$\mathbf{U}_{b;l} = \mathbf{Q}\mathbf{U}_{b;d}(\mathbf{Q} \otimes \mathbf{Q}) \qquad \text{(XI-25)}$$

The Second Bonds in Branches 2 and 3. The statistical weight matrix for the second bond in branch 2 must incorporate the first- and second-order interactions in the fragment depicted in Fig. XI-5. The first-order interactions are given by **D**. The second-order interactions are

$$[\mathbf{V}_{2;2;d}]_{ij} = [\mathbf{V}]_{ij}[\mathbf{Q}_d\mathbf{V}]_{ij} \qquad \text{(XI-26)}$$

where the second factor on the right-hand side incorporates the second-order interactions between $A_{3;2}$ and $A_{1;3}$.

$$\mathbf{U}_{2;2;d} = \begin{bmatrix} \tau_2 & \sigma\omega & \sigma\psi \\ \tau_2 & \sigma\psi & \sigma\omega \\ 1 & \sigma\psi\omega & \sigma\psi\omega \end{bmatrix} \qquad \text{(XI-27)}$$

Proceding in a similar fashion, the statistical weight matrix for the second bond in branch 3 is

$$\mathbf{U}_{2;3;d} = ([\mathbf{V}]_{ij}[\mathbf{Q}_d^T\mathbf{V}]_{ij})\mathbf{D} = \begin{bmatrix} \tau_2 & \sigma\psi & \sigma\omega \\ 1 & \sigma\psi\omega & \sigma\psi\omega \\ \tau_2 & \sigma\omega & \sigma\psi \end{bmatrix} \qquad \text{(XI-28)}$$

If the branch has an l pseudoasymmetric center, the statistical weight matrices for the second bond in branches 2 and 3 are given by

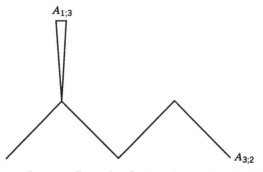

Figure XI-5. Atoms that contribute the first- and second-order interactions in the statistical weight matrix for the second bond in branch 2.

TABLE XI-1. The Special Statistical Weight Matrices Near an Isolated Trifunctional Branch Point[a]

Matrix	$\mathbf{V}$ Followed (on the Next Line) by $\mathbf{D}$
$\mathbf{U}_{n_1;1;d}$	$[\mathbf{V}]_{ij}[\mathbf{V}\mathbf{Q}_d]_{ij}$ $\mathbf{D}\mathbf{Q}_d^T\mathbf{D}\mathbf{Q}_d$
$\mathbf{U}_{b;d}$	$[\mathbf{V}\otimes[1\ \ 1\ \ 1]]_{ij}[[1\ \ 1\ \ 1]\otimes(\mathbf{Q}_d^T\mathbf{V})]_{ij}\cdot$ $[[1\ \ 1\ \ 1]^T\otimes(\mathbf{Q}_d^R[\mathbf{I}_3\otimes(\mathbf{V}\mathbf{Q}_d)])]_{ij}$ $(\mathbf{D}\mathbf{Q}_d\mathbf{D}\mathbf{Q}_d^T)\otimes(\mathbf{D}\mathbf{Q}_d^T\mathbf{D}\mathbf{Q}_d)$
$\mathbf{U}_{2;2;d}$	$[\mathbf{V}]_{ij}[\mathbf{Q}_d\mathbf{V}]_{ij}$ $\mathbf{D}$
$\mathbf{U}_{2;3;d}$	$[\mathbf{V}]_{ij}[\mathbf{Q}_d^T\mathbf{V}]_{ij}$ $\mathbf{D}$

[a]Statistical weight matrices near an l pseudoasymmetric center are generated from these expressions using $\mathbf{Q}$ and Eqs. (XI-9), (XI-25), (XI-29), and (XI-30).

$$\mathbf{U}_{2;2;l} = \mathbf{Q}\mathbf{U}_{2;2;d}\mathbf{Q} \qquad (XI\text{-}29)$$

$$\mathbf{U}_{2;3;l} = \mathbf{Q}\mathbf{U}_{2;3;d}\mathbf{Q} \qquad (XI\text{-}30)$$

Bonds in branches 2 and 3 that are more remote from the branch point will use the statistical weight matrix in Eq. (XI-2) if the treatment is confined to first- and second-order interactions. The special statistical weight matrices required by the presence of a single trifunctional branch point are $\mathbf{U}_{n_1;1;d}, \mathbf{U}_{b;d}, \mathbf{U}_{2;2;d}$, and $\mathbf{U}_{2;3;d}$, or their counterparts for l replacing d, which are produced using $\mathbf{Q}$. The construction of these matrices is summarized in Table XI-1. The only information required is $\mathbf{D}, \mathbf{I}_3, \mathbf{Q}_d, \mathbf{V}$, and $[1\ \ 1\ \ 1]$.

In Section XI.1.B the statistical weight matrices for the bonds at the branch point and for the internal bonds in the three branches will be combined to generate Z. Before proceding with that task, we will define the statistical weight matrices for bonds near a tetrafunctional branch point.

Tetrafunctional Branch Point. The prototype of an alkane with a tetrafunctional branch point is neopentane, for which all four n_i are 1. Here we consider an alkane with larger n_i, treated with first- and second-order interactions. Since the configuration at the tetrafunctional branch point is assumed to be tetrahedral, only a single set of statistical weight matrices is needed.[4] The only information required for the assembly of the special statistical weight matrices for the bonds near a tetrafunctional branch point is $\mathbf{D}, \mathbf{I}_3, \mathbf{Q}_d, \mathbf{V}$, and $[1\ \ 1\ \ 1]$.

[4]Mattice, W. L.; Carpenter, D. K. *Macromolecules* **1976**, *9*, 53.

The statistical weight matrices for the last bond in branch 1, and the second bonds in branches 2, 3, and 4, are

$$\mathbf{U}_{n_1;1} = ([\mathbf{V}]_{ij}[\mathbf{VQ}_d]_{ij}[\mathbf{VQ}_d^T]_{ij}) \text{ diag } (1,1,1) = \begin{bmatrix} \tau_2 & \tau_2 & \tau_2 \\ \psi\omega & \psi\omega & \psi\omega \\ \psi\omega & \psi\omega & \psi\omega \end{bmatrix} \quad \text{(XI-31)}$$

$$\mathbf{U}_{2;k} = ([\mathbf{V}]_{ij}[\mathbf{Q}_d\mathbf{V}]_{ij}[\mathbf{Q}_d^T\mathbf{V}]_{ij}) \text{ diag } (1,\sigma,\sigma)$$

$$= \begin{bmatrix} \tau_2 & \sigma\psi\omega & \sigma\psi\omega \\ \tau_2 & \sigma\psi\omega & \sigma\psi\omega \\ \tau_2 & \sigma\psi\omega & \sigma\psi\omega \end{bmatrix} \quad k = 2,3,4 \quad \text{(XI-32)}$$

The rectangular matrix corresponding to $\mathbf{U}_b$ above is of dimensions 3×3^3, has the rows indexed by the state of the last bond in branch 1, and the columns indexed by the states of the first bonds in branches 2, 3, and 4. The t, g^+, and g^- states at the first bond in branch 2 index the first, second, and third blocks of nine columns, respectively. Within each of these blocks of nine columns, the t, g^+, and g^- states at the first bond in branch 3 index the first, second, and third block of three columns, respectively. Within each of these block of three columns, the first, second, and third columns are indexed by the t, g^+, and g^- states at the first bond in branch 4. The matrix of first-order interactions, $\mathbf{D}_d$, is simply diag $(1, 1, \cdots, 1)$. Its dimensions are $3^3 \times 3^3$.

The rectangular matrix $\mathbf{U}_b, 3 \times 3^3$, must contain the second-order interactions of all six combinations of pairs of atoms that are two bonds removed from the branch point. The second-order interactions of the atoms in branches i and j appear in $\mathbf{V}_{ij}$, and

$$[\mathbf{V}_b]_{ij} = [\mathbf{V}_{12}]_{ij}[\mathbf{V}_{13}]_{ij}[\mathbf{V}_{14}]_{ij}[\mathbf{V}_{23}]_{ij}[\mathbf{V}_{24}]_{ij}[\mathbf{V}_{34}]_{ij} \quad \text{(XI-33)}$$

The matrix of second-order interactions of the pair of atoms in the "main chain" containing branches 1 and 2 is

$$\mathbf{V}_{12} = \mathbf{V} \otimes [1 \quad 1 \quad 1] \otimes [1 \quad 1 \quad 1] \quad \text{(XI-34)}$$

Assuming that branches 3 and 4 were numbered so that a d pseudoasymmetric center is defined by the main chain and branch 3, and an l pseudoasymmetric center is defined by the main chain and branch 4, the remaining second-order interactions are properly incorporated with

$$\mathbf{V}_{13} = [1 \quad 1 \quad 1] \otimes (\mathbf{Q}_d^T\mathbf{V}) \otimes [1 \quad 1 \quad 1] \quad \text{(XI-35)}$$

$$\mathbf{V}_{14} = [1 \quad 1 \quad 1] \otimes [1 \quad 1 \quad 1] \otimes (\mathbf{Q}_d\mathbf{V}) \quad \text{(XI-36)}$$

$$\mathbf{V}_{23} = \begin{bmatrix} 1 \\ 1 \\ 1 \end{bmatrix} \otimes (\mathbf{Q}_d^R[\mathbf{I}_3 \otimes (\mathbf{V}\mathbf{Q}_d)]) \otimes [1 \quad 1 \quad 1] \qquad \text{(XI-37)}$$

$$\mathbf{V}_{24} = \begin{bmatrix} 1 \\ 1 \\ 1 \end{bmatrix} \otimes [[1 \quad 1 \quad 1] \otimes (\mathbf{Q}_d^T\mathbf{V}\mathbf{Q}_d^T)]^R \qquad \text{(XI-38)}$$

$$\mathbf{V}_{34} = \begin{bmatrix} 1 \\ 1 \\ 1 \end{bmatrix} \otimes [1 \quad 1 \quad 1] \otimes (\mathbf{Q}_d^R[\mathbf{I}_3 \otimes (\mathbf{V}\mathbf{Q}_d)]) \qquad \text{(XI-39)}$$

Only nine of the 81 elements in $\mathbf{U}_b$ do not contain at least one factor of ω, as a consequence of the numerous short-range interactions near a tetrafunctional branch point.

B. Z for Tri- and Tetrafunctional Stars

The conformational partition function for a trifunctional polyethylene star with $n_2 = n_3$ is generated from the statistical weight matrices described in the previous section as[2]

$$Z = \mathbf{U}_1\mathbf{U}^{n_1-2}\mathbf{U}_{n_1;1}\mathbf{U}_b(\mathbf{U}_{2;2} \otimes \mathbf{U}_{2;3})(\mathbf{U} \otimes \mathbf{U})^{n_2-3}(\mathbf{U}_n \otimes \mathbf{U}_n) \qquad \text{(XI-40)}$$

where the matrices of the appropriate chirality are used for $\mathbf{U}_{n_1;1}, \mathbf{U}_b, \mathbf{U}_{2;2}$, and $\mathbf{U}_{2;3}$. The final result for Z is, of course, independent of the chirality of the trifunctional branch point. The terminal matrices, denoted by $\mathbf{U}_n$, are the columns specified by Eq. (IV-24). The direct product is used for incorporation of branches 2 and 3 into Z because the order for indexing the rows and columns is $tt, tg^+, tg^-, g^+t, g^+g^+, g^+g^-, g^-t, g^-g^+, g^-g^-$, with the first index being for the bond in branch 2, and the second index for the bond in branch 3. This order of indexing is the same as the order for indexing the columns in $\mathbf{U}_b$. In each $\mathbf{U} \otimes \mathbf{U}$, the first $\mathbf{U}$ is contributed by a bond in branch 2, and the second $\mathbf{U}$ is contributed by the bond in branch 3 that is the equivalent distance from the branch point. Thus each element in $\mathbf{U} \otimes \mathbf{U}$ can unambiguously be associated with the states at four bonds: bonds i and $i-1$ in branch 2, and bonds i and $i-1$ in branch 3. The dimensions are 9×9 for the internal $\mathbf{U} \otimes \mathbf{U}$, and 9×1 for $\mathbf{U}_{n_2} \otimes \mathbf{U}_{n_3}$.

Repetitive application of the *theorem on direct products*[5,6] renders Eq. (XI-40) into the alternative form[2]

$$Z = \mathbf{U}_1 \mathbf{U}^{n_1-2} \mathbf{U}_{n_1;1} \mathbf{U}_b[(\mathbf{U}_{2;2} \mathbf{U}^{n_2-3} \mathbf{U}_n) \otimes (\mathbf{U}_{2;3} \mathbf{U}^{n_3-3} \mathbf{U}_n)] \qquad \text{(XI-41)}$$

This form shows that the branch point is handled by an expansion in dimensions (at $\mathbf{U}_b$), and the contributions from the last two branches are appended to the expanded matrix, using the direct product. The separate appearance of $\mathbf{U}^{n_1-2}, \mathbf{U}^{n_2-3}$, and $\mathbf{U}^{n_3-3}$ in Eq. (XI-41) also shows that this formulation is not restricted to those stars in which the three branches have the same number of bonds. The formulation requires $n_1 > 1, n_2 > 2$, and $n_3 > 2$, but no two n_i need be identical. Stars with $n_i = 2, i = 2, 3$ are handled by the substitution of $\mathbf{I}_3$ for $\mathbf{U}_{2;i} \mathbf{U}^{n_i-3}$.

A completely analogous application to the tetrafunctional star yields

$$Z = \mathbf{U}_1 \mathbf{U}^{n_1-2} \mathbf{U}_{n_1;1} \mathbf{U}_b[(\mathbf{U}_{2;2} \mathbf{U}^{n_2-3} \mathbf{U}_n) \otimes (\mathbf{U}_{2;3} \mathbf{U}^{n_3-3} \mathbf{U}_n) \otimes (\mathbf{U}_{2;4} \mathbf{U}^{n_4-3} \mathbf{U}_n)]$$

$$\text{(XI-42)}$$

where $\mathbf{U}_{n_1;1}, \mathbf{U}_b$, and $\mathbf{U}_{2,j}$ ($j = 2, 3, 4$) are defined by Eq. (XI-31) through Eq. (XI-39).

The bond and pair a priori probabilities for occupation of rotational isomeric states are extracted from Z for the stars in a manner that is completely analogous to the application to the linear chains. All that is required is the replacement of the appropriate $\mathbf{U}_i$ by either $\mathbf{U}'_{\eta;i}$ or $\mathbf{U}'_{\xi\eta;i}$. Another probability may be of potential interest at the branch point. In the case of the trifunctional star one can define $\mathbf{U}'_{\zeta\xi\eta}$ as the form adopted by $\mathbf{U}_b$ when every element save one is rendered null. The nonzero element defines a combination of rotational isomeric states for the three interdependent bonds to the atom at the trifunctional branch point. The indices ζ, ξ, and η denote the states at these three bonds. Use of this $\mathbf{U}'_{\zeta\xi\eta}$ yields a probability for three bonds, $p_{\zeta\xi\eta}$.

The effect of a trifunctional branch point on $p_{\eta;i}$ is presented in Fig. XI-6. The perturbation in $p_{\eta;i}$ is much larger near the branch than near the free end. The probability for a t placement is reduced at the bonds emanating at the branch point, and it is enhanced at the neighbors of these bonds.

---

[5] Flory, P. J.; Jernigan, R. L. *J. Chem. Phys.* **1965**, *42*, 3509.
[6] If $\mathbf{A}$ is conformable for multiplication onto $\mathbf{B}$, and $\mathbf{C}$ is conformable for multiplication onto $\mathbf{D}$, the theorem on direct products states that

$$(\mathbf{A} \otimes \mathbf{C})(\mathbf{B} \otimes \mathbf{D}) = (\mathbf{AB}) \otimes (\mathbf{CD})$$

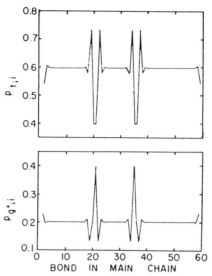

Figure XI-6. $p_{t;i}$ and $p_{g^+;i}$ for a polyethylene chain of 60 bonds, with branches of 4 bonds each attached at atoms 20 and 35 in this chain. Reprinted with permission from Mattice, W. L. *Macromolecules* **1975**, *8*, 644. Copyright 1975 American Chemical Society.

C. Unperturbed Dimensions

The unperturbed dimensions of greatest interest are the mean square unperturbed end-to-end distances for the linear chains found in the star, and the mean square radius of gyration for the entire molecule. The calculations are illustrated below for the case of the trifunctional star.[4]

Mean Square Unperturbed End-to-End Distance of a Branch. The mean-square end-to-end distance of branch i is denoted here by $\langle r_{bi}^2 \rangle_0$. It is obtained for the trifunctional star as

$$\langle r_{b1}^2 \rangle_0 = Z^{-1} \mathscr{G}_{[1} \cdot \mathscr{G}^{n_1-2} \mathscr{G}_{n_1;1]} \mathbf{U}_b [(\mathbf{U}_{2;2} \mathbf{U}^{n_2-3} \mathbf{U}_n) \otimes (\mathbf{U}_{2;3} \mathbf{U}^{n_3-3} \mathbf{U}_n)] \quad \text{(XI-43)}$$

$$\langle r_{b2}^2 \rangle_0 = Z^{-1} \mathbf{U}_1 \mathbf{U}^{n_1-2} \mathbf{U}_{n_1;1} \mathscr{G}_{[b} [(\mathscr{G}_{2;2} \cdot \mathscr{G}^{n_2-3} \cdot \mathscr{G}_{n]}) \otimes (\mathbf{U}_{2;3} \mathbf{U}^{n_3-3} \mathbf{U}_n)] \quad \text{(XI-44)}$$

$$\langle r_{b3}^2 \rangle_0 = Z^{-1} \mathbf{U}_1 \mathbf{U}^{n_1-2} \mathbf{U}_{n_1;1} \mathscr{G}_{[b} [(\mathbf{U}_{2;2} \mathbf{U}^{n_2-3} \mathbf{U}_n) \otimes (\mathscr{G}_{2;3} \cdot \mathscr{G}^{n_3-3} \cdot \mathscr{G}_{n]})] \quad \text{(XI-45)}$$

where the $\mathscr{G}$ are formulated from the statistical weight matrix and the 5×5 generator matrix $\mathbf{G}$, in a manner completely analogous to the computation of

$\langle r^2 \rangle_0$ for the linear chain. The symbol "[" has been placed in the subscript for each $\mathscr{G}$ that uses only the top row of $\mathbf{G}$, and the symbol "]" has been added when the $\mathscr{G}$ uses only the last column of $\mathbf{G}$. In applications to trifunctional polyethylene stars with $n_1 = n_2 = n_3$, exactly the same result is obtained from Eq. (XI-43), Eq. (XI-44), and Eq. (XI-45).

The presence of the trifunctional branch point only modifies the conformations of the bonds very near one end of the subchain the mean square unperturbed end-to-end distance of which is $\langle r_{bi}^2 \rangle_0$. Application of Eqs. (XI-43), (XI-44), and (XI-45) to polyethylene yields mean square unperturbed dimensions for the branches that are nearly indistinguishable from those for the linear chain with n_i bonds when $n_i > \sim 10$.[4] For trifunctional stars of the sizes that are usually considered, the mean square unperturbed dimensions of a branch will be indistinguishable from $\langle r^2 \rangle_0$ for the linear molecule with n_i bonds.

The mean square unperturbed end-to-end distance for the chain consisting of branches 1 and 2, or 1 and 3, denoted $\langle r_{c12}^2 \rangle_0$ or $\langle r_{c13}^2 \rangle_0$, is obtained as

$$\langle r_{c12}^2 \rangle_0 = Z^{-1} \mathscr{G}_{[1} \mathscr{G}^{n_1-2} \mathscr{G}_{n_1;1} \mathscr{G}_b[(\mathscr{G}_{2;2} \mathscr{G}^{n_2-3} \mathscr{G}_n]) \otimes (\mathbf{U}_{2;3} \mathbf{U}^{n_3-3} \mathbf{U}_n)]$$

(XI-46)

$$\langle r_{c13}^2 \rangle_0 = Z^{-1} \mathscr{G}_{[1} \mathscr{G}^{n_1-2} \mathscr{G}_{n_1;1} \mathscr{G}_b[(\mathbf{U}_{2;2} \mathbf{U}^{n_2-3} \mathbf{U}_n) \otimes (\mathscr{G}_{2;3} \mathscr{G}^{n_3-3} \mathscr{G}_n])]$$

(XI-47)

Special care must be used in the construction of the $\mathbf{T}$ that are inserted in $\mathscr{G}_{n_1;1}$ in Eq. (XI-47). Recall that the definitions of the t, g^+, and g^- states at the last bond in branch 1 used the chain comprising branches 1 and 2. In the calculation of $\langle r_{c13}^2 \rangle_0$, the ϕ inserted in each $\mathbf{T}$ must be selected to give the proper geometry for the chain comprised of branches 1 and 3.

The results obtained from Eq. (XI-46) and Eq. (XI-47) are independent of the chirality of the branch point. Identical results are obtained from Eqs. (XI-46) and (XI-47) for polyethylene stars with $n_2 = n_3$. The mean square unperturbed end-to-end distance for the chain consisting of branches 2 and 3 can be obtained by beginning anew with another formulation of Z in which a different branch is selected as "branch 1." Then either branch 2 or 3 of the original formulation will become branch 1 in the new formulation, permitting use of either Eq. (XI-46) or Eq. (XI-47). This awkward procedure can be avoided if any two of the branches contain the same number of bonds, and one of these branches is selected initially as "branch 1." Then $\langle r_{c23}^2 \rangle_0$ will be equal to either $\langle r_{c12}^2 \rangle_0$ or $\langle r_{c13}^2 \rangle_0$.

Figure XI-6 shows that a trifunctional branch perturbs the conformation for the bonds in the vicinity of the branch point. These bonds are located in the interior of the chain comprising branches 1 and $i, i = 2, 3$. When n_1 and n_i are small, the perturbation produces a reduction in the mean square unperturbed dimensions, by disrupting the preferred trajectory of an unperturbed linear polyethylene chain. Nevertheless, we anticipate $\langle r_{c1i}^2 \rangle_0$ will approach $\langle r^2 \rangle_0$ for the linear chain with $n_1 + n_i$ bonds, in the limit as $n_1 + n_i \to \infty$. This expectation is supported by the calculations for polyethylene. The mean square unperturbed dimensions for the two situations become nearly indistinguishable when n is on the order of 10^2.[4]

Mean Square Unperturbed Radius of Gyration. The squared radius of gyration for a specified conformation of a trifunctional star is obtained from the squared distances between all pairs of atoms, using Eq. (I-4). The sum of the squares of the distances between all pairs of atoms can be expressed as

$$\sum_{\text{star}} \langle r_{ij}^2 \rangle_0 = \sum_{\text{chains}} \langle r_{ij}^2 \rangle_0 - \sum_{\text{branches}} \langle r_{ij}^2 \rangle_0 \qquad \text{(XI-48)}$$

Here the first sum on the right-hand side contains all of the r_{ij}^2 in the chains consisting of all possible pairs of branches. The atoms that must be considered in each of these chains are the ones that constitute the chain the mean-square end-to-end distances of which were denoted by $\langle r_{c12}^2 \rangle_0, \langle r_{c13}^2 \rangle_0$, and $\langle r_{c23}^2 \rangle_0$. During the evaluation of the first sum on the right-hand side of Eq. (XI-48), there is a double-counting of the r_{ij}^2 within each branch, because each branch contributes to the $\langle r_{ij}^2 \rangle_0$ for two different chains. The second term in Eq. (XI-48) corrects for this double-counting.

The terms contributed to the first summation by the chains composed of branches 1 and 2, and of branches 1 and 3, are obtained from Eq. (XI-46) and Eq. (XI-47), respectively, with each $\mathcal{G}$ replaced by $\mathcal{H}$. The terms contributed by the chain composed of branches 2 and 3 are obtained by new (but equivalent) formulation of Z in which a different branch is selected as "branch 1." If any two of the branches contain the same number of bonds, this awkward procedure can be avoided by selecting as "branch 1" one of the branches that contains the same number of bonds as another branch. If all three branches contain the same number of bonds, each of the three chains makes exactly the same contribution to the first sum on the right-hand side of Eq. (XI-48). The terms in the last sum in Eq. (XI-48) are obtained from Eq. (XI-43), Eq. (XI-44), and Eq. (XI-45) with the substitution of $\mathcal{H}$ for $\mathcal{G}$. The mean square unperturbed radius of gyration is obtained by division of the completed sum in Eq. (XI-48) by the square of the number of atoms, $(n + 1)^2$.

For the special case of an unperturbed trifunctional star in which all branches contain the same number of bonds, the result becomes

$$\langle s^2 \rangle_0 = \frac{3(A - B)}{Z(n + 1)^2} \tag{XI-49}$$

$$A = \mathscr{H}_{[1}.\mathscr{H}^{n_1-2}.\mathscr{H}_{n_1;1}.\mathscr{H}_b[(\mathscr{H}_{2;2}.\mathscr{H}^{n_2-3}.\mathscr{H}_n]) \otimes (\mathbf{U}_{2;3}\mathbf{U}^{n_3-3}\mathbf{U}_n)] \tag{XI-50}$$

$$B = \mathscr{H}_{[1}.\mathscr{H}^{n_1-2}.\mathscr{H}_{n_1;1]}\mathbf{U}_b[(\mathbf{U}_{2;2}\mathbf{U}^{n_2-3}\mathbf{U}_n) \otimes (\mathbf{U}_{2;3}\mathbf{U}^{n_3-3}\mathbf{U}_n)] \tag{XI-51}$$

For a tetrafunctional star in which all branches are identical, an equivalent treatment yields

$$\langle s^2 \rangle_0 = \frac{6A - 8B}{Z(n + 1)^2} \tag{XI-52}$$

$$A = \mathscr{H}_{[1}.\mathscr{H}^{n_1-2}.\mathscr{H}_{n_1;1}.\mathscr{H}_b[(\mathscr{H}_{2;2}.\mathscr{H}^{n_2-3}.\mathscr{H}_n]) \otimes (\mathbf{U}_{2;3}\mathbf{U}^{n_3-3}\mathbf{U}_n)$$
$$\otimes (\mathbf{U}_{2;4}\mathbf{U}^{n_4-3}\mathbf{U}_n)] \tag{XI-53}$$

$$B = \mathscr{H}_{[1}.\mathscr{H}^{n_1-2}.\mathscr{H}_{n_1;1]}\mathbf{U}_b[(\mathbf{U}_{2;2}\mathbf{U}^{n_2-3}\mathbf{U}_n) \otimes (\mathbf{U}_{2;3}\mathbf{U}^{n_3-3}\mathbf{U}_n)$$
$$\otimes (\mathbf{U}_{2;4}\mathbf{U}^{n_4-3}\mathbf{U}_n)] \tag{XI-54}$$

Illustrative results for tetrafunctional stars in which all branches contain the same number of bonds are depicted in Fig. XI-7. The dimensions are reported as the dimensionless ratio g, defined in Eq. (XI-1). Application of random flight statistics yields[1,7]

$$g = \frac{1}{n^3} \sum_j (3nn_j^2 - 2n_j^3) \tag{XI-55}$$

where the sum is over all j branches. This expression yields $g = 5/8$ for a tetrafunctional star in which all branches contain the same number of bonds. That result is obtained from RIS theory in the limit as $n \to \infty$, as is apparent in Fig. XI-7. The number of bonds required for convergence to this limit, and the size and direction of the deviation from the limit at small n, depend on the nature of the short range interactions in the chain.[4]

Often the branch site is constructed from a different moiety than the monomer that appears in the branches. For example, triglycerides formed from saturated fatty acids can be viewed as trifunctional polyethylene stars in which the branch site is $-COOCH_2-CH(OOC-)-CH_2OOC-$, and the branches are

[7]Orofino, T. A. *Polymer* **1961**, 2, 305.

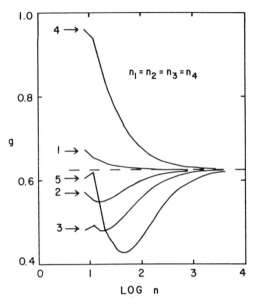

Figure XI-7. Values of g for tetrafunctional stars with branches of the same length and statistical weights of (1) $\sigma = \tau_1 = \tau_2 = \psi = \omega = 1$, (2) $\sigma = \tau_1 = 0.54, \tau_2 = \psi = \omega = 1$, (3) $\sigma = \tau_1 = 0.54, \tau_2 = \psi = 1, \omega = 0.088$, (4) $\sigma = \tau_1 = 10, \tau_2 = \psi = \omega = 1$, and (5) $\sigma = \tau_1 = 10, \tau_2 = \psi = 1, \omega = 0$. The dashed line indicates the limit of 5/8. Reprinted with permission from Mattice, W. L.; Carpenter, D. K. *Macromolecules* **1976**, *9*, 53. Copyright 1976 American Chemical Society.

$-(CH_2)_{n_i}H.$[8] Such molecules can be handled with the formalism presented above by appropriate construction of the statistical weight matrices for the bonds in the vicinity of the branch point.

2. CROSSLINKS AND GRAFTS

From the standpoint of rotational isomeric state theory, two *crosslinked chains* can be viewed as a molecule with two trifunctional branch points, as there are only a small number of bonds in the segment between the atoms at the two branch points. This short segment often contains different types of bonds than those found in the two chains. As an example, consider two poly(amino acid) chains of the composition X_a-Cys-X_b, where Cys denotes the cysteinyl residue, and X denotes some other amino acid residue. When the sulfhydryl groups at two such Cys residues are joined via a disulfide bond, the portion of the polymer

---

[8]Mattice, W. L.; Saiz, E. *J. Am. Chem. Soc.* **1978**, *100*, 6308. Mattice, W. L. *J. Am. Chem. Soc.* **1979**, *101*, 732.

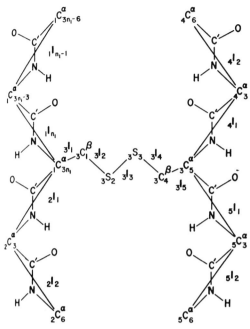

Figure XI-8. Representation of two poly(amino acid) chains crosslinked via a disulfide bond between two cysteinyl residues. Hydrogen atoms bonded to carbon, and side chains other than those in the Cys residues, are omitted. Reprinted with permission from Mattice, W. L. *J. Am. Chem. Soc.* **1977**, *99*, 2324. Copyright 1977 American Chemical Society.

near the crosslink has the structure depicted in Fig. XI-8. In this case, there are five bonds in the segment between the two trifunctional branch points.

By a generalization of the procedures employed in the description of a trifunctional star, the conformational partition function for two crosslinked chains, based on first- and second-order interactions, can be written as

$$Z = U_1 U^{n_1-2} U_{n_1;1} U_b (A \otimes B) \tag{XI-56}$$

$$A = U_{2;2} U^{n_2-3} U_n \tag{XI-57}$$

$$B = U_{2;3} U^{n_3-3} U_{n_3;3} U_b [(U_{2;4} U^{n_4-3} U_n) \otimes (U_{2;5} U^{n_5-3} U_n)] \tag{XI-58}$$

Here one of the crosslinked chains is denoted by branches 1 and 2, the other chain is denoted by branches 4 and 5, and the crosslink itself is branch 3. Usually n_3 would be much smaller than the other n_i, because the crosslink is often much shorter than the chains that are crosslinked, and crosslinking often occurs

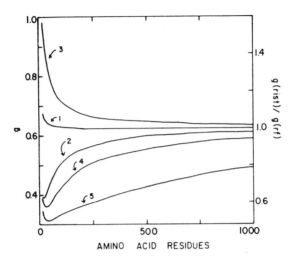

Figure XI-9. Values of g for crosslinked polyglycine (curve 1), poly(L-alanine) (curve 2), alternating D-Ala-L-Ala (curve 3), and two models for poly(L-proline) (curves 4 and 5), where the two chains are of the same length, and are crosslinked via the residue in the middle of each chain. Reprinted with permission from Mattice, W. L. *Macromolecules* **1977**, *10*, 511. Copyright 1977 American Chemical Society.

at sites remote from either end of the chains. Usually the composition of the system will be such that the serial indices need to be retained on several of the terms that are written more concisely above with notation such as U^n.

Illustrative results for unperturbed crosslinked polyglycine, poly(L-alanine), and poly(L-proline) are depicted in Fig. XI-9. The conformational partition function for the crosslink itself was formulated in the manner described above, and shown to be consistent with the optical activity exhibited by the disulfide bond in this crosslink.[9] Both chains are of the same degree of polymerization, and they are crosslinked via L-cysteinyl residues situated in the middle of each chain. For this system, the value of g predicted by Eq. (XI-55) approaches $\frac{5}{8}$ as the degree of polymerization of the chains approaches infinity, because the molecule can be approximated by a tetrafunctional star if n_3 is much smaller than the other n_i. The results from RIS theory approach this limit as the size of the chains increases, but significant deviations remain for poly(L-alanine) and poly(L-proline) when the chains contain a few hundred residues.[10] Crosslinked polyglycine reaches the limiting value at smaller degrees of polymerization.

The conformational partition function for a chain with a few *grafts*, of specific degrees of polymerization, and attached to the main chain at specific sites, can be generated by a simple elaboration of the treatment of two chains joined

[9]Mattice, W. L. *J. Am. Chem. Soc.* **1977**, *99*, 2324.
[10]Oka, M.; Nakajima, A. *Polym. J.* **1977**, *9*, 573.

by a crosslink. The expression in Eq. (XI-56) can be used as is for a chain that has two grafts. The main chain might be represented by the odd-numbered branches; and the grafts, by the two even-numbered branches. Since the attachment sites for the two grafts might be widely separated, it is likely that n_3 may be comparable in size with (or even larger than) the other n_i. But the size of n_3 does not enter the form of Eq. (XI-58), so long as $n_3 > 2$. Extension to a chain with more grafts (represented by even-numbered branches) yields an expression of the form

$$Z = \mathbf{A}_1[\mathbf{B}_2 \otimes (\mathbf{A}_3[\mathbf{B}_4 \otimes (\mathbf{A}_5[\mathbf{B}_6 \otimes \mathbf{A}_7])])] \qquad \text{(XI-59)}$$

where $\mathbf{A}_1 = \mathbf{U}_1\mathbf{U}^{n_1-2}\mathbf{U}_{n_1;i}\mathbf{U}_b$, internal $\mathbf{A}_i$ are $\mathbf{U}_{2;i}\mathbf{U}^{n_i-3}\mathbf{U}_{n_i;i}\mathbf{U}_b$, and the final $\mathbf{A}_i$, as well as each $\mathbf{B}_i$, is of the form $\mathbf{U}_{2;i}\mathbf{U}^{n_i-3}\mathbf{U}_{n_i;i}$. There is no limitation on the number of grafts that can be appended. The particular formalism adopted in Eq. (XI-59) assumes that first- and second-order interactions are adequate for a description of the short-range interactions, and there are at least three bonds between the sites of attachment of successive grafts.

3. MONOMERS WITH ARTICULATED SIDE CHAINS

Many polymers contain side chains with internal degrees of freedom. A simple example is poly(1-butene), where an ethyl group is appended to alternate atoms on a polyethylene backbone. The short-range interactions of the methyl group with the backbone depend on the torsion at the C—C bond at which the side group is attached to the main chain. This torsion is denoted by χ_1. For present purposes, polymers formed from monomers with articulated side chains can be viewed as a special case of a graft copolymer, in which the "grafts," or side chains, are attached to the main chain at regular intervals, and contain many fewer bonds than the main chain.

A. Conformational Partition Functions

If the homopolymer can be treated successfully using only first- and second-order interactions, and there are at least three bonds in the main chain between the sites of attachment of successive side chains, Z can be obtained by a simple elaboration of Eq. (XI-59). Each of the $\mathbf{B}_i$ represents the contribution made by a side chain to Z. Each of the internal $\mathbf{A}_i$ denotes the contribution by the segment of three or more bonds in the main chain between successive sites of attachment for the side chains. If the attachment sites are chiral, and Z is written using the *dl* notation, there will be two different expressions for $\mathbf{B}$, and four different expressions for the internal $\mathbf{A}$. The first and last $\mathbf{A}$ are determined by the end groups in the chain.

Vinyl Polymers with Articulated Side Chains. The vinyl polymers constitute an important class of polymers in which there are only two bonds in the main chain between the sites of attachment of the branches. The first atoms in two successive branches can participate in a second-order interaction. The formulation of Z that was described earlier in this chapter accounted for all first- and second-order interactions of a side chain with the main chain, but it did not include any short-range interactions of one side chain with another. The missing second-order interactions were presented in Chapter VIII, which described Z for vinyl polymers with nonarticulated side chains. A description of Z for vinyl polymers with articulated side chains can be obtained by an elaboration of the procedures in Chapter VIII, where the articulated side chain is introduced using methodology appropriate for the description of a branched molecule.[11,12] The restriction of the formulation of Z to inclusion of all first- and second-order interactions will not necessarily be sufficient. Crowding near the backbone produced by the high density of side chains raises the question of whether short range interactions of higher order may make an important contribution to Z, as occurs in poly(4-methyl-1-pentene) and poly[(S)-4-methyl-1-hexene].[13] Here we will focus on the development for those cases where first- and second-order interactions are sufficient.

Consider the vinyl polymer with x side chains to be a special case of a polymer with $2x + 1$ branches. The side chains contribute x branches (sections), the segments of the main chain between successive side chains contribute $x - 1$ sections (each with two bonds), and the segments of the main chain at the ends contribute 2 branches (sections). The x side chains will be indexed by even numbers, and the remaining $x + 1$ branches (sections) contributed by the main chain will be indexed by odd numbers. The expression for Z can be written as

$$Z = \mathbf{A}_1 \mathbf{X}_2 \mathbf{X}_4 \cdots \mathbf{X}_{2x-2} \mathbf{U}_b \mathbf{A}_{2x+1} \tag{XI-60}$$

where each $\mathbf{X}_i$ incorporates the contribution from statistical weight matrices for the bonds in a side chain (branch, or section, i) and the segment of the main chain between section i and the next side-chain carrying section.

$$\mathbf{X}_i = \mathbf{U}_b[(\mathbf{U}^{n_i-1}) \otimes \mathbf{U}_{2;i+1}] \tag{XI-61}$$

Here $\mathbf{U}_b$ is the rectangular $\nu \times \nu^2$ matrix that contains the statistical weights for the rotational isomers at the first bond in sections i and $i + 1$, $\mathbf{U}^{n_i-1}$ is the product of the statistical weight matrices for all remaining bonds in the side chain (section i), and $\mathbf{U}_{2;i+1}$ is the statistical weight matrix for the second bond

[11] Abe, A. *J. Am. Chem. Soc.* **1968**, *90*, 2205; *Polym. J.* **1970**, *1*, 232; *Macromolecules* **1977**, *10*, 34.
[12] Mattice, W. L. *Macromolecules* **1977**, *10*, 1171.
[13] Wittwer, H.; Suter, U. W. *Macromolecules* **1985**, *18*, 403.

in the segment of the main chain between two sections. The dimensions of $\mathbf{U}^{n_i-1}$ and $\mathbf{U}_{2;i+1}$ are $\nu \times 1$ and $\nu \times \nu$, respectively, and $[(\mathbf{U}^{n_i-1}) \otimes \mathbf{U}_{2;i+1}]$ is of dimensions $\nu^2 \times \nu$. The matrix $\mathbf{X}_i$ is relatively compact, $\nu \times \nu$. It arises from the multiplication of two larger matrices, the first of dimensions, $\nu \times \nu^2$, and the second of dimensions $\nu^2 \times \nu$. If $\nu = 3$, the $\mathbf{X}_i$ of Eq. (XI-61) is of the same dimensions as the products $\mathbf{U}'\mathbf{U}_m$ and $\mathbf{U}'\mathbf{U}_r$, formulated from the statistical weight matrices defined in Eqs. (VIII-14), (VIII-15), and (VIII-22).

As an illustrative example, consider a *meso* diad in which the side chain contains two bonds, with atoms denoted by X_1 and X_2, and atom X_1 bonded directly to the backbone, as depicted in Fig. XI-10. Then, using statistical weights of the same type as those adopted in Chapter VIII, plus first-order statistical weights of diag $(1, 1, \tau_X)$ at the C—X_1 bond, and ω_{X_2} for the "pentane-effect" second-order interaction of X_2 with a carbon atom in the main chain, we have

$$
\mathbf{U}_b = \begin{bmatrix} \eta\omega_{X_2} & 1 & \tau & \eta\omega_{X_2} & \omega_{X_2} & \tau\omega_{X_2} & \eta\tau_X & \tau_X & \tau\tau_X\omega_{X_2} \\ \eta\omega_{X_2} & \omega & \tau & \eta & \omega & \tau & \eta\tau_X & \tau_X\omega & \tau\tau_X\omega_{X_2} \\ \eta\omega_{X_2} & 1 & \tau\omega & \eta & 1 & \tau\omega & \eta\tau_X\omega_{X_2} & \tau_X\omega_{X_2} & \tau\tau_X\omega\omega_{X_2}^2 \end{bmatrix}
$$

$$\tag{XI-62}$$

$$
\mathbf{U}^{n_i-1} = \begin{bmatrix} 1 \\ 1 \\ 1 \end{bmatrix}
$$

$$\tag{XI-63}$$

$$
\mathbf{U}_{2;i+1} = \mathbf{U}_m
$$

$$\tag{XI-64}$$

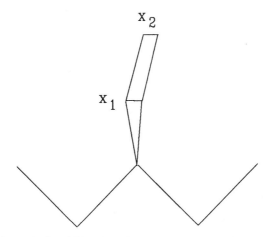

Figure XI-10. An articulated X_1—X_2 side chain in a vinyl polymer, and the atoms involved in the first- and second-order interactions denoted by τ_X and ω_{X_2}

where the matrix denoted by $U_{2;i+1}$ is identical with the U_m defined in Eq. (VIII-15), during the discussion of the *meso* diad in a vinyl polymer with a nonarticulated side chain. The statistical weights that arise from the fact that the branch is articulated, namely, τ_X and ω_{X_2}, are found in U_b. The desired product, $U_b[(U^{n_i-1}) \otimes U_{2;i+1}]$, is a 3×3 matrix that can be generated identically as the product of two 3×3 matrices:

$$U_b[(U^{n_i-1}) \otimes U_{2;i+1}]_m = \begin{bmatrix} \eta\tau^* & 1 & \tau f_1 \\ \eta & \omega(2-\tau^*) & \tau f_2 \\ \eta f_1 & f_2 & \tau\omega f_4 \end{bmatrix} U_m \qquad \text{(XI-65)}$$

Here the first statistical weight matrix on the right-hand side is a modification of the U_p that was defined in Eq. (VIII-14). The modification is the introduction of statistical weights denoted by τ^*, f_1, f_2, and f_4. These statistical weights are determined by τ_X and ω_{X_2} as[12]

$$\tau^* = \frac{\tau_X + 2\omega_{X_2}}{1 + \tau_X + \omega_{X_2}} \qquad \text{(XI-66)}$$

$$f_1 = \frac{1 + \omega_{X_2} + \tau_X\omega_{X_2}}{1 + \tau_X + \omega_{X_2}} \qquad \text{(XI-67)}$$

$$f_2 = \frac{2 + \tau_X\omega_{X_2}}{1 + \tau_X + \omega_{X_2}} \qquad \text{(XI-68)}$$

$$f_4 = \frac{2 + \tau_X\omega_{X_2}^2}{1 + \tau_X + \omega_{X_2}} \qquad \text{(XI-69)}$$

If $\tau_X = \omega_{X_2} = 1$, U_p of Eq. (VIII-14) is recovered. If the diad under discussion were *racemo* rather than *meso*, the only change required on the right-hand side of Eq. (XI-65) would be the substitution of U_r for U_m.

Extension to a longer side chain, with three bonds, can be accomplished by modification of the definitions of τ^*, f_1, f_2, and f_4. Representing a side chain with three bonds by -X_1-X_2-X_3, each τ_X in these definitions is multiplied by a factor c, defined as[12]

$$c = \frac{1 + 2\sigma_{X_3}\omega_{X_3}}{1 + \sigma_{X_3} + \sigma_{X_3}\omega_{X_3}} \qquad \text{(XI-70)}$$

where the statistical weights appearing in c arise from first- and second-order interactions of atom X_3 with carbon atoms in the main chain.

The procedures described here show that vinyl polymers with articulated side chains can be treated with a formalism very similar to the one described in Chapter VIII for vinyl polymers with nonarticulated side chains, provided no interactions higher than second-order are necessary. All that is required is a modification of seven of the elements in the statistical weight matrix denoted by U_p.

If the side chain participates in highly repulsive short-range interactions with the main chain, such that $\tau_X \to 0$ and $\omega_{X_2} \to 0$, the limiting values for the factors that arise from these interactions are $\tau^* \to 0$, $f_1 \to 1$, $f_2 \to f_4 \to 2$. Under these conditions the limiting form for the statistical weight matrix denoted by U_p is

$$\lim_{\tau_X \to 0, \omega_{X_2} \to 0} U_p = \begin{bmatrix} 0 & 1 & \tau \\ \eta & 2\omega & 2\tau \\ \eta & 2 & 2\tau\omega \end{bmatrix} \qquad \text{(XI-71)}$$

Further insight into the consequences of introduction of an articulated side chain can be obtained by examination of typical values expected for τ^*, f_1, f_2, and f_4. Illustrative results are presented in Table XI-2, as computed for several simple side chains at 300 K. The illustrative results presented in Table XI-2, as well as the limiting form in Eq. (XI-71), suggest the most important new parameter is likely to be τ^*. The three f_i do not change by more than a factor of 2, but τ^* changes by an order of magnitude in Table XI-2, and becomes zero in Eq. (XI-71). The value of τ^* is more sensitive to the interactions produced by the articulated side chain than are the f_i because the defining expressions for the f_i contain a leading term of 1 or 2 in both the numerator and denominator, but for τ^* the numerator contains no term that is simply an integer. Alternatively

TABLE XI-2. Illustrative τ^*, f_1, f_2, and f_4 at 300 K (Energies in kJ mol^{-1})

Side Chain	$E_{\tau X}$	$E_{\omega X_2}$	$E_{\sigma X_3}$	$E_{\omega X_3}$	τ^*	f_1	f_2	f_4
$-X_1-X_2$	0	0			1	1	1	1
$-X_1-X_2$	∞	0			1	1	1	1
$-X_1-X_2$	0	∞			0.5	0.5	1	1
$-X_1-X_2$	∞	∞			0	1	2	2
$CH_2CH_3^a$	2.1	8.4			0.34	0.72	1.4	1.4
$CH_2CH_2CH_3^a$	2.1	8.4	2.1	8.4	0.28	0.78	1.5	1.5
OCH_3^b	3.8	∞			0.18	0.82	1.6	1.6
$OCH_2CH_3^b$	3.8	∞	3.8	∞	0.15	0.85	1.7	1.7
$CH_2OCH_3^b$	-0.8	1.4	3.8	∞	0.84	0.82	0.98	0.88

[a] Abe, A.; Jernigan, R. L.; Flory, P. J. *J. Am. Chem. Soc.* **1966**, *88*, 631.
[b] Abe, A.; Mark, J. E. *J. Am. Chem. Soc.* **1976**, *98*, 6468.

stated, atom X_2 in an $-X_1-X_2$ side chain *must* participate in a short-range interaction weighted by either τ_X or ω_{X_2} if the two main-chain bonds to the atom at the branch point both adopt t states. In other combinations of states at these two bonds in the main chain, there is at least one state at the $C—X_1$ bond that places X_2 in a position where it can avoid these short-range interactions.

If $\tau^* < 1$, as is often the case, simultaneous occupancy of t states for both main-chain bonds to a branch point becomes less likely when the side chain is articulated. The tt sequence at the two bonds in the main chain will be suppressed if the side chain is articulated, and $E_{\tau X} > 0, E_{\omega X_2} > 0$. The importance of the suppression of the tt sequences, which produces τ^*, was recognized early in the development of the rotational isomeric state description of vinyl polymers.[14]

If the conformational partition function for the vinyl polymer is developed using the dl notation, instead of *meso* and *racemo* diads, the replacements for the first matrix on the right-hand side of Eq. (XI-65) are

$$\mathbf{U}_d = \begin{bmatrix} \eta\tau^* & 1 & \tau f_1 \\ \eta f_1 & f_2 & \tau\omega f_4 \\ \eta & \omega(2-\tau^*) & \tau f_2 \end{bmatrix} \tag{XI-72}$$

$$\mathbf{U}_l = \mathbf{Q}\mathbf{U}_d\mathbf{Q} \tag{XI-73}$$

and the second matrix on the right-hand side of Eq. (XI-65) is replaced by $\mathbf{U}_{dd}, \mathbf{U}_{dl}, \mathbf{U}_{ld}$, or $\mathbf{U}_{ll}$ from Chapter VIII.

The treatment of articulated side chains is greatly simplified if the conformation of the side chain is not altered by the conformation of the main chain. This situation occurs when the main chain has only one conformation, as in the α helix of poly(α-amino acids). Then the conformation of the side chain may depend on the conformation of neighboring side chains, but the main chain provides a constant set of interactions.[15]

B. Unperturbed Dimensions for the Main Chain

Several applications of RIS theory to the description of $\langle r^2 \rangle_0$ for polymers in which each monomer contributes an articulated side chain have appeared in the literature. Illustrative examples are collected in Table XI-3.

When the structure of the side chains, or the density of their attachments to the main chain, produces numerous short-range interactions that include important interactions higher than second-order, methods based on specific enumeration of all of the relevant short-range interactions may become unwieldy. It may be preferable to determine the numeric values of the elements in the statistical weight matrices by integration of the conformational energy surface,

[14]Flory, P. J.; Mark, J. E.; Abe, A. *J. Am. Chem. Soc.* **1966**, *88*, 639.
[15]Yamazaki, T.; Abe, A.; Ono, H.; Toriumi, H. *Biopolymers* **1989**, *28*, 1959.

TABLE XI-3. Illustrative Applications to $\langle r^2 \rangle_0$ for Homopolymers

Monomeric Unit	R	Dimensions of $\mathbf{U}$	Ref.
CH_2CHR	Illustrative[a]	3×3	b
CH_2CHR	$(CH_2)_zCH_3$, $z = 1, 2$	5×5	c
CH_2CHR	$CH_2CH(CH_3)_2$	5×5	c
CH_2CHR	$CH_2CH(CH_3)CH_2CH_3$	10×10	c
CH_2CHR	OH	3×3	d
CH_2CHR	OCH_3	3×3	e
CH_2CHR	$OCH(CH_3)C_2H_5$	3×3	e
CH_2CHR	$OCH_2CH(CH_3)C_2H_5$	3×3	e
CH_2CHR	$(C{=}O)C(CH_3)_3$	2×2	f
$OSiR_2$	$(CH_2)_zCH_3$	3×3	g
$OSiR_2$	$(CH_2)_zCH_3$	9×9	h
$OSiR_2$	$CH(CH_3)_2$, $C(CH_3)_3$, $CH(CH_3)C_2H_5$, $CH_2CH(CH_3)_2$	9×9	h
OCH_2CHR	C_2H_5, $CH(CH_3)_2$, $C(CH_3)_3$	3×3	i

[a]Illustrative calculations that explore the consequences for various assignments of the energetics and geometry.
[b]Abe, A. *Polym. J.* **1970**, *1*, 232. Mattice, W. L. *Macromolecules* **1977**, *10*, 1171.
[c]Wittwer, H.; Suter, U. W. *Macromolecules* **1985**, *18*, 403.
[d]Wolf, R. M.; Suter, U. W. *Macromolecules* **1984**, *17*, 669.
[e]Abe, A. *Macromolecules* **1977**, *10*, 34.
[f]Suter, U. W. *J. Am. Chem. Soc.* **1979**, *101*, 6481. Guest, J. A.; Matsuo, K.; Stockmayer, W. H.; Suter, U. W. *Macromolecules* **1980**, *13*, 560.
[g]Mattice, W. L. *Macromolecules* **1978**, *11*, 517.
[h]Neuburger, N. A.; Bahar, I.; Mattice, W. L. *Macromolecules* **1992**, *25*, 2447.
[i]Abe, A.; Hirano, T.; Tsuji, K.; Tsuruta, T. *Macromolecules* **1979**, *12*, 1100.

as done for poly(4-methyl-1-pentene) and poly[(S)-methyl-1-hexene],[13] or by integration of a molecular dynamics trajectory of sufficient length to sample all of conformational space, as done for the poly(dialkylsiloxanes).[16] The copious short-range interactions that force application of these integration schemes may also necessitate an expansion in the dimensions of the statistical weight matrices.[13,16]

Experimentally, the unperturbed dimensions of isotactic poly(1-alkenes) are sensitive to the structure of the side chain.[17] The values of C_∞ tend to increase with the size of the side chain. The few data available suggest a similar trend for isotactic poly[oxy(1-alkylethylenes)].[18] Calculations suggest a much different behavior for the syndiotactic poly[oxy(1-alkylethylenes)], with the syndiotactic chain with *tert*-butyl side chains predicted to have a C_∞ of only 0.9. The calculated unperturbed dimensions of the symmetrically substituted

[16]Neuburger, N. A.; Bahar, I.; Mattice, W. L. *Macromolecules* **1992**, *25*, 2447.
[17]Results from several groups, on five isotactic poly(1-alkenes), have been summarized by Wittwer and Suter.[13]
[18]Data summarized by Abe, A.; Hirano, T.; Tsuji, K.; Tsuruta, T. *Macromolecules* **1979**, *12*, 1100.

poly(dialkylsiloxanes), where tacticity is not an issue, are less sensitive to the details of the structure of the alkyl groups,[16] perhaps in part due to the high intrinsic flexibility of the siloxane backbone. This prediction is supported by the observation that the characteristic ratio of poly(diethylsiloxane) is measured to be little different from the result for poly(dimethylsiloxane).[19]

PROBLEMS

Answers for Problems XI-1, XI-4 through XI-8, and XI-10 can be found in Appendix B.

XI-1. How do the first- and second-order interactions involving the oxygen atom appear in the statistical weight matrices for a molecule in which a —CH_2—O—$(CH_2)_x CH_3$ branch ($x \gg 1$) is attached to a carbon atom near the middle of a long polyethylene chain?

XI-2. Convince yourself that application of the statistical weight matrices described in this chapter yields Z for 5-(n-propyl)decane that is independent of the chirality of the branch point, and independent of the arbitrary selection of a branch as "branch 1."

XI-3. Applying the theorem on direct products, verify that Eq. (XI-40) is identical with Eq. (XI-41).

XI-4. For a polyethylene star with $f = 3$, what are the conformations for the three bonds at the branch point that produce the largest $p_{\zeta\xi\eta}$?

XI-5. Write out, element by element, the 3×27 form of U_b for an alkane with a tetrafunctional branch point, including all first- and second-order interactions.

XI-6. For a polyethylene star, are the probabilities for *trans* placements at the bonds in the chain composed of branches 1 and 2 disturbed more by a trifunctional branch point or by a tetrafunctional branch point?

XI-7. Specify the values of ϕ that appear in each of the **G** in $\mathscr{G}_{n_1;1}$ in Eq. (XI-47), for both chiralities for the branch point.

XI-8. Sketch the architecture of a polyethylene, showing the connections of the branches, when the conformational partition function is given by

$$Z = U_1 U^8 U_{10;1} U_b [A \otimes B]$$

[19] Mark, J. E.; Chiu, D. S.; Su, T.-K. *Polymer* **1978**, *19*, 407.

$$\mathbf{A} = \mathbf{U}_{2;2}\mathbf{U}^7\mathbf{U}_{10;2}\mathbf{U}_b[(\mathbf{U}_{2;4}\mathbf{U}^2\mathbf{U}_{n_4}) \otimes (\mathbf{U}_{2;5}\mathbf{U}\mathbf{U}_{n_5})]$$

$$\mathbf{B} = \mathbf{U}_{2;3}\mathbf{U}^5\mathbf{U}_{8;3}\mathbf{U}_b[(\mathbf{U}_{2;7}\mathbf{U}^9\mathbf{U}_{n_7}) \otimes (\mathbf{U}_{2;8}\mathbf{U}^{20}\mathbf{U}_{n_8})]$$

XI-9. Verify the definition of τ^* in Eq. (XI-66).

XI-10. What types of short-range interactions are required of an articulated side chain if it is to produce $\tau^* > 1$? How likely are these interactions with real side chains?

XI-11. An $-X_1-X_2$ side chain in a vinyl polymer becomes "nonarticulated" if one of the ν states at the $C-X_1$ bond is populated to the exclusion of the other $\nu - 1$ states. For what values of $E_{\tau X}$ and $E_{\omega X_2}$ will the probability for one of these states exceed 0.95?

XI-12. If the first- and second-order interactions of X_2 with the atoms in the backbone of a vinyl polymer with an articulated $-X_1-X_2$ side chain produce a decrease in C_∞ for the isotactic polymer, will they necessarily also produce a decrease in C_∞ for the syndiotactic polymer?

XII Some Special Issues with Copolymers

Rotational isomeric state (RIS) theory has been employed in previous chapters primarily for homopolymers. Issues that concern the sequence of monomers in the chain were usually suppressed.[1] This chapter is concerned specifically with copolymers of monomer A and monomer B. Homopolymers comprising constitutional units that render their description as copolymers convenient are also included here; typical examples would be vinyl polymers with head–tail (head-to-tail) isomerism. The conformational partition function for the copolymer of specific sequence is represented by Eq. (IV-22), where no two of the U_i need necessarily be identical. Copolymers in which the monomer units are not arranged in a specific sequence present a few new issues. New statistical weight matrices, incorporating short-range interactions not found in either homopolymer, may be necessary at the bonds near the junctions of two different monomers.

Two important classes of copolymers of specific sequence are provided by proteins and nucleic acids. The expressions developed earlier in this book apply, with the reservations that serial indices be maintained on the statistical weight matrices, and that the range of the interactions is adequate to encompass all interactions that determine the conformation.

1. "RANDOM" COPOLYMERS WITH INTERDEPENDENT BONDS

The literature contains many examples of RIS descriptions of random copolymers with interdependent bonds. Representative examples that illustrate several issues are collected in Table XII-1. The issues raised in the analysis of the unperturbed dimensions of copolymers can be divided into two groups. One group addresses the composition and configuration, and the other group addresses the evaluation of $\langle r^2 \rangle_0$ for any chain defined in the first.

A. Covalent Structure

An issue common to all modeling of copolymers is the distribution of monomer units along the chain. For binary copolymers, such as copolymers of ethylene

---

[1] But they did occur in the determination of the stereochemical sequence in vinyl polymers.

TABLE XII-1. Treatments of $\langle r^2 \rangle_0$ for Representative Random Copolymers

Monomer A	Monomer B	Ref.
Ethylene	Propylene	*a*
Ethylene	Vinyl chloride	*b*
Ethylene	Vinyl bromide	*c*
Propylene	Vinyl chloride	*d*
Vinyl choride[e]	Vinyl chloride[e]	*f*
Vinylidene fluoride[e]	Vinylidene fluoride[e]	*f*
Ethylene	Carbon monoxide	*g*
Ethylene	1-Butene	*h*

[a]Mark, J. W. *J. Chem. Phys.* **1972**, *57*, 2541.
[b]Mark, J. E. *Polymer* **1973**, *14*, 553.
[c]Tonelli, A. E. *Macromolecules* **1982**, *15*, 290.
[d]Mark, J. E. *J. Polym. Sci., Polym. Phys. Ed.* **1973**, *11*, 1375.
[e]Pseudocopolymers with head-to-head, as well as head-to-tail, monomer placements.
[f]Wang, S.; Mark, J. E. *Comput. Polym. Sci.* **1991**, *1*, 188.
[g]Wittwer, H.; Pino, P.; Suter, U. W. *Macromolecules* **1988**, *21*, 1262.
[h]Mattice, W. L. *Macromolecules* **1986**, *19*, 2303.

and propylene,[2] this aspect of the composition can be handled by use of *reactivity ratios*.[3] The conditional probabilities that a monomer of type α will be followed by a monomer of type $\beta, q_{\alpha\beta}$, are given by

$$q_{AA} = 1 - q_{AB} = \frac{r_A X_A}{r_A X_A + X_B} \tag{XII-1}$$

$$q_{BB} = 1 - q_{BA} = \frac{r_B X_B}{r_B X_B + X_A} \tag{XII-2}$$

where the r_α are the reactivity ratios defined in terms of the rate constants, $k_{\alpha\beta}$, for the addition of a monomer of type β to a growing chain with a monomer of type α at its reactive end,

$$r_A = \frac{k_{AA}}{k_{AB}} \tag{XII-3}$$

$$r_B = \frac{k_{BB}}{k_{BA}} \tag{XII-4}$$

[2]Mark, J. E. *J. Chem. Phys.* **1972**, *57*, 2541.
[3]Flory, P. J. *Principles of Polymer Chemistry*, Cornell University Press, Ithaca, NY, **1953**, Chapter 5.

and X_α is the mole fraction of monomer α in the mixture of the two monomers. For very long chains, in a system where the X_α are constants, the fraction of the monomer units in the chain that are of type A is[3]

$$f_A = \frac{r_A X_A^2 + X_A X_B}{r_A X_A^2 + 2 X_A X_B + r_B X_B^2} \tag{XII-5}$$

Specification of f_A and $r_A r_B$ determines the values of q_{AA} and q_{BB}, that can be used in conjunction with a sequence of random numbers in order to produce the sequence of monomer units in a representative chain.

The stereochemistry of the attachment of the methyl group in each propylene unit is handled in a manner similar to that employed with the vinyl polymers. If the stereochemistry can be taken to be Bernoullian, the probability for a *meso* "diad," p_m, is used in conjunction with a random number in order to assign the stereochemistry of each propylene unit added to the chain. A random number is also used to assign the sequence of the units in pseudocopolymers with head-to-head, as well as head-to-tail, placements.

B. Unperturbed Dimensions

The value of Z for a chain of x monomer units, and also the value of $\langle r^2 \rangle_0$ for this chain, will in general depend on the sequence of the monomer units. It will also depend on the stereochemical composition if one or both monomers contributes bonds with an asymmetric rotation potential energy function. However, once the covalent structure of a representative chain has been specified, both Z and $\langle r^2 \rangle_0$ can be evaluated using the methods described in earlier chapters. Implementation requires the statistical weight matrices appropriate for both homopolymers, and additional statistical weight matrices for the bonds near the junctions of two monomer units with different covalent structures.

In the approximation that $\nu = 3$, the statistical weight matrix for the homopolymer of ethylene is given by Eq. (IV-19), and the statistical weight matrices for the homopolymer of propylene, using the dl notation, are given by Eqs. (VIII-5), (VIII-7), (VIII-9), (VIII-10), (VIII-18), and (VIII-19). Incorporation of the first- and second order interactions that occur in copolymers of ethylene and propylene requires additional statistical weight matrices for the bonds written with long dashes in the sequences $CH_2CH_2—CH(CH_3)CH_2$ and $CH(CH_3)CH_2—CH_2CH_2$. Using the subscript e to represent an ethylene unit, and subscripts d and l to represent d- and l-propylene units, the statistical weight matrix for $CH_2CH_2—CH(CH_3)CH_2$ can be written as[2]

$$\mathbf{U}_{ed} = \begin{bmatrix} \eta & \tau & 1 \\ \eta & \tau\omega & \omega \\ \eta\omega & \tau\omega & 1 \end{bmatrix} \tag{XII-6}$$

$$\mathbf{U}_{el} = \mathbf{Q}\mathbf{U}_{ed}\mathbf{Q} \tag{XII-7}$$

and the statistical weight matrix for $CH(CH_3)CH_2—CH_2CH_2$ is[4]

$$\mathbf{U}_{de} = \begin{bmatrix} \eta/\tau & \omega & 1 \\ \eta/\tau & 1 & \omega \\ \eta/\tau & \omega & \omega \end{bmatrix} \tag{XII-8}$$

$$\mathbf{U}_{le} = \mathbf{Q}\mathbf{U}_{de}\mathbf{Q} \tag{XII-9}$$

In these matrices, the ω in the 2,3 and 3,2 elements arise from the second-order interaction of groups in the main chain, and the remaining two ω in each matrix arise from the second-order interaction of a methyl group in a propylene unit with a group in the main chain.

Averages of $\langle r^2 \rangle_0$ must be generated over a suitable number of independently generated chains, obtained using the same f_A and $r_A r_B$. This procedure is comparable with the generation of C_n for an atactic vinyl polymer by the average of the $\langle r^2 \rangle_0$ over a suitable number of independently generated chains, each with the desired p_m and n, but different stereochemical sequences.

The characteristic ratios will, in general, depend on the overall composition (f_A), the reactivity ratios $(r_A r_B)$, and the stereochemical composition (p_m), as illustrated in Figs. XII-1 and XII-2 for copolymers of ethylene and propylene.[2] When $r_A r_B = \infty$, the system contains a mixture of the two homopolymers. In this limit, C_n is a linear function of the overall composition. Relatively little deviation from this straight line is seen at other $r_A r_B$ if the stereochemistry has $p_m = 0.50$ (Fig. XII-2), but strong deviations are observed when the stereochemistry is highly isotactic (Fig. XII-1). The appearance of the minimum in C_n at intermediate overall composition, when the system is highly isotactic and $r_A r_B$ is small, has been rationalized as follows:[2] Random insertion of ethylene units into an isotactic sequence of propylene units disrupts its preferred 3_1 helix, and random insertion of propylene units into a sequence of ethylene units disrupts its preferred sequence of *trans* placements. Both types of disruption reduce $\langle r^2 \rangle_0$, thereby causing the minimum in C_n.

Copolymers of propylene and vinyl chloride present slightly different issues insofar as the statistical weight matrices are concerned. Here each monomer is of the form $CH_2=CHX$, with X being either Cl or CH_3. If each "X" is taken as

---

[4]Equation (XII-8) implies $\tau/\eta = \sigma$ as the relationship between the statistical weights for the first-order interactions. In his treatment of these copolymers, Mark[2] wrote the statistical weight matrix for bonds in polyethylene as

$$\mathbf{U} = \begin{bmatrix} 1 & \tau/\eta & \tau/\eta \\ 1 & \tau/\eta & \tau\omega/\eta \\ 1 & \tau\omega/\eta & \tau/\eta \end{bmatrix}$$

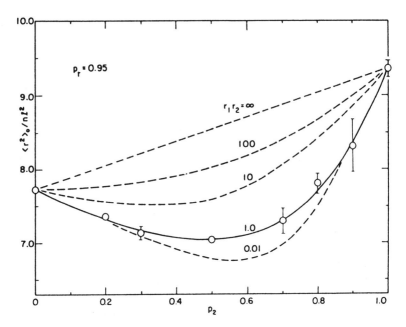

Figure XII-1. C_n for highly isotactic ($p_m = 0.95$) ethylene–propylene copolymers of 100 units, as a function of the mole fraction of propylene. The values of $r_A r_B$ are noted for each curve, and Mark's p_r is our p_m. Reprinted with permission from Mark, J. E. *J. Chem. Phys.* **1972**, *57*, 2541. Copyright 1972 American Physical Society.

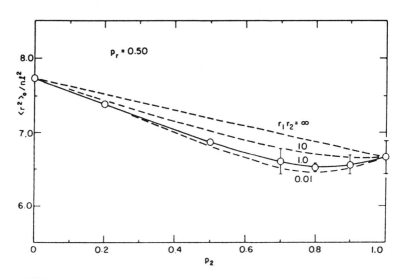

Figure XII-2. C_n for atactic ($p_m = 0.50$) ethylene–propylene copolymers of 100 units, as a function of the mole fraction of propylene. The values of $r_A r_B$ are noted for each curve, and Mark's p_r is our p_m. Reprinted with permission from Mark, J. E *J. Chem. Phys.* **1972**, *57*, 2541. Copyright 1972 American Physical Society.

a single unit, and $\nu = 3$ is a satisfactory approximation for the copolymers, the statistical weight matrices for all of the bonds take the form for vinyl polymers, as described in the first section of Chapter VIII (Section VIII.1). Appropriate identifiers are needed for each statistical weight, for proper identification of the interacting groups. For example, the second-order interaction denoted by ω_{XX} in Chapter VIII will now come in three forms: $\omega_{Cl,Cl}$, ω_{Cl,CH_3}, and ω_{CH_3,CH_3}. The first and last of these ω values are also required for treatment of the homopolymers, but ω_{Cl,CH_3} is unique to the copolymers. Distinction between these three ω_{XX} may have little influence on the results for C_n if they arise from repulsive second-order interactions of comparable energies.[5]

Cases with Mixed Head-to-Head and Head-to-Tail Units. Polymers with mixed head-to-head and head-to-tail units are not copolymers in the usual sense. The chain is derived from a single chemical species. Nevertheless, insofar as the formulation of Z is concerned, the chains with mixed head-to-head and head-to-tail units present very similar issues to those that arise with the real copolymers of two chemically different monomer units that were described above. Evaluation of the effect of head-to-head placements on C_n requires first the generation of chains with specific sequences of monomer units. Computation of $\langle r^2 \rangle_0$ for each chain will require statistical weight matrices, and statistical weights, in addition to those employed for the chain composed exclusively of head-to-tail units. In the case of poly(vinyl chloride), for example, two chlorine atoms participate in the second-order interaction denoted in Chapter VIII by ω_{XX} if the chain is exclusively head-to-tail. When head-to-head placements are present, two chlorine atoms can also participate in a first-order interaction in the fragment $-CHCl{-}CHCl-$.[6]

Ethylene–Carbon Monoxide Copolymers. Copolymers of ethylene and carbon monoxide are a simple example of a copolymer in which the two monomer units contribute different numbers of bonds to the main chain. Furthermore, the profound differences in the conformational restrictions produced by the two monomer units require that some of the statistical weight matrices must be rectangular but not square.[7] The origin of the rectangular statistical weight matrices is apparent in the conformational energy surface for a molecule as simple as 2-pentanone, as depicted in Fig. XII-3. Three distinct rotational isomers are evident at the $CH_2{-}CH_2$ bond, as expected from the conformational analysis of n-alkanes. However, the $CO{-}CH_2$ bond is characterized by a broad region of low energy that does not exhibit an energy barrier between the torsion angles usually associated with t and with $g^{\pm}$. Accurate representation of this broad region of low conformational energy for the $CO{-}CH_2$ bond is difficult if only three rotational isomeric states

[5]Mark, J. E. *J. Polym. Sci., Polym. Phys. Ed.* **1973**, *11*, 1375.
[6]Wang, S.; Mark, J. E. *Comput. Polym. Sci.* **1991**, *1*, 188.
[7]Wittwer, H.; Pino, P.; Suter, U. W. *Macromolecules* **1988**, *21*, 1262.

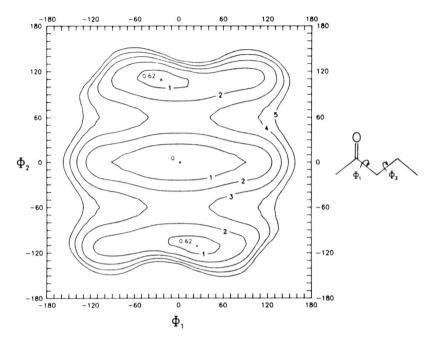

Figure XII-3. Conformational energy surface for 2-pentanone. Contours are drawn at 1, 2, 3, 4, and 5 kcal mol^{-1} relative to the minimum. The figure uses $\phi = 0°$ in the *trans* state. Reprinted with permission from Wittwer, H.; Pino, P.; Suter, U. W. *Macromolecules* **1988**, *21*, 1262. Copyright 1988 American Chemical Society.

are employed. A more accurate approach for the CO—CH$_2$ bond splits the customary t state into three states, denoted by t^-, t^0, and t^+, with torsion angles in the ranges $120° < \phi < 160°$, $160° < \phi < 200°$, $200° < \phi < 240°$, respectively. The $g^{\pm}$ states share equally the remaining 240° range for ϕ. Thus $\nu = 5$ at the CO—CH$_2$ bonds, but ν retains its customary value of 3 at the CH$_2$—CH$_2$ bonds.

For copolymers of ethylene and carbon monoxide in which there are at least two ethylene units between each pair of carbon monoxide units, the statistical weights required for the three successive bonds in the CH$_2$—CO—CH$_2$—CH$_2$ sequence are of the form

$$
\mathbf{U}_{CH_2CO} =
\begin{array}{c}
\\ t \\ g^+ \\ g^-
\end{array}
\begin{array}{ccccc}
t^- & t^0 & t^+ & g^+ & g^- \\
\left[\begin{array}{ccccc}
\kappa_{ab} & 1 & \kappa_{ab} & \sigma_a & \sigma_a \\
\kappa_{ab} & 1 & \kappa_{ab} & \sigma_a & \sigma_a\omega_a \\
\kappa_{ab} & 1 & \kappa_{ab} & \sigma_a\omega_a & \sigma_a
\end{array} \right]
\end{array}
\qquad \text{(XII-10)}
$$

$$\mathbf{U}_{COCH_2} = \begin{array}{c} \\ t^- \\ t^0 \\ t^+ \\ g^+ \\ g^- \end{array} \begin{array}{ccccc} t^- & t^0 & t^+ & g^+ & g^- \\ \begin{bmatrix} \lambda & 1 & \lambda & \sigma_a & \sigma_a \\ \lambda & 1 & \lambda & \sigma_a & \sigma_a \\ \lambda & 1 & \lambda & \sigma_a & \sigma_a \\ \lambda & 1 & \lambda & \sigma_a & \sigma_a \\ \lambda & 1 & \lambda & \sigma_a & \sigma_a \end{bmatrix} \end{array} \qquad \text{(XII-11)}$$

$$\mathbf{U}_{CH_2CH_2} = \begin{array}{c} \\ t^- \\ t^0 \\ t^+ \\ g^+ \\ g^- \end{array} \begin{array}{ccc} t & g^+ & g^- \\ \begin{bmatrix} 1 & \sigma_b & \sigma_b \\ 1 & \sigma_b & \sigma_b \\ 1 & \sigma_b & \sigma_b \\ 1 & \sigma_b & \sigma_b\omega_{ab} \\ 1 & \sigma_b\omega_{ab} & \sigma_b \end{bmatrix} \end{array} \qquad \text{(XII-12)}$$

The serial product $\mathbf{U}_{CH_2CO}\mathbf{U}_{COCH_2}\mathbf{U}_{CH_2CH_2}$ is of dimensions 3×3. Hence it is conformable for multiplication with $\mathbf{U}$ of the type employed for bonds in polyethylene.

If only a single ethylene unit separates two carbon monoxide units, as in the alternating copolymers, the statistical weight matrices retain the dimensions defined in Eqs. (XII-10), (XII-11), and (XII-12), but the statistical weights appearing in these matrices are different, as a result of direct interaction of the two closely spaced carbon monoxide units.[7]

The calculated C_∞ are not very sensitive to the carbon monoxide content, increasing by only about 15% as the carbon monoxide content increases from 0 to 50%.[7] This prediction is consistent with the experimental estimate of $C_\infty = 8.5 \pm 1.5$ at 25°C for highly alternating copolymers.[8]

Ethylene–1-Alkene Copolymers. Statistical weight matrices of varying dimensions are also required for the formulation of Z for random copolymers of ethylene and a 1-alkene containing four or more carbon atoms. The 1-alkene then contributes an articulated side chain. When these copolymers have a composition that is predominantly ethylene, with only a few percent of the monomers being the 1-alkene, they are the "linear low-density polyethylene" of commerce. In this range of composition, a useful approximation is that at least one ethylene unit occurs between successive 1-alkene units. Then the conformational partition function, including all first- and second-order interactions, for a chain of specified sequence can be formulated using the statistical weight matrices described in the previous chapter. For purposes of the formulation of Z, the molecule can be viewed as a graft copolymer in which the grafts are short and of the same chemistry as the main chain.

---

[8]Wittwer, H. dissertation, ETH Zürich 8117, **1986**.

Calculations for random ethylene–1-butene copolymers show that C_∞ is reduced by the random incorporation of 1-butene into polyethylene.[9] The depression in C_∞ reaches 10% when there is an ethyl group bonded to one out of every 12–14 carbon atoms in the main chain. The reduction in unperturbed dimensions occurs because the short branch disrupts the tendency for propagation of sequences of t placements. The disturbance in the $p_{t;i}$ in the main chain is depicted in Fig. XI-6, for the case where the 1-alkene is 1-hexene.

2. "RANDOM" COPOLYMERS WITH INDEPENDENT BONDS

Very significant simplifications in the formalism are possible if the copolymer is constructed from bonds that are independent. The unperturbed ensembles of an important class of copolymers have been treated with this approximation, within the framework of a virtual bond model. If a group of homopolymers can be treated as being composed of independent virtual bonds [as was the case for the poly(α-amino acids) in Chapter IX], one might anticipate use of a similar model for the copolymers. In this section we first describe the formalism, and then present illustrative applications.

A. Matrix Formalism

The averaged end-to-end vector for a chain with independent bonds of constant length is

$$\langle \mathbf{r} \rangle_0 = \mathbf{l}_1 + \langle \mathbf{T}_1 \rangle \mathbf{l}_2 + \langle \mathbf{T}_1 \rangle \langle \mathbf{T}_2 \rangle \mathbf{l}_3 + \cdots + \langle \mathbf{T}_1 \rangle \langle \mathbf{T}_2 \rangle \cdots \langle \mathbf{T}_{n-1} \rangle \mathbf{l}_n \quad \text{(XII-13)}$$

If bond angles are also constant, $\mathbf{T}_1$ does not require angle brackets, as was the case in Eq. (VI-39). The independence of the bonds allows replacement of each $\langle \mathbf{T}_i \cdots \mathbf{T}_{i+j} \rangle$ by $\langle \mathbf{T}_i \rangle \cdots \langle \mathbf{T}_{i+j} \rangle$. The development in Chapter VI imagined the case where the variables that must be considered in the evaluation of $\langle \mathbf{T}_i \rangle$ were the same for bond i in each chain in the unperturbed ensemble. If some of the chains contain monomer A at bond i, and others contain monomer B at the same position, with complete independence of the bonds in both situations, then

$$\langle \mathbf{T}_i \rangle = X_A \langle \mathbf{T}_A \rangle + (1 - X_A) \langle \mathbf{T}_B \rangle \quad \text{(XII-14)}$$

where X_A denotes the mole fraction of the monomer units at position i that are of type A. The homopolymers are recovered when $X_A = 1$ or 0. When these requirements are met by the copolymer, the values of $\langle \mathbf{r} \rangle_0$ and $\langle r^2 \rangle_0$ are obtained by use of $\langle \mathbf{T}_i \rangle$ in Eqs. (VI-40) and (VI-56).

[9]Mattice, W. L. *Macromolecules* **1986**, *19*, 2303.

A subtle aspect of the independence of bonds can arise in the copolymers. One might have a situation where the torsion represented by ϕ_i is independent of the torsion represented by ϕ_{i-1}, but the nature of the torsion potential energy function depends on whether the chemistry of the preceding unit is monomer A or monomer B. Thus the torsion potential energy function at bond i in monomer A may be either $E_{AA}(\phi_i)$ or $E_{BA}(\phi_i)$, which might be quite different (as illustrated schematically in Fig. XII-4), but in neither case depends on ϕ at the preceeding bond. This dependence on the sequence of the units in the chain can be incorporated by the computation of $\langle r^2 \rangle_0$ as

$$[X_A \langle G_A \rangle (1 - X_A) \langle G_B \rangle]_1 \begin{bmatrix} P_{same} \langle G_{AA} \rangle & (1 - P_{same}) \langle G_{AB} \rangle \\ (1 - P_{same}) \langle G_{BA} \rangle & P_{same} \langle G_{BB} \rangle \end{bmatrix}^{n-2} \cdot \begin{bmatrix} G_A \\ G_B \end{bmatrix}_n$$

$$(XII-15)$$

where X_A denotes the probability that the first bond in a chain is of type A, and P_{same} denotes the probability that bond i and $i - 1$ are derived from the same type of monomer. Each of the $\langle G \rangle$ in the 10×10 matrix contains the $\langle T \rangle$ that is appropriate for the combinations of bonds at positions $i - 1$ and i that is denoted by its subscript. The $\langle T \rangle$ used for a bond of type A can depend on whether its predecessor is of type A or B, but in both cases ϕ_i does not depend on ϕ_{i-1}.

B. Amino Acid Copolymers

If the virtual bonds in a poly(α-amino acid) are independent, a copolymer of the D and L amino acids can be treated using Eq. (XII-15). In the case of a copolymer composed of L-alanyl and D-alanyl units, the averaged transformation matrix that appears in $\langle G_A \rangle, \langle G_{BA} \rangle$, and $\langle G_{AA} \rangle$ might be the one in Eq.

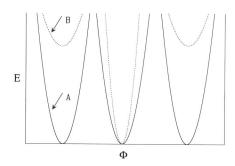

Figure XII-4. Conformational energy *vs.* ϕ at virtual bond A when the previous virtual bond is of type A or B.

(IX-32), and the averaged transformation matrix in $\langle \mathbf{G_B} \rangle, \langle \mathbf{G_{AB}} \rangle$, and $\langle \mathbf{G_{BB}} \rangle$ would be the one in which the signs have been reversed at the 1,3, 2,3, 3,1, and 3,2 elements.

$$\langle \mathbf{T}_{\text{D-Ala}} \rangle = \begin{bmatrix} 0.51 & 0.20 & -0.59 \\ -0.046 & -0.61 & -0.21 \\ -0.65 & 0.23 & -0.30 \end{bmatrix} \qquad \text{(XII-16)}$$

The dependence of C_∞ on P_{same} is depicted in Fig. XII-5. The system contains an equimolar mixture of the two sterically pure homopolymers when $P_{\text{same}} = 1$. In this limit C_∞ is identical with the result for either of the homopolymers. Much smaller dimensions are obtained for the completely random copolymer, with $P_{\text{same}} = 0.5$. The L-alanyl residues prefer conformations in the upper left quadrant of the ϕ, ψ conformational energy map (Fig. IX-8), but the D-alanyl residues prefer conformations in the lower right quadrant of the map. The preference for a quadrant of the ϕ, ψ conformational energy map undergoes an exact alternation as one proceeds along the chain in the alternating copolymer, $P_{\text{same}} = 0$.

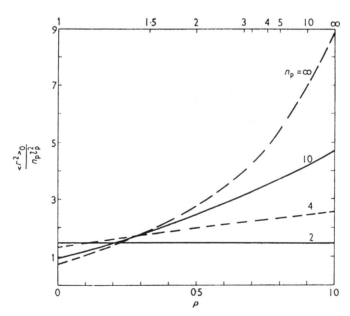

Figure XII-5. C_∞ as a function of P_{same} ($= p$) for racemic copolymers of alanine. From Miller, W. G.; Brant, D. A.; Flory, P. J. *J. Mol. Biol.* **1967**, *23*, 67. Copyright © John Wiley & Sons 1967. Reprinted by permission of John Wiley & Sons, Inc.

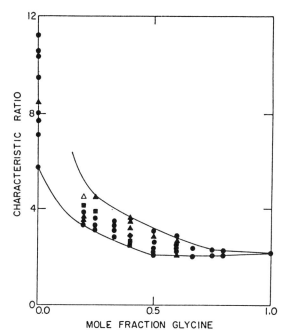

Figure XII-6. C_∞ as a function of the mole fraction of glycine for sequential copolymers of glycine, L-alanine, and L-proline. Reprinted with permission from Mattice, W. L.; Mandelkern, L. *Biochemistry* **1971**, *10*, 1934. Copyright 1971 American Chemical Society.

Random copolymers of glycine and L-alanine can be treated using Eq. (XII-14), but Eq. (XII-15) must be employed if the copolymer contains L-proline. The steric interactions generated by the pyrrolidine ring in an L-prolyl residue at position $i + 1$ modify the conformational energy surface of the residue at position i.[10] Therefore the $\langle T \rangle$ used for a glycyl, L-alanyl, or L-prolyl residue at position i depends on whether the following residue is or is not L-prolyl. This distinction must be introduced into the calculation of the unperturbed dimensions of protein random coils[11] and sequential copolypeptides that contain L-proline.[12]

When the glycyl residue is an important component of random copolypeptides, it will exert a strong control over C_∞ because its conformational energy surface (Fig. IX-7) is unaffected by inversion through the origin.[11] Figure XII-6 depicts the computed C_∞ for sequential copolymers of L-alanine, L-proline, and

---

[10]Schimmel, P. R.; Flory, P. J. *J. Mol. Biol.* **1968**, *34*, 105.

[11]Miller, W. G.; Goebel, C. V. *Biochemistry* **1968**, *7*, 3925. Mattice, W. L. *Macromolecules* **1977**, *10*, 516.

[12]Mattice, W. L.; Mandelkern, L. *Biochemistry* **1971**, *10*, 1934.

glycine with all possible repeating sequences of three, four, or five residues. If the copolymer contains a high content of glycine, C_∞ is defined within rather narrow limits, independent of the nature of the remaining residues in the chain. The same data do not exhibit nearly as precise a trend when plotted against the content of L-alanine (Fig. XII-7) or L-proline.

The treatment of the helix $\rightleftharpoons$ coil transitions in α-amino acid copolymers may require the introduction of important interactions between nonidentical amino acid residues that are separated by several other amino acid residues. For example, the C^α atoms in residues i and $i + 4$ are separated by only ~0.6 nm in an α helix, permitting easy interaction of their side chains. These interactions may stabilize the α helix if residues i and $i + 4$ bear charges of opposite sign, as would be the case with L-glutamate and L-lysine in an aqueous solution with a pH near 7. The side chains on these residues are long enough so that they can interact in the α helix even when they are separated by two turns, as at positions i and $i + 7$.

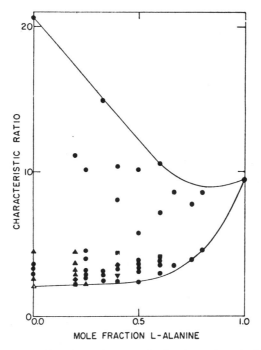

Figure XII-7. C_∞ as a function of the mole fraction of L-alanine for sequential copolymers of glycine, L-alanine, and L-proline. Reprinted with permission from Mattice, W. L.; Mandelkern, L. *Biochemistry* **1971**, *10*, 1934. Copyright 1971 American Chemical Society.

PROBLEMS

Appendix B contains answers for Problems XII-3, XII-9 through XII-12, XII-14, and XII-15.

XII-1. How many distinguishable statistical weight matrices are required for the evaluation of Z for poly(vinylidene fluoride) if the chain contains both head-to-head and head-to-tail units, and the treatment is confined to consideration of first- and second-order interactions? (See Tonelli[13] for a set of statistical weight matrices used for this purpose.)

XII-2. How many distinguishable statistical weight matrices are required for the evaluation of Z for poly(vinyl chloride) if the chain contains both head-to-head and head-to-tail units, and the treatment is confined to consideration of first- and second-order interactions? (See Wang and Mark.[6])

XII-3. Set up the statistical weight matrices required for the study of random copolymers of ethylene and oxymethylene. When applied to the strictly alternating copolymer, your formulation should reduce to the one described for poly(trimethylene oxide) in Chapter VII.

XII-4. As shown in the discussion in Chapter VII, polyoxymethylene and polyoxyethylene have quite different values of C_∞. How does C_∞ depend on composition for random copolymers of oxymethylene and oxyethylene?

XII-5. Formulate the U required for computation of C_∞ for random copolymers of ethylene oxide and propylene oxide.

XII-6. How might one formulate Z for a polybutadiene in which the microstructure is 55% *trans*, 35% *cis*, 10% vinyl?

XII-7. An alternating copolymer of ethylene and propylene can be obtained by complete hydrogenation of polyisoprene. How might one formulate Z for chains that result from *partial* hydrogenation of polyisoprene? What is the prediction for the manner in which $\langle r^2 \rangle_0$ changes during the reaction?

XII-8. Assuming equal reactivities, how does C_∞ depend on X for copolymers in which the molar ratio of ethylene oxide, terephthalate, and isophthalate units is $1 : X : 1 - X$?

XII-9. How does C_∞ for racemic copolymers of L-lactyl and D-lactyl units depend on the tendency for propagation of sequences of monomer units with the same chirality?

---

[13]Tonelli, A. E. *Macromolecules* **1976**, *9*, 547.

XII-10. How does C_∞ depend on the mole fraction of glycine in random copolymers of L-alanine and glycine? Why isn't C_∞ a linear function of the composition?

XII-11. If a chain contains equal numbers of glycyl and L-alanyl residues, will C_∞ be higher if the sequence is random or alternating?

XII-12. If the error in a measurement of C_∞ is ±5%, what level of randomly placed D-alanyl residues can be incorporated into poly(L-alanine) before they have an effect on C_∞ that is comparable with the experimental uncertainty?

XII-13. Suggest a formulation of Z for the helix $\rightleftharpoons$ coil transition in a polyalanine containing mole fraction X_D of D-alanyl residues and mole fraction $1 - X_D$ of L-alanyl residues, $X_D \ll 1$. Is the presence of the minor component more likely to affect C_∞ when s for the L-alanyl residue is less than 1, or when it is greater than 1?

XII-14. Find C for a tapered racemic copolymer of 1000 alanyl units when the probability for an L-alanyl unit at position 1 is 0, its probability at position 1000 is 1, and the probability at intermediate positions is $(i - 1)/999$. How does this result for C compare with the results for the strictly alternating copolymer, and the completely random copolymer of the same composition?

XII-15. Find C for the five non-overlapping subchains of 200 units for the tapered copolymer in Problem XII-14. How can one rationalize the difference in the unperturbed dimensions of the subchains?

XIII More Generator Matrices
from l and T

Chapter VI described matrices used for generation of $\mathbf{r}, r^2$, and s^2 for a chain in a specified conformation. The generator matrices $\mathbf{A}_i, \mathbf{G}_i$, and $\mathbf{H}_i$ were formulated from l_i, θ_i, and ϕ_i. When combined with the statistical weight matrices to form $\mathscr{A}_i, \mathscr{G}_i$, and $\mathscr{H}_i$, they provide a rapid and exact evaluation of $\langle \mathbf{r} \rangle_0, \langle r^2 \rangle_0$, and $\langle s^2 \rangle_0$ for the rotational isomeric state (RIS) model of a chain with interdependent bonds. If the bonds are independent, the averages can be achieved with an even simpler scheme, where every element in each $\mathbf{A}_i, \mathbf{G}_i$, and $\mathbf{H}_i$ is replaced by its average.

Several other generator matrices, appropriate for evaluation of other conformation-dependent physical properties of a chain, can be constructed. Some of these matrices employ the same information $(\mathbf{l}_i, \mathbf{T}_i)$ that is used in $\mathbf{A}_i, \mathbf{G}_i$, and $\mathbf{H}_i$. The formulation of other of these matrices requires additional information. The new generator matrices that use only $\mathbf{l}_i$ and $\mathbf{T}_i$ are the subject of this chapter. The generator matrices that incorporate additional information (e.g., dipole moment, anisotropic part of the polarizability tensor) will be described in Chapter XIV. In both cases, the objective is a matrix expression for a conformation-dependent physical property, constructed so that all pertinent information relevant to bond i is found in the generator matrix for that bond, with the matrices for the n bonds appearing in the same order in the serial product as do the n bonds in the chain.

The method for efficiently computing the average over all conformations in the ensemble remains the one described in Chapter VI, using supermatrices that are completely analogous to $\mathscr{A}_i, \mathscr{G}_i$, and $\mathscr{H}_i$ when the bonds are interdependent, and replacing each generator matrix by its average if they are independent.

1. HIGHER MOMENTS OF r, s, AND r

The *higher even moments* of $r, \langle r^{2p} \rangle_0, p > 1$, are used to provide information about the *shape* of the distribution function for the end-to-end distance. The *breadth* of the distribution function for r^2, for example, is measured by $\langle r^4 \rangle_0 / \langle r^2 \rangle_0^2$. Knowledge of all of the even moments, $\langle r^{2p} \rangle_0, p = 1, 2, \ldots$, would completely specify the shape of the distribution function for r^2. Complete knowledge of all $\langle r^{2p} \rangle_0$ is not readily obtainable in practice, but the lower even moments are computed from generator matrices for the r^{2p} and the averaging procedures described in Chapter VI.

A. Construction of the Generator Matrices

The generator matrices for the $r^{2p}, p > 1$, are constructed from the generator matrix for r^2.[1] If the objective is r^4, we start with

$$r^4 = r^2 \otimes r^2 \tag{XIII-1}$$

which merely states that the product of two scalars can be obtained by using the rule for forming the *direct product* of two matrices.[2] Inserting the generator matrix expression for r^2 yields

$$r^4 = (\mathbf{G}_1 \mathbf{G}_2 \cdots \mathbf{G}_n) \otimes (\mathbf{G}_1 \mathbf{G}_2 \cdots \mathbf{G}_n) \tag{XIII-2}$$

As it stands, this expression is not usable for facile computation of $\langle r^4 \rangle_0$ because information about bond i occurs in two copies of $\mathbf{G}_i$, which are found at two widely separated positions. This difficulty can be alleviated by n successive applications of the *theorem on direct products*,[3] which permits the successive rewriting of Eq. (XIII-2) as

$$r^4 = (\mathbf{G}_1 \otimes \mathbf{G}_1)[(\mathbf{G}_2 \cdots \mathbf{G}_n) \otimes (\mathbf{G}_2 \cdots \mathbf{G}_n)]$$

$$\vdots$$

$$r^4 = (\mathbf{G}_1 \otimes \mathbf{G}_1)(\mathbf{G}_2 \otimes \mathbf{G}_2) \cdots (\mathbf{G}_n \otimes \mathbf{G}_n) \tag{XIII-3}$$

All information about bond i—namely, l_i, θ_i, and ϕ_i—is now found in a single

---

[1] Flory, P. J. *Macromolecules* **1974**, *7*, 381.
[2] If $\mathbf{A}$ and $\mathbf{B}$ are matrices of dimensions $i \times j$ and $k \times l$, then $\mathbf{A} \otimes \mathbf{B}$ is of dimensions $ik \times jl$. This matrix is conveniently constructed as ij blocks, each the size of $\mathbf{B}$, namely, $k \times l$. The ijth block in $\mathbf{A} \otimes \mathbf{B}$ is $A_{ij}\mathbf{B}$. When $i = j = k = l = 1$, $\mathbf{A} \otimes \mathbf{B} = A_{11}B_{11}$, which is just the product of two scalars.
[3] The theorem on direct products states that $(\mathbf{A} \otimes \mathbf{B})(\mathbf{C} \otimes \mathbf{D}) = (\mathbf{AC}) \otimes (\mathbf{BD})$, provided $\mathbf{A}$ and $\mathbf{C}$, and $\mathbf{B}$ and $\mathbf{D}$, are conformable for forming the products $\mathbf{AC}$ and $\mathbf{BD}$.

matrix, $\mathbf{G}_i \otimes \mathbf{G}_i$. The generator matrices for the internal bonds in Eq. (XIII-3) are of dimensions 25×25. The matrices for bonds 1 and n are a row vector and column vector, respectively, of 25 elements. The same information can be generated with matrices of a more compact form, as will be described in Section XIII.1.B.

Several of the generator matrices described in this chapter will make use of the direct product of a matrix with itself. A more concise notation for these self-direct products is $\mathbf{A}^{\times j}$, where $\mathbf{A}^{\times 2} \equiv \mathbf{A} \otimes \mathbf{A}, \mathbf{A}^{\times 3} \equiv \mathbf{A} \otimes \mathbf{A} \otimes \mathbf{A}$, and so on. With this notation, the expression for r^4 becomes

$$r^4 = (r^2)^{\times 2} = \mathbf{G}_1^{\times 2} \mathbf{G}_2^{\times 2} \cdots \mathbf{G}_n^{\times 2} \tag{XIII-4}$$

The final expression in Eq. (XIII-3) can be combined with the statistical weight matrices in the manner described in Chapter VI, for rapid computation of $\langle r^4 \rangle_0$. The supermatrices for the internal bonds are $(\mathbf{U}_i \otimes \mathbf{I}_{25}) \| \mathbf{G}_i^{\times 2} \|$ if the unreduced form of $\mathbf{G}^{\times 2}$ is used. If $\mathbf{G}^{\times 2}$ has been reduced, $\mathbf{I}_{15}$ replaces $\mathbf{I}_{25}$. This information, along with $\langle r^2 \rangle_0$, can be used for evaluation of $\langle r^4 \rangle_0 / \langle r^2 \rangle_0^2$, which is a measure of the width of the distribution function for r^2.

The averaging of r^4 to obtain $\langle r^4 \rangle_0$ for a chain with *independent* bonds, and with ϕ_i as the only variable, is achieved as

$$\langle r^4 \rangle_0 = \mathbf{G}_1^{\times 2} \langle \mathbf{G}_2^{\times 2} \rangle \cdots \langle \mathbf{G}_{n-1}^{\times 2} \rangle \mathbf{G}_n^{\times 2} \tag{XIII-5}$$

The averaged generator matrices for the internal bonds cannot be replaced by $\langle \mathbf{G}_i \rangle^{\times 2}$. The correct generator matrix, $\langle \mathbf{G}_i^{\times 2} \rangle$, will contain terms with $\langle \cos \phi_i \rangle, \langle \sin \phi_i \rangle, \langle \cos^2 \phi_i \rangle, \langle \sin^2 \phi_i \rangle$, and $\langle \cos \phi_i \sin \phi_i \rangle$. The incorrect $\langle \mathbf{G}_i \rangle^{\times 2}$ would supply $\langle \cos \phi_i \rangle^2$ instead of $\langle \cos^2 \phi_i \rangle, \langle \sin \phi_i \rangle^2$ instead of $\langle \sin^2 \phi_i \rangle$, and $\langle \cos \phi_i \rangle \langle \sin \phi_i \rangle$ instead of $\langle \cos \phi_i \sin \phi_i \rangle$.

The procedure illustrated above can be extended to higher even moments. For example, r^6 can be written as

$$r^6 = (r^2)^{\times 3} = r^2 \otimes r^2 \otimes r^2 \tag{XIII-6}$$

$$r^6 = \mathbf{G}_1^{\times 3} \mathbf{G}_2^{\times 3} \cdots \mathbf{G}_n^{\times 3} \tag{XIII-7}$$

where the generator matrices for the internal bonds are now of dimensions

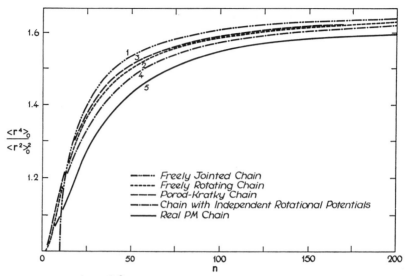

Figure XIII-1. $\langle r^4 \rangle_0 / \langle r^2 \rangle_0^2$ for different RIS approximations to an unperturbed linear polyethylene. Reprinted with permission from Jernigan, R. L.; Flory, P. J. *J. Chem. Phys.* **1969**, *50*, 4178. Copyright 1969 American Physical Society.

$5^3 \times 5^3$.[4] Extension to the higher even moments of the radius of gyration, s^4, s^6, ..., is obvious.

$$s^4 = s^2 \otimes s^2 = (s^2)^{\times 2} \tag{XIII-8}$$

$$s^4 = \frac{1}{(n+1)^4} \mathbf{H}_1^{\times 2} \mathbf{H}_2^{\times 2} \cdots \mathbf{H}_n^{\times 2} \tag{XIII-9}$$

Before reduction, $\mathbf{H}_i^{\times p}$ for internal bonds are of dimensions $7^p \times 7^p$.

These generator matrices can be used to demonstrate that flexible chains usually have a distribution function for s^2 that is narrower than the distribution function for r^2: $\langle s^4 \rangle_0 / \langle s^2 \rangle_0^2 < \langle r^4 \rangle_0 / \langle r^2 \rangle_0^2$. As $n \to \infty$, these dimensionless ratios for flexible chains will approach the limits presented in Table I-1. Departures from these limiting values can be found at finite n. The nature of the departure depends on n and on the conformations adopted by the chain, as depicted in Fig. XIII-1.[5]

---

[4]In general, $\mathbf{G}_i^{\times p}$ for the internal bonds are of dimensions $5^p \times 5^p$. They can be reduced by the method described in Section XIII.1.B.

[5]Jernigan, R. L.; Flory, P. J. *J. Chem. Phys.* **1969**, *50*, 4178.

The analogous manipulation of the generator matrices for **r** is different only in that the results are not scalars, because **r** is a vector. The expression

$$\mathbf{r}^{\times 2} = \mathbf{A}_1^{\times 2}\mathbf{A}_2^{\times 2} \cdots \mathbf{A}_n^{\times 2} \qquad (\text{XIII-10})$$

defines a vector of nine elements, which can be written in terms of the $x, y,$ and z components of **r** as

$$\mathbf{r}^{\times 2} = \begin{bmatrix} r_x^2 \\ r_x r_y \\ r_x r_z \\ r_y r_x \\ r_y^2 \\ r_y r_z \\ r_z r_x \\ r_z r_y \\ r_z^2 \end{bmatrix} \qquad (\text{XIII-11})$$

The sum of the first, fifth, and ninth elements is r^2. The column vector contains three pairs of elements ($r_x r_y$ and $r_y r_x, r_x r_z$ and $r_z r_x, r_y r_z$ and $r_z r_y$) that are numerically identical. The nine elements in the column vector can also be arranged in the matrix form of a second-order tensor, which is symmetric because of the arrangement of these three pairs of identical elements:

$$\mathbf{r}\mathbf{r}^T = \begin{bmatrix} r_x^2 & r_x r_y & r_x r_z \\ r_y r_x & r_y^2 & r_y r_z \\ r_z r_x & r_z r_y & r_z^2 \end{bmatrix} \qquad (\text{XIII-12})$$

The trace of $\mathbf{r}\mathbf{r}^T$ is r^2.

B. Reduction in Dimensions of the Generator Matrices

Equations (XIII-11) and (XIII-12) contain three pairs of *redundant elements*. Elimination of one member of each pair would reduce the size of the representation of $\mathbf{r}^{\times 2}$ to a column of six distinguishable elements. In general, $\mathbf{r}^{\times p}$ has 3^p elements, but elimination of the redundancies leaves only $(p + 2)!/p!2!$ distinguishable elements.[1] This result suggests that the generator matrices used for computation of all distinguishable elements in $\mathbf{r}^{\times p}$ need not be as large as $4^p \times 4^p$.

The procedure for reduction in the dimensions of the generator matrices was described initially by Nagai.[6] The implementation described here is due to Flory and Abe.[1,7] It will be illustrated with $A_i^{\times 2}$. Incorporating the fact that the third element in the first row of T_i is null, A_i for an internal bond is

$$
A_i = \begin{bmatrix} T_{11} & T_{12} & 0 & l \\ T_{21} & T_{22} & T_{23} & 0 \\ T_{31} & T_{32} & T_{33} & 0 \\ 0 & 0 & 0 & 1 \end{bmatrix}_i \tag{XIII-13}
$$

The 16×16 representation of $A_i^{\times 2}$, with rows and columns indexed by two digits selected from 1–4, in the order 11, 12, ... , 44, is presented in Table XIII-1.

Corresponding rows are those indexed by the same digits, but in *permuted order*. Corresponding columns are defined in a like manner. There are six pairs of corresponding rows in $A_i^{\times 2}$ (12 and 21, 13 and 31, 14 and 41, 23 and 32, 24 and 42, 34 and 43), and six pairs of corresponding columns. The condensation is achieved by adding corresponding rows for each of the six pairs. The corresponding columns thereby become identical. The first member of each set of redundant rows and columns is retained, and the rest are deleted. When applied to the matrix in Table XIII-1, the dimensions decrease from 16×16 to 10×10 (or from $4^p \times 4^p$ to $(p + 3)!/p!3! \times (p + 3)!/p!3!$), and lead to the matrix presented in Table XIII-2. The condensed form of $A_1^{\times 2}$ is of dimensions 6×10, and the condensed $A_n^{\times 2}$ is a column vector of 10 elements.

Further condensation of the $\mathscr{A}_i^{\times 2}$, constructed from the reduced $A_{\eta;i}^{\times 2}$, can be achieved if the chain has a symmetric torsion potential energy function. The method for achieving this additional reduction is the one described in Section IV.5. The advantage gained by the condensations is illustrated in Table XIII-3 for the case of $\langle r^{2p} \rangle_0$ for a chain of bonds subject to nearest neighbor interdependence and a symmetric threefold rotation potential energy function. With no reduction, the square order is $3 \cdot 5^p$. The size is reduced to $3[(p + 4)!]/(p!4!)$ if the reduction described in this section is employed. A further reduction of $\frac{2}{3}$ is achieved if the statistical weight matrix is also reduced.

If one wished to treat a chain with $\nu = 3$ without resort to square matrices of order larger than 10^3, $\langle r^6 \rangle_0$ would be the largest accessible even moment, in the absence of any condensation. With the condensation described in this section, one gains access to $\langle r^{14} \rangle_0$. The additional condensation achieved by reducing U then provides $\langle r^{16} \rangle_0$. Flory and Abe[7] describe a further condensation that would bring $\langle r^{18} \rangle_0$ within range.

[6]Nagai, K. *J. Chem. Phys.* **1963**, *38*, 924. Nagai, K.; Ishikawa, T. *J. Chem. Phys.* **1966**, *45*, 3128. Nagai, K. *J. Chem. Phys.* **1968**, *48*, 5646.
[7]Flory, P. J.; Abe, Y. *J. Chem. Phys.* **1971**, *54*, 1351.

TABLE XIII-1. $A_i^{\times 2}$ for an Internal Bond

	11	12	13	14	21	22	23	24	31	32	33	34	41	42	43	44
11	T_{11}^2	$T_{11}T_{12}$	0	$T_{11}l$	$T_{12}T_{11}$	T_{12}^2	0	$T_{12}l$	0	0	0	0	lT_{11}	lT_{12}	0	l^2
12	$T_{11}T_{21}$	$T_{11}T_{22}$	$T_{11}T_{23}$	0	$T_{12}T_{21}$	$T_{12}T_{22}$	$T_{12}T_{23}$	0	0	0	0	0	lT_{21}	lT_{22}	lT_{23}	0
13	$T_{11}T_{31}$	$T_{11}T_{32}$	$T_{11}T_{33}$	0	$T_{12}T_{31}$	$T_{12}T_{32}$	$T_{12}T_{33}$	0	0	0	0	0	lT_{31}	lT_{32}	lT_{33}	0
14	0	0	0	T_{11}	0	0	0	T_{12}	0	0	0	0	0	0	0	l
21	$T_{21}T_{11}$	$T_{21}T_{12}$	0	$T_{21}l$	$T_{22}T_{11}$	$T_{22}T_{12}$	0	$T_{22}l$	$T_{23}T_{11}$	$T_{23}T_{12}$	0	$T_{23}l$	0	0	0	0
22	T_{21}^2	$T_{21}T_{22}$	$T_{21}T_{23}$	0	$T_{22}T_{21}$	T_{22}^2	$T_{22}T_{23}$	0	$T_{23}T_{21}$	$T_{23}T_{22}$	T_{23}^2	0	0	0	0	0
23	$T_{21}T_{31}$	$T_{21}T_{32}$	$T_{21}T_{33}$	0	$T_{22}T_{31}$	$T_{22}T_{32}$	$T_{22}T_{33}$	0	$T_{23}T_{31}$	$T_{23}T_{32}$	$T_{23}T_{33}$	0	0	0	0	0
24	0	0	0	T_{21}	0	0	0	T_{22}	0	0	0	T_{23}	0	0	0	0
31	$T_{31}T_{11}$	$T_{31}T_{12}$	0	$T_{31}l$	$T_{32}T_{11}$	$T_{32}T_{12}$	0	$T_{32}l$	$T_{33}T_{11}$	$T_{33}T_{12}$	0	$T_{33}l$	0	0	0	0
32	$T_{31}T_{21}$	$T_{31}T_{22}$	$T_{31}T_{23}$	0	$T_{32}T_{21}$	$T_{32}T_{22}$	$T_{32}T_{23}$	0	$T_{33}T_{21}$	$T_{33}T_{22}$	$T_{33}T_{23}$	0	0	0	0	0
33	T_{31}^2	$T_{31}T_{32}$	$T_{31}T_{33}$	0	$T_{32}T_{31}$	T_{32}^2	$T_{32}T_{33}$	0	$T_{33}T_{31}$	$T_{33}T_{32}$	T_{33}^2	0	0	0	0	0
34	0	0	0	T_{31}	0	0	0	T_{32}	0	0	0	T_{33}	0	0	0	0
41	0	0	0	0	0	0	0	0	0	0	0	0	T_{11}	T_{12}	0	l
42	0	0	0	0	0	0	0	0	0	0	0	0	T_{21}	T_{22}	T_{23}	0
43	0	0	0	0	0	0	0	0	0	0	0	0	T_{31}	T_{32}	T_{33}	0
44	0	0	0	0	0	0	0	0	0	0	0	0	0	0	0	1

TABLE XIII-2. Condensed Form of $A_i^{\times 2}$ for an Internal Bond

	11	12 + 21	13 + 31	14 + 41	22	23 + 32	24 + 42	33	34 + 43	44
11	T_{11}^2	$T_{11}T_{12}$	0	$T_{11}l$	T_{12}^2	0	$T_{12}l$	0	0	l^2
12 + 21	$2T_{11}T_{21}$	$T_{11}T_{22}+T_{21}T_{12}$	$T_{11}T_{23}$	$T_{21}l$	$2T_{12}T_{22}$	$T_{12}T_{23}$	$T_{22}l$	0	$T_{23}l$	0
13 + 31	$2T_{11}T_{31}$	$T_{11}T_{32}+T_{31}T_{12}$	$T_{11}T_{33}$	$T_{31}l$	$2T_{12}T_{32}$	$T_{12}T_{33}$	$T_{32}l$	0	$T_{33}l$	0
14 + 41	0	0	0	T_{11}	0	0	T_{12}	0	0	$2l$
22	T_{21}^2	$T_{21}T_{22}$	$T_{21}T_{23}$	0	T_{22}^2	$T_{22}T_{23}$	0	T_{23}^2	0	0
23 + 32	$2T_{21}T_{31}$	$T_{21}T_{32}+T_{31}T_{22}$	$T_{21}T_{33}+T_{31}T_{23}$	0	$2T_{22}T_{32}$	$T_{22}T_{33}+T_{32}T_{23}$	0	$2T_{23}T_{33}$	0	0
24 + 42	0	0	0	T_{21}	0	0	T_{22}	0	T_{23}	0
33	T_{31}^2	$T_{31}T_{32}$	$T_{31}T_{33}$	0	T_{32}^2	$T_{32}T_{33}$	0	T_{33}^2	0	0
34 + 43	0	0	0	T_{31}	0	0	T_{32}	0	T_{33}	0
44	0	0	0	0	0	0	0	0	0	1

i

328

TABLE XIII-3. Square Orders of Supermatrices for Internal Bonds When the Chain Has a Symmetric Threefold Torsion Potential Energy Function with Pairwise Interdependent Bonds[a]

p	No Condensation	Reduce r^{2p}, but Not U[b]	Reduce Both U and r^{2p}[c]
1	15	15	10
2	75	45	30
3	375	105	70
4	1,875	210	140
5	9,375	378	252
6	46,875	630	420
7	234,375	990	660
8	1,171,875	1485	990

[a] Adapted from Flory, P. J.; Abe, Y. *J. Phys. Chem.* **1971**, *54*, 1351.
[b] Using the reduced forms of the generator matrices for r^{2p} for individual conformations (Section XIII.1) and the 3×3 form of U.
[c] Using the reduced forms of the generator matrices for r^{2p} for individual conformations (Section XIII.1), and the additional reduction associated with the 2×2 form of U (Sections IV.5 and VI.6).

2. MIXED MOMENTS FROM r AND s

The *correlation coefficient* for two variables, x and y, is customarily defined as

$$\rho_{xy} = \left(\frac{\langle xy \rangle}{\langle x \rangle \langle y \rangle} - 1 \right) \left[\left(\frac{\langle x^2 \rangle}{\langle x \rangle^2} - 1 \right) \left(\frac{\langle y^2 \rangle}{\langle y \rangle^2} - 1 \right) \right]^{-1/2} \qquad \text{(XIII-14)}$$

Its range is $-1 \le \rho_{xy} \le 1$. If we seek the correlation between r^2 and s^2 for all conformations in the RIS model, then $\langle x \rangle$ and $\langle y \rangle$ (or $\langle r^2 \rangle_0$ and $\langle s^2 \rangle_0$) are obtainable by the methods described in Chapter VI, and $\langle x^2 \rangle$ and $\langle y^2 \rangle$ (or $\langle r^4 \rangle_0$ and $\langle s^4 \rangle_0$) are obtained by the methods presented in the first Section XIII.1. That still leaves $\langle xy \rangle$, which is $\langle r^2 s^2 \rangle_0$ in the present example. This mixed moment is obtained by the obvious simple extension of the methods developed in the previous section. Following Eq. (XIII-1)

$$r^2 s^2 = r^2 \otimes s^2 \qquad \text{(XIII-15)}$$

Insertion of the generator matrix expressions for r^2 and s^2 from Chapter VI, followed by n applications of the theorem on direct products, yields

$$r^2 s^2 = \frac{1}{(n+1)^2} (\mathbf{G}_1 \otimes \mathbf{H}_1)(\mathbf{G}_2 \otimes \mathbf{H}_2) \cdots (\mathbf{G}_n \otimes \mathbf{H}_n) \qquad \text{(XIII-16)}$$

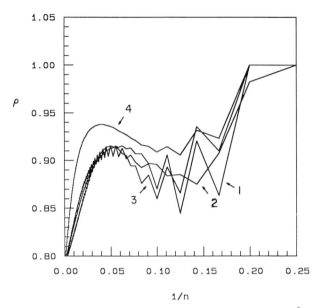

Figure XIII-2. Local maxima and minima in ρ for the correlation of r^2 and s^2 for RIS models of (1) polyphosphate, (2) polyoxymethylene, (3) polydimethylsiloxane, and (4) polydimethylsilane. Reprinted with permission from Neuburger, N. A.; Mattice, W. L. in *Computer Simulation of Polymers*, Roe, R. J., ed., Prentice-Hall, Englewood Cliffs, NJ, **1990**, p. 341. Copyright 1990 Prentice-Hall.

where the internal $\mathbf{G}_i \otimes \mathbf{H}_i$ are of dimensions 35×35. This matrix is not susceptible to reduction by the procedure described in the previous section, because it is the direct product of two different matrices, rather than the self-direct product of a single matrix.

The manipulation illustrated by Eq. (XIII-15) is of broad application. It provides an easy route to the product of any two conformation-dependent physical properties if each property individually can be expressed as the serial product of n generator matrices, one for each bond. Because this manipulation is general, it will be seen again later in this chapter. This manipulation was used to advantage in the organization of the sums required for the evaluation of the exact closed-form expression for $\langle r^2 s^2 \rangle_0$ [see Eq. (I-15)] for the freely jointed chain with $1 < N < \infty$.[8]

In the limit as $n \to \infty$, $\langle r^2 s^2 \rangle_0 / \langle r^2 \rangle_0 \langle s^2 \rangle_0$ approaches $\frac{4}{3}$, and $\rho_{r^2 s^2}$ approaches $(\frac{5}{8})^{1/2}$.[8] Much different values can be obtained with chains of finite n, as depicted in Fig. XIII-2.[9] There is a tendency for the conformations with large r^2 to also have large s^2, leading to $\rho_{r^2 s^2} > 0$. The fully extended chain, with

[8] Mattice, W. L.; Sienicki, K. *J. Chem. Phys.* **1989**, *90*, 1956.
[9] Neuburger, N. A.; Mattice, W. L. in *Computer Simulation of Polymers*, Roe, R. J., ed., Prentice-Hall, Englewood Cliffs, NJ, **1990**, p. 341.

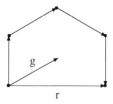

Figure XIII-3. The vectors **r** and **g** for a short chain of $n + 1$ (=5) identical atoms.

squared end-to-end distance r^2_{max}, has squared radius of gyration s^2_{max}, which will be $r^2_{max}/12$ if n is large. But the correlation is not perfect, as can be instantly appreciated by consideration of those nearly cyclic conformations for which r^2 is close to zero. These conformations can have a variety of different values of s^2, depending of the manner in which the ring is closed. The range for s^2 for the cyclic conformations is from the very small value found in the cyclic globular state to the much larger value of $r^2_{max}/48$, found when the fully extended chain reverses direction at the bond in its middle, thereby bringing the two ends together.

The parsing capability of the programming language "C" is particularly well suited to efficient formulation of the large generator matrices used in the computation of the higher moments and mixed moments.[10]

3. THE CENTER-OF-MASS VECTOR

Figure XIII-3 depicts a short chain consisting of $n + 1$ atoms of identical mass. The atoms are connected sequentially by bond vectors, $\mathbf{l}_i$. The vector from A_0 to A_n is the end-to-end vector, **r**, which can be computed via Eq. (VI-12). The figure also depicts another vector, **g**, which is drawn from A_0 to the center of mass of the $n + 1$ identical atoms. This vector is the *center-of-mass vector*. Its size and direction can be computed with the same information used for computation of **r**.[11]

A. Generator Matrix

When the chain atoms are of the same mass, the center-of-mass vector can be written in terms of the vectors from A_0 to A_i as

$$\mathbf{g} = \frac{1}{n+1} \sum_{i=0}^{n} \mathbf{r}_{0i} \qquad \text{(XIII-17)}$$

[10]Galiatsatos, V.; Mattice, W. L. *J. Comput. Chem.* **1990**, *11*, 396.
[11]Flory, P. J.; Yoon, D. Y. *J. Chem. Phys.* **1974**, *61*, 5358.

This sum can be generated by a slight elaboration of the method described in Chapter VI for the generation of **r**. Consider the product of the following two matrices for bonds 1 and 2:[12]

$$[\mathbf{A}_1 \quad \mathbf{l}_1] \begin{bmatrix} \mathbf{A}_2 & \mathbf{A}_{2]} \\ \mathbf{0} & 1 \end{bmatrix} = [\mathbf{A}_1\mathbf{A}_2 \quad \mathbf{A}_1\mathbf{A}_{2]} + \mathbf{l}_1] \qquad \text{(XIII-18)}$$

The first matrix on the left-hand side of Eq. (XIII-18), as well as the matrix on the right-hand side, is of dimensions 3×5, but it can be written as a row of two blocks. The second block in $[\mathbf{A}_1 \quad \mathbf{l}_1]$ is the sum in Eq. (XIII-17) when $n = 1$; the second block in the matrix on the right-hand side of Eq. (XIII-18) is this sum when $n = 2$. In general, the desired sum is the second block in

$$[\mathbf{A}_1 \quad \mathbf{l}_1] \begin{bmatrix} \mathbf{A}_2 & \mathbf{A}_{2]} \\ \mathbf{0} & 1 \end{bmatrix} \cdots \begin{bmatrix} \mathbf{A}_n & \mathbf{A}_{n]} \\ \mathbf{0} & 1 \end{bmatrix}$$

which yields

$$\mathbf{g} = \frac{1}{n+1} [\mathbf{A}_1 \quad \mathbf{l}_1] \begin{bmatrix} \mathbf{A}_2 & \mathbf{A}_{2]} \\ \mathbf{0} & 1 \end{bmatrix} \cdots \begin{bmatrix} \mathbf{A}_{n-1} & \mathbf{A}_{n-1]} \\ \mathbf{0} & 1 \end{bmatrix} \begin{bmatrix} \mathbf{A}_{n]} \\ 1 \end{bmatrix} \qquad \text{(XIII-19)}$$

In terms of the transformation matrices and bond vectors, the 5×5 generator for the internal bonds is

$$\begin{bmatrix} \mathbf{T} & \mathbf{1} & \mathbf{1} \\ \mathbf{0} & 1 & 1 \\ \mathbf{0} & \mathbf{0} & 1 \end{bmatrix}_i, \qquad 1 < i < n$$

and the special forms for the first and last bonds are

$$[\mathbf{T}_1 \quad \mathbf{l}_1 \quad \mathbf{l}_1]$$

$$\begin{bmatrix} \mathbf{l}_n \\ 1 \\ 1 \end{bmatrix}$$

---

[12]The symbol $\mathbf{A}_{i]}$ denotes the last column of $\mathbf{A}_i$.

B. Relationship between $\langle \mathbf{g} \rangle_0$ and $\langle \mathbf{r} \rangle_0$

Figure XIII-3 shows that $\mathbf{r}$ and $\mathbf{g}$ will usually differ in specified conformations. For highly extended conformations, $\mathbf{r} > \mathbf{g}$, but for cyclic conformations, $\mathbf{r} <$ $\mathbf{g}$. Averaging of $\mathbf{g}$ over all conformations is achieved by use of the methods described in Chapter VI for generation of $\langle \mathbf{r} \rangle_0$, with the substitution of the 3 × 5, 5 × 5, and 5 × 1 generator matrices defined above.

A comparison of the behavior of $\langle \mathbf{r} \rangle_0$ and $\langle \mathbf{g} \rangle_0$ for unperturbed linear polyethylene chains is depicted in Fig. XIII-4.[13] The Z components of both $\langle \mathbf{r} \rangle_0$ and $\langle \mathbf{g} \rangle_0$ are zero because the torsion potential energy function is symmetric. The figure shows that $\langle \mathbf{g} \rangle_0 \rightarrow \langle \mathbf{r} \rangle_0$ as $n \rightarrow \infty$ for linear polyethylene. This result is general for linear chains of equally weighted atoms.[13] Define a vector ρ_i as the vector from $\langle \mathbf{r} \rangle_0$ to $\mathbf{r}_i$ in a specified conformation:

$$\rho_i = \mathbf{r}_{0i} - \langle \mathbf{r} \rangle_0 \qquad \text{(XIII-20)}$$

In general, ρ_i depends on the conformation selected, and is rarely null. The average of ρ_i over all conformations will not vanish if i is small.[14] Combination of Eqs. (XIII-17) and (XIII-20) yields

$$\mathbf{g} = \langle \mathbf{r} \rangle_0 + \frac{1}{n+1} \sum_{i=0}^{n} \rho_i \qquad \text{(XIII-21)}$$

$$\langle \mathbf{g} \rangle_0 = \langle \mathbf{r} \rangle_0 + \frac{1}{n+1} \sum_{i=0}^{n} \langle \rho_i \rangle_0 \qquad \text{(XIII-22)}$$

Since $\langle \rho_i \rangle$ vanishes as $i \rightarrow \infty$, the limiting form from Eq. (XIII-22) is

$$\lim_{n \rightarrow \infty} \langle \mathbf{g} \rangle_0 = \langle \mathbf{r} \rangle_0 \qquad \text{(XIII-23)}$$

The approach to this limit is illustrated in Fig. XIII-4 for the case of an unperturbed linear polyethylene chain.

The situation becomes more complicated for star-branched polymers composed of atoms of identical mass. Here $\langle \mathbf{g} \rangle_0$ is well defined as soon as a particular branch is selected as "branch 1," but $\langle \mathbf{r} \rangle_0$ is undefined until the branch that completes the chain is also specified. If $n_1 \rightarrow \infty$ in the branched molecule, all of the $\langle \mathbf{r} \rangle_0$ become identical, equal to $\langle \mathbf{g} \rangle_0$, and equal to the results obtained

---

[13] Yoon, D. Y.; Flory, P. J. *J. Chem. Phys.* **1974**, *61*, 5366.

[14] For example, the definition in Eq. (XIII-20) yields $\langle \rho_0 \rangle = -\langle \mathbf{r} \rangle_0$. But $\langle \rho_i \rangle$ will vanish as $i \rightarrow \infty$, because the free end of a very long flexible chain has a spherically symmetric distribution about the point defined by $\langle \mathbf{r} \rangle_0$.

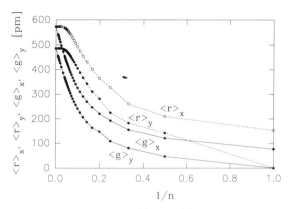

Figure XIII-4. X and Y components of $\langle \mathbf{r} \rangle_0$ and $\langle \mathbf{g} \rangle_0$ for unperturbed linear polyethylene chains.

for the linear chain of n_1 bonds.[15] But if n_1 is small, the $\langle \mathbf{r} \rangle_0$ may depend on which branches make up the chain, and none of the $\langle \mathbf{r} \rangle_0$ need to be the same as $\langle \mathbf{g} \rangle_0$.

4. ASYMMETRY OF THE DISTRIBUTION OF CHAIN ATOMS

The asymmetry of the distribution of $n + 1$ identical chain atoms about the center of mass in a specified conformation can be deduced from the ratios of the principal moments of the *radius-of-gyration tensor*.[11] If we seek an average over all conformations, the type of information obtained depends on the stage at which the averaging is done. One approach might be to average corresponding elements in the radius-of-gyration tensors, and then transform the coordinate system for the averaged tensor to a principal-axis system, thereby finding the principal moments of the averaged tensor. An alternative approach, yielding different information, first finds the principal moments for the radius-of-gyration tensor in each conformation, and then averages corresponding principal moments over all conformations. The sum of the averages of the corresponding principal moments is identical with the trace of the averaged tensor, because both procedures recover $\langle s^2 \rangle_0$. The asymmetry will be different, because the two procedures employ different definitions of the asymmetry.

Both procedures require the ability to generate the radius-of-gyration tensor for any arrangement of the $n + 1$ atoms. We start with that common feature of the calculation, and then proceed to the different methods for averaging over all conformations.

[15]Mattice, W. L. *Macromolecules* **1977**, *10*, 1182.

A. The Radius-of-Gyration Tensor

The location of the center of mass is specified by $\mathbf{g}$, and the position of A_i is given by $\mathbf{r}_{0i}$. The vector $\mathbf{s}_i$ from the center of mass to the position of A_i is

$$\mathbf{s}_i = \mathbf{r}_{0i} - \mathbf{g} \qquad \text{(XIII-24)}$$

and the nine elements of the radius-of-gyration tensor, arranged in the form of a column vector, are generated from the $\mathbf{s}_i$ as

$$\mathbf{S} = \frac{1}{n+1} \sum_{i=0}^{n} \mathbf{s}_i^{\times 2} \qquad \text{(XIII-25)}$$

The squared radius of gyration, s^2, is the sum of the first, fifth, and ninth elements of the column vector (or the trace of the 3×3 representation of the same nine elements). Combination of the two equations yields[11]

$$\mathbf{S} = \frac{1}{n+1} \sum_{i=0}^{n} (\mathbf{r}_{0i}^{\times 2} - \mathbf{r}_{0i} \otimes \mathbf{g} - \mathbf{g} \otimes \mathbf{r}_{0i} + \mathbf{g}^{\times 2}) \qquad \text{(XIII-26)}$$

The second, third, and fourth terms can be written more concisely by use of the definition for $\mathbf{g}$ in Eq. (XIII-17)

$$\mathbf{S} = \frac{1}{n+1} \sum_{i=0}^{n} \mathbf{r}_{0i}^{\times 2} - \mathbf{g} \otimes \mathbf{g} - \mathbf{g} \otimes \mathbf{g} + \mathbf{g}^{\times 2} = \frac{1}{n+1} \sum_{i=0}^{n} \mathbf{r}_{0i}^{\times 2} - \mathbf{g}^{\times 2}$$

$$\text{(XIII-27)}$$

The $\mathbf{r}_{0i}^{\times 2}$ and $\mathbf{g}^{\times 2}$ are easily computed from the generator matrix expressions for $\mathbf{r}_{0i}$ (Chapter VI) and $\mathbf{g}$ (Section XIII.3), using the methods described at the beginning of the chapter.

If desired, the 3×3 representation of $\mathbf{S}$ for any conformation can be rendered into diagonal form by similarity transforms that convert from the local coordinate system defined for bond 1 to the principal axis system for $\mathbf{S}$:

$$\mathbf{S}_{\text{diag}} = \text{diag} (L_1^2, L_2^2, L_3^2) \qquad \text{(XIII-28)}$$

where $L_1^2 \geq L_2^2 \geq L_3^2$. As stated above, $s^2 = L_1^2 + L_2^2 + L_3^2$. The shape of the chain can be measured in various ways using the principal moments of this tensor. One measurement employs the ratios of the principal moments, $(L_2^2/L_1^2$

and L_3^2/L_1^2).[16] Another group of measurements is derived from the diagonal elements of the traceless tensor, defined as

$$\hat{\mathbf{S}} = \mathbf{S}_{\text{diag}} - \frac{s^2}{3}\mathbf{I}_3 \tag{XIII-29}$$

Following the treatment used for the polarizability tensor,[17] the traceless tensor is written as

$$\hat{\mathbf{S}} = b \text{ diag} \left(\tfrac{2}{3}, -\tfrac{1}{3}, -\tfrac{1}{3}\right) + c \text{ diag} \left(0, \tfrac{1}{2}, -\tfrac{1}{2}\right) \tag{XIII-30}$$

by which the *asphericity b* and the *acylindricity c* are defined as[18]

$$b = L_1^2 - \tfrac{1}{2}(L_2^2 + L_3^2) \tag{XIII-31}$$

$$c = L_2^2 - L_3^2 \tag{XIII-32}$$

For shapes of tetrahedral or higher symmetry, $b = 0$, and for other shapes, $b > 0$. For shapes of cylindrical symmetric, $c = 0$, and for other shapes, $c > 0$. The *relative shape anisotropy* κ^2, defined such that $0 \le \kappa^2 \le 1$, with the lower limit being achieved for a structure of tetrahedral or higher symmetry, and the upper limit for a linear array, is

$$\kappa^2 = \frac{3}{2}\left(\frac{\text{tr}(\hat{\mathbf{S}}\hat{\mathbf{S}})}{\text{tr}\,\mathbf{S}^2}\right) = \frac{b^2 + (3/4)c^2}{s^4} \tag{XIII-33}$$

B. Averaging of the Radius-of-Gyration Tensor

Internal Coordinate System. If $\mathbf{r}_{0i}^{\times2}$ and $\mathbf{g}^{\times2}$ are averaged using generator matrices formulated as described in Chapter VI, we can obtain an averaged radius-of-gyration tensor as

$$\langle \mathbf{S} \rangle_0 = \frac{1}{n+1} \sum_{i=0}^{n} \langle \mathbf{r}_{0i}^{\times2} \rangle_0 - \langle \mathbf{g}^{\times2} \rangle_0 \tag{XIII-34}$$

[16]Solc, K.; Stockmayer, W. H. *J. Chem. Phys.* **1971**, *54*, 2756. Solc, K. *J. Chem. Phys.* **1971**, *55*, 355.
[17]Smith, R. P.; Mortensen, E. M. *J. Chem. Phys.* **1960**, *32*, 502.
[18]Theodorou, D. N.; Suter, U. W. *Macromolecules* **1985**, *18*, 1206.

The averaged tensor can be written in the form

$$\langle \mathbf{S} \rangle_0 = \begin{bmatrix} \langle X^2 \rangle & \langle XY \rangle & \langle XZ \rangle \\ \langle YX \rangle & \langle Y^2 \rangle & \langle YZ \rangle \\ \langle ZX \rangle & \langle ZY \rangle & \langle Z^2 \rangle \end{bmatrix} \tag{XIII-35}$$

The result must reduce to a simpler form

$$\langle \mathbf{S} \rangle_0 = \begin{bmatrix} \langle X^2 \rangle & \langle XY \rangle & 0 \\ \langle YX \rangle & \langle Y^2 \rangle & 0 \\ 0 & 0 & \langle Z^2 \rangle \end{bmatrix} \tag{XIII-36}$$

if the bonds are subject to a symmetric torsion potential energy function. The mean square radius of gyration is obtained as the trace, $\langle s^2 \rangle = \langle X^2 \rangle + \langle Y^2 \rangle + \langle Z^2 \rangle$.

The averaged tensor can be expressed in its principal-axis system by an appropriate similarity transformation, with principal moments that we shall write here as P_i^2.

$$\langle \mathbf{S} \rangle_{0,\,\text{principal axis}} = \begin{bmatrix} P_1^2 & 0 & 0 \\ 0 & P_2^2 & 0 \\ 0 & 0 & P_3^2 \end{bmatrix} \tag{XIII-37}$$

For flexible chains in the limit as $n \to \infty$, the diagonal elements in this tensor are given by

$$P_1^2 = P_2^2 = P_3^2 = \frac{\langle s^2 \rangle_0}{3}, \qquad n \to \infty \tag{XIII-38}$$

The distribution deduced from the principal moments of $\langle \mathbf{S} \rangle_0$ is spherically symmetric, in the limit where $n \to \infty$, because the averaging was performed before conversion of the tensor into a principal-axis system. Performance of the averaging at this stage in the calculation will suppress the asymmetries found in the instantaneous conformations of the chain.

Average Asymmetry of Instantaneous Conformations. In order to retain information about the average asymmetry of instantaneous conformations, we should transform **S** *for each conformation* into its principal-axis system. After doing so, we might write the diagonalized tensor as in Eq. (XIII-28), where the indexing for the principal moments is chosen so that $L_1^2 \geq L_2^2 \geq L_3^2$. Then corresponding principal moments can be averaged over all conformations in the ensemble (if n is small) or over a representative subset of conformations,

TABLE XIII-4. Asymptotic Limits, as $n \to \infty$, for Measures of Average Asymmetry of Individual Conformations of Unperturbed Chains

Property	Limit
$\langle L_2^2 \rangle_0 / \langle L_1^2 \rangle_0$	$\sim 0.23^{a,b}$
$\langle L_3^2 \rangle_0 / \langle L_1^2 \rangle_0$	$\sim 0.08^{a,b}$
$\langle b \rangle_0 / \langle s^2 \rangle_0$	$\sim 0.66^b$
$\langle c \rangle_0 / \langle s^2 \rangle_0$	$\sim 0.11^b$
$\langle \kappa^2 \rangle_0$	$\sim 0.41^b$

[a]Solc, K.; Stockmayer, W. H. *J. Chem. Phys.* **1971**, *54*, 2756. Solc, K. *J. Chem. Phys.* **1971**, *55*, 355; *Macromolecules* **1973**, *6*, 378, 796. Yoon, D. Y.; Flory, P. J. *J. Chem. Phys.* **1974**, *61*, 5366. Kranbuehl, D. E.; Verdier, P. J. *J. Chem. Phys.* **1977**, *67*, 361. Mattice, W. L. *Macromolecules* **1980**, *13*, 506.
[b]Theodorou, D. N.; Suter, U. W. *Macromolecules* **1985** *18*, 1206.

generated using the Monte Carlo methods described in Section VI.7 (if n is too large for discrete enumeration of all conformations). The sum of these averaged principal moments is indistinguishable from the sum of the P_i^2:

$$\sum_{i=1}^{3} \langle L_i^2 \rangle_0 = \sum_{i=1}^{3} P_i^2 = \langle s^2 \rangle_0 \qquad \text{(XIII-39)}$$

However, the $\langle L_i^2 \rangle_0$ are not identical, even in the limit as $n \to \infty$, but instead decrease in the sequence $\langle L_1^2 \rangle_0 > \langle L_2^2 \rangle_0 > \langle L_3^2 \rangle_0$, because the average is always over corresponding principal moments derived from the individual conformations. Information about the average asymmetry of individual conformations, as represented by dimensionless ratios obtained from pairs of the $\langle L_i^2 \rangle_0$, is retained. For unperturbed linear chains, whether studied by RIS models[13,18,19] or on cubic lattices,[16,20] the measures of the average asymmetry converge to specific limits as $n \to \infty$, as shown in Table XIII-4. All of these measures can assume different values at small n, with the actual result depending on the local structure and short-range interactions.

The limiting values of $\langle L_2^2 \rangle_0 / \langle L_1^2 \rangle_0$ and $\langle L_3^2 \rangle_0 / \langle L_1^2 \rangle_0$ are different for different molecular architectures, as determined both by chains on a cubic lattice[21] and by

[19]Mattice, W. L. *Macromolecules* **1980**, *13*, 506.
[20]Kranbuehl, D. E.; Verdier, P. J. *J. Chem. Phys.* **1977**, *67*, 361.
[21]Solc, K. *Macromolecules* **1973**, *6*, 378.

TABLE XIII-5. Dimensionless Ratios for Tri- and Tetrafunctional Stars with Arms Containing the Same Number of Bonds, and for Macrocycles, All in the Limit as $n \to \infty$[a]

Architecture	$\langle L_2^2 \rangle_0 / \langle L_1^2 \rangle_0$	$\langle L_3^2 \rangle_0 / \langle L_1^2 \rangle_0$
Trifunctional star	0.334 (0.326)	0.118 (0.116)
Tetrafunctional star	0.412 (0.39)	0.161 (0.153)
Macrocycle	0.362 (0.37)	0.154 (0.155)

[a]Numbers outside parentheses are from a cubic lattice, as reported in Solc, K. *Macromolecules* **1973**, *6*, 378. Numbers in parentheses are from RIS chains, as reported in Mattice, W. L. *Macromolecules* **1980**, *13*, 506.

rotational isomeric state chains,[19] as shown in Table XIII-5. The conformations of the stars and macrocycles are less asymmetric than the conformations of the linear chains, as shown by the values of $\langle L_i^2 \rangle_0 / \langle L_1^2 \rangle_0, i = 2, 3$, being closer to 1. Tetrafunctional stars are less asymmetric than trifunctional stars, as expected.

The asymmetry of the distribution of chain atoms changes dramatically when a poly(α-amino acid) undergoes the helix $\rightleftharpoons$ coil transitions.[22] If the degree of polymerization is high, the unperturbed random coil will have values of $\langle L_2^2 \rangle_0 / \langle L_1^2 \rangle_0$ and $\langle L_3^2 \rangle_0 / \langle L_1^2 \rangle_0$ given by Table XIII-4. As the helix $\rightleftharpoons$ coil transition is induced by increasing the value of the propagation parameter s, the chain approaches the conformation of a rod, leading to $L_2^2 / L_1^2 = L_3^2 / L_1^2 \sim 0$ in the limit as $s \to \infty$. The behavior of $\langle L_2^2 \rangle_0 / \langle L_1^2 \rangle_0$ and $\langle L_3^2 \rangle_0 / \langle L_1^2 \rangle_0$ at intermediate helix content depends on the mechanism by which the transition takes place. The dominant intermediate species might be a single helix with disordered ends, two helices connected by a flexible segment, or several (or many) independent helices connected by flexible segments, depending on the values of $\hat{\sigma}, s$, and n_p (see Chapter X). Each of these folding pathways produces recognizably different patterns in the manner in which $\langle L_2^2 \rangle_0 / \langle L_1^2 \rangle_0$ and $\langle L_3^2 \rangle_0 / \langle L_1^2 \rangle_0$ depend on s. In contrast, the different pathways would be much less readily distinguishable by examination of the ratios of the P_i^2. All of these mechanisms for the helix $\rightleftharpoons$ coil transition proceed via species with floppy chain ends, and the floppy chain end, if of sufficient size, will produce $P_1^2 = P_2^2 = P_3^2$, independent of how many rod-like segments may be present along the chain. A sufficiently long floppy chain end will cause these rod-like segments to occur with equal probability in all directions in the local coordinate system defined for bond 1, and the P_i^2 will be indistinguishable from one another.

[22]Mattice, W. L. *Macromolecules* **1980**, *13*, 904.

A Monte Carlo calculation used to generate the $\langle L_i^2 \rangle_0$ can be analyzed in more detail by inspection of the densities of the chain atoms in the principal axis system. When viewed in this manner, the maximum density for a random coil is not at the origin of the principal-axis system, but instead at two positions displaced from the origin by equal distance along the axis for L_1^2, as depicted in Fig. XIII-5.

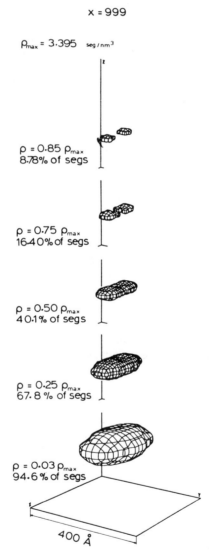

$$x = 999$$

$\rho_{max} = 3.395$ seg/nm³

z

$\rho = 0.85\,\rho_{max}$
8.78% of segs

$\rho = 0.75\,\rho_{max}$
16.40% of segs

$\rho = 0.50\,\rho_{max}$
40.1% of segs

$\rho = 0.25\,\rho_{max}$
67.8% of segs

$\rho = 0.03\,\rho_{max}$
94.6% of segs

400 Å

Figure XIII-5. Surfaces of equal average segment density for unperturbed atactic ($p_m = 0.50$) polypropylene chains with $x = 999$. Reprinted with permission from Theodorou, D. N.; Suter, U. W. *Macromolecules* **1985**, *18*, 1206. Copyright 1985 American Chemical Society.

5. MACROCYCLIZATION EQUILIBRIUM

The *macrocyclization equilibrium constant* K_x is the equilibrium constant for the reaction

$$\text{linear}_{x+y} \; \rightleftharpoons \; \text{cyclic}_x + \text{linear}_y$$

where linear$_i$ denotes a linear chain of DP_i, while cyclic$_i$ represents the corresponding end-linked macrocycle. For $x + y \gg 1$ and $x < y$, $K_x \approx [\text{cyclic}_x]$, and depends mainly on the probability that the two ends of the chain will approach one another to within a displacement difference **dr**, which is very close to zero.[23] The probability for this juxtaposition of the chain ends is denoted by $W(0)\mathbf{dr}$. If the chain obeys Gaussian statistics, $W(0)$ is given by $(3/2\pi\langle r^2\rangle_0)^{3/2}$, Eq. (II-10). In general, $W(0) = \tilde{Z}_{r=0}/Z$, using $\tilde{Z}_{r=0}$ to denote the conformational partition function for the cyclic chains (and therefore $\tilde{Z}_{r=0} \ll Z$). Improved approximations to $W(0)$ for chains of finite x can be obtained with dimensionless ratios formulated using the higher moments of the squared end-to-end distance:[24]

$$W(\mathbf{0}) = \left(\frac{3}{2\pi\langle r^2\rangle_0} \right)^{3/2} h(0)$$

where $h(0)$ denotes the scalar Hermite series

$$h(0) = 1 + 3 \cdot 5 g_4 + 3 \cdot 5 \cdot 7 g_6 + \cdots \tag{XIII-40}$$

with coefficients given by

$$g_4 = \left(\frac{-1}{2^3} \right) \left(1 - \frac{3\langle r^4\rangle_0}{5\langle r^2\rangle_0^2} \right) \tag{XIII-41}$$

$$g_6 = -\left(\frac{1}{2^3 \cdot 3!} \right) \left[3\left(1 - \frac{3\langle r^4\rangle_0}{5\langle r^2\rangle_0^2} \right) - \left(1 - \frac{3^2\langle r^6\rangle_0}{5 \cdot 7\langle r^2\rangle_0^3} \right) \right] \tag{XIII-42}$$

and so on. These coefficients can be computed using the generator matrices described in the first section of this chapter. The approximation for the macrocyclization equilibrium constant, K_x, is

[23]Jacobson, H.; Stockmayer, W. H. *J. Chem. Phys.* **1950**, *18*, 1600.
[24]Jernigan, R. L.; Flory, P. J. *J. Chem. Phys.* **1969**, *50*, 4185.

$$K_x = \left(\frac{3}{2\pi\langle r^2\rangle_0}\right)^{3/2}\left(\frac{h(0)}{\sigma_{cx}\mathscr{L}}\right) \tag{XIII-43}$$

where σ_{cx} is the symmetry number for the macrocycle of x units and $\mathscr{L}$ is Avogadro's number.

When the value of x becomes sufficiently small, the correction introduced by $h(0)$ is no longer adequate for an accurate calculation of K_x. The conformations must now be weighted with respect to not only $\mathbf{r}$ but also whether the new bond angle formed by cyclization is within an appropriate tolerance of the preferred bond angle for the three atoms involved. This problem is illustrated with Fig. XIII-6. We imagine a chain consisting of n atoms, A_0 to A_{n-1}, followed by two "hypothetical" atoms, A_n and A_{n+1}. Ring closure of the "real" chain (A_0 to A_{n-1}) with correct bond angle at A_0 requires (1) that atoms A_0 and A_n superimpose ($\mathbf{r} = \mathbf{0}$) as well as (2) that atoms A_1 and A_{n+1} superimpose, that is, that

$$\gamma = \cos(\theta^{\text{required}} - \theta_{A_{n-1}A_0A_1}) \tag{XIII-44}$$

be unity.

An improved description of the macrocyclization replaces Eq. (XIII-43)

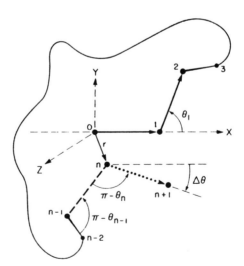

Figure XIII-6. Atoms and bonds involved in the definition of γ. θ_i in this figure represents the complement of bond angle i. Reprinted with permission from Flory, P. J.; Suter, U. W.; Mutter, M. *J. Am. Chem. Soc.* **1976**, *98*, 5733. Copyright 1976 American Chemical Society.

with[25]

$$K_x = \left(\frac{3}{2\pi \langle r^2 \rangle_0} \right)^{3/2} \left(\frac{2\Gamma_0(1)h(0)}{\sigma_{cx} \mathscr{L}} \right) \qquad \text{(XIII-45)}$$

where $\Gamma_0(1)\delta\gamma$ is the probability that the cosine of the difference in the angles is 1, within the range $\delta\gamma$, given that $r = 0$. The evaluation via Legendre polynomials yields a doubly infinite expansion:

$$2\Gamma_0(1) = \sum_{k=0}^{\infty} (2k + 1) \langle P_k \rangle_{r=0} \qquad \text{(XIII-46)}$$

with

$$\langle P_k \rangle_{r=0} = \frac{1}{h(0)} \left[f_{k;0} - \frac{3}{2} f_{k;2} + \left(\frac{3}{2} \right)^2 \frac{1}{2!} f_{k;4} - \left(\frac{3}{2} \right)^3 \frac{1}{3!} f_{k;6} + \cdots \right]$$

$$\text{(XIII-47)}$$

All but the first of the $f_{k;j}$ require the averages over the unperturbed ensemble of products that involve γ and r. Thus

$$f_{k;0} = \langle P_k \rangle \qquad \text{(XIII-48)}$$

and the next member of the series is

$$f_{k;2} = \frac{\langle P_k r^2 \rangle_0}{\langle r^2 \rangle_0} - \langle P_k \rangle \qquad \text{(XIII-49)}$$

which is followed by

$$f_{k;4} = \frac{\langle P_k r^4 \rangle_0}{\langle r^2 \rangle_0^2} - \frac{10}{3} \frac{\langle P_k r^2 \rangle_0}{\langle r^2 \rangle_0} + \frac{5}{3} \langle P_k \rangle \qquad \text{(XIII-50)}$$

and so forth. For each term $\langle P_k \rangle_{r=0}$ in Eq. (XIII-46) a set of terms $f_{k;2p}$ is needed, defined by the Legendre polynomials of γ of order k [$P_0 = 1, P_1 = \gamma, P_2 = \frac{1}{2}(3\gamma^2 - 1)$, etc.], and their products with $r^2, r^4, \ldots r^{2p}$, averaged without restriction on end-to-end distance or bond angles at the termini. For example, evalu-

[25] Flory, P. J.; Suter, U. W.; Mutter, M. *J. Am. Chem. Soc.* **1976**, *98*, 5733.

ation of $\langle \gamma r^2 \rangle_0$ and $\langle \gamma \rangle_0$ is required for $f_{1;2}$; and of $\langle \gamma^2 r^4 \rangle_0, \langle \gamma^2 r^2 \rangle_0$, and $\langle \gamma^2 \rangle_0$, for $f_{2;4}$.

The generator matrix for $\gamma^m r^{2p}$ is obtained by the method described in Section XIII.2. In a specified conformation, γ is generated as

$$\gamma = [1 \quad 0 \quad 0] \mathbf{T}_1 \cdots \mathbf{T}_n \begin{bmatrix} 1 \\ 0 \\ 0 \end{bmatrix} \tag{XIII-51}$$

and the mth power of γ is generated as

$$\gamma^m = [1 \quad 0 \quad 0]^{\times m} \mathbf{T}_1^{\times m} \cdots \mathbf{T}_n^{\times m} \begin{bmatrix} 1 \\ 0 \\ 0 \end{bmatrix}^{\times m} \tag{XIII-52}$$

The generator matrix for $\gamma^m r^{\times p}$ is obtained as $\gamma^m \otimes r^{\times p}$, which, for internal bonds, is $(\mathbf{T}^{\times m} \otimes \mathbf{G}^{\times p})_i$.[26]

PROBLEMS

Appendix B contains the answers to Problems XIII-1, XIII-2, and XIII-8 through XIII-10.

XIII-1. How many nonzero elements occur in $\langle \mathbf{r}^{\times p} \rangle$ for a chain with a threefold symmetric torsion potential energy function?

XIII-2. What are the condensed forms of $\mathbf{A}_1^{\times 2}$ and $\mathbf{A}_n^{\times 2}$?

XIII-3. By computation of $\langle r^2 \rangle_0$ and $\langle r^4 \rangle_0$ as a function of n for chains of your choice, convince yourself that $\langle r^4 \rangle_0 / \langle r^2 \rangle_0^2 \to 5/3$ as $n \to \infty$, if the chain is flexible.

XIII-4. Figure VI-6 presents non-Gaussian distribution functions for some short chains, generated by Monte Carlo methods. Use generator matrix methods to compute $\langle r^4 \rangle_0 / \langle r^2 \rangle_0^2$ for the same chains. Compare the results for the short chains with the limiting values in Table I-1.

XIII-5. In an unperturbed polyethylene chain of 100 bonds, one might expect a very low correlation between r^2 for the subchains consisting of bonds 1–50 and bonds 51–100. Determine ρ for the r^2 for these two subchains.

[26]Suter, U. W.; Mutter, M.; Flory, P. J. *J. Am. Chem. Soc.* **1976**, *98*, 5740.

XIII-6. What happens to the correlation coefficient if we compare the squared dimensions for two subchains that partially overlap, such as subchains composed of bonds 1–50 and bonds 26–75 in an unperturbed polyethylene of 100 bonds? Will the value of the correlation coefficient depend on whether we use r^2 or s^2 as the measure of the squared dimensions for these subchains? Does it depend linearly on the fraction of the bonds that is common to both subchains?

XIII-7. Calculate $\langle g \rangle_0$ as a function of n for a chain of identical atoms in which all bond lengths are 154 pm, all bond angles are 112°, and each internal bond is subject to free rotation. (The result is depicted in Fig. 2 of Yoon and Flory.[13])

XIII-8. For a flexible chain represented as A—$(B$—$A)_{x-1}$—B, which of the following quantities are necessarily independent of the direction we arbitrarily select for indexing the bonds along the chain? Which become independent of the direction of indexing in the limit as $x \to \infty$?

(a) $\langle g \rangle_0$ (b) $\langle r_{0n}^{\times 2} \rangle_0$ (c) $\langle S \rangle_0$

(d) P_1^2 (e) $\langle L_1^2 \rangle_0$ (f) P_2^2/P_1^2

(g) $\langle L_2^2 \rangle_0 / \langle L_1^2 \rangle_0$ (h) $P_1^2 + P_2^2 + P_3^2$ (i) $\langle L_1^2 \rangle_0 + \langle L_2^2 \rangle_0 + \langle L_3^2 \rangle_0$

XIII-9. A diblock copolymer consists of 100 bonds. All bonds are of length l, and all bond angles are tetrahedral. Bonds 2–10 have access to three independent, equally weighted rotational isomeric states, with $\phi_i = 180°, \pm 60°$. All other internal bonds have access to only one state, with $\phi_i = 180°$. Why do the ratios of the principal moments of $\langle S \rangle_0$ depend strongly on the direction chosen for indexing the bonds in the chain?

XIII-10. For typical flexible chains of 10–20 bonds, will the correction to the macrocyclization equilibrium constant introduced by $\Gamma_0(1)$ tend to increase, or to decrease, the estimate of K_x?

XIV Generator Matrices Using Information in Addition to l_i and T_i

Chapters VI and XIII described generator matrices formulated from l_i, ϕ_i, and θ_i. No additional information was required for computation of $r^{\times p}$, $g^{\times p}$, r^{2p}, and s^{2p} (where $p = 1, 2, \ldots$), as well as mixed moments such as $r^2 s^2$ and $\gamma^m r^{2p}$. These properties, which are determined completely by n bond lengths, $n - 1$ bond angles, and $n - 2$ torsion angles, provide a great deal of information about a specified conformation of the chain.

Other conformation-dependent physical properties can be handled by similar methods, but only after incorporation into the generator matrices of more information about the units that constitute the chain. Several such generator matrices are presented in this chapter. We first describe generator matrices that are a very simple extension of the methods used for computation of r and r^2. Those generator matrices provided a means for manipulating vectors that were called l_i. They can, of course, be applied in likewise fashion to other properties of the chain that can be expressed as a sum of vectors rigidly attached to the local coordinate systems defined for the component bonds. The dipole moment vector of polar chains is therefore susceptible to treatment by methods used for the treatment of r.

Properties that depend on the sum of tensors, rather than vectors, rigidly fixed in the local coordinate systems are also described in this chapter. By this means one can calculate properties that depend on the anisotropy of the polarizability of the molecule, through the anisotropy of tensors associated with each of the local coordinate systems. The *valence-optical scheme* is employed in the construction of these generator matrices.

1. THE DIPOLE MOMENT VECTOR AND MEAN SQUARE DIPOLE MOMENT

In Chapter VI the end-to-end vector, r, was expressed as a sum of bond vectors, l_i. The l_i were defined so that only a single component, the one along the x axis for the local coordinate system of bond i, was nonzero. However, the matrix used for transformation of this vector from its expression in the local coordinate system of bond i into the local coordinate system of bond $i - 1$ handles all three components of l_i correctly. Therefore that procedure can be used for evaluation

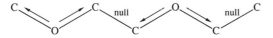

Figure XIV-1. Dipole moment vectors for the C—O and C—C bonds in polyoxyethylene.

of the sum of any type of vector that is unambiguously defined in the coordinate systems for the individual bond. The most important application, other than the one already presented for **r**, is for evaluation of the *dipole moment vector*, **μ**, and the *squared dipole moment*, μ^2, in polar chains.[1]

A. Dipole Moment Vectors for Individual Bonds

The dipole moment vectors associated with the individual bonds are denoted by $\mathbf{m}_i$, which differ from the $\mathbf{l}_i$ in that any, all, or none of the elements can be nonzero. For example, in the case of polyoxyethylene, a useful approximation is to consider that the dipole moment vector for the chain is the sum of bond dipole moment vectors associated with, and parallel with, the C—O bonds. As shown in Fig. XIV-1, the three types of $\mathbf{m}_i$ required for this chain, each expressed in the local coordinate system to which it is affixed, are

$$\mathbf{m}_{CO} = \begin{bmatrix} -m_{OC} \\ 0 \\ 0 \end{bmatrix} \qquad \mathbf{m}_{OC} = \begin{bmatrix} m_{OC} \\ 0 \\ 0 \end{bmatrix} \qquad \mathbf{m}_{CC} = \begin{bmatrix} 0 \\ 0 \\ 0 \end{bmatrix} \qquad \text{(XIV-1)}$$

where m_{OC} denotes the magnitude of the bond dipole moment vector for the O—C bond. Components of $\mathbf{m}_i$ can be positive or negative, because the $\mathbf{m}_i$ are not oriented in an unique pattern, in contrast with $\mathbf{l}_i$. Some of the $\mathbf{m}_i$ will be null in the local frames of reference if the chain contains bonds that contribute a negligible dipole moment.

Simple examples of chains in which some of the $\mathbf{m}_i$ contain three nonzero elements are vinyl polymers with a polar nonarticulated substituent X, such as a halogen atom. When expressed in the coordinate system of the C^α—X bond, $\mathbf{m}_{CX}$ is $[m_{CX} \quad 0 \quad 0]^T$. It can also be expressed in the coordinate system of the C^α—C bond in the main chain, because the substituent "X" is assumed to be rigidly attached to the main chain. In the coordinate system of the C^α—C bond, the bond dipole moment vector takes either of two forms:

[1]Flory, P. J. *Macromolecules* **1974**, *7*, 381.

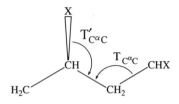

Figure XIV-2. The transformations carried out by $\mathbf{T}_{C^\alpha-C}$ and $\mathbf{T}'_{C^\alpha-C}$ in poly-$(-CHX-CH_2)$.

$$
\mathbf{m}_{CX,d} = \begin{bmatrix} x \\ y \\ z \end{bmatrix} \qquad \mathbf{m}_{CX,l} = \begin{bmatrix} x \\ y \\ -z \end{bmatrix} \tag{XIV-2}
$$

depending on the stereochemistry of the attachment of X to C^α. In the coordinate system of the C^α—C bond, the dipole moment vector for the C^α—X bond is $\mathbf{T}'_{C^\alpha-C}\mathbf{m}_{CX}$, where the matrix for transformation of $\mathbf{m}_{CX}$ from its own coordinate system into the coordinate system of the C^α—C bond in the backbone (see Fig. XIV-2) is[2]

$$
\mathbf{T}'_{C^\alpha-C} = \begin{bmatrix} \cos\theta_X & \sin\theta_X & 0 \\ -\sin\theta_X\cos\beta & \cos\theta_X\cos\beta & -\sin\beta \\ -\sin\theta_X\sin\beta & \cos\theta_X\sin\beta & \cos\beta \end{bmatrix} \tag{XIV-3}
$$

where β is the dihedral angle between the chain skeleton and X ($\beta = \angle X-C^\alpha-C-C^\alpha$) and is $\sim \pm 120°$, the sign depending on the stereochemistry of the attachment of the side chain to C^α (– for a d pseudoasymmetric center, and + for an l pseudoasymmetric center), and the exact value depending on the angles for the bonds centered on C^α (exactly $\pm 120°$ if all bond angles centered on C^α are tetrahedral), and θ_X is the bond angle $\angle X-C^\alpha-C$. All components of this transformation matrix are completely determined by bond angles centered on C^α, all of which are taken to be constants. The fact that $\mathbf{T}'_{C^\alpha-C}$ is independent of the conformation is consistent with the description of "X" as being rigidly attached to the main chain. This transformation matrix does depend on the configuration, through the sign of β.

B. Generator Matrices

The generator matrices used for the computation of $\boldsymbol{\mu}$ and μ^2 are constructed from $\mathbf{m}_i$ and $\mathbf{T}_i$ by an obvious extension of the matrices described in Chapter VI for the computation of $\mathbf{r}$ and r^2.

---

[2]The prime differentiates this matrix from another matrix, $\mathbf{T}_{C^\alpha-C}$, that transforms from the coordinate system of a C—C^α bond into the coordinate system of the preceding C^α—C bond, with both bonds in the main chain.

Dipole Moment Vector. The dipole moment vector, $\boldsymbol{\mu}$, expressed in the coordinate system for bond 1, is generated as

$$\boldsymbol{\mu} = \mathbf{B}_1 \mathbf{B}_2 \cdots \mathbf{B}_n \qquad \text{(XIV-4)}$$

where the generator matrices are obtained from $\mathbf{A}_1$, $\mathbf{A}_i$, and $\mathbf{A}_n$ (Table VI-1) by replacement of $\mathbf{l}_i$ with $\mathbf{m}_i$:

$$\mathbf{B}_1 = [\mathbf{T} \quad \mathbf{m}]_1 \qquad \text{(XIV-5)}$$

$$\mathbf{B}_i = \begin{bmatrix} \mathbf{T} & \mathbf{m} \\ \mathbf{0} & 1 \end{bmatrix}_i, \qquad 1 < i < n \qquad \text{(XIV-6)}$$

$$\mathbf{B}_n = \begin{bmatrix} \mathbf{m} \\ 1 \end{bmatrix}_n \qquad \text{(XIV-7)}$$

Only by coincidence will $\boldsymbol{\mu}$ and $\mathbf{r}$ point in the same direction. For example, in the all-*trans* conformation of a syndiotactic poly(vinyl chloride) chain, $\mathbf{r}$ is perpendicular to $\boldsymbol{\mu}$, as illustrated in Fig. XIV-3. In general, chains that lack a distinguishable direction along the backbone will have $\boldsymbol{\mu}$ perpendicular to $\mathbf{r}$ when the chain is fully extended. Chains that do have a distinguishable direction, as in the poly(α-amino acids), where the amino terminus is distinguishable from the carboxyl terminus, will in general have a component of $\boldsymbol{\mu}$ parallel to $\mathbf{r}$ in the fully extended conformation.

The Square of the Dipole Moment. The square of the dipole moment, μ^2, is obtained via the expected modification of $\mathbf{G}_1$, $\mathbf{G}_i$, and $\mathbf{G}_n$ (Table VI-1),

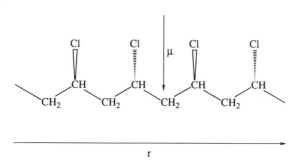

Figure XIV-3. Directions of $\boldsymbol{\mu}$ and $\mathbf{r}$ for a syndiotactic poly(vinyl chloride) chain in the all-*trans* conformation.

$$\mu^2 = \mathbf{D}_1 \mathbf{D}_2 \cdots \mathbf{D}_n \qquad (\text{XIV-8})$$

where now the generator matrices are given by

$$\mathbf{D}_1 = [1 \quad 2\mathbf{m}^T\mathbf{T} \quad m^2]_1 \qquad (\text{XIV-9})$$

$$\mathbf{D}_i = \begin{bmatrix} 1 & 2\mathbf{m}^T\mathbf{T} & m^2 \\ 0 & \mathbf{T} & \mathbf{m} \\ 0 & 0 & 1 \end{bmatrix}_i \qquad 1 < i < n \qquad (\text{XIV-10})$$

$$\mathbf{D}_n = \begin{bmatrix} m^2 \\ \mathbf{m} \\ 1 \end{bmatrix}_n \qquad (\text{XIV-11})$$

In some chains, the conformations that produce the largest r^2 will also produce the largest μ^2. Consider, for example, poly(A–B) with $\theta_{ABA} = \theta_{BAB} = \theta > 90°$ and a nonzero $\mathbf{m}_{AB}$ parallel with $\mathbf{l}_{AB}$. The all-*trans* conformation has $r^2 = [nl_{AB}\sin(\theta/2)]^2$ and $\mu^2 = [nm_{AB}\cos(\theta/2)]^2$ (ignoring end effects), both of which are proportional to n^2. On the other hand, in poly(A–B–A) with $\theta_{AAB} = \theta_{ABA} = \theta > 90°$, $l_{AA} = l_{AB} = l$, and a nonzero $\mathbf{m}_{AB}$ parallel with $\mathbf{l}_{AB}$, the all-*trans* conformation still has $r^2 = [nl\sin(\theta/2)]^2$, but that chain has $\mu^2 = 4m_{AB}^2\sin^2(\theta/2)$ or $\mu^2 = 0$, depending on whether the number of monomer units is odd or even, respectively.

C. The Mean and the Mean Square Dipole Moment

Averaging of $\boldsymbol{\mu}$ and μ^2 over all conformations in the unperturbed ensemble is handled in exactly the same way as the averaging of $\mathbf{r}$ and r^2 (see Chapter VI):

$$\langle \boldsymbol{\mu} \rangle_0 = \frac{1}{Z} \mathscr{B}_1 \mathscr{B}_2 \cdots \mathscr{B}_n \qquad (\text{XIV-12})$$

$$\langle \mu^2 \rangle_0 = \frac{1}{Z} \mathscr{D}_1 \mathscr{D}_2 \cdots \mathscr{D}_n \qquad (\text{XIV-13})$$

where $\mathscr{B}_i$ is constructed in the same manner as $\mathscr{A}_i$, with the substitution of $\mathbf{B}_i$ for $\mathbf{A}_i$, and $\mathscr{D}_i$ is constructed in the same manner as $\mathscr{G}_i$, with the substitution of $\mathbf{D}_i$ for $\mathbf{G}_i$.

The Characteristic Ratio of the Dipole Moment. In analogy with the characteristic ratio for the mean square end-to-end distance, we can also define a dimensionless characteristic ratio for the mean square dipole moment. In practice, two different dimensionless ratios are used, with the difference lying in the term in the denominator:[3]

$$D_n = \frac{\langle \mu^2 \rangle_0}{\sum_{i=1}^{n} m_i^2} \qquad \text{(XIV-14)}$$

$$D_x = \frac{\langle \mu^2 \rangle_0}{\sum_{i=1}^{x} \mu_i^2} \qquad \text{(XIV-15)}$$

The first definition has in its denominator the sum of the squares of the lengths of all bond dipole moment vectors, and the second definition substitutes the sum of the squares of the dipole moment vectors for each monomer unit. The two definitions are equivalent for some chains [e.g., poly(vinyl chloride) in the approximation that the dipole moment vector is completely dominated by the C—Cl bond], but nonequivalent for other chains [e.g., polyoxyethylene, where the contribution to the denominator from each monomer unit can be either $2m_{CO}^2$ or $4\cos^2(\theta_{COC}/2)\, m_{CO}^2$, depending on which definition is adopted].

Both D_n and C_n depend on n when n is small, but they approach limiting values, denoted by D_∞ and C_∞, respectively, as the number of bonds increases without limit. Usually D_n experiences most of its change at smaller values of n than does C_n. Therefore calculations with relatively short chains are often sufficient for an excellent approximation to D_∞. The program listed in Appendix C calculates D_x and D_∞ for cases where the $\mathbf{m}_i$ are fixed in the skeletal frames of reference, when the appropriate $\mathbf{m}_i$ are given as input.

Typical flexible chains have $C_n > 1$, as has been documented extensively in earlier chapters. The contributing causes are bond angles predominantly greater than 90°, and short-range interactions that usually discriminate against extremely compact local conformations. In contrast, flexible polymers with $\mathbf{m}_i$ closely spaced along the chain usually have $D_n < 1$, as shown by the selected examples in Table XIV-1.[4] Flexible chains tend to avoid local conformations that cause a series of $\mathbf{m}_i$ to be parallel. Also the two A—B bond dipoles in an A—B—A fragment produce a net squared dipole of $4m^2 \cos^2(\theta/2)$, which is smaller than $2m^2$ if $\theta > 90°$. Since bond angles are usually greater than 90°, $\mathbf{m}_{AB}$ tends to partially cancel $\mathbf{m}_{BA}$ in A—B—A.

The stereochemical composition can have a strong influence on $\langle \mu^2 \rangle_0$ in polar

---

[3]Sometimes the result is reported instead as $\langle \mu^2 \rangle_0/x$, where x is the number of monomer units having a group dipole moment. This ratio does have units, because the denominator is a pure number. It is equal to D_n only in the special case where the sum of the squares of the group dipoles is equal to x, expressed in the units used for $\langle \mu^2 \rangle_0$. The analogous ratio for the unperturbed dimensions would be $\langle r^2 \rangle_0/n$, which has units of length squared.

[4]If the group dipoles are separated by a large number of flexible units, such as $(CH_2)_x$ with large x, D_∞ will usually be close to 1, because the flexible spacer eliminates any correlations in the directions of the $\mathbf{m}_i$ at either end. For the same reason, $D_n \to 1$ as $x \to \infty$ in X–$(CH_2)_x$–X.

TABLE XIV-1. $D_n, n \to \infty$ for Selected Chains[a]

Polymer	D_∞^b	Ref. Exp	Ref. RIS
Poly(*tert*-butyl vinyl ketone), $p_m = 0.5$–0.6	0.18^c	d	d
Poly(dimethylsiloxane)	0.30-0.38	e	f
Poly(oxyethylene)	0.60	g	h
Poly(*p*-chlorostyrene), $p_m \sim 0.5$	~ 0.7	i	i
Poly[(S)-oxy(1-isopropylethylene)]	0.83	j	k

[a]The program given in Appendix C computes D_x and D_∞ for the configurationally regular linear chains, namely, poly(dimethylsiloxane) and poly(oxyethylene).
[b]Or D_n at large n, where the difference in D_∞ and D_n is trivial.
[c]$\langle \mu^2 \rangle / x\mu_1^2$, where μ_1 is the dipole moment of pinacolone.
[d]Guest, J. A.; Matsuo, K.; Stockmayer, W. H.; Suter, U. W. *Macromolecules* **1980**, *13*, 560.
[e]Dasgupta, S.; Smyth, C. P. *J. Chem. Phys.* **1967**, *47*, 2911. Sutton, C.; Mark, J. E. *J. Chem. Phys.* **1971**, *54*, 5011. Liao, S. C.; Mark, J. E. *J. Chem. Phys.* **1973**, *59*, 3825.
[f]Mark, J. E. *J. Chem. Phys.* **1968**, *49*, 1398.
[g]Marchal, J.; Benoit, H. *J. Chim. Phys. Phys.-Chim. Biol.* **1955**, *52*, 818; *J. Polym. Sci.* **1957**, *23*, 223. Bak, K.; Elefante, G.; Mark, J. E. *J. Phys. Chem.* **1967**, *71*, 4007.
[h]Mark, J. E.; Flory, P. J. *J. Am. Chem. Soc.* **1966**, *88*, 3702. Abe, A.; Mark, J. E. *J. Am. Chem. Soc.* **1976**, *98*, 6468.
[i]Mark, J. E. *J. Chem. Phys.* **1972**, *56*, 458.
[j]Hirano, T.; Khanh, P. H.; Tsuji, K.; Sato, A.; Tsuruta, T.; Abe, A.; Shimozawa, T.; Kotera, A.; Yamaguchi, N.; Kitahara, S. *Polym. J.* **1979**, *11*, 905.
[k]Abe, A.; Hirano, T.; Tsuji, K.; Tsuruta, T. *Macromolecules* **1979**, *12*, 1100.

vinyl polymers, as illustrated in Fig. XIV-4 for poly(vinyl chloride) and poly(*p*-chlorostyrene).[5] Poly(vinyl chloride) has its largest mean square dipole moment in the syndiotactic regime, because the *racemo* chain has a high probability for propagation of sequences of *trans* placements, and the components of the C—Cl dipole in the plane of the backbone, and perpendicular to the chain axis, are additive in this conformation. Poly(*p*-chlorostyrene), on the other hand, has its largest mean square dipole moment in the isotactic regime, because the *meso* chain has a strong probability for propagation of *tg* helices (that have an additive component in the helix axis).

In atactic copolymers of a polar vinyl monomer and a nonpolar monomer, calculations suggest that $\langle \mu^2 \rangle_0$ is often a monotonic function of the mole fraction of polar monomer, such as, styrene–*p*-chlorostyrene copolymers,[6] propylene–vinyl chloride copolymers,[7] and ethylene–vinyl chloride copolymers.[8] Semialternating copolymers of ethylene and carbon monoxide also have

[5]Mark, J. E. *J. Chem. Phys.* **1972**, *56*, 458.
[6]Mark, J. E. *J. Am. Chem. Soc.* **1972**, *94*, 6645.
[7]Mark, J. E. *J. Polym. Sci., Polym. Phys. Ed.* **1973**, *11*, 1375.
[8]Mark, J. E. *Polymer* **1973**, *14*, 553.

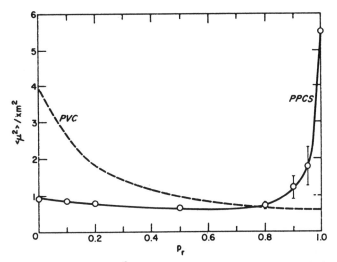

Figure XIV-4. Dependence of $\langle\mu^2\rangle_0$ on the probability for a *meso* diad (denoted by p_r in the figure) for poly(vinyl chloride) (PVC) and poly(p-chlorostyrene) (PPCS). Reproduced with permission from Mark, J. E. *J. Chem. Phys.* **1972**, *56*, 458. Copyright 1972 American Physical Society.

$\langle\mu^2\rangle_0$ that exhibit a monotonic dependence on carbon monoxide content.[9] The dependence of $\langle\mu^2\rangle_0$ on composition may become non-monotonic if the product of the reactivity ratios, $r_A r_B$, is very small [e.g., p-methylstyrene–p-chlorostyrene copolymers,[5] Fig. XIV-5], or if the sequences containing only the polar unit have a strong preference for either *meso* or *racemo* diads (e.g., styrene–p-chlorostyrene, propylene–vinyl chloride, and ethylene–vinyl chloride copolymers).[6–8]

Dipoles in Articulated Side Chains. Polar polymers that have mean square dipole moments arising entirely from $\mathbf{m}_i$ colinear with bonds in the main chain, or rigidly attached to these bonds [as in the poly(vinyl halides)], are treated by the methods outlined above. Polymers with more complicated polar side chains also yield to those methods if the polar side chain strongly prefers a single conformation, as though it were rigidly attached to the backbone. In a third class of cases, the polar side chain may have multiple conformations that are susceptible to the same simple treatment, as in poly(methyl acrylate)[10] and poly(ethyl methacrylate);[11] here two conformations of the ester group, separated by 180° in the torsion angle at the bond that attaches the side chain to the

[9]Wittwer, H.; Pino, P.; Suter, U. W. *Macromolecules* **1988**, *21*, 1262.
[10]Saiz, E.; Tarazona, M. P. *Macromolecules* **1983**, *16*, 1128.
[11]Kuntman, A.; Bahar, I.; Baysal, B. M. *Macromolecules* **1990**, *23*, 4959.

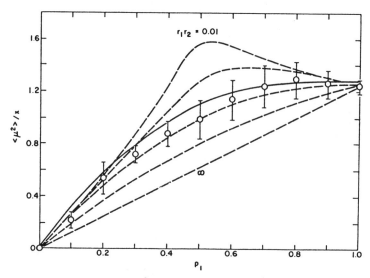

Figure XIV-5. Dependence of $\langle\mu^2\rangle_0$ on the probability for a p-chlorostyrene unit (denoted by p_1 in the figure) for atactic copolymers of poly(p-methylstyrene) and poly(p-chlorostyrene). The value of $r_A r_B$ for the dashed curves is 0.01, 0.1, 1, 10, and ∞, reading from top to bottom. The solid curve shows the experimental dependence. Reprinted with permission from Mark, J. E. *J. Am. Chem. Soc.* **1972**, *94*, 6645. Copyright 1972 American Chemical Society.

polymer, are simply averaged under the assumption that the conformation of the side chain is not coupled with the conformation of the backbone.

Many polar polymers, however, have $\mathbf{m}_i$ located in articulated side chains, such that $\mathbf{m}_i$ is not rigidly affixed to any skeletal frames of reference. Examples are vinyl polymers with $-(CH_2)_x-X$ and $-(CH_2)_x-X-(CH_2)_y CH_3$ side chains, where the C—X bond is polar. The RIS treatment of $\langle\mu^2\rangle_0$ for polymers in this class has been described by Abe.[12,13]

As an example, consider a poly(n-alkyl vinyl ether), where the side chain has the structure $-O-(CH_2)_y-CH_3$, with the dipole moment vector of the chain being completely dominated by the dipoles contributed by the ether groups. Let each C—O bond contribute a bond dipole moment vector of magnitude m, which for this polymer is 1.07 Debye (1 Debye = 3.336×10^{-30} Cm).[13] Each ether group is represented by two such vectors, the one associated with the C^α—O bond (where C^α is the carbon atom in the main chain that is bonded to the ether oxygen atom), and the other associated with the O—C^s bond (where C^s is a carbon atom in the side chain), as depicted in Fig. XIV-6. Their vector sum, expressed in the local coordinate system of the C^α—O bond, is

[12] Abe, A. *J. Polym. Sci., Symp.* **1976**, *54*, 135.
[13] Abe, A. *Macromolecules* **1977**, *10*, 34.

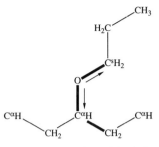

Figure XIV-6. The dipole moment vectors in a side chain of a poly(n-alkyl vinyl ether). The highlighted bonds are discussed in the text.

$$\mathbf{m}_{COC} = -\mathbf{m} + \mathbf{T}_{C^\alpha - O}\mathbf{m} \qquad \text{(XIV-16)}$$

$$\mathbf{m} = \begin{bmatrix} m \\ 0 \\ 0 \end{bmatrix} \qquad \text{(XIV-17)}$$

The vector associated with the C^α—O bond is rigidly attached to the backbone. It can be handled in the same manner as the vector associated with the C—X bond in a poly(vinyl halide). The vector associated with the O—C^s bond, in contrast, is not rigidly attached to the backbone, but can instead adopt different orientations with respect to the backbone, depending on the rotational isomeric state adopted at the C^α—O bond and incorporated into $\mathbf{T}_{C^\alpha - O}$. The conventional symbol for the torsion angle at C^α—O is χ_1. The states of this bond are involved in the indexing of the columns in the rectangular statistical weight matrix denoted by $\mathbf{U}_b$, as exemplified by Eq. (XI-10). In the coordinate system of the C^α—C bond, the dipole moment vector for the ether group is

$$\mathbf{m}_{COC;\varsigma} = -\mathbf{T}'_{C^\alpha - C}\mathbf{m} + \mathbf{T}'_{C^\alpha - C}\mathbf{T}_{C^\alpha - O;\varsigma}\mathbf{m} \qquad \text{(XIV-18)}$$

which depends on the rotational isomeric state ς at the C^α—O bond, through $\mathbf{T}_{C^\alpha - O;\varsigma}$.

The $\mathscr{D}$ matrices for the two bonds in the monomer unit, with proper incorporation of its articulated side chain, and corresponding to the statistical weight matrices denoted by $\mathbf{U}_b$ and the term in square brackets in [see Eq. (XI-61)]

$$\mathbf{X}_i = \mathbf{U}_b[(\mathbf{U}_{2;i} \cdots \mathbf{U}_{n_i;i}) \otimes \mathbf{U}_{2;i+1}] \qquad \text{(XIV-19)}$$

appear in the order $\mathscr{D}_b\mathscr{D}_a$, the former of dimensions 15×45, the latter of dimensions 45×15, with their product being of dimensions 15×15, as expected for a linear chain with $\nu = 3$.

$$\mathscr{D}_b = (\mathbf{U}_b \otimes \mathbf{I}_5) \, \text{diag} \, (\mathbf{D}_{tt}, \mathbf{D}_{tg^+}, \mathbf{D}_{tg^-}, \mathbf{D}_{g^+t}, \mathbf{D}_{g^+g^+}, \mathbf{D}_{g^+g^-}, \mathbf{D}_{g^-t}, \mathbf{D}_{g^-g^+}, \mathbf{D}_{g^-g^-})$$

$$\text{(XIV-20)}$$

$$\mathscr{D}_a = ([(\mathbf{U}_{2;i} \cdots \mathbf{U}_{n_i;i}) \otimes \mathbf{U}_{2;i+1}] \otimes \mathbf{I}_5)(\mathbf{I}_3 \otimes \|\mathbf{D}\|) \qquad \text{(XIV-21)}$$

The subscripts on $\mathbf{D}_{\zeta\eta}$ in Eq. (XIV-20) denote the states at bonds C^α—O and C^α—C, respectively; the former are used to formulate $\mathbf{m}_{COC;\zeta}$ via Eq. (XIV-18), and the latter are used to specify the $\mathbf{T}$ that appear in each $\mathbf{D}_{\zeta\eta}$:

$$\mathbf{D}_{\zeta\eta} = \begin{bmatrix} 1 & 2\mathbf{m}_{COC;\zeta}^T \mathbf{T}_\eta & m_{COC}^2 \\ 0 & \mathbf{T}_\eta & \mathbf{m}_{COC;\zeta} \\ 0 & 0 & 1 \end{bmatrix} \qquad \text{(XIV-22)}$$

The subscript ζ on m_{COC}^2 can be removed, because this scalar is the same for all ζ. The $\mathbf{D}$ in Eq. (XIV-21) are

$$\mathbf{D} = \begin{bmatrix} 1 & 0 & 0 \\ 0 & \mathbf{T} & 0 \\ 0 & 0 & 1 \end{bmatrix} \qquad \text{(XIV-23)}$$

where the state used for formulation of the $\mathbf{T}$ is the one at the C—C^α bond highlighted in Fig. XIV-6.

Calculations performed for $\langle \mu^2 \rangle_0 / x$ for poly(alkyl vinyl ethers) show the highest values for the *meso* polymer, with a monotonic decrease to the minimum values for the *racemo* polymer.[13] For *meso*-dominated atactic polymers, $p_m \sim 0.7$, the values of $\langle \mu^2 \rangle_0 / x$ are in the range 1.0 ± 0.3 Debye^2. These values are in reasonable agreement with measurements reported for moderately isotactic poly(isopropyl vinyl ether)[14] and poly(isobutyl vinyl ether).[14,15] The squared dipole moment for each monomer unit in Abe's calculation was 1.5 Debye^2. Since the calculated values of $\langle \mu^2 \rangle_0 / x$ are smaller than this number, there is a correlation between the bond dipole moment vectors, with that correlation acting to reduce the mean square dipole moment of the chain.

D. Correlations with Other Properties

Since it is clear from the preceding section that μ^2 need not be proportional to r^2, there is interest in assessing whether μ^2 is correlated with r^2 for the

[14]Luisi, P. L.; Chiellini, E.; Franchini, P. F.; Orienti, M. *Makromol. Chem.* **1968**, *112*, 197.
[15]Takeda, M.; Imamura, Y.; Okamura, S.; Higashimura, T. *J. Chem. Phys.* **1960**, *33*, 631. Pohl, H. A.; Zabusky, H. H. *J. Phys. Chem.* **1962**, *66*, 1390.

chains that contribute strongest to the unperturbed ensemble, and whether $\langle \mu^2 \rangle$ is correlated with $\langle r^2 \rangle$ when the weighting of the chains in the ensemble is altered by the introduction of the excluded volume effect.

Correlation Coefficients. The correlation between μ^2 and r^2 for all conformations in the unperturbed ensemble is measured by the correlation coefficient, which now takes the form

$$\rho_{r^2,\mu^2} = \left(\frac{\langle r^2 \mu^2 \rangle_0}{\langle r^2 \rangle_0 \langle \mu^2 \rangle_0} - 1 \right) \left[\left(\frac{\langle r^4 \rangle_0}{\langle r^2 \rangle_0^2} - 1 \right) \left(\frac{\langle \mu^4 \rangle_0}{\langle \mu^2 \rangle_0^2} - 1 \right) \right]^{-1/2} \quad \text{(XIV-24)}$$

where r^2, r^4, and μ^2 are generated by Eq. (VI-23), (XIII-4), and (XIV-8), respectively. The remaining two terms in Eq. (XIV-24) are obtained through

$$\mu^4 = \mu^2 \otimes \mu^2 \quad \text{(XIV-25)}$$

$$r^2 \mu^2 = r^2 \otimes \mu^2 \quad \text{(XIV-26)}$$

Correlation coefficients evaluated in this fashion can be positive, negative, or zero for short chains, but they are often zero for long chains. Results for several *model* chains that all have the l_i, θ_i, ϕ_i, and $\mathbf{U}_i$ for polyethylene, and with dipole moment vectors affixed to the bonds in different positions, are presented in Table XIV-2.

Average Dot Product of the End-to-End and Dipole Moment Vectors. The average of the dot product of $\mathbf{r}$ and $\boldsymbol{\mu}$ plays an important role in the expression [Eq. (II-32)], obtained by Nagai and Ishikawa[16] and by Doi[17] for the influence of excluded volume on the mean square dipole moment. The generator matrix for this term is a slight elaboration of the generator matrix used for r^2 and μ^2, both of which are self-dot products of these vectors. In the present case, we seek

$$\mathbf{r} \cdot \boldsymbol{\mu} = \left(\sum_{i=1}^{n} \mathbf{l}_i \right) \cdot \left(\sum_{i=1}^{n} \mathbf{m}_i \right) \quad \text{(XIV-27)}$$

which can be evaluated with a trivial modification of the generator matrices described previously, and the relationship[16]

[16]Nagai, K.; Ishikawa, T. *Polym. J.* **1971**, 2, 416.
[17]Doi, M. *Polym. J.* **1972**, 3, 252.

TABLE XIV-2. Correlation Coefficients for Several Chains with Conformations of Unperturbed Polyethylene, and with Dipole Moment Vectors Affixed in Different Patterns

		ρ_{r^2,μ^2}	
$\mathbf{m}_i$, Even i	$\mathbf{m}_i$, Odd i	Small n	$n \to \infty$
$\begin{bmatrix} m \\ 0 \\ 0 \end{bmatrix}$	$\begin{bmatrix} m \\ 0 \\ 0 \end{bmatrix}$	1	1
$\begin{bmatrix} 0 \\ m \\ 0 \end{bmatrix}$	$\begin{bmatrix} 0 \\ m \\ 0 \end{bmatrix}$	1	1
$\begin{bmatrix} 0 \\ 0 \\ m \end{bmatrix}$	$\begin{bmatrix} 0 \\ 0 \\ m \end{bmatrix}$	Negative	0
$\begin{bmatrix} m \\ 0 \\ 0 \end{bmatrix}$	$\begin{bmatrix} -m \\ 0 \\ 0 \end{bmatrix}$	Positive	0
$\begin{bmatrix} 0 \\ m \\ 0 \end{bmatrix}$	$\begin{bmatrix} 0 \\ -m \\ 0 \end{bmatrix}$	Positive	0
$\begin{bmatrix} 0 \\ 0 \\ m \end{bmatrix}$	$\begin{bmatrix} 0 \\ 0 \\ -m \end{bmatrix}$	Positive	0

$$\langle \mathbf{r} \cdot \boldsymbol{\mu} \rangle_0 = \tfrac{1}{2}[2\langle (\mathbf{r} + \boldsymbol{\mu}) \cdot (\mathbf{r} + \boldsymbol{\mu}) \rangle_0 - \langle r^2 \rangle_0 - \langle \mu^2 \rangle_0] \qquad \text{(XIV-28)}$$

The generator matrices used for finding the mean square length (mean self-dot product) of the vector $\mathbf{r}$ or the vector $\boldsymbol{\mu}$ can also be used to find the mean self-dot product of the vector $\mathbf{r} + \boldsymbol{\mu}$. The form of the generator matrix for the internal bonds is

$$\begin{bmatrix} 1 & 2(\mathbf{l} + \mathbf{m})^T\mathbf{T} & (\mathbf{l} + \mathbf{m})^T(\mathbf{l} + \mathbf{m}) \\ \mathbf{0} & \mathbf{T} & \mathbf{l} + \mathbf{m} \\ 0 & \mathbf{0} & 1 \end{bmatrix}_i \qquad 1 < i < n \qquad \text{(XIV-29)}$$

which, with Eq. (XIV-28), $\langle r^2 \rangle_0$, and $\langle \mu^2 \rangle_0$, gives $\langle \mathbf{r} \cdot \boldsymbol{\mu} \rangle_0$.

An alternative path provides a direct route to $\langle \mathbf{r} \cdot \boldsymbol{\mu} \rangle_0$, without the necessity for also calculating $\langle r^2 \rangle_0$ and $\langle \mu^2 \rangle_0$. A larger generator matrix (of dimensions 8×8, instead of 5×5) must be used. This generator matrix evaluates Eq. (XIV-27) directly as

$$\mathbf{r} \cdot \boldsymbol{\mu} = \sum_{i=1}^{n} l_i m_i + \sum_{i=1}^{n-1} \sum_{j=i+1}^{n} \mathbf{l}_i \cdot \mathbf{m}_j + \sum_{i=1}^{n-1} \sum_{j=i+1}^{n} \mathbf{m}_i \cdot \mathbf{l}_j \qquad \text{(XIV-30)}$$

This equation differs from Eq. (VI-16) only in that there are two double sums, rather than one, in order to account for the fact that $\mathbf{l}_i \cdot \mathbf{m}_j \neq \mathbf{l}_j \cdot \mathbf{m}_i$. The extra double sum requires an additional row and column in the blocked forms of the generator matrices, which become

$$[1 \quad \mathbf{l}^T \mathbf{T} \quad \mathbf{m}^T \mathbf{T} \quad lm]_1 \qquad \text{(XIV-31)}$$

$$\begin{bmatrix} 1 & \mathbf{l}^T \mathbf{T} & \mathbf{m}^T \mathbf{T} & lm \\ \mathbf{0} & \mathbf{T} & \mathbf{0} & \mathbf{m} \\ \mathbf{0} & \mathbf{0} & \mathbf{T} & \mathbf{l} \\ \mathbf{0} & \mathbf{0} & \mathbf{0} & 1 \end{bmatrix}_i, \qquad 1 < i < n \qquad \text{(XIV-32)}$$

$$\begin{bmatrix} lm \\ \mathbf{m} \\ \mathbf{l} \\ 1 \end{bmatrix}_n \qquad \text{(XIV-33)}$$

for the first bond, internal bonds, and last bond, respectively. Averaging over all chains in the unperturbed ensemble simply requires combination of these matrices with the appropriate $\mathbf{U}_i$. When extrapolated to $n \to \infty$, $\langle \mathbf{r} \cdot \boldsymbol{\mu} \rangle_0^2$ is frequently much smaller than $\langle r^2 \rangle_0 \langle \mu^2 \rangle_0$. For such chains, $\langle \mu^2 \rangle_0$ is insensitive to the excluded volume effect, in the limit where $n \to \infty$.

2. NMR CHEMICAL SHIFTS AND COUPLING CONSTANTS

Evaluation of the vicinal spin–spin coupling constants and the chemical shifts is based on the a priori probabilities, because the dominant effects are of short range. Frequently they depend on three-bond fragments, the conformational distribution of which is described by a bond probability denoted by $p_{\eta;i}$.[18]

[18] Gutowsky, H. S.; Belford, G. G.; McMahon, P. B. *J. Chem. Phys.* **1962**, *36*, 3353. Bovey, F. A.; Hood III, F. B.; Anderson, B. W.; Snyder, L. C. *J. Chem. Phys.* **1965**, *42*, 3900. Moritani, T.; Fujiwara, Y. *J. Chem. Phys.* **1973**, *59*, 1175.

The vicinal spin–spin coupling constant, $^3J_{HH}$, usually for two ^1H nuclei separated by three bonds, as in $^1H\text{--}A\text{—}B\text{--}^1H$, depends on the torsion angle at the A—B bond. The average value involves averages of trigonometric functions of this torsion angle [see Eq. (II-28)], with the exact form depending on the chemistry of the fragment. The averages of the appropriate trigonometric functions are easily generated from the $p_{\eta;i}$ for the A—B bond.

Evaluation of the γ effect on the chemical shift has been used extensively in the analysis of ^{13}C NMR spectra of polymers.[19] The γ-*gauche* interaction produces an upfield chemical shift relative to that obtained in the *trans* arrangement. The upfield shifts, γ_{ab}, for nuclei a and b when bond A—B in a–A—B–b is in a *gauche* state, are averaged using the $p_{\eta;i}$ for the A—B bond. This approach can be used for analysis of the interactions between similar nuclei (e.g., a = b = C) or different nuclei (e.g., a = C, b = Cl).[20]

3. OPTICAL ROTATORY POWER

Poly(α-olefins) can exhibit a large optical rotatory power, which may be strongly dependent on the temperature, if the side chain contains a chiral center.[21] The optical activities of these polyolefins,[22] as well as of several other polymers[10,23] and small molecules,[24] is susceptible to rationalization by combination of empirical rules for the molar optical rotation[25] with the averaging capability of RIS theory. The usual approach places the focus on three-bond interactions, as was also the case for the vicinal spin–spin coupling constants and chemical shifts described in the previous section. In the treatment of the molar rotation, however, the contributions from all bonds in the chain are additive.

A. Polymers without Articulated Side Chains

Let the ν_i rotational isomeric states at bond i make contributions of $[\Delta M]_{\lambda,1,i}$, $[\Delta M]_{\lambda,2,i}, \cdots [\Delta M]_{\lambda,\nu,i}$ to the molar optical rotation, $[M]_\lambda^T$, when this bond is

[19]Tonelli, A. E.; Schilling, F. C. *Acct. Chem. Res.* **1981**, *14*, 233. Tonelli, A. E. *NMR Spectroscopy and Polymer Microstructure. The Conformational Connection*, VCH, New York, **1989**.

[20]Tonelli, A. E.; Schilling, F. C. *Macromolecules* **1984**, *17*, 1946.

[21]Pino, P.; Lorenzi, G. P. *J. Am. Chem. Soc.* **1960**, *82*, 4745. Pino, P.; Lorenzi, G. P.; Lardicci, L.; Ciardelli, F. *Vysokomolekul. Soedin.* **1961**, *3*, 1597. Nozakura, S.; Takeuchi, S.; Yuki, H.; Murahashi, S. *Bull. Chem. Soc. Japan* **1961**, *34*, 1673. Pino, P.; Ciardelli, F.; Lorenzi, G. P.; Montagnoli, G. *Makromol. Chem.* **1963**, *61*, 207. Goodman, M.; Clark, K. J.; Stake, M. A.; Abe, A. *Makromol. Chem.* **1964**, *72*, 131.

[22]Abe, A. *J. Am. Chem. Soc.* **1968**, *90*, 2205.

[23]Abe, A. *J. Am. Chem. Soc.* **1970**, *92*, 1136.

[24]Colle, R.; Suter, U. W.; Luisi, P. L. *Tetrahedron* **1981**, *37*, 3727. Azzena, U.; Luisi, P. L.; Suter, U. W.; Gladiali, S. *Helv. Chim. Acta* **1981**, *64*, 2821.

[25]Whiffen, D. H. *Chem. Ind.* (*London*) **1956**, 964. Brewster, J. H. *J. Am. Chem. Soc.* **1959**, *81*, 5475. Brewster, J. H. *Top. Stereochem.* (Eliel, E. L.; Allinger, N. L., eds.) **1967**, *2*, 1.

in states 1, 2, $\cdots$ ν, Eq. (II-55). It is assumed that the $[\Delta M]_{\lambda,i}$ values are not affected by the conformations adopted by other bonds, and that the molar optical rotation of a conformation of the chain is the sum of the $[\Delta M]_{\lambda,\eta,i}$ values for all of the internal bonds. Then the average contribution of bond i, $[M]^T_{\lambda;i}$ is given by

$$[M]^T_{\lambda;i} = \frac{1}{Z} \mathbf{U}_1 \mathbf{U}_2 \cdots \mathbf{U}_{i-1} \mathbf{U}_i \|\mathbf{M}\|_{\lambda,i} \mathbf{U}_{i+1} \cdots \mathbf{U}_{n-1} \mathbf{U}_n \qquad \text{(XIV-34)}$$

where $\|\mathbf{M}\|_{\lambda,i}$ is the diagonal matrix with the $[\Delta M]_\lambda$ on the main diagonal:

$$\|\mathbf{M}\|_{\lambda,i} = \begin{bmatrix} [\Delta M]_1 & 0 & \cdots & 0 \\ 0 & [\Delta M]_2 & \cdots & 0 \\ \vdots & \vdots & & \vdots \\ 0 & 0 & \cdots & [\Delta M]_\nu \end{bmatrix}_{\lambda,i} \qquad \text{(XIV-35)}$$

The observable $[M]^T_\lambda$ is simply the sum of all $[M]^T_{\lambda;i}$ [Eq. (XIV-34)]; the summation is similar to the one that must be evaluated for computation of p_η using the $n-2$ values of $p_{\eta;i}$, and hence the matrix required for the evaluation is similar to the $\hat{\mathbf{U}}_{\eta;i}$ defined in Eq. (V-28):

$$[M]^T_\lambda = \frac{1}{Z} [\mathbf{U} \quad \mathbf{0}]_1 \begin{bmatrix} \mathbf{U} & \mathbf{U}\|\mathbf{M}\| \\ \mathbf{0} & \mathbf{U} \end{bmatrix}_{\lambda,2} \cdots \begin{bmatrix} \mathbf{U} & \mathbf{U}\|\mathbf{M}\| \\ \mathbf{0} & \mathbf{U} \end{bmatrix}_{\lambda,n-1} \begin{bmatrix} \mathbf{0} \\ \mathbf{U} \end{bmatrix}_n$$

$$\text{(XIV-36)}$$

This equation yields $[M]^T_\lambda = 0$ for linear polyethylene, as it must. In that chain the probability of each conformation is the same as the probability of its mirror image, and the matrix $\|\mathbf{M}\|_{\lambda,i}$ takes the form

$$\|\mathbf{M}\|_{\lambda,i} = \begin{array}{c} \\ t \\ g^+ \\ g^- \end{array} \begin{array}{c} \overset{\displaystyle t \quad\quad g^+ \quad\quad g^-}{} \\ \begin{bmatrix} 0 & 0 & 0 \\ 0 & [\Delta M]_g & 0 \\ 0 & 0 & -[\Delta M]_g \end{bmatrix}_i \end{array} \qquad \text{(XIV-37)}$$

for each internal bond. Large molar rotations can be generated, however, by polyolefins that have chiral centers in the side chains.

B. Polymers with Articulated Side Chains

Polymers with side chains that are rigidly attached to the main chain are easily handled by a simple extension of the methods described above. All that is necessary is to include the $[\Delta M]_\lambda$ from the bonds in the side chain in the $[\Delta M]_{\lambda;i}$ for the bond in the main chain to which the side chain is attached.

When the side chain has access to multiple conformations that interact with the bonds in the main chain, as in poly(alkyl vinyl ethers) or poly(α-olefins), the evaluation of $[M]_\lambda^T$ can utilize the formalism for treating articulated side chains. Consider, for example, the fragment of poly((S)-5-methyl-1-octene) depicted in Fig. XIV-7. Following Eq. (XI-61), the monomer unit highlighted with heavy links contributes a factor of

$$X_i = U_b[(U_{2;i}U_{3;i}U_{4;i}U_{5;i}) \otimes U_{2;i+1}]$$
(XIV-38)

to Z for this chain. The contribution by this same set of bonds to $[M]_\lambda^T$ is given by a 9×27 matrix followed by a 27×9 matrix

$$\begin{bmatrix} U_b & U_{b,M,i} & U_{b,M,i+1} \\ 0 & U_b & 0 \\ 0 & 0 & U_b \end{bmatrix} \cdot \begin{bmatrix} Y_i \otimes U_{2;i+1} & Y_{M,i} \otimes U_{2;i+1} & Y_i \otimes U_{2;i+1} \| M \|_{\lambda,2,i+1} \\ 0 & Y_i \otimes U_{2;i+1} & 0 \\ 0 & 0 & Y_i \otimes U_{2;i+1} \end{bmatrix}$$
(XIV-39)

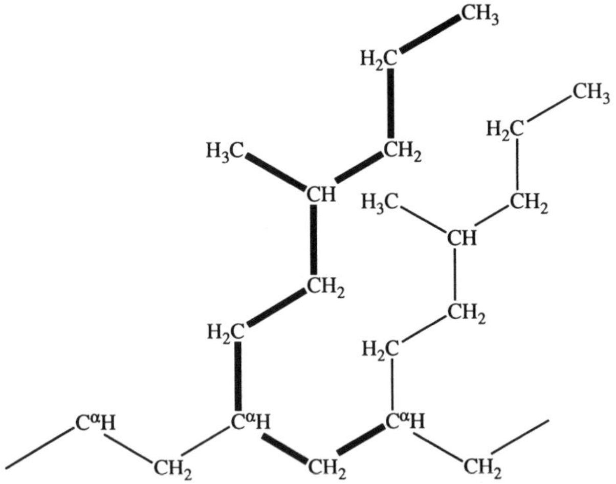

Figure XIV-7. Fragment of poly((S)-5-methyl-1-octene). The bonds drawn with heavy lines are described in the text.

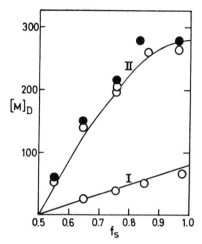

Figure XIV-8. Measured (points) and calculated (lines) molar rotations for (I) poly((R)(S)-5-methyl-1-heptene) and (II) poly((R)(S)-4-methyl-1-hexene). f_S denotes the fraction of S monomeric units in chains of random arrangement of monomeric units. Reprinted with permission from Abe, A. *J. Am. Chem. Soc.* **1970**, *92*, 1136. Copyright 1970 American Chemical Society.

where

$$\mathbf{U}_{b,M,i} = \mathbf{U}_b(\mathbf{I}_3 \otimes \|\mathbf{M}\|)_{\lambda,i} \tag{XIV-40}$$

$$\mathbf{U}_{b,M,i+1} = \mathbf{U}_b(\|\mathbf{M}\| \otimes \mathbf{I}_3)_{\lambda,i+1} \tag{XIV-41}$$

$$\mathbf{Y}_i = \mathbf{U}_{2;i}\mathbf{U}_{3;i}\mathbf{U}_{4;i}\mathbf{U}_{5;i} \tag{XIV-42}$$

$$\mathbf{Y}_{M,i} = [\mathbf{U} \quad \mathbf{U}\|\mathbf{M}\|]_{\lambda,2,i} \begin{bmatrix} \mathbf{U} & \mathbf{U}\|\mathbf{M}\| \\ \mathbf{0} & \mathbf{U} \end{bmatrix}_{\lambda,3,i} \begin{bmatrix} \mathbf{U} & \mathbf{U}\|\mathbf{M}\| \\ \mathbf{0} & \mathbf{U} \end{bmatrix}_{\lambda,4,i} \begin{bmatrix} \mathbf{0} \\ \mathbf{U} \end{bmatrix}_{5,i}$$

$$\tag{XIV-43}$$

The matrices in Eqs. (XIV-40) and (XIV-41) incorporate the contributions of the first bond in the branch, and the bond in the main chain that follows the site of attachment of the branch, respectively. Equation (XIV-42) is merely a compact notation for the product of the statistical weight matrices for the remaining bonds in the side chain, and Eq. (XIV-43) incorporates their contributions to the molar rotation.

Figure XIV-8 presents a comparison between measured[26] molar rotations at

[26]Pino, P.; Ciardelli, F.; Montagnoli, G.; Pieroni, O. *J. Polym. Sci., Part B* **1967**, *5*, 307.

the sodium D-line and calculations via methods equivalent to those described here.[23] The molar rotation increases with the excess of side chains of one chirality over side chains with the other chirality, and, for a given stereochemical composition, is larger for the side chain that places the chiral center nearer the main chain.

4. THE MOLECULAR OPTICAL ANISOTROPY

Depolarized Rayleigh scattering depends on the anisotropy of the polarizability tensor for the molecule, via Eq. (II-50) and Eq. (II-51). Rotational isomeric state theory, in conjunction with the *valence optical scheme*, is used for the evaluation of $\langle \gamma^2 \rangle_0$, the *molecular optical anisotropy*. The valence optical scheme is adopted for the construction of the anisotropic part of the polarizability tensor, $\hat{\boldsymbol{\alpha}}$, for the molecule in a particular conformation. The trace of $\hat{\boldsymbol{\alpha}}^2$ is required[27] for calculation of γ^2 for the same conformation, via Eq. (II-51). Implementation of the valence optical scheme requires that the chain can conceptually be decomposed into units with an $\hat{\boldsymbol{\alpha}}_i$, expressed in its own coordinate system, that is independent of the conformation of the molecule.[28] The $\hat{\boldsymbol{\alpha}}_i$ for all of the component units are then summed, with due regard for the orientation of their individual coordinate systems, in order to obtain $\hat{\boldsymbol{\alpha}}$ (and hence γ^2) for the molecule in that conformation. This information is then combined with Z in order to achieve the average over all conformations.

A. Transformation of a Second-Rank Tensor

In analogy with the representations of $\mathbf{r}^{\times 2}$ and $\mathbf{r}^T \mathbf{r}$ in Eqs. (XIII-11) and (XIII-12), it will be convenient to employ different representations for the nine elements in the symmetric second order tensor, $\hat{\boldsymbol{\alpha}}$, which can be written variously as

$$\hat{\boldsymbol{\alpha}} = \begin{bmatrix} \hat{\alpha}_{11} & \hat{\alpha}_{12} & \hat{\alpha}_{13} \\ \hat{\alpha}_{21} & \hat{\alpha}_{22} & \hat{\alpha}_{23} \\ \hat{\alpha}_{31} & \hat{\alpha}_{32} & \hat{\alpha}_{33} \end{bmatrix} \qquad \text{(XIV-44)}$$

$$\hat{\boldsymbol{\alpha}}^R = [\hat{\alpha}_{11} \quad \hat{\alpha}_{12} \quad \hat{\alpha}_{13} \quad \hat{\alpha}_{21} \quad \hat{\alpha}_{22} \quad \hat{\alpha}_{23} \quad \hat{\alpha}_{31} \quad \hat{\alpha}_{32} \quad \hat{\alpha}_{33}] \qquad \text{(XIV-45)}$$

---

[27] Although $\hat{\boldsymbol{\alpha}}$ is traceless (*i.e.*, tr $\hat{\boldsymbol{\alpha}} = 0$), $\hat{\boldsymbol{\alpha}}^2$ is not.

[28] This requirement is not fulfilled equally by all chains. It may be troublesome in chains with inductive effects that are conformation-dependent.

$$\hat{\boldsymbol{\alpha}}^C = \begin{bmatrix} \hat{\alpha}_{11} \\ \hat{\alpha}_{12} \\ \hat{\alpha}_{13} \\ \hat{\alpha}_{21} \\ \hat{\alpha}_{22} \\ \hat{\alpha}_{23} \\ \hat{\alpha}_{31} \\ \hat{\alpha}_{32} \\ \hat{\alpha}_{33} \end{bmatrix} \qquad (\text{XIV-46})$$

The latter two representations express the elements of the tensor in a manner that gives the appearance of a row vector or its transpose. These two representations will be particularly advantageous in what follows, because they will let us easily modify generator matrices used previously for handling vectors (such as $\mathbf{r}$ and $\boldsymbol{\mu}$) so that they can handle $\hat{\boldsymbol{\alpha}}$. Before proceeding along that line, however, we must first examine the formalism for transforming a tensor expressed in the coordinate system of bond $i + 1$ into its representation in the coordinate system of bond i.

Let $\mathbf{t}^{i+1}$ denote a second-rank tensor expressed in the coordinate system of bond $i + 1$, and $\mathbf{t}^i$ denote that same tensor, now expressed in the coordinate system of bond i. All of the information required for the transformation of $\mathbf{t}^{i+1}$ into $\mathbf{t}^i$ is contained in the transformation matrix, $\mathbf{T}_i$:

$$\mathbf{t}^i = \mathbf{T}_i \mathbf{t}^{i+1} \mathbf{T}_i^T \qquad (\text{XIV-47})$$

If $\mathbf{t}^{i+1}$ is expressed in the 3×3 form of Eq. (XIV-44), with elements denoted by t_{kl}^{i+1}, then $\mathbf{t}^i$ is also obtained in 3×3 form, with elements denoted by t_{mn}^i. Carrying out the multiplication as specified by Eq. (XIV-47) yields

$$t_{mn}^i = \sum_{k=1}^{3} \sum_{l=1}^{3} T_{mk;i} T_{nl;i} t_{kl}^{i+1} \qquad (\text{XIV-48})$$

Using instead the column representation of the tensors, Eq. (XIV-46), we obtain

$$[t_{3(m-1)+n}^i]^C = \sum_{k=1}^{3} \sum_{l=1}^{3} T_{mk;i} T_{nl;i} [t_{3(k-1)+l}^{i+1}]^C \qquad (\text{XIV-49})$$

Row $3(m - 1) + n$ of $\mathbf{T}_i^{\times 2}$ has exactly the nine pairs of elements as specified in Eq. (XIV-49), appearing in the proper order for meshing with the correct element from $\mathbf{t}^{i+1,C}$; thus, that row vector is

$$[T_{m1}T_{n1} \quad T_{m1}T_{n2} \quad T_{m1}T_{n3} \quad T_{m2}T_{n1} \quad T_{m2}T_{n2} \quad T_{m2}T_{n3} \quad T_{m3}T_{n1} \quad T_{m3}T_{n2} \quad T_{m3}T_{n3}]_i$$

$$(XIV-50)$$

If we are willing to work with the row vector and column vector forms of the tensor, Eqs. (XIV-45) and (XIV-46), a replacement for Eq. (XIV-47) is[29]

$$\mathbf{t}^{i,C} = \mathbf{T}_i^{\times 2} \mathbf{t}^{i+1,C} \tag{XIV-51}$$

Some insight into whether it will be preferable to use the row vector and column vector forms of the tensor, or the 3×3 form, is immediately gained by considering one further transformation. If $\mathbf{t}^{i+1}$ is transformed from its representation in the coordinate system of bond $i + 1$ into its representation in the coordinate system of bond i, and then transformed again into the coordinate system of bond $i - 1$, the choice lies between

$$\mathbf{t}^{i-1} = \mathbf{T}_{i-1} \mathbf{T}_i \mathbf{t}^{i+1} \mathbf{T}_i^T \mathbf{T}_{i-1}^T \tag{XIV-52}$$

and

$$\mathbf{t}^{i,C} = \mathbf{T}_{i-1}^{\times 2} \mathbf{T}_i^{\times 2} \mathbf{t}^{i+1,C} \tag{XIV-53}$$

Special attention is directed to the manner in which information about bond $i - 1$ appears in the two expressions. When the column representation is used for the tensor, all such information occurs at a single location, in the $\mathbf{T}_{i-1}^{\times 2}$ immediately after the equal sign. A much different situation is obtained if the tensor is expressed in 3×3 form. Then the information about bond $i - 1$ appears in two widely separated parts of the equation, as the first and last matrix in the serial product on the right-hand side of Eq. (XIV-52). Looking forward to the averaging procedure by which Z is combined with generator matrices, we prefer an expression where all information about bond $i - 1$ occurs in a single location. This requirement will cause us to often construct the generator matrix using the transformation of the tensors with elements expressed as a column vector, instead of using the tensor in its 3×3 form.

---

[29] Jernigan, R. L.; Flory, P. J. *J. Chem. Phys.* **1967**, *47*, 1999. The general relationships for three conformable matrices, **A**, **B**, and **C**, are

$$(\mathbf{ABC})^C = (\mathbf{A} \otimes \mathbf{C}^T)\mathbf{B}^C$$
$$(\mathbf{ABC})^R = \mathbf{B}^R(\mathbf{A}^T \otimes \mathbf{C})$$

B. Generator Matrix

The optical anisotropy of a specified conformation can be expressed using either of the following equivalent representations for the anisotropic part of the polarizability tensor.

$$\gamma^2 = \tfrac{3}{2} \operatorname{tr} \hat{\boldsymbol{\alpha}}^2 = \tfrac{3}{2} \hat{\boldsymbol{\alpha}}^R \hat{\boldsymbol{\alpha}}^C \qquad \text{(XIV-54)}$$

With the trivial exception of the leading factor of $\tfrac{3}{2}$, the latter representation can be treated in the manner used for generating $\mathbf{r}^T \mathbf{r} = r^2$. Instead of

$$r^2 = 2 \sum_{k=1}^{n-1} \sum_{j=k+1}^{n} \mathbf{l}_k^T \mathbf{T}_k \mathbf{T}_{k+1} \cdots \mathbf{T}_{j-1} \mathbf{l}_j + \sum_{k=1}^{n} l_k^2 \qquad \text{(VI-18)}$$

we now have

$$\gamma^2 = \frac{3}{2} \left(2 \sum_{k=1}^{n-1} \sum_{j=k+1}^{n} \hat{\boldsymbol{\alpha}}_k^R \mathbf{T}_k^{\times 2} \mathbf{T}_{k+1}^{\times 2} \cdots \mathbf{T}_{j-1}^{\times 2} \hat{\boldsymbol{\alpha}}_j^C + \sum_{k=1}^{n} \hat{\boldsymbol{\alpha}}_k^R \hat{\boldsymbol{\alpha}}_k^C \right) \qquad \text{(XIV-55)}$$

Proceeding along the line that led to $\mathbf{G}_i$, we obtain[30]

$$\mathbf{P}_i = \begin{bmatrix} 1 & 2\hat{\boldsymbol{\alpha}}^R \mathbf{T}^{\times 2} & \hat{\boldsymbol{\alpha}}^R \hat{\boldsymbol{\alpha}}^C \\ \mathbf{0} & \mathbf{T}^{\times 2} & \hat{\boldsymbol{\alpha}}^C \\ 0 & \mathbf{0} & 1 \end{bmatrix}_i \qquad 1 < i < n \qquad \text{(XIV-56)}$$

$$\mathbf{P}_1 = \begin{bmatrix} 1 & 2\hat{\boldsymbol{\alpha}}^R \mathbf{T}^{\times 2} & \hat{\boldsymbol{\alpha}}^R \hat{\boldsymbol{\alpha}}^C \end{bmatrix}_1 \qquad \text{(XIV-57)}$$

$$\mathbf{P}_n = \begin{bmatrix} \hat{\boldsymbol{\alpha}}^R \hat{\boldsymbol{\alpha}}^C \\ \hat{\boldsymbol{\alpha}}^C \\ 1 \end{bmatrix}_n \qquad \text{(XIV-58)}$$

$$\gamma^2 = \tfrac{3}{2} \mathbf{P}_1 \mathbf{P}_2 \cdots \mathbf{P}_n \qquad \text{(XIV-59)}$$

The dimensions of the internal $\mathbf{P}_i$ are 11×11. Reduction to 8×8 can be affected, using the same procedure described in the preceding chapter for the

[30]Flory, P. J. *J. Chem. Phys.* **1972**, *56*, 862.

reduction in the dimensions of $\mathbf{A}_i^{\times 2}$.[31] According to that procedure, the reduced forms of $\hat{\boldsymbol{\alpha}}^R$, $\hat{\boldsymbol{\alpha}}^C$, and $\mathbf{T}^{\times 2}$ are

$$\hat{\boldsymbol{\alpha}}^R = [\hat{\alpha}_{11} \quad \hat{\alpha}_{12} \quad \hat{\alpha}_{13} \quad \hat{\alpha}_{22} \quad \hat{\alpha}_{23} \quad \hat{\alpha}_{33}]$$

$$\hat{\boldsymbol{\alpha}}^C = \begin{bmatrix} \hat{\alpha}_{11} \\ 2\hat{\alpha}_{12} \\ 2\hat{\alpha}_{13} \\ \hat{\alpha}_{22} \\ 2\hat{\alpha}_{23} \\ \hat{\alpha}_{33} \end{bmatrix}$$

$$\mathbf{T}^{\times 2} = \begin{bmatrix} T_{11}^2 & T_{11}T_{12} & T_{11}T_{13} & T_{12}^2 & T_{12}T_{13} & T_{13}^2 \\ 2T_{11}T_{21} & T_{11}T_{22}+T_{21}T_{12} & T_{11}T_{23}+T_{21}T_{13} & 2T_{12}T_{22} & T_{12}T_{23}+T_{22}T_{13} & 2T_{13}T_{23} \\ 2T_{11}T_{31} & T_{11}T_{32}+T_{31}T_{12} & T_{11}T_{33}+T_{31}T_{13} & 2T_{12}T_{32} & T_{12}T_{33}+T_{32}T_{13} & 2T_{13}T_{33} \\ T_{21}^2 & T_{21}T_{22} & T_{21}T_{23} & T_{22}^2 & T_{22}T_{23} & T_{23}^2 \\ 2T_{21}T_{31} & T_{21}T_{32}+T_{31}T_{22} & T_{21}T_{33}+T_{31}T_{23} & 2T_{22}T_{32} & T_{22}T_{33}+T_{32}T_{23} & 2T_{23}T_{33} \\ T_{31}^2 & T_{31}T_{32} & T_{31}T_{33} & T_{32}^2 & T_{32}T_{33} & T_{33}^2 \end{bmatrix}$$

In the present case, the reduction takes advantage of the fact that $\hat{\boldsymbol{\alpha}}$ is a symmetric tensor, with three pairs of redundant elements in the unreduced forms of $\hat{\boldsymbol{\alpha}}^C$ and $\hat{\boldsymbol{\alpha}}^R$. Combination of these pairs of redundant elements reduces the size of the middle block in $\mathbf{P}_i$ from 9×9 to 6×6, with corresponding reductions in $\hat{\boldsymbol{\alpha}}^R \mathbf{T}^{\times 2}$ and $\hat{\boldsymbol{\alpha}}^C$.

Combination of $\mathbf{P}_i$ with $\mathbf{U}_i$ can be achieved in the usual way.

$$\mathscr{P}_i = (\mathbf{U}_i \otimes \mathbf{I}_P)\|\mathbf{P}_i\| \qquad \text{(XIV-60)}$$

According to the valence optical scheme, $\hat{\boldsymbol{\alpha}}^R$ and $\hat{\boldsymbol{\alpha}}^C$ are the same in each $\mathbf{P}_i$ in $\|\mathbf{P}_i\|$. However, if one had information that justified using a $\hat{\boldsymbol{\alpha}}^R$ and $\hat{\boldsymbol{\alpha}}^C$ that depended on the state at bond i, that information could be incorporated into $\|\mathbf{P}_i\|$. The formalism would also permit letting $\hat{\boldsymbol{\alpha}}^R$ and $\hat{\boldsymbol{\alpha}}^C$ depend on the state of the previous bond.[32]

For flexible chains, $\langle \gamma^2 \rangle_0/n$ usually reaches its asymptotic limit at small n. Illustrative results are presented in Table XIV-3.

[31] Nagai, K. *J. Chem. Phys.* **1963**, *38*, 924. Nagai, K.; Ishikawa, T. *J. Chem. Phys.* **1966**, *45*, 3128. Nagai, K. *J. Chem. Phys.* **1968**, *48*, 5646. Flory, P. J.; Abe, Y. *J. Chem. Phys.* **1971**, *54*, 1351. Flory, P. J. *Macromolecules* **1974**, *7*, 381. Galiatsatos, V.; Mattice, W. L. *J. Comput. Chem.* **1990**, *11*, 396.
[32] See footnote 11 in Chapter VI.

TABLE XIV-3. $\langle \gamma^2 \rangle_0/x$ (in 10^{-60} m^6 = Å^6) for Selected Polymers

Polymer	$\langle \gamma^2 \rangle_0/x$		
	Calc.	Exp.	Ref.
Poly(1,3-dioxolane)	~0.4	—	a
Polypropylene, atactic	1-2	—	b
Polystyrene, atactic	~34	31	c
	37 ± 5	32 ± 1	d
Polystyrene, isotactic	30	—	d
Polystyrene, syndiotactic	97	—	d
Poly(p-chlorostyrene), atactic	70 ± 10	63	e
Poly(p-bromostyrene), atactic	130 ± 20	111	e
Polycarbonate of Bisphenol A	~120	117	f
	$111(x = 11)$	$120 \pm 5(\bar{x}_n = 10.7)$	g

[a] Riande, E.; Saiz, E.; Mark, J. E. *Macromolecules* **1980**, *13*, 448. Calculated values over a range of head-to-head and head-to-tail compositions.
[b] Tonelli, A. E.; Abe, Y.; Flory, P. J. *Macromolecules* **1970**, *3*, 303.
[c] Konishi, T.; Yoshizaki, T.; Shimada, J.; Yamakawa, H. *Macromolecules* **1989**, *22*, 1921.
[d] Suter, U. W.; Flory, P. J. *J. Chem. Soc., Faraday Trans. II* **1977**, *73*, 1521.
[e] Saiz, E.; Suter, U. W.; Flory, P. J. *J. Chem. Soc., Faraday Trans. II* **1977**, *73*, 1538.
[f] Floudas, G.; Lappas, A.; Fytas, G.; Meier, G. *Macromolecules* **1990**, *23*, 1747.
[g] Erman, B.; Wu, D.; Irvine, P. A.; Marvin, D. C.; Flory, P. J. *Macromolecules* **1982**, *15*, 670.

5. BIREFRINGENCE

Birefringence can be produced in macromolecules by applying a mechanical stress, electric field, or magnetic field. The term that dominates the birefringence is $\langle \mathbf{r}^T \hat{\alpha} \mathbf{r} \rangle_0$, $\langle \boldsymbol{\mu}^T \hat{\alpha} \boldsymbol{\mu} \rangle_0$, or $\langle \mathrm{tr}\, \hat{\chi} \hat{\alpha} \rangle_0$, as described in Chapter II.[33] As expected from the prior treatments of $\mathbf{r}$ and $\boldsymbol{\mu}$, the generator matrices for $\mathbf{r}^T \hat{\alpha} \mathbf{r}$ and $\boldsymbol{\mu}^T \hat{\alpha} \boldsymbol{\mu}$ are similar in form.

A. Notation for Products of Vectors and Symmetric Tensors

The treatments of the stress-optical coefficient, Eq. (II-45), and the molar Kerr constant, Eq. (II-48), require evaluation of the products of a vector, $\mathbf{v}$, and a symmetric tensor, $\mathbf{t}$, which are of the form $\mathbf{v}^T \mathbf{t} \mathbf{v}$. Before proceeding, it is useful to remember that $\mathbf{v}^T \mathbf{t} \mathbf{v}$ can be written variously as[29,34]

---

[33] In polymers with very small dipole moments, the electrically induced birefringence may be more dependent on the term in $\langle \gamma^2 \rangle_0$, rather than on the term in $\langle \boldsymbol{\mu}^T \hat{\alpha} \boldsymbol{\mu} \rangle_0$.

[34] Each of the terms in Eq. (XIV-61) yields the scalar

$$v_{A,1}v_{B,1}t_{11} + 2v_{A,1}v_{B,2}t_{12} + 2v_{A,1}v_{B,3}t_{13} + v_{A,2}v_{B,2}t_{22} + 2v_{A,2}v_{B,3}t_{23} + v_{A,3}v_{B,3}t_{33}$$

$$\mathbf{v}_A^T \mathbf{t} \mathbf{v}_B = \mathbf{v}_B^T \mathbf{t} \mathbf{v}_A = (\mathbf{v}_A^T \otimes \mathbf{v}_B^T) \mathbf{t}^C = (\mathbf{v}_B^T \otimes \mathbf{v}_A^T) \mathbf{t}^C = \mathbf{t}^R (\mathbf{v}_A \otimes \mathbf{v}_B) = \mathbf{t}^R (\mathbf{v}_B \otimes \mathbf{v}_A)$$

$$(XIV\text{-}61)$$

In construction of the generator matrix for the stress-optical coefficient and molar Kerr constant, we will use whichever form in Eq. (XIV-61) is best suited to the purpose of the moment.

B. Stress-Induced: The Stress-Optical Coefficient

The stress-optical coefficient, Eq. (II-45), depends on the ratio $\langle \mathbf{r}^T \hat{\alpha} \mathbf{r} \rangle_0 / \langle r^2 \rangle_0$. The new information required here is a generator matrix for $\mathbf{r}^T \hat{\alpha} \mathbf{r}$. If all vectors and tensors are expressed in the same coordinate system, and if we can retain the assumption of the validity of the valence optical scheme, this information is generated as a triple sum

$$\mathbf{r}^T \hat{\alpha} \mathbf{r} = \sum_{h=1}^{n} \sum_{j=1}^{n} \sum_{k=1}^{n} \mathbf{l}_h^T \hat{\alpha}_j \mathbf{l}_k \qquad (XIV\text{-}62)$$

where the components of every $\mathbf{l}_h^T \hat{\alpha}_j \mathbf{l}_k$ are expressed in the same coordinate system when the product is formed. Each term is the product of two vectors with a tensor. The terms in the triple sum are conveniently separated into eight types depending on the relationship between the indices h, j, and k, as summarized in Table XIV-4.

The simplest case is where $h = j = k$, because then the vectors and tensor expressed in the coordinate system of their own bond are necessarily in the

TABLE XIV-4. Evaluation of Terms Required for the Stress-Optical Coefficient

Relationship between Subscripts	Equation	Add Factor of 2?	Blocks of $\mathbf{K}_i$ Used
$h = j = k$	(XIV-63)	No	1,1; 1,6; 6,6
$j < h = k$	(XIV-64)	No	1,3; 3,3; 3,6
$h = k < j$	(XIV-65)	No	1,4; 4,4; 4,6
$h < j = k$	(XIV-66)	Yes	1,2; 2,2; 2,6
$h = j < k$	(XIV-67)	Yes	1,5; 5,5; 5,6
$h < j < k$	(XIV-68)	Yes	1,2; 2,2; 2,5; 5,5; 5,6
$j < h < k$	(XIV-69)	Yes	1,3; 3,3; 3,5; 5,5; 5,6
$h < k < j$	(XIV-70)	Yes	1,2; 2,2; 2,4; 4,4; 4,6

same coordinate system:

$$\sum_{h=j=k=1}^{n} \hat{\alpha}_j^R \mathbf{l}_h^{\times 2} \tag{XIV-63}$$

If only $\mathbf{l}^T$ and $\mathbf{l}$ are contributed by the same bond, with $\hat{\alpha}_j$ from another bond located earlier in the chain, we obtain

$$\sum_{j=1}^{n-1} \sum_{h=k=j+1}^{n} \hat{\alpha}_j^R \mathbf{T}_j^{\times 2} \mathbf{T}_{j+1}^{\times 2} \cdots \mathbf{T}_{h-1}^{\times 2} \mathbf{l}_h^{\times 2} \tag{XIV-64}$$

but if $\hat{\alpha}_j$ is located later in the chain, we have instead

$$\sum_{h=k=1}^{n-1} \sum_{j=h+1}^{n} \mathbf{l}_h^{T,\times 2} \mathbf{T}_h^{\times 2} \mathbf{T}_{h+1}^{\times 2} \cdots \mathbf{T}_{j-1}^{\times 2} \hat{\alpha}_j^C \tag{XIV-65}$$

For the remaining terms, $\mathbf{l}^T$ and $\mathbf{l}$ are contributed by different bonds. The terms can be paired, with members of each pair obtained by reversal of h and k. It will be convenient to simply write an expression for one of the pairs, and then account for the other member by multiplication of that expression by a leading factor of 2. If $\hat{\alpha}_j\mathbf{l}$ is contributed by the same bond, but $\mathbf{l}^T$ is from a bond with a smaller index, we have

$$\sum_{h=1}^{n-1} \sum_{j=k=h+1}^{n} \mathbf{l}_h^T \mathbf{T}_h \mathbf{T}_{h+1} \cdots \mathbf{T}_{j-1} \hat{\alpha}_j \mathbf{l}_k \tag{XIV-66}$$

If $\mathbf{l}^T \hat{\alpha}_j$ is contributed by the same bond, and $\mathbf{l}$ is from a bond that appears later in the sequence, we obtain

$$\sum_{h=j=1}^{n-1} \sum_{k=h+1}^{n} \hat{\alpha}_j^R (\mathbf{l}_h \otimes \mathbf{T}_j) \mathbf{T}_{h+1} \cdots \mathbf{T}_{k-1} \mathbf{l}_k \tag{XIV-67}$$

For the remaining terms, three different bonds are involved for $\mathbf{l}^T$, $\hat{\alpha}_j$, and $\mathbf{l}$. The three distinct combinations of indices (all of which require a leading factor of 2 to catch the partner) are

$$\sum_{h=1}^{n-2}\sum_{j=h+1}^{n-1}\sum_{k=j+1}^{n} \mathbf{l}_h^T \mathbf{T}_h \mathbf{T}_{h+1} \cdots \mathbf{T}_{j-1}\hat{\boldsymbol{\alpha}}_j \mathbf{T}_j \cdots \mathbf{T}_{k-1}\mathbf{l}_k \qquad \text{(XIV-68)}$$

$$\sum_{j=1}^{n-2}\sum_{h=j+1}^{n-1}\sum_{k=h+1}^{n} \hat{\boldsymbol{\alpha}}_j^R \mathbf{T}_j^{\times 2}\mathbf{T}_{j+1}^{\times 2} \cdots \mathbf{T}_{h-1}^{\times 2}(\mathbf{l}_h \otimes \mathbf{T}_h)\mathbf{T}_{h+1} \cdots \mathbf{T}_{k-1}\mathbf{l}_k \qquad \text{(XIV-69)}$$

$$\sum_{h=1}^{n-2}\sum_{k=h+1}^{n-1}\sum_{j=k+1}^{n} \mathbf{l}_h^T \mathbf{T}_h \mathbf{T}_{h+1} \cdots \mathbf{T}_{k-1}(\mathbf{I}_3 \otimes \mathbf{l}_k^T)\mathbf{T}_{k+1}^{\times 2} \cdots \mathbf{T}_{j-1}^{\times 2}\hat{\boldsymbol{\alpha}}_j^C \qquad \text{(XIV-70)}$$

All of these sums, with added factors of 2 where appropriate, are generated with a matrix of dimensions 26×26, which appears as a 6×6 matrix when written in blocked form:[1]

$$\mathbf{K}_i = \begin{bmatrix} 1 & 2\mathbf{l}^T\mathbf{T} & \hat{\boldsymbol{\alpha}}^R\mathbf{T}^{\times 2} & \mathbf{l}^{T,\times 2}\mathbf{T}^{\times 2} & 2\hat{\boldsymbol{\alpha}}^R(\mathbf{l}\otimes\mathbf{T}) & \hat{\boldsymbol{\alpha}}^R\mathbf{l}^{\times 2} \\ 0 & \mathbf{T} & 0 & (\mathbf{I}_3\otimes\mathbf{l}^T)\mathbf{T}^{\times 2} & \hat{\boldsymbol{\alpha}}\mathbf{T} & \hat{\boldsymbol{\alpha}}\mathbf{l} \\ 0 & 0 & \mathbf{T}^{\times 2} & 0 & 2\mathbf{l}\otimes\mathbf{T} & \mathbf{l}^{\times 2} \\ 0 & 0 & 0 & \mathbf{T}^{\times 2} & 0 & \hat{\boldsymbol{\alpha}}^C \\ 0 & 0 & 0 & 0 & \mathbf{T} & \mathbf{l} \\ 0 & 0 & 0 & 0 & 0 & 1 \end{bmatrix}_i \quad 1 < i < n$$

$$\text{(XIV-71)}$$

The specific elements utilized in the evaluation of each of the sums are identified in Table XIV-4. The terminal matrices, $\mathbf{K}_1$ and $\mathbf{K}_n$, are the first row and last column, respectively, of $\mathbf{K}_i$. With this notation

$$\mathbf{r}^T\hat{\boldsymbol{\alpha}}\mathbf{r} = \mathbf{K}_1\mathbf{K}_2 \cdots \mathbf{K}_n \qquad \text{(XIV-72)}$$

The averaging to obtain $\langle\mathbf{r}^T\hat{\boldsymbol{\alpha}}\mathbf{r}\rangle_0$ is accomplished by the usual procedure that couples $\mathbf{K}_i$ with $\mathbf{U}_i$ to form $\mathscr{K}_i$.

A few representative values of Γ_2 are presented in Table XIV-5. In some cases, as illustrated by atactic poly(methylphenylsiloxane), agreement between calculation and experiment is satisfactory. The comparison of calculation and experiment is disappointing in other cases, as illustrated by poly(methyl acrylate), perhaps because the calculated value of Γ_2 is sensitive to the uncertainties associated with the formulation of $\hat{\boldsymbol{\alpha}}$.

Chains in which the direction of greatest polarizability is perpendicular to the chain axis tend to have $\Gamma_2 < 0$. The phenyl and ester groups in poly(methylphenylsiloxane), polystyrene, and poly(methyl acrylate) are the

TABLE XIV-5. $\Gamma_2 = 9\langle r^T \hat{\alpha} r \rangle_0 / 10 \langle r^2 \rangle_0$ (in 10^{-30} m^3 = Å^3) for Selected Polymers

Polymer	Calc.	Exp.	Ref.
		Γ_2	
Poly(methylphenylsiloxane)			a
Isotactic	−4	—	
Atactic	−7 to −8	−7	
Syntiotactic	−13	—	
Poly(methyl acrylate)			b
Isotactic	−3.0	—	
Atactic	−3.3	−0.5	
Syntiotactic	−7.9	—	
Polystyrene, atactic	−9	—	c
Polypropylene, atactic	~1	—	c

^{a} Llorente, M. A.; Fernández de Pierola, I.; Saiz, E. *Macromolecules* **1985**, *18*, 2663.
^{b} Saiz, E.; Riande, E.; Mark, J. E. *Macromolecules* **1984**, *17*, 899.
^{c} Abe, Y.; Tonelli, A. E.; Flory, P. J. *Macromolecules* **1970**, *3*, 294. Calculated values.

most polarizable groups in those polymers, and these groups are attached to the chain such that the axis of their greatest polarizability is perpendicular to the chain. Thus the negative sign of their Γ_2 is easily rationalized. For polypropylene, in contrast, it is the C—C bonds that are the most polarizable units. The direction of their greatest polarizability is along the C—C bonds, which, of course, makes the direction of greatest polarizability coincident with the chain axis. Hence Γ_2 for this polymer, while small, is positive.

C. Electrically Induced: The Molar Kerr Constant

The electrically induced birefringence measured with small electric fields is usually reported as the *molar Kerr constant*, $\langle _m K \rangle_0$, defined in Eq. (II-46). It is often reported as the ratio of the molar Kerr constant to the number of units in the chain, $\langle _m K \rangle_0 / x$. As shown in Eq. (II-48), the conformational averages required for a theoretical interpretation of $\langle _m K \rangle_0$ are $\langle \mu^T \hat{\alpha} \mu \rangle_0$ and $\langle \gamma^2 \rangle_0$. The latter was described above in conjunction with the molecular optical anisotropy, and $\langle \mu^T \hat{\alpha} \mu \rangle_0$ is the electric dipole analog of $\langle r^T \hat{\alpha} r \rangle_0$ discussed in conjunction with the stress-optical coefficient. The generator matrix for $\mu^T \hat{\alpha} \mu$ can be obtained directly from $\mathbf{K}_i$, Eq. (XIV-71), by substitution of $\mathbf{m}_i$ for each $\mathbf{l}_i$.[1]

Most polymers for which $\langle _m K \rangle_0$ is of interest are polar. In these polymers, the term in $\langle \mu^T \hat{\alpha} \mu \rangle_0$ in Eq. (II-48) dominates the term in $\langle \gamma^2 \rangle_0$. For polymers with weak dipole moments and strong anisotropies, as in polystyrene, the term in $\langle \gamma^2 \rangle_0$ becomes more important. Of course, if the polymer has a vanishingly

TABLE XIV-6. Illustrative Calculations of $\langle_m K \rangle_0 / x$ (in 10^{-22} m^5 statV^{-2} mol^{-1}) for Selected Polymers[a,b]

| Polymer | Tacticity | $\langle_m K \rangle_0 / x$ | Contribution from | |
			$\langle \mu^T \hat{\alpha} \mu \rangle_0$	$\langle \gamma^2 \rangle_0$
Polypropylene	Syndiotactic	49	0	49
	Atactic	45	0	45
	Isotactic	39	0	39
Polystyrene	Syndiotactic	116	-18	134
	Atactic	127	-1	128
	Isotactic	107	-4	113
Poly(vinyl chloride)	Syndiotactic	$-13,426$	$-13,500$	74
	Atactic	71	45	26
	Isotactic	306	287	19

[a]Tonelli, A. E. *Macromolecules* **1977**, *10*, 153. The author has suggested in a more recent publication (Khanarian, G.; Cais, R. E.; Kometani, J. M.; Tonelli, A. E. *Macromolecules* **1982**, *15*, 866) that different formulations of $\hat{\alpha}_i$ might be more appropriate than those used to calculate the results presented in this table. However, these results are an excellent illustration of the major trends from the two contributions to $\langle_m K \rangle_0$, and for that reason we use them here.
[b]The fourth and fifth columns present separately the contributions to $\langle_m K \rangle_0$ from the two terms in Eq. (II-48).

small dipole moment, the term in $\langle \gamma^2 \rangle_0$ is the sole contributor, but $\langle_m K \rangle_0$ will usually be small, as in polypropylene. A contribution from the dipolar term is necessary if the molar Kerr constant is to be negative. The term contributed by $\langle \gamma^2 \rangle_0$ is always positive (since the bias in orientation to the electric field favors the most polarizable direction parallel to the field), but the other term takes the sign of $\langle \mu^T \hat{\alpha} \mu \rangle_0$. These effects are apparent in the calculations presented in Table XIV-6.

D. Magnetically Induced: The Cotton–Mouton Constant

Theoretical evaluation of the Cotton–Mouton constant, Eq. (II-49), requires calculation of $\langle \text{tr} \, \hat{\alpha} \hat{\chi} \rangle_0$, where $\hat{\chi}$ is the magnetic susceptibility analog of $\hat{\alpha}$. The generator matrix is formulated using the assumption that the valence optical scheme can be extended to the magnetic susceptibilities, $\hat{\chi} = \sum_i \hat{\chi}_i$.

The trace of the product of $\hat{\alpha}$ and $\hat{\chi}$ (both being symmetric) is generated as $\hat{\alpha}^R \hat{\chi}^C$, which has the appearance of the dot product of two different vectors:

$$\text{tr} \, \hat{\alpha} \hat{\chi} = \sum_{i=1}^{n} \hat{\alpha}_i^R \hat{\chi}_i^C + \sum_{i=1}^{n-1} \sum_{j=i+1}^{n} \hat{\alpha}_i^R \hat{\chi}_j^C + \sum_{i=1}^{n-1} \sum_{j=i+1}^{n} \hat{\chi}_i^R \hat{\alpha}_j^C \qquad \text{(XIV-73)}$$

Simple substitution of $\hat{\chi}_i^C$ for $\hat{\alpha}_i^C$, and $\hat{\alpha}_i^R \hat{\chi}_i^C$ for $\hat{\alpha}_i^2$, in the generator matrix for γ^2, Eq. (XIV-56), would miss the terms contributed by the last double sum in Eq. (XIV-73). The proper generator matrix can be constructed using the previous form for the generation of the dot product of $\mathbf{r}$ and $\boldsymbol{\mu}$, Eq. (XIV-32):

$$\mathbf{C}_i = \begin{bmatrix} 1 & \hat{\alpha}^R \mathbf{T}^{\times 2} & \hat{\chi}^R \mathbf{T}^{\times 2} & \hat{\alpha}^R \hat{\chi}^C \\ 0 & \mathbf{T}^{\times 2} & 0 & \hat{\chi}^C \\ 0 & 0 & \mathbf{T}^{\times 2} & \hat{\alpha}^C \\ 0 & 0 & 0 & 1 \end{bmatrix}_i, \qquad 1 < i < n \qquad \text{(XIV-74)}$$

$$\mathbf{C}_1 = [1 \quad \hat{\alpha}^R \mathbf{T}^{\times 2} \quad \hat{\chi}^R \mathbf{T}^{\times 2} \quad \hat{\alpha}^R \hat{\chi}^C]_1 \qquad \text{(XIV-75)}$$

$$\mathbf{C}_n = \begin{bmatrix} \hat{\alpha}^R \hat{\chi}^C \\ \hat{\chi}^C \\ \hat{\alpha}^C \\ 1 \end{bmatrix}_n \qquad \text{(XIV-76)}$$

$$\text{tr } \hat{\alpha} \hat{\chi} = \mathbf{C}_1 \mathbf{C}_2 \cdots \mathbf{C}_n \qquad \text{(XIV-77)}$$

The conformational average, $\langle \text{tr } \hat{\alpha} \hat{\chi} \rangle_0$ ($\equiv \text{tr} \langle \hat{\alpha} \hat{\chi} \rangle_0$), is generated by the usual combination of $\mathbf{C}_i$ with $\mathbf{U}_i$.

PROBLEMS

Appendix B contains the answers for Problems XIV-1 and XIV-2.

XIV-1. What is the limit for $\langle \mu^2 \rangle / 2m^2$ for Cl—(CH$_2$)$_n$—Cl as $n \to \infty$? At what value of n is the result within 5% of this limit?

XIV-2. The $\mathbf{m}$ for the C—Cl bond highlighted in boldface should be incorporated into the $\mathbf{D}$ matrix for which of the C—C bonds in –CHCl–CH$_2$**CCl**HCH$_2$–CHCl–?

XIV-3. Which of the polymers described in Table VII-3 has the largest D_∞?

XIV-4. Which of the polymers described in Table VII-5 is most likely to have D_∞ near 1?

XIV-5. If a fraction f of the C—Cl bonds in poly(vinyl chloride) of high molecular weight is replaced at random with C—H bonds, how do $\langle \mu^2 \rangle_0$ and D depend on f? Does the result also depend on p_m?

XIV-6. Assume that $\hat{\alpha}_i$ for the C—C and C—H bonds in polyethylene are cylindrically symmetric, with the differences in the polarizabilities along the axes denoted by $\Delta \alpha_{CC}$ and $\Delta \alpha_{CH}$, respectively. Formulate

$\hat{\alpha}_i$ for an internal methylene unit in a linear polyethylene, including the contribution from the two C—H bonds.

XIV-7. Consider a simple chain, poly(A), in which each $\hat{\alpha}_i$ is

$$\hat{\alpha}_i = \mathrm{diag}\left(\Delta\alpha_{AA}, \frac{-\Delta\alpha_{AA}}{2}, \frac{-\Delta\alpha_{AA}}{2} \right)_i$$

with bonds subject to a symmetric threefold rotation potential-energy function, pairwise interdependencies, and rotational isomers with $\phi_i =$ 180°, $\pm\phi_g$. For chains long enough so that end effects are not important, how should θ_i, ϕ_g, σ, and ω be chosen so that the stress-optical coefficient will be

(a) Large and positive?

(b) Large and negative?

(c) Close to zero?

XIV-8. Consider a poly(A—B), in which each $\hat{\alpha}_i$ is

$$\hat{\alpha}_i = \mathrm{diag}\left(\Delta\alpha_{AB}, \frac{-\Delta\alpha_{AB}}{2}, \frac{-\Delta\alpha_{AB}}{2} \right)_i$$

and each $\mathbf{m}_i$ is

$$\mathbf{m}_i = \begin{bmatrix} \pm m \\ 0 \\ 0 \end{bmatrix}_i$$

with bonds subject to a symmetric threefold rotation potential-energy function, pairwise interdependencies, and rotational isomers with $\phi_i =$ 180°, $\pm\phi_g$. For chains long enough so that end effects are not important, how should θ_{ABA}, θ_{BAB}, ϕ_g, σ, ω_{AA}, and ω_{AB} be chosen so that $\langle \boldsymbol{\mu}^T \hat{\alpha} \boldsymbol{\mu} \rangle_0$ will be

(a) Large and positive?

(b) Large and negative?

(c) Close to zero?

XIV-9. For the chain in Problem XIV-8, take $m = 1$ Debye and $\Delta\alpha_{AB} = 1$ Å^3. Which of the two terms in Eq. (II-48) is likely to be the more important, the one with $\langle \boldsymbol{\mu}^T \hat{\alpha} \boldsymbol{\mu} \rangle_0$ or the one with $\langle \mathrm{tr}\, \hat{\alpha}^2 \rangle_0$?

XIV-10. Give the dimensions of the matrices required for polyoxyethylene in order to calculate the correlation coefficients for

(a) r^2 and μ^2 (b) r^2 and γ^2 (c) r^2 and $\boldsymbol{\mu}^T \hat{\alpha} \boldsymbol{\mu}$

(d) γ^2 and $\mathbf{r}^T \hat{\alpha} \mathbf{r}$ (e) $\mathbf{r}^T \hat{\alpha} \mathbf{r}$ and $\boldsymbol{\mu}^T \hat{\alpha} \boldsymbol{\mu}$

XV Additional Information from Statistical Weight Matrices

This chapter collects several miscellaneous topics that require additional modifications in the formulation of Z, or manipulations of statistical weight matrices. The emphasis in previous chapters has been on the treatment of chains under conditions where the conformation is controlled by short-range interactions. A few interactions of moderate to long range were addressed, most notably in conjunction with the intramolecular antiparallel sheet $\rightleftharpoons$ coil transition in Chapter X. More general methods exist for incorporation of the influence of *long-range interactions*. These methods are described in the first section of the present chapter.

In media that are immobile on the time scale defined by the lifetime of a singlet excited state (the nanosecond time scale for many chromophores), the *fluorescence* of a system with multiple chromophores is sensitive to their spatial distribution in a readily analyzable fashion. If those chromophores are all attached to a chain molecule, the knowledge of the distribution of conformations contained in Z has implications for the fluorescence. That topic is discussed here with respect to intramolecular excimer emission and nonradiative singlet energy transfer (Förster transfer).

In conjunction with the discussion of vinyl polymers in Chapter VIII, it was noted that the dependence of Z on stereochemical composition implied that chains with various stereochemical sequences would occur in different populations if they could be equilibrated with one another. That process, known as *epimerization to stereochemical equilibrium*, is also treated in this chapter.

Introduction of time into the conformational partition function permits examination of local intramolecular dynamics using a methodology that is extremely efficient with regard to computation. This branch of RIS, know as *dynamic rotational isomeric state theory*, is introduced in the final section of the chapter.

1. INCORPORATION OF LONG-RANGE INTERACTIONS

Most of the rotational isomeric state models described in previous chapters have employed first- and second-order interactions. Interactions of longer range are not required for successful interpretation of conformation-dependent physical properties of many flexible chains under Θ conditions. Exceptions exist, of course, as exemplified by specific interactions of longer range in the treatments of the helix $\rightleftharpoons$ coil and sheet $\rightleftharpoons$ coil transitions in Chapter X. However, in no case was there an attempt to incorporate all of the long-range interactions into the description of the conformations of the chain.

These long-range interactions have been incorporated in a few instances by two different methods, both of which are described here. The principal applications of both lies in the generation of representative chains by Monte Carlo methods. The methods differ with regard to whether they change the weight of a chain after it has been generated, or modify the conditional probabilities as the chain is being generated.

A. Reweighting for Incorporation of Long-Range Interactions

The last section in Chapter VI described a Monte Carlo method by which a sample of $\mathcal{N}$ representative unperturbed chains can be generated using the $p_{\eta;2}$ and $q_{\xi\eta;i}$, $2 < i < n$, determined by first- and second-order interactions. If the energy of the interaction between atoms A_i and A_j is denoted by E_{ij}, the short-range energies, E_{SR}, considered in that Monte Carlo process are represented by

$$E_{SR} = \sum_{i=0}^{n-j} \sum_{j=3}^{4} E_{i,i+j} \tag{XV-1}$$

The total energy, contributed by all pairs, the separation of which is not fixed by the assumption of constant bond lengths and bond angles, is

$$E = \sum_{i=0}^{n-j} \sum_{j=3}^{n} E_{i,i+j} \tag{XV-2}$$

Let us define the energy of the long-range interactions, E_{LR}, as $E - E_{SR}$:

$$E_{LR} \equiv E - E_{SR} = \sum_{i=0}^{n-j} \sum_{j=5}^{n} E_{i,i+j} \tag{XV-3}$$

More generally, E_{LR} is defined as all pairwise interaction energies that were not included in the probabilities deduced from Z, and used in the generation of the representative sample of $\mathcal{N}$ unperturbed chains.

The term $\mathcal{N}$ is the number of representative unperturbed chains in the sample. Every one of these chains is weighted equally in the extraction of the averages of unperturbed conformation-dependent properties from the sample, as in Eq. (VI-88) for $\langle r^{2p} \rangle_{0,MC}$. For the second moment, we obtain

$$\langle r^2 \rangle_{0,\text{MC}} = \frac{1}{\mathcal{N}} \sum_{m=1}^{\mathcal{N}} r_m^2 \tag{XV-4}$$

where m indexes the $\mathcal{N}$ representative unperturbed chains. The estimate of the effect of the long-range interactions on the average conformation-dependent physical properties *for exactly the same set of chains* can be generated by

$$\langle r^2 \rangle_{\text{MC}} = \frac{1}{\sum_{m=1}^{\mathcal{N}} w_m} \sum_{m=1}^{\mathcal{N}} w_m r_m^2 \tag{XV-5}$$

where the w_m are the weighting factors that arise from the imposition of the long-range interactions. The unperturbed sample is recovered when all $w_m = 1$. The energy of the long-range interactions was specified in Eq. (XV-3), and

$$w_m = \exp \frac{-E_{m,\text{LR}}}{kT} \tag{XV-6}$$

The energies appearing in Eq. (XV-6) should be appropriate for the environment in which the macromolecule is found; that is, they should be the appropriate potential of mean force with respect to the solvent and chemical structure, $\hat{U}(\{\phi\})$, Eq. (III-4).

Often very simple approximations are used for $\hat{U}(\{\phi\})$ in order to obtain an estimate of how a property might respond to the excluded volume effect. One of the simplest approximations is that each term in Eq. (XV-3) contributes a specified nonzero energy if $r_{ij} < r_{\text{max}}$, where r_{max} is some cutoff, and contributes nothing if $r_{ij} > r_{\text{max}}$. If the specified energy divided by kT is represented by E', and n_m is the number of "long-range" r_{ij} that fall below the cutoff in conformation m, we have

$$w_m = \exp(-E' n_m) \tag{XV-7}$$

Having evaluated n_m for all of the conformations in the sample, the strength of the excluded volume effect can be manipulated, within the framework of the simple model, by changing E'. Expansion of $\langle r^2 \rangle$ for the chains in the sample will be obtained with $E' > 0$, and $E' < 0$ will produce collapse of $\langle r^2 \rangle$ for the same chains. The atoms behave as hard spheres, insofar as the long-range interactions are concerned, when $E' = \infty$.

This approach has been used to examine how $\langle \mu^2 \rangle$ responds to the excluded volume effect in RIS models for chains with a finite number of bonds,[1] and the

[1]Mattice, W. L.; Carpenter, D. K. *Macromolecules* **1984**, *16*, 625. Mattice, W. L. *J. Chem. Phys.* **1987**, *87*, 5512; *Macromolecules* **1988**, *21*, 3320.

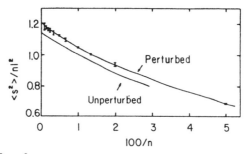

Figure XV-1. $\langle s^2 \rangle / nl^2$ for unperturbed polyethylene chains, and for polyethylene chains attached at one end to a flat impenetrable surface. Reprinted with permission from Mattice, W. L.; Napper, D. H. *Macromolecules* **1981**, *14*, 1066. Copyright 1981 American Chemical Society.

manner in which the expansion factor for a subchain depends on its location within the main chain.[2] The $\mathscr{G}_i$ for computation of $\langle r^2 \rangle_0$ and $\langle r_{ij}^2 \rangle_0$ for unperturbed polyethylene can be modified so that it mimics, in a rapid computation, the values of $\langle r^2 \rangle$, and the position and size dependence of the $\langle r_{ij}^2 \rangle$, when the chain is subject to the excluded volume effect.[2]

The method is easily adapted to the incorporation of interactions that arise from sources other than the interaction between the atoms in the chain. It has been employed for modification of the conformations in the sample so that it places the chains in a restrictive environment, as when one end of the chain is attached to a flat impenetrable interface.[3] This perturbation produces a small expansion of the mean-square radius of gyration, as depicted in Fig. XV-1. It produces more impressive changes in the x component (perpendicular to the surface) of $\langle \mathbf{r} \rangle$ and $\langle \mathbf{g} \rangle$, both of which increase without limit, with $r_x > g_x$, as depicted in Fig. XV-2.

Care must be exercised if the environment becomes so restrictive that only a few, or only one, of the conformations *from the RIS model that described the unperturbed chain* can meet the restriction. An example is provided by the confinement of the chain by a very narrow channel. In severely constrained environments, degrees of freedom that could be ignored in the model for the unperturbed chain may now become of importance.[4] Exclusion of a consideration of these fluctuations when only a single conformation from the usual RIS model will meet the restrictions is legitimate only if the rotational isomers are extremely well defined, such that ϕ cannot depart from the discrete set used in the formulation of Z, and if θ is also confined precisely to the value used in the model. For a chain such as poly(1,4-*trans*-butadiene), where the torsion

[2]Mattice, W. L. *Macromolecules* **1981**, *14*, 1491.

3Mattice, W. L.; Napper, D. H. *Macromolecules* **1981**, *14*, 1066.

[4]A similar situation arises in the treatment of unconfined "stiff" chains, such as poly(L-proline) and poly(*cis*-benzobisoxazole), as described in Chapter IX, where fluctuations that are not considered in the formulation of Z for the usual unperturbed chains become of vital importance.

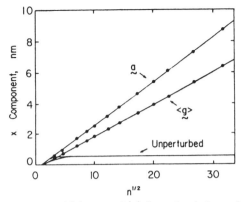

Figure XV-2. x components of $\langle \mathbf{r} \rangle \equiv \mathbf{a}$ and $\langle \mathbf{g} \rangle$ for polyethylene chains attached at one end to a flat impenetrable surface, compared with the same components for the unperturbed chains. Reprinted with permission from Mattice, W. L.; Napper, D. H. *Macromolecules* **1981**, *14*, 1066. Copyright 1981 American Chemical Society.

potential energy function for the CH—CH$_2$ bond has rather shallow minima at the $s^{\pm}$ states, it is imperative to include the fluctuations if the polymer is to be successfully modeled in a narrow channel.[5]

The method is easily adapted to the examination of the consequences of the interaction of selected A_i and A_j, $j \ll i$, in the unrestricted chain. If the only atoms included in the "long-range" interaction are A_0 and A_n, and E' for the interaction of these two atoms becomes increasingly negative, the conformation-dependent properties deduced from Eq. (XV-5) are driven toward those expected for macrocycles.

With the use of somewhat more computational resources, E' can be replaced by a function that depends on r_{ij} in a continuous manner, as in the expansion of star-like polyethylenes with similarly charged ends.[6] Here the only terms included in E_{LR} are those arising from the repulsive interaction of the groups at the ends. The energy of this interaction was taken to be inversely proportional to r_{0n}.

B. Modification of the $q_{\xi\eta}$ by Incorporation of Long-Range Interactions

The previous section used the $p_{\eta;2}$ and $q_{\xi\eta;i}$, $2 < i < n$, to generate representative unperturbed chains, and then modified the weights of these chains by incorporating additional interactions. In the present section, an approximation is employed where the method instead alters the $q_{\xi\eta;i}$ as a representative chain is grown, so that they reflect the consequences of the interaction of the new atom,

[5] Dodge, R.; Mattice, W. L. *Macromolecules* **1991**, *24*, 2709; Zhan, Y.; Mattice, W. L. *Macromolecules* **1992**, *25*, 1554; *J. Chem. Phys.* **1992**, *96*, 3279.
[6] Mattice, W. L.; Skolnick, J. *Macromolecules* **1981**, *14*, 1463.

A_{i+1}, being added to the chain with atoms that were placed earlier, but that do not participate in short-range interactions (those interactions incorporated in Z) with A_{i+1}. When the treatment uses first- and second-order interactions, the additional interactions that must be considered are those of A_{i+1} with A_j, $0 \leq j < i - 3$. This approach was developed in conjunction with the generation of representative chains packed in an amorphous glass at bulk density.[7]

When a chain is generated by assignment of consecutive torsion angles from ϕ_2 through ϕ_{n-1}, the assignment of all except ϕ_2 is done with absolute knowledge of the conformation of the preceding bond. The assignment of the conformation at bond i ($1 < i < n$) establishes the position of A_{i+1} and the substituents (if any) bonded directly to A_i. When i is in the range $2 < i < n$, the ensemble is unperturbed, and the bonds are subject to pairwise interdependencies, the $q_{\xi\eta;i}$ defined by Eq. (V-36) contain the information necessary for the correct assignment of the state at bond i. This information is insufficient if the chain is subject to interactions of longer range than those incorporated in the Z that was used for generation of the $q_{\xi\eta;i}$. A remedy would be to modify the $p_{\eta;2}$ and all of the $q_{\xi\eta;i}$, $2 < i < n$, so that each incorporates the consequences of all of the longer-range interactions that were not incorporated in Z. An approximation to this procedure is, at each stage in the generation of the representative chain, to modify $q_{\xi\eta;i}$ so that it incorporates the consequences of the long-range interactions of A_{i+1} and the substituents (if any) on A_i with the portion of the chain that has already been generated. The modified conditional probabilities, denoted by $q'_{\xi\eta;i}$, are calculated as

$$q'_{\xi\eta;i} = q_{\xi\eta;i} \left[\frac{\exp(-E_{\eta;i;\text{LR}}/RT)}{\sum_{\eta_i} q_{\xi\eta;i} \exp(-E_{\eta;i;\text{LR}}/RT)} \right] \qquad \text{(XV-8)}$$

where ξ is the known state about bond $i - 1$, and $E_{\eta;i;\text{LR}}$ (see also Fig. XV-3) is the interaction of A_{i+1} and the substituents (if any) on A_i with all atoms already placed in the representative chain, the interactions of which with A_{i+1} and the substituents on A_i was not incorporated into Z as a first- or second-order interaction. The $q_{\xi\eta;i}$ of the unperturbed chains are recovered if all of these long-range interactions are of exactly the same size in the assignment of each of the ν_i potential rotational isomeric states at bond i. However, if $E_{\eta;i;\text{LR}}$ depends on the assignment of η_i, the $q'_{\xi\eta;i}$ will bias the selection of η_i away from those conformations where $E_{\eta;i;\text{LR}}$ is most repulsive. If $E_{\eta_a;i;\text{LR}} - E_{\eta_b;i;\text{LR}} \gg kT$, state η_a is completely supressed at bond i in the representative chain being generated, even though it might have been allowed in that chain if the chain were unperturbed.

A consequence of this method is that a directionality in the conformational preferences is imposed by the method itself, even when no such bias is required

[7]Theodorou, D. N.; Suter, U. W. *Macromolecules* **1985**, *18*, 1467.

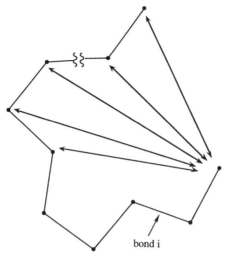

Figure XV-3. Interactions that contribute to the $E_{\eta;i;\mathrm{LR}}$ in Eq. (XV-8).

by the fundamental physics. The origin of this directionality can be seen by consideration of the application to a polyethylene chain with very large n, and identical groups at the ends. The $p_{\eta;2}$ and $q_{\xi\eta;i}$, $2 < i < n$, for the unperturbed chain will produce an ensemble of representative chains in which the ends remain indistinguishable. Average conformation-dependent properties evaluated for the subchain consisting of bonds 1 through j are the same as those evaluated for the subchain consisting of bonds $n - j + 1$ through n. The perturbation produced by the excluded volume effect does not produce a distinction between the two ends, but instead modifies in an identical manner the average conformation-dependent properties of the subchains of bonds 1 through j and bonds $n - j + 1$ through n. However, application of Eq. (XV-8) causes the two ends to become distinct. The end that is generated first "feels" the perturbation more weakly than does the end that is generated last. If, for example, one focuses on the third bond from each end of the chain, the unmodified $q_{\xi\eta;3}$ are used to assign the state at bond 3, but the modified (perhaps strongly modified) $q'_{\xi\eta;n-2}$ are used to assign the state at bond $n - 2$.

The principal use of Eq. (XV-8) has been in the generation of the "initial guess" of chains packed at amorphous bulk density. The "initial guess" is then refined by energy minimization, sometimes accompanied by molecular dynamics, in order to relax strongly repulsive contacts.

2. EXCIMERS AND FÖRSTER TRANSFER

Rotational isomeric state theory as it has been described in previous chapters can be employed for rationalization of fluorescence measurements provided two requirements are met. Since RIS theory provides a description of the static conformations of an unperturbed sample, this method will not be appropriate if the fluorescence is responding to the *rates* of interconversion of the conformations in the ensemble. The viscosity of the medium must be sufficiently high that these conformational transitions are suppressed on the time scale of the experiment. That time scale is determined by the fluorescence lifetime of the chromophore. Even when the molecular framework, as defined by the nuclei, is static, there is the possibility that the excitation itself might migrate along the chain, from a chromophore bonded to A_i to a chemically identical chromophore bonded to A_{i+j}. This *energy migration* along the chain is also assumed to be absent in the usual applications of RIS theory to the rationalization of fluorescence spectra.

Given the requirements presented above, the probability for formation of an intramolecular excimer by two chromophores at the ends of the chain can be assessed by a simple modification of the analysis of macrocyclization equilibrium. The problems can be made similar by consideration of Fig. XV-4. The description of the chain in Fig. XV-4 defines positions for three nonexistent atoms, in order to convert the problem into one of macrocyclization equilibrium. These three "atoms" are A_0, A_1, and A_n. Atoms A_1 and A_n are at the centers of the two chromophores, which in this case are the benzene rings. Bonds 2 and n are virtual bonds that extend between the chain atom to which the benzene ring is attached and A_1 or A_n. The zeroth atom, A_0 is placed at a site that should be occupied by A_n if the two rings are to form an excimer. There are

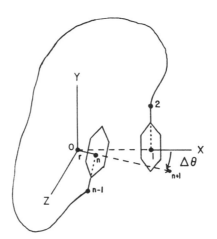

Figure XV-4. Depiction of a polymer with chromophores (drawn as hexagons) at its ends. The chromophores are close to the conformation required for formation of an excimer. Reprinted with permission from Pannikottu, A.; Mattice, W. L. *Macromolecules* **1990**, *23*, 867. Copyright 1990 American Chemical Society.

two possible sites for A_0, on either side of the plane described by the benzene ring that contains A_1. The length of l_1 is ~350 pm, and θ_1 (i.e., the angle $\angle A_0-A_1-A_2$) is 90°, so that the direction of $\mathbf{l}_1$ is perpendicular to the benzene ring. If all bonds in the chain are subject to symmetric rotation potential energy functions, excimer formation is equally probable at the two sites, and it will suffice to evaluate the probability for one of them. If the bonds are not subject to symmetric rotation potential energy functions, it may be necessary to evaluate separately the probabilities for both sites.

With the definition of A_0, A_1, and A_n as in Fig. XV-4, the evaluation proceeds via $W_n(0)$ with the Hermite expansion, if it is sufficient to simply examine the probability that A_n is superimposed on A_0, or via $\Gamma_0(1)$ if one also wishes to enforce the proper angle between $\mathbf{l}_1$ and $\mathbf{l}_n$, as in Chapter XIII. The incorporation of $\Gamma_0(1)$ is important for end-labeled alkanes containing fewer than 20 CH_2—CH_2 bonds.[8]

When the planar chromophore is described by an ellipse, rather than a circle, and the flexible spacer is short, it may become desirable to include also a term[9] to account for the preferred values of the "torsion angle" defined by $\mathbf{l}_n$, $\mathbf{l}_1$, and $\mathbf{l}_2$. For short chains, this exercise is most easily accomplished by either a discrete enumeration of all of the rotational isomeric states or a sampling of those states by Monte Carlo methods.

The treatment of the efficiency of Förster transfer is best done by Monte Carlo methods, where in each representative chain the efficiency of Förster transfer can be evaluated from the separation of the chromophores and R_0 via Eq. (II-63), or from their separation, relative orientation (and hence κ^2), and Eq. (II-64) in conjunction with Eq. (II-63).

3. EPIMERIZATION TO STEREOCHEMICAL EQUILIBRIUM

The conformational partition functions for vinyl polymers were introduced in Chapter VIII. In general, Z was observed to depend on the stereochemical sequence of the diads in the chain. The applications envisioned in Chapter VIII were to systems in which the stereochemical sequence of each chain was fixed, such that the chain was characterized by a single value of Z. Systems for which $0 < p_m < 1$ were studied by generation of a representative set of chains, each with its own stereochemical sequence and Z. If the interest was in C_∞, $\langle r^2 \rangle_0$ was calculated for each of the representative chains, and the results from the representative chains were averaged to obtain the characteristic ratio.

In the present application, the system contains a catalyst that rapidly interconverts *meso* and *racemo* diads, but without otherwise affecting the properties of the system. The Z for the individual chains will then determine the stereochemical sequences observed after equilibration of the diads. The system is

[8]Pannikottu, A.; Mattice, W. L. *Macromolecules* **1990**, *23*, 867.
[9]Flory, P. J.; Suter, U. W.; Mutter, M. *J. Am. Chem. Soc.* **1976**, *98*, 5733.

described by a new partition function, denoted by $\mathscr{Z}$, that is the sum of the Z for all stereochemical sequences.[10]

A. The Sum of Partition Functions, $\mathscr{Z}$

Let the conformational partition function for a vinyl polymer with x units be written as

$$Z = \mathbf{U}_1 \mathbf{U}_1^{(2)} \mathbf{U}_2^{(2)} \cdots \mathbf{U}_{x-1}^{(2)} \mathbf{U}_n \tag{XV-9}$$

where each $\mathbf{U}_k^{(2)}$ is the product of two statistical weight matrices, for the two bonds between successive C^α. For the case where the chain is a homopolymer, the internal $\mathbf{U}_k^{(2)}$ have either of two structures:

$$\mathbf{U}_m^{(2)} = \mathbf{U}_p \mathbf{U}_m \tag{XV-10}$$

$$\mathbf{U}_r^{(2)} = \mathbf{U}_p \mathbf{U}_r \tag{XV-11}$$

The structures of $\mathbf{U}_1^{(2)}$ and $\mathbf{U}_{x-1}^{(2)}$ will depend on the nature of the end groups.[10,11]

If each of the $x-1$ diads can be *meso* or *racemo*, there are 2^{x-1} stereochemical sequences accessible to the chain. The sum of the Z values for these 2^{x-1} stereochemical sequences is denoted by $\mathscr{Z}$, which is generated as

$$\mathscr{Z} = \mathbf{U}_1 \mathscr{U}_1 \mathscr{U}_2 \cdots \mathscr{U}_{x-1} \mathbf{U}_n \tag{XV-12}$$

$$\mathscr{U}_k = \mathbf{U}_m^{(2)} + \mathbf{U}_r^{(2)} \tag{XV-13}$$

For a homopolymer, all $\mathscr{U}_k$ with $1 < k < x - 1$ are identical. For methyl-terminated polypropylene, $\mathscr{U}_1$ and $\mathscr{U}_{x-1}$ require modification.[10]

B. Stereochemical Composition at Equilibrium

The stereochemical composition of the chain at equilibrium is extracted from $\mathscr{Z}$ by methods that are similar to those used to obtain the probability for the occupancy of rotational isomeric states from Z. Those methods were described in Chapter V. The probability for a *meso* diad, p_m, is obtained by a method

---

[10]Flory, P. J. *J. Am. Chem. Soc.* **1967**, *89*, 1798. Suter, U. W.; Flory, P. J. *Macromolecules* **1975**, *8*, 765.

[11]If the chain is terminated by groups equivalent to R, as in methyl-terminated polypropylene, $\mathbf{U}_1^{(2)}$ and $\mathbf{U}_{x-1}^{(2)}$ each has a different structure, because that unit does not contain an asymmetric carbon atom. In the formalism developed here, it is assumed that R is distinguishable from the terminating groups. Statistical weights for second-order interactions must be suppressed in $\mathbf{U}_2$.

similar to the one used to generate p_η, Eq. (V-33). The crucial matrix for generation of p_η was a $2\nu \times 2\nu$ matrix denoted by $\hat{\mathbf{U}}_{\eta;i}$. Here the crucial role is also played by a $2\nu \times 2\nu$ matrix, denoted by $\hat{\mathcal{U}}_{m;k}$:

$$p_m = \frac{1}{x-1} \frac{1}{\mathcal{Z}} [\mathbf{U}_1 \quad \mathbf{0}] \hat{\mathcal{U}}_{m;1} \cdots \hat{\mathcal{U}}_{m;x-1} \begin{bmatrix} \mathbf{0} \\ \mathbf{U}_n \end{bmatrix} \qquad \text{(XV-14)}$$

where $\hat{\mathcal{U}}_{m;k}$ is defined as

$$\hat{\mathcal{U}}_{m;k} = \begin{bmatrix} \mathcal{U} & \mathbf{U}_m^{(2)} \\ \mathbf{0} & \mathcal{U} \end{bmatrix}_k \qquad \text{(XV-15)}$$

The block in the upper right of $\hat{\mathcal{U}}_{m;k}$ contains all of the statistical weights that are appropriate if this diad is *meso*, but rejects the contributions from chains in which this diad is *racemo*. The $\mathbf{U}'_{\eta;i}$ in the upper right of the blocked form of $\hat{\mathbf{U}}_{\eta;i}$ played a similar role in the calculation of p_η in Chapter V, retaining the statistical weights called on when bond i is in state η, but rejecting the contributions in which bond i was in another state. The probability for a *racemo* diad is extracted from the same expression, with the block in the upper right of the $2\nu \times 2\nu$ matrix replaced by $\mathbf{U}_r^{(2)}$. It is also obtained as $1 - p_m$.

The generation of the probabilities for triads utilizes the same concept, but the matrices become larger. The probability for an isotactic (*meso, meso*) triad, denoted by p_{mm}, is

$$p_{mm} = \frac{1}{x-2} \frac{1}{\mathcal{Z}} [\mathbf{U}_1\mathbf{U}_{m;1}^{(2)} \quad \mathbf{U}_1\mathbf{U}_{r;1}^{(2)} \quad \mathbf{0} \quad \mathbf{0}] \hat{W}_{mm;2} \cdots \hat{W}_{mm;x-1} \begin{bmatrix} \mathbf{0} \\ \mathbf{0} \\ \mathbf{U}_n \\ \mathbf{U}_n \end{bmatrix}$$

$$\text{(XV-16)}$$

$$\hat{W}_{mm;k} = \begin{bmatrix} \mathbf{U}_m^{(2)} & \mathbf{U}_r^{(2)} & \mathbf{U}_m^{(2)} & \mathbf{0} \\ \mathbf{U}_m^{(2)} & \mathbf{U}_r^{(2)} & \mathbf{0} & \mathbf{0} \\ \mathbf{0} & \mathbf{0} & \mathbf{U}_m^{(2)} & \mathbf{U}_r^{(2)} \\ \mathbf{0} & \mathbf{0} & \mathbf{U}_m^{(2)} & \mathbf{U}_r^{(2)} \end{bmatrix}_k \qquad \text{(XV-17)}$$

For heterotactic and syndiotactic triads, the substitutions for $\hat{W}_{mm;k}$ are

$$
\hat{W}_{mr,k} = \begin{bmatrix} U_m^{(2)} & U_r^{(2)} & 0 & U_r^{(2)} \\ U_m^{(2)} & U_r^{(2)} & U_m^{(2)} & 0 \\ 0 & 0 & U_m^{(2)} & U_r^{(2)} \\ 0 & 0 & U_m^{(2)} & U_r^{(2)} \end{bmatrix}_k
$$
(XV-18)

$$
\hat{W}_{rr,k} = \begin{bmatrix} U_m^{(2)} & U_r^{(2)} & 0 & 0 \\ U_m^{(2)} & U_r^{(2)} & 0 & U_r^{(2)} \\ 0 & 0 & U_m^{(2)} & U_r^{(2)} \\ 0 & 0 & U_m^{(2)} & U_r^{(2)} \end{bmatrix}_k
$$
(XV-19)

with $p_{mm} + p_{mr} + p_{rr} = 1$.

The procedure described above for triads can be extended to longer sequences.[12] It is convenient to write matrices of the types presented in Eqs. (XV-17)–(XV-19) as

$$
\hat{W}_q = \begin{bmatrix} W & W'_q \\ 0 & W \end{bmatrix}
$$
(XV-20)

where q denotes the stereochemical sequences for which the probability will be calculated. For a sequence of y units ($y = 3$ for the triads that were the subject of the previous paragraph)

$$
W = \begin{bmatrix} 1 \\ 1 \end{bmatrix} \otimes I_{y-2} \otimes [U_m^{(2)} \quad U_r^{(2)}]
$$
(XV-21)

Its dimensions are $2^{y-2}\nu \times 2^{y-2}\nu$. The prime appended to the last block in the top row of $\hat{W}_q$ denotes a modification of W in which selected blocks have been replaced by 0, as illustrated in Eq. (XV-17), (XV-18), and (XV-19) for the cases where q specifies selected triads. This usage of the prime is analogous to its use in U'_η in Chapter V. The probability of sequence q is

$$
P_q = \frac{1}{x - y + 1} \frac{1}{\mathscr{Z}} [U_1 \quad 0] [W_1 W_2 \cdots W_{y-2} \quad 0] \hat{W}_q^{x-y-1} \begin{bmatrix} 0 \\ \vdots \\ 0 \\ U_n \\ \vdots \\ U_n \end{bmatrix}
$$
(XV-22)

[12] Suter, U. W. *Macromolecules* **1980**, *14*, 523.

The q for the desired sequence is represented by a number ψ in binary form, using 0 for m and 1 for r, so that the pentad $rmrm$ is denoted by $\psi = 1010_{(2)} = 10_{(10)}$. This number defines two natural numbers by the equations

$$g_R = 1 + \left\{ \frac{\psi}{2} \right\} \tag{XV-23}$$

$$g_C = 1 + (\psi \text{ modulo } 2^{y-2}) \tag{XV-24}$$

where $\{a\}$ denotes the largest natural number not exceeding a, and a modulo $b \equiv a - b\{a/b\}$ is the remainder of a after division by b. Define two matrices of order $2^{y-2} \times 2^{y-2}$ as

$$\mathbf{V}_R = \text{diag}(\delta_{ig_R}) \equiv [\delta_{ij}\delta_{ig_R}] \tag{XV-25}$$

$$\mathbf{V}_C = \text{diag}(\delta_{ig_C}) \equiv [\delta_{ij}\delta_{ig_C}] \tag{XV-26}$$

where δ is the Kronecker delta ($\delta_{kl} = 1$ if $k = l$, and is zero otherwise). Then W'_q is given by

$$W'_q = (\mathbf{V}_R \otimes \mathbf{I}_\nu)W(\mathbf{V}_C \otimes \mathbf{I}_\nu) \tag{XV-27}$$

using the expression for W in Eq. (XV-21). The expression for W'_q that is used in Eq. (XV-22) must be the sum of the W'_q obtained by two applications of Eq. (XV-27) if q implicitly comprises two different stereochemical sequences, as in $mmmr$ (includes $rmmm$). This procedure applies to triads and to all longer sequences. Explicit expressions for the W'_q required for all tetrads and pentads have been given by Suter.[12]

Application of this method to polystyrene[12] yields values of p_m, p_{mm}, p_{mr}, and p_{rr} that are within 0.02 of the experimental results for the polymer at 100°C.[13] A favorable comparison between experiment and theory has also been obtained for polypropylene, where the comparison can be extended to pentads, but the treatment of the chain ends is slightly different than that envisoned in the equations presented above.[14] The terminal groups and the side chains are both methyl groups in polypropylene, which reduces the number of pseudoasymmetric centers by two.

[13] Shepherd, L.; Chen, T. K.; Harwood, H. J. *Polym. Bull.* **1979**, *1*, 445.
[14] Suter, U. W.; Neuenschwander, P. *Macromolecules* **1980**, *14*, 528.

4. INCORPORATION OF TIME INTO Z

All applications of RIS theory described in previous sections have dealt with ensemble averages for the static properties of chains at equilibrium. The RIS formalism can also be used to examine the local intrachain dynamics of the chains. The procedure that allows the adaptation of RIS theory for the description of the dynamics of the transitions between rotational isomeric states was initially introduced by Jernigan,[15] and extended by Bahar and coworkers.[16–19]

The version described here is computationally efficient, and is developed in a manner that presents close analogies with the classic RIS formalism introduced in Chapters IV–VI. The focus on the extraction from Z of $p_{\xi\eta;i}$ in the classic treatment is replaced by the evaluation of $p_{\xi\eta;\xi^0\eta^0;i}(\tau)$, which denotes the joint probability that bonds $i-1$ and i will occupy states $\xi^0\eta^0$ at time 0, and states $\xi\eta$ a time τ later. The geometry of the chain can be incorporated to calculate orientation autocorrelation functions, defined as

$$F_j(\tau) = \langle \mathbf{m}_j^T(0)\mathbf{m}_j(\tau)\rangle \qquad \text{(XV-28)}$$

$$G_j(\tau) = \tfrac{3}{2}\langle [\mathbf{m}_j^T(0)\mathbf{m}_j(\tau)]^2\rangle - \tfrac{1}{2} \qquad \text{(XV-29)}$$

where $\mathbf{m}_j$ is a unit vector rigidly affixed to bond j in the chain. The angle brackets in Eqs. (XV-28) and (XV-29) denote the ensemble average over time-delayed joint states observed with an interval of time τ. All of the initial and final states for the rotatable bonds in the chain must be considered in the calculation of this average. Dielectric relaxation depends on $F_j(\tau)$, and $G_j(\tau)$ is involved in nmr, esr (electron spin resonance), and fluorescence depolarization experiments. We will formulate the dynamic RIS model so that it can be used to calculate $\langle \mathbf{m}_j^T(0)\mathbf{m}_j(\tau)\rangle$ and $\langle [\mathbf{m}_j^T(0)\mathbf{m}_j(\tau)]^2\rangle$, and thereby $F_j(\tau)$ and $G_j(\tau)$.

A. The Transition Partition Function and Joint Probabilities

The conformational partition function for a chain of n pairwise interdependent bonds is written as

$$Z = \mathbf{U}_1\mathbf{U}_2\mathbf{U}_3 \cdots \mathbf{U}_{n-1}\mathbf{U}_n \qquad \text{(XV-30)}$$

[15] Jernigan, R. L. in *Dielectric Properties of Polymers* Karasz, F. E., ed., Plenum, New York, **1972**, p. 99.
[16] Bahar, I.; Erman, B. *Macromolecules* **1987**, *20*, 1368.
[17] Bahar, I.; Erman, B. *J. Chem. Phys.* **1988**, *33*, 1228. Bahar, I.; Erman, B.; Monnerie, L. *Macromolecules* **1989**, *22*, 431; *Macromolecules* **1989**, *22*, 2396; *Adv. Polym. Sci.*, **1994**, *116*, 145.
[18] Bahar, I. *J. Chem. Phys.* **1989**, *91*, 6525.
[19] Bahar, I.; Mattice, W. L. *Macromolecules* **1990**, *23*, 2719.

The dynamic counterpart, the *transition partition function* denoted by $Z(\tau)$, is written in an analogous fashion in Eq. (XV-31) for a given time interval, τ, for a chain of n bonds subject to pairwise interdependencies:

$$Z(\tau) = \mathbf{V}_1 \mathbf{V}_2(\tau) \mathbf{V}_3(\tau) \cdots \mathbf{V}_{n-1}(\tau) \mathbf{V}_n \qquad (\text{XV-31})$$

The terminal row and column vectors, $\mathbf{V}_1$ and $\mathbf{V}_n$, respectively, contain ν^2 elements, all of which have the value of 1. The time dependence resides in the matrices for the internal bonds. The diagonal matrix for the second bond is formulated from $p_{\eta;\eta^0;2}(\tau)$, which is the joint probability that bond 2 will be in state η^0 at time 0, and in state η a time interval τ later, evaluated in the approximation that the bonds are independent. For a chain with a symmetric threefold rotation potential energy function and pairwise interdependent bonds this diagonal matrix takes the form

$$\mathbf{V}_2(\tau) = \text{diag}\,(p_{t;t}, p_{t;g^+}, p_{t;g^-}, p_{g^+;t}, p_{g^+;g^+}, p_{g^+;g^-}, p_{g^-;t}, p_{g^-;g^+}, p_{g^-;g^-})_2(\tau)$$

$$(\text{XV-32})$$

The matrix for all remaining internal bonds is formulated from ratios of $p_{\xi\eta;\xi^0\eta^0;i}(\tau)$ and $p_{\xi;\xi^0;i-1}(\tau)$. This ratio is the dynamic counterpart of the conditional probability, $q_{\xi\eta;i}$, defined in Eq. (V-36). Again assuming a chain with a symmetric threefold rotation potential energy function and pairwise interdependent bonds, we obtain

$$\mathbf{V}_i(\tau) = \begin{bmatrix} v_{tt;tt} & v_{tt;tg^+} & v_{tt;tg^-} & \cdots & v_{tg^-;tg^-} \\ v_{tt;g^+t} & v_{tt;g^+g^+} & v_{tt;g^+g^-} & \cdots & v_{tg^-;g^+g^-} \\ v_{tt;g^-t} & v_{tt;g^-g^+} & v_{tt;g^-g^-} & \cdots & v_{tg^-;g^-g^-} \\ \vdots & \vdots & \vdots & & \vdots \\ v_{g^-t;g^-t} & v_{g^-t;g^-g^+} & v_{g^-t;g^-g^-} & \cdots & v_{g^-g^-;g^-g^-} \end{bmatrix}_i (\tau) \qquad (\text{XV-33})$$

where each element is formulated as

$$v_{\xi\eta;\xi^0\eta^0;i}(\tau) = \frac{p_{\xi\eta;\xi^0\eta^0;i}(\tau)}{p_{\xi;\xi^0;i-1}(\tau)} \qquad (\text{XV-34})$$

The $p_{\xi;\xi^0;i-1}(\tau)$ are evaluated ignoring the interdependence of the bonds, and the interdependence of the transitions between rotational isomeric states is incorporated via the $p_{\xi\eta;\xi^0\eta^0;i}(\tau)$.

The joint probabilities that appear in Eq. (XV-34) are obtained from a consideration of the dynamics of the rotational isomeric state transitions in small fragments of the polymer. The rate for a transition between rotational isomeric states at a specified bond is usually expressed in Kramers' high friction limit as[20]

$$r = \frac{(\gamma\gamma^*)^{1/2}}{2\pi\zeta} \exp{(-E_a/RT)} \qquad \text{(XV-35)}$$

where E_a is the activation energy for the transition, γ and γ^* depend on the curvature of the reaction coordinate in the initial state and at the transition state, and ζ is a friction coefficient.[16] Assignment of ζ, which depends on the effective viscosity of the medium, determines the time scale of the dynamics. It is usually assigned by comparison of the calculated dynamics with an experimental measurement.

If n-alkanes are treated in the approximation that the bonds are independent, transitions between the t and g states must be considered according to the kinetic scheme $g^+ \rightleftharpoons t \rightleftharpoons g^-$, but direct $g^{\pm} \rightleftharpoons g^{\mp}$ can be ignored, because of the much larger E_a for direct transitions between the g states. This conclusion follows from consideration of the dependence of the conformational energy on ϕ_2 in n-butane, as depicted in Fig. III-3. A 3×3 matrix with the joint probabilities for occupation of state η at time τ (with the state at time τ indexing the rows, in the order t, g^+, g^-), given state η^0 at time 0 (with the state at time 0 indexing the columns, in the same order as the indexing of the rows), is obtained as[15,16]

$$\begin{bmatrix} p_{t;t^0} & p_{t;g^{+0}} & p_{t;g^{-0}} \\ p_{g^+;t^0} & p_{g^+;g^{+0}} & p_{g^+;g^{-0}} \\ p_{g^-;t^0} & p_{g^-;g^{+0}} & p_{g^-;g^{-0}} \end{bmatrix}_i (\tau) = \mathbf{B} \exp{(\Lambda_i \tau)} \mathbf{B}^{-1} \operatorname{diag}{(p_{t^0}, p_{g^{+0}}, p_{g^{-0}})_i}$$

$$\text{(XV-36)}$$

where the final matrix in Eq. (XV-36) contains along the diagonal the probabilities for t, g^+, and g^- states at bond i at zero time, and Λ_i is a diagonal matrix formulated from the rates:

$$\mathbf{B}\Lambda_i\mathbf{B}^{-1} = \mathbf{A}_i^{(1)} = \begin{bmatrix} -2r_1 & r_{-1} & r_{-1} \\ r_1 & -r_{-1} & 0 \\ r_1 & 0 & -r_{-1} \end{bmatrix}_i \qquad \text{(XV-37)}$$

If the conformational energy surface controlling the dynamics is shifted from the one for n-butane (Fig. III-3) to the one for n-pentane (Fig. III-5) the dynamics will have a pairwise interdependence. On the basis of the sizes of the activation energies deduced from Fig. III-5, the important transitions for the pair of bonds are those listed in Table XV-1. The matrix on the left-hand side of

[20]Kramers, H. A. *Physica* **1940**, 7, 284.

TABLE XV-1. Symbols Used for the Rates of Transitions between Rotational Isomers

Transition	Forward Rate	Reverse Rate
$tt \rightleftharpoons tg^{\pm}$ and $tt \rightleftharpoons g^{\pm}t$	r_1	r_{-1}
$tg^{\pm} \rightleftharpoons g^{\pm}g^{\pm}$ and $g^{\pm}t \rightleftharpoons g^{\pm}g^{\pm}$	r_2	r_{-2}
$tg^{\pm} \rightleftharpoons g^{\mp}g^{\pm}$ and $g^{\pm}t \rightleftharpoons g^{\pm}g^{\mp}$	r_3	r_{-3}

Eq. (XV-36) is now of dimensions 9×9, as are $\mathbf{A}_i^{(2)}$ and the diagonal matrix with the probabilities at zero time. With the rows (and columns) indexed in the order tt, tg^+, tg^-, g^+t, g^+g^+, g^+g^-, g^-t, g^-g^+, g^-g^-, the rows denoting the state at time τ, and the columns denoting the state at time 0, $\mathbf{A}_i^{(2)}$ takes the form[16]

$$
\mathbf{A}_i^{(2)} =
\begin{bmatrix}
-4r_1 & r_{-1} & r_{-1} & r_{-1} & 0 & 0 & r_{-1} & 0 & 0 \\
r_1 & -a & 0 & 0 & r_{-2} & 0 & 0 & r_{-3} & 0 \\
r_1 & 0 & -a & 0 & 0 & r_{-3} & 0 & 0 & r_{-2} \\
r_1 & 0 & 0 & -a & r_{-2} & r_{-3} & 0 & 0 & 0 \\
0 & r_2 & 0 & r_2 & -2r_{-2} & 0 & 0 & 0 & 0 \\
0 & 0 & r_3 & r_3 & 0 & -2r_{-3} & 0 & 0 & 0 \\
r_1 & 0 & 0 & 0 & 0 & 0 & -a & r_{-3} & r_{-2} \\
0 & r_3 & 0 & 0 & 0 & 0 & r_3 & -2r_{-3} & 0 \\
0 & 0 & r_2 & 0 & 0 & 0 & r_2 & 0 & -2r_{-2}
\end{bmatrix}
$$

$$(XV-38)$$

where $a = r_2 + r_3 + r_{-1}$. Equation (XV-36) with $\mathbf{A}_i^{(1)}$, or its counterpart with $\mathbf{A}_i^{(2)}$, provides the $p_{\eta; \eta^0; i-1}(\tau)$ and $p_{\xi\eta; \xi^0\eta^0; i}(\tau)$ required for the $v_{\xi\eta; \xi^0\eta^0; i}(\tau)$, Eq. (XV-34).

Let $p^*_{\eta; \eta^0; i}(\tau)$ denote the joint probability for state η at bond i at time τ, given state η^0, taking into account the interdependence of the transitions between rotational isomeric states. This probability is extracted from $Z(\tau)$ in a manner similar to the extraction of $p_{\eta; i}$ from Z:

$$
p^*_{\eta; \eta^0; i}(\tau) = \frac{1}{Z(\tau)} \mathbf{V}_1 \mathbf{V}_2(\tau) \cdots \mathbf{V}_{i-1}(\tau) \mathbf{V}'_{\eta; \eta^0; i}(\tau) \mathbf{V}_{i+1}(\tau) \cdots \mathbf{V}_{n-1}(\tau) \mathbf{V}_n
$$

$$(XV-39)$$

where $\mathbf{V}'_{\eta; \eta^0; i}(\tau)$ differs from $\mathbf{V}_i(\tau)$ in that all elements are replaced by 0 unless they have the desired state (η^0) at time 0, and the desired state (η) at time τ.[19]

B. Orientation Autocorrelation Functions

The approach to $F(\tau)$ will be similar to the approach to $\langle \mathbf{r} \rangle_0$ in Chapter VI. The intermediate objective is the construction of a set of matrices, one for each bond, that provide the contribution to $F(\tau)$ from a specified initial conformation at time 0, and a specified final conformation a time τ later. These matrices play the role for $F(\tau)$ that is analogous to the role played by the generator matrix $\mathbf{A}_i$ for $\langle \mathbf{r} \rangle_0$. The matrices will then be combined with $Z(\tau)$ in order to obtain $F(\tau)$, in the same manner in which the $\mathbf{A}_i$ were combined with Z to obtain $\langle \mathbf{r} \rangle_0$ in Chapter VI.[19]

Let $F_{j,\kappa\lambda}$ denote one of the terms that must be averaged to obtain $F_j(\tau)$, just as $\mathbf{r}$ denotes one of the end-to-end vectors that is averaged to obtain $\langle \mathbf{r} \rangle_0$. For $F_{j,\kappa\lambda}$ one of the many conformations is selected for time 0, and another conformation (or perhaps the same one) is selected for time τ. These two conformations are indexed by κ and λ, respectively. With these restrictions on the conformations, Eq. (XV-28) takes the form

$$F_{j,\kappa\lambda} = \mathbf{m}_{j,\kappa}^T \mathbf{m}_{j,\lambda} \tag{XV-40}$$

where the angle brackets are no longer necessary, since there is only one initial conformation, and one final conformation. The vectors in Eq. (XV-40) can be expressed in the coordinate system of the first bond through premultiplication by $\mathbf{T}_1 \mathbf{T}_2 \cdots \mathbf{T}_{j-1}$, for conformation κ or λ, as appropriate:

$$F_{j,\kappa\lambda} = (\mathbf{T}_1 \mathbf{T}_2 \cdots \mathbf{T}_{j-1} \mathbf{m}_j)_\kappa^T (\mathbf{T}_1 \mathbf{T}_2 \cdots \mathbf{T}_{j-1} \mathbf{m}_j)_\lambda \tag{XV-41}$$

Equation XV-41 is not in a convenient form, because information about bond i, $1 \leq i \leq j$ occurs at two widely separated positions on the right-hand side. This problem can be eliminated[19] using relationships between matrices that were described by Jernigan and Flory.[21] For any three conformable matrices

$$(\mathbf{ABC})^R = \mathbf{B}^R (\mathbf{A}^T \otimes \mathbf{C}) \tag{XV-42}$$

Application of this relationship, with $\mathbf{A} = (\mathbf{T}_1 \mathbf{T}_2 \cdots \mathbf{T}_{j-1} \mathbf{m})_\kappa^T$, $\mathbf{B} = \mathbf{I}_3$, and $\mathbf{C} = (\mathbf{T}_1 \mathbf{T}_2 \cdots \mathbf{T}_{j-1} \mathbf{m})_\lambda$ yields

$$F_{j,\kappa\lambda} = \mathbf{I}_3^R [(\mathbf{T}_1 \mathbf{T}_2 \cdots \mathbf{T}_{j-1} \mathbf{m}_j)_\kappa \otimes (\mathbf{T}_1 \mathbf{T}_2 \cdots \mathbf{T}_{j-1} \mathbf{m}_j)_\lambda] \tag{XV-43}$$

After j successive applications of the theorem on direct products, all information about bond i occurs at a single location in the expression:

---

[21] Jernigan, R. L.; Flory, P. J. *J. Chem. Phys.* **1967**, *47*, 1999.

$$F_{j,\kappa\lambda} = \mathbf{I}_3^R[(\mathbf{T}_{1,\kappa} \otimes \mathbf{T}_{1,\lambda})(\mathbf{T}_{2,\kappa} \otimes \mathbf{T}_{2,\lambda}) \cdots (\mathbf{T}_{j-1,\kappa} \otimes \mathbf{T}_{j-1,\lambda})(\mathbf{m}_j \otimes \mathbf{m}_j)]$$

$$(\text{XV-44})$$

The terms in parentheses in Eq. (XV-44) cannot be written as $\mathbf{T}_i^{\times 2}$, as occurs in the $\mathbf{G}_i^{\times 2}$ used for computation of r^4 in Chapter XIII. The notation $\mathbf{T}_i^{\times 2}$ is the direct product of $\mathbf{T}_i$ with itself, but Eq. (XV-44) requires the direct product of $\mathbf{T}_i$ in the initial conformation, κ, with the transformation matrix for the same bond in the final conformation, λ. These two transformation matrices need not be the same.

Averaging of $F_{j,\kappa\lambda}$ over all pairs of conformations, with due regard for the likelihood that they will be separated in time by τ, is obtained by a straightforward extension of the method described in Chapter VI for the computation of the statistical mechanical average of static properties. Using $\mathscr{F}_i$ to denote the matrices required for this purpose, we have[19]

$$F_j(\tau) = \frac{1}{Z(\tau)} \mathscr{F}_1 \mathscr{F}_2(\tau) \cdots \mathscr{F}_{j-1}(\tau) \mathscr{F}_j(\tau) \mathbf{V}_{j+1}(\tau) \cdots \mathbf{V}_{n-1}(\tau)\mathbf{V}_n \quad (\text{XV-45})$$

$$\mathscr{F}_1 = [1 \quad 1 \quad 1 \quad 1 \quad 1 \quad 1 \quad 1 \quad 1 \quad 1] \otimes \mathbf{I}_3^R \quad (\text{XV-46})$$

$$\mathscr{F}_i(\tau) = (\mathbf{V}_i(\tau) \otimes \mathbf{I}_9)\|\mathbf{T}_{\eta^0;i} \otimes \mathbf{T}_{\eta;i}\|, \quad 1 < i < j \quad (\text{XV-47})$$

$$\mathscr{F}_j = \mathbf{V}_j(\tau) \otimes (\mathbf{m}_j \otimes \mathbf{m}_j) \quad (\text{XV-48})$$

with the nine combinations of the direct product of the transformation matrices on the main diagonal of the blocked form of $\|\mathbf{T}_{\eta^0;i} \otimes \mathbf{T}_{\eta;i}\|$:

$$\|\mathbf{T}_{\eta^0;i} \otimes \mathbf{T}_{\eta;i}\| = \begin{bmatrix} \mathbf{T}_t \otimes \mathbf{T}_t & \mathbf{0} & \mathbf{0} & \cdots & \mathbf{0} \\ \mathbf{0} & \mathbf{T}_t \otimes \mathbf{T}_{g^+} & \mathbf{0} & \cdots & \mathbf{0} \\ \mathbf{0} & \mathbf{0} & \mathbf{T}_t \otimes \mathbf{T}_{g^-} & \cdots & \mathbf{0} \\ \vdots & \vdots & \vdots & & \vdots \\ \mathbf{0} & \mathbf{0} & \mathbf{0} & \cdots & \mathbf{T}_{g^-} \otimes \mathbf{T}_{g^-} \end{bmatrix}$$

$$(\text{XV-49})$$

The $\mathscr{F}_i(\tau)$, $1 < i < j$, are of dimensions 81×81 if there are three rotational isomeric states per bond.

The second orientation autocorrelation function requires $\langle[\mathbf{m}_j^T(0)\mathbf{m}_j(\tau)]^2\rangle$. Define $G_{j,\kappa\lambda}$ as one of the possible contributions to this average, in a manner completely analogous to the definition of $F_{j,\kappa\lambda}$. The relationship between these two terms is simple: $G_{j,\kappa\lambda} = F_{j,\kappa\lambda}^2$, or, more usefully

$$G_{j,\kappa\lambda} = F_{j,\kappa\lambda} \otimes F_{j,\kappa\lambda} \qquad (XV\text{-}50)$$

just as was done in Chapter XIII to express r^4 as $r^2 \otimes r^2$. Starting from Eq. (XV-44) for $F_{j,\kappa\lambda}$, and using j successive applications of the theorem on direct products, we obtain

$$G_{j,\kappa\lambda} = (\mathbf{I}_3^R)^{\times 2}[(\mathbf{T}_{1,\kappa} \otimes \mathbf{T}_{1,\lambda})^{\times 2} \cdots (\mathbf{T}_{j-1,\kappa} \otimes \mathbf{T}_{j-1,\lambda})^{\times 2}(\mathbf{m}_j \otimes \mathbf{m}_j)^{\times 2}]$$

$$(XV\text{-}51)$$

The desired average is obtained as

$$\langle[\mathbf{m}_j^T(0)\mathbf{m}_j(\tau)]^2\rangle = \frac{1}{Z(\tau)} \,\mathscr{G}_1\mathscr{G}_2(\tau) \cdots \mathscr{G}_j(\tau)\mathbf{V}_{j+1}(\tau) \cdots \mathbf{V}_{n-1}(\tau)\mathbf{V}_n$$

$$(XV\text{-}52)$$

$$\mathscr{G}_1 = [1 \quad 1 \quad 1 \quad 1 \quad 1 \quad 1 \quad 1 \quad 1 \quad 1] \otimes (\mathbf{I}_3^R)^{\times 2} \qquad (XV\text{-}53)$$

$$\mathscr{G}_i(\tau) = (\mathbf{V}_i(\tau) \otimes \mathbf{I}_{81})\|(\mathbf{T}_{\eta^0;i} \otimes \mathbf{T}_{\eta;i})^{\times 2}\|, \qquad 1 < i < j \qquad (XV\text{-}54)$$

$$\mathscr{G}_j = \mathbf{V}_j(\tau) \otimes (\mathbf{m}_j \otimes \mathbf{m}_j)^{\times 2} \qquad (XV\text{-}55)$$

The $\mathscr{G}_i(\tau)$, $1 < i < j$, are of dimensions 729×729 if there are three rotational isomeric states per bond.

C. Application to Polyoxymethylene

The type of problem susceptible to treatment by the dynamic RIS formalism described above is illustrated by the helix $\rightleftharpoons$ coil transition in polyoxymethylene. As was described in Chapter VII, the internal bonds in this chain have a strong preference for g placements, and pairs of consecutive internal bonds strongly avoid occupancy of g placements of opposite sign. At low temperatures, chains with a small number of bonds prefer the two conformations in which every internal bond adopts a g placement of the same sign. The chain describes a helix, either right- or left-handed, as depicted in Fig. VII-4. Let p_h denote the probability that *all* internal bonds in the chain adopt a g placement of the same sign. At sufficiently low temperature, $p_h \sim 1$. There will be an onset of competing conformations as the temperature increases, and in a narrow temperature interval p_h falls from ~ 1 to a much smaller number. The temperature range in which this disordering occurs depends on n. The dependence of p_h on the temperature and n, described as the thermal melting of the g helices, was studied by classic RIS theory by Curro et al.[22]

If the temperature is chosen so that it is in the range where p_h shows its sharp decrease, the conformation of a chain becomes dependent on time in a manner that reflects the mechanism by which the helices become disordered. Let $p_{h;h^0}(\tau)$ denote the probability that a chain with all internal bonds in g states of the same sign at time 0, is also in this same state a time τ later. If τ, T, and n are sufficiently small, this number will be very close to one, because the static analysis yields $p_h \sim 1$ under these conditions. That result is indeed obtained from the dynamic RIS analysis of polyoxymethylene,[23] as depicted in Fig. XV-5.

Insight into the mechanism of the melting of the helices is obtained by selecting a particular chain, and following $p_{h;h^0}(\tau)$ as a function of τ under isothermal conditions. The behavior of the chain with $n = 40$ at three temperatures is depicted in Fig. XV-6. The system is more sluggish at the lower temperatures, as expected. More interesting is the appearance of a bimodal relaxation, with the two processes separated by several orders of magnitude in time.

The physical origin of the faster process, the time scale on which it occurs, and the value of $p_{h;h^0}(\tau)$ at the plateau that is achieved after completion of the fast process, can all be rationalized with a very simple model for the melting.[23] Assume that each helix with n bonds and *gauche* placements of the same sign participates first in an equilibrium with the $n-2$ structures in which one of the internal bonds occupies a t placement.

$$g^+g^+g^+ \cdots g^+ \rightleftharpoons \begin{bmatrix} tg^+g^+ & \cdots & g^+ \\ g^+tg^+ & \cdots & g^+ \\ g^+g^+t & \cdots & g^+ \\ & \vdots & \\ g^+g^+g^+ & \cdots & t \end{bmatrix} \tag{XV-56}$$

[22]Curro, J. G.; Schweizer, K. S.; Adolf, D.; Mark, J. E. *Macromolecules* **1986**, *19*, 1739.
[23]Bahar, I.; Mattice, W. L. *Macromolecules* **1991**, *24*, 877.

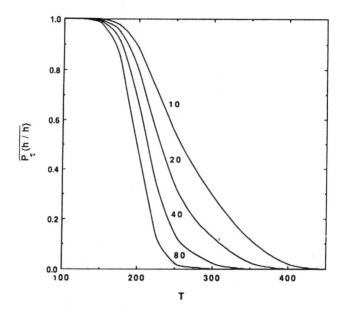

Figure XV-5. Values of $p_{h;h^0}(\tau)$ as a function of temperature (in K) for polyoxymethylene chains of 10, 20, 40, and 80 bonds, using $\tau = 10$ ns. Reprinted with permission from Bahar, I.; Mattice, W. L. *Macromolecules* **1991**, *24*, 877. Copyright 1991 American Chemical Society.

The rates for the forward and reserve processes are given by r_{-2} and r_2, as defined in the second line of Table XV-1. This initial process is followed, on a much slower time scale, by further disruption of the once-disrupted helices. Thus the model assumes a once-disrupted helix is more likely to revert to the perfect helix than to become more completely disrupted.

The statistical weights for the helix and for each of the once-disrupted helices are σ^{n-2} and σ^{n-3}, respectively. If the fast process corresponds to the equilibration between the intact helices and the once-disrupted helices, the value of $p_{h;h^0}(\tau)$ at the plateau is

$$p_{h;h^0}(\tau_{\text{plateau}}) = \frac{\sigma}{\sigma + n - 2} \qquad \text{(XV-57)}$$

This result provides an excellent estimate of the height of the plateau when the plateau is well defined, as it is for the lowest temperature considered in Fig. XV-6. The relaxation from the complete helix to the collection of once-disrupted helices is also described by a very simple model, because the values of r_2 and r_{-2} are assumed to be independent of the index i for the bond undergoing the transition in Eq. (XV-56).

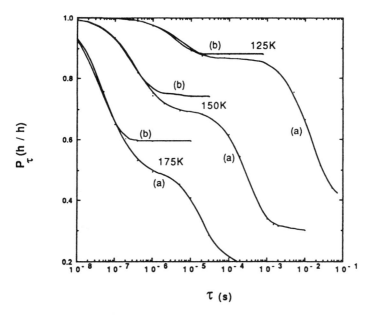

Figure XV-6. Values of $p_{h;h^0}(\tau)$, as a function of τ, for the entire ensemble of polyoxymethylene chains of 40 bonds. The temperatures are 125, 150, and 175 K. Curves denoted by (*a*) are from the full dynamic rotational isomeric state model, and curves denoted by (*b*) are for a simple model of the faster of the two processes, using $E_\sigma = -5.86$ kJ mol^{-1} and an activation energy of 14.6 kJ mol^{-1} for r_2. Reprinted with permission from Bahar, I.; Mattice, W. L. *Macromolecules* **1991**, *24*, 877. Copyright 1991 American Chemical Society.

$$p_{h;h^0}(\tau) = p_{h;h^0}(\tau_{\text{plateau}}) + [1 - p_{h;h^0}(\tau_{\text{plateau}})]\exp(-[r_2 + (n - 2)r_{-2}]\tau)$$

$$(\text{XV-58})$$

The time dependence specified by this equation is shown by curves (*b*) in Fig. XV-6. It captures the time dependence of the fast process rather well, particularly at the lower temperatures. Thus the physical origin of the bimodal relaxation, obtained directly from the dynamic RIS analysis, is explained by a simple model.

The manner in which the bimodal relaxation appears in $F(\tau)$ depends on the manner in which **m** is oriented in the local coordinate system of the bond to which it is rigidly attached.[23] The time scales of the bimodal relaxation are not affected by the nature of the attachment, nor is the value of $F(\tau)$ at the plateau, but the final value of $F(\tau)$ attained after completion of the slower process depends strongly on the orientation of **m**, as is shown in Fig. XV-7.

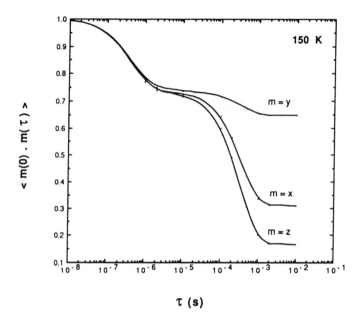

Figure XV-7. $F_j(\tau)$ at 150 K, as a function of τ, for **m** attached to a central bond in a polyoxymethylene chain of 40 bonds. Reprinted with permission from Bahar, I.; Mattice, W. L. *Macromolecules* **1991**, *24*, 877. Copyright 1991 American Chemical Society.

PROBLEMS

XV-1. The RIS model for a polyethylene chain will produce a collapse in dimensions if $E' < 0$ in Eq. (XV-7). When the chain collapses, does it remain disordered, or does it crystallize?

XV-2. Figure XV-2 depicts an environment in which $\langle r \rangle$ and $\langle g \rangle$ remain different as $n \to \infty$, in contrast with the result for the unperturbed chain. Will $\langle r \rangle$ remain different from $\langle g \rangle$ at large n if

(a) Long-range interactions are introduced by Eq. (XV-8)?

(b) Atoms A_0 and A_n bear charges of opposite sign?

(c) Atoms A_0 and A_n bear charges of the same sign?

XV-3. At stereochemical equilibrium, what is p_m for a long poly(vinyl bromide) chain?

XV-4. From the values of $\partial \ln p_m / \partial w$, where w is a statistical weight, can you make any generalizations about whether the composition at stereochemical equilibrium is more sensitive to the first- or second-order interactions in typical vinyl polymers?

APPENDIX A
Definitions of Generally Used Symbols

Greek

α	Polarizability. Statistical weight. Expansion factor for the end-to-end distance. Angle of rotation of plane-polarized light.
$\boldsymbol{\alpha}$	Polarizability tensor.
$\tilde{\boldsymbol{\alpha}}$	Anisotropic part of the polarizability tensor.
α_μ	Expansion factor for the dipole moment.
β	Statistical weight. Dihedral angle.
Γ_2	$\equiv 3\Delta a/5$.
γ	Statistical weight. Cosine of the difference between two angles. $R_E/\langle s^2 \rangle^{1/2}$.
γ^2	Molecular anisotropy.
Δa	Segmental polarizability anisotropy.
δ	Statistical weight for a second-order interaction or a bend. Chemical shift.
ϵ	Static dielectric constant.
ζ	Statistical weight for a first-order interaction. Denotes a domain.
η	Index for a state at bond i. Statistical weight for a first-order interaction. Viscosity. Angle between a virtual bond and the C'—C^α bond.
$[\eta]$	Intrinsic viscosity.
Θ	A solvent, or T, at which the chain is unperturbed by long-range interactions.
θ	Bond angle. Scattering angle.
θ_c	Complement of the bond angle, $\pi - \theta$.
κ	Index for conformations.
κ^2	Shape anisotropy. Orientation factor in Förster transfer.
Λ	Diagonal matrix of eigenvalues.
λ	Eigenvalue. Wavelength.
μ	Dipole moment.

401

$\boldsymbol{\mu}$	Dipole moment vector.
ν	Number of rotational isomers at a bond. Exponent for the chain dimensions. Frequency.
ξ	Index for a state at bond $i-1$. Statistical weight for a second-order interaction. Angle between a virtual bond and the $N-C^\alpha$ bond (in amides) or $O-C^\alpha$ bond (in esters).
ρ	Statistical weight. The vector $\mathbf{r} - \langle\mathbf{r}\rangle_0$. Correlation coefficient. Density.
σ	Statistical weight for a first-order interaction.
$\hat{\sigma}$	With s, the statistical weight for initiation of a helix.
σ_c	Symmetry number.
τ	Statistical weight. Axial stress. Fluorescence lifetime. Time.
Φ	Viscosity function.
ϕ	Torsion angle. Volume fraction.
χ	Torsion angle at a bond in a side chain.
$\tilde{\chi}$	Anisotropy of the magnetic susceptibility.
ψ	Statistical weight for a second-order interaction. Torsion angle.
ω	Statistical weight for a second-order interaction.
$\hat{\omega}$	Statistical weight for helix–helix interaction.

Latin

$\mathbf{A}$	4×4 generator matrix for $\mathbf{r}$. Matrix used in diagonalization.
$\|\mathbf{A}\|$	$4\nu \times 4\nu$ supermatrix with the $\mathbf{A}_\eta$ on the diagonal.
$\mathscr{A}$	Supermatrix $(\mathbf{U} \otimes \mathbf{I}_4)\|\mathbf{A}\|$.
A	Atom.
A_2	Second virial coefficient of the chemical potential.
a	Persistence length. van der Waals constant for attractions.
$\mathbf{B}$	4×4 generator matrix for the dipole moment vector.
$\|\mathbf{B}\|$	$4\nu \times 4\nu$ supermatrix with the $\mathbf{B}_\eta$ on the diagonal.
$\mathscr{B}$	Supermatrix $(\mathbf{U} \otimes \mathbf{I}_4)\|\mathbf{B}\|$.
B	Stress-optical coefficient.
b	Asphericity. van der Waals constant for repulsions.
$\mathbf{C}$	20×20 generator matrix for the Cotton–Mouton constant.
$\|\mathbf{C}\|$	$20\nu \times 20\nu$ supermatrix with the $\mathbf{C}_\eta$ on the diagonal.
$\mathscr{C}$	Supermatrix $(\mathbf{U} \otimes \mathbf{I}_{20})\|\mathbf{C}\|$.
$_m C$	Cotton–Mouton constant.
C_E	$\equiv N/V_E$.
C_n	Characteristic ratio $\equiv \langle r^2\rangle_0/nl^2$.
$C_{s,n}$	Characteristic ratio for the radius of gyration, $\langle s^2\rangle_0/nl^2$.
c	cis. Mass concentration. Acylindricity.
$\mathbf{D}$	Diagonal matrix of first-order interactions. 5×5 generator matrix for μ^2.
$\|\mathbf{D}\|$	$5\nu \times 5\nu$ supermatrix with the $\mathbf{D}_\eta$ on the diagonal.
$\mathscr{D}$	Supermatrix $(\mathbf{U} \otimes \mathbf{I}_5)\|\mathbf{D}\|$.

D_x Characteristic dipole moment ratio, $\langle \mu^2 \rangle_0 / xm^2$.

d Pseudoasymmetric center. Distance between scatterer and detector. Light path.

$\mathscr{E}$ Electric field.

E Energy. Efficiency of Förster transfer.

f Scattering factor. Statistical weight for articulated side chain. Functionality of a branch point.

F First orientation autocorrelation functions.

$\mathscr{F}$ Supermatrix with elements of $\mathbf{U}$ expanded by a generator matrix $\mathbf{F}$.

G Second orientation autocorrelation functions.

$\mathbf{G}$ 5×5 generator matrix for r^2.

$\|\mathbf{G}\|$ $5\nu \times 5\nu$ supermatrix with the $\mathbf{G}_\eta$ on the diagonal.

$\mathscr{G}$ Supermatrix $(\mathbf{U} \otimes \mathbf{I}_5)\|\mathbf{G}\|$.

$\mathbf{g}$ Center-of-mass vector.

g *gauche*. Ratio of $\langle s^2 \rangle$ for a branched and a linear molecule with the same n. Coefficient in Hermite series.

$\mathbf{H}$ 7×7 generator matrix for s^2.

$\|\mathbf{H}\|$ $7\nu \times 7\nu$ supermatrix with the $\mathbf{H}_\eta$ on the diagonal.

$\mathscr{H}$ Supermatrix $(\mathbf{U} \otimes \mathbf{I}_7)\|\mathbf{H}\|$.

H Enthalpy.

h Planck's constant.

$h(0)$ Hermite series.

$\mathbf{I}_\nu$ Identity matrix of dimensions $\nu \times \nu$.

I Number of units allowed in a strand of an antiparallel sheet.

I_0 Intensity of incident radiation.

I_E Intensity of the fluorescence from an excimer.

I_M Intensity of the fluorescence from a monomer.

i Scattering intensity, integer for indexing.

$\tilde{i}$ $(-1)^{1/2}$.

J Coupling constant.

j Integer for indexing.

$\mathbf{K}$ 26×26 generator matrix for $\mathbf{r}^T \hat{\alpha} \mathbf{r}$ or $\boldsymbol{\mu}^T \hat{\alpha} \boldsymbol{\mu}$.

$\|\mathbf{K}\|$ $26\nu \times 26\nu$ supermatrix with the $\mathbf{K}_\eta$ on the diagonal.

$\mathscr{K}$ Supermatrix $(\mathbf{U} \otimes \mathbf{I}_{26})\|\mathbf{K}\|$. Scattering constant.

K Constant in the Mark–Houwink relationship. Number of scattering centers.

$_m K$ Molar Kerr constant.

K_x Macrocyclization equilibrium constant.

k Boltzmann's constant. Rate constant. Integer for indexing.

$\mathbf{L}$ Vector for a random-flight step.

$\mathscr{L}$ Avogadro's number.

L Length of a random-flight step. Length of a rod.

L_i^2 Principal moment of $\mathbf{S}$.

$\mathbf{l}$ vector for a bond.

l Length of a bond. Pseudoasymmetric center. Integer for indexing.

M Molecular weight.

M_x Molecular weight of a monomeric unit.

$[M]$ Molar rotation.

m   Local dipole moment vector.

m Local dipole moment. Mass. Molar concentration.

$\mathscr{N}$ Number of chains generated in a Monte Carlo simulation.

N Number of particles. Number of random-flight steps

n Number of bonds.

$\tilde{n}$ Refractive index.

P 11×11 generator matrix for γ^2.

$\|\mathbf{P}\|$ $11\nu \times 11\nu$ supermatrix with the $\mathbf{P}_\eta$ on the diagonal.

$\mathscr{P}$ Supermatrix $(\mathbf{U} \otimes \mathbf{I}_{11})\|\mathbf{P}\|$.

P_i^2 Principal moment of $\langle \mathbf{S} \rangle$.

$P(Q)$ Scattering function.

p Probability. Polarization factor. Integer for indexing.

$p_{\eta;i}$ Probability of state η at bond i.

p_η Probability of state η in a chain.

$p_{\xi\eta;i}$ Probability of state ξ at bond $i-1$ and state η at bond i.

$p_{\xi\eta}$ Probability of state ξ and state η at two successive bonds in a chain.

p_m Probability for a *meso* diad.

Q   Scattering vector. The matrix

$$\begin{bmatrix} 1 & 0 & 0 \\ 0 & 0 & 1 \\ 0 & 1 & 0 \end{bmatrix}$$

$\mathbf{Q}_d$ The matrix

$$\begin{bmatrix} 0 & 0 & 1 \\ 1 & 0 & 0 \\ 0 & 1 & 0 \end{bmatrix}$$

Q Classical partition function. Length of scattering vector. Quantum yield for fluorescence.

q Conditional probability. Charge.

R   Row vector.

$\mathbf{R}^N$ Configuration of N centers.

R Gas constant. Radius of a sphere.

R_0 Förster radius.

ΔR_λ Bond or group refractivity.

r   End-to-end vector.

$\mathbf{r}_i$ Position of A_i.

$\mathbf{r}_{ij}$	Vector from A_i to A_j.
r	End-to-end distance.
r_i	Reactivity ratio.
r_{ij}	Distance between A_i and A_j.
$r_{\max}$	Contour length.
S	Entropy.
$\mathbf{S}$	Radius-of-gyration tensor.
$\mathbf{s}_i$	Vector from the center of mass (or charge) to mass (or charge) i.
s	Radius of gyration. *skew.* Statistical weight for propagation of a helix.
s_i	Distance from the center of mass (or charge) to mass (or charge) i.
$\mathbf{T}$	Transformation matrix.
T	Temperature. As a superscript, denotes the transpose of a matrix.
$\mathbf{T}_i$	Transformation matrix for bond i.
$\mathbf{t}$	A tensor.
t	*trans.* Statistical weight for propagation of an antiparallel sheet.
$\mathbf{U}$	Statistical weight matrix.
$\mathscr{U}$	Supermatrix used for epimerization to stereochemical equilibrium.
$U(\mathbf{R}^N)$	Total potential energy.
$\overline{U}(\mathbf{R}^N)$	Potential of mean force with respect to solvent.
$\hat{U}(\mathbf{R}^N)$	Potential of mean force with respect to solvent and chemical structure.
u	Statistical weight.
u_{ij}	Element in row i, column j, of $\mathbf{U}$.
$\mathbf{V}$	Matrix with statistical weights for second-order interactions.
$\mathbf{V}_i(\tau)$	Transition probability matrix.
V	Volume.
V_E	$\equiv (4/3)\pi R_E^3$.
$\overline{V}$	Partial molar volume.
$\mathbf{v}$	A vector.
υ	$\equiv \sqrt{Q^2 \langle s^2 \rangle}$.
$\mathbf{W}$	Matrix with statistical weights for third-order interactions.
$W(x)$	Distribution function for x.
w	Statistical weight. Weight fraction of solute.
$\mathbf{X}$	Matrix used in the reduction of $\mathscr{F}$.
$\mathbf{X}^0$	Matrix used in the reduction of $\mathbf{U}$.
X	Mole fraction.
x	Number of monomer units.
$\mathbf{Y}$	Matrix used in the reduction of $\mathscr{F}$.
$\mathbf{Y}^0$	Matrix used in the reduction of $\mathbf{U}$.
Z	Conformational partition function for a chain (also $Z_{\{\phi\}}$).
Z_N	Configuration partition function.
$Z(\tau)$	Transition partition function.

$\mathscr{Z}$	Sum of conformational partition functions, used for epimerization to stereochemical equilibrium.
z	Conformational partition function for a bond. Excluded volume parameter.

Subscripts

∞	Limit as $n \to \infty$.
θ	Angle.
Θ	Θ state.
0	Θ state, incident beam, zero angle, term in a preexponential factor.
$\parallel$	Parallel component.
$\perp$	Perpendicular component.
E	Equivalent sphere.
G	Center of mass.
d	d pseudoasymmetric center.
l	l pseudoasymmetric center.
m	*meso.*
n	Chain of n bonds.
r	*racemo.*
s	Solvent.
v	in vacuo.
x	Monomeric unit.
x, y, z	Components along the Cartesian axes.

Other

$\langle \cdots \rangle$	Denotes a conformational ensemble average over the enclosed quantity.

APPENDIX B
Answers to Selected Problems

Chapter I

I-1. If you wonder whether a particular (large) value of N has values of $\langle r^4 \rangle_0 / \langle r^2 \rangle_0^2$ and $\langle s^4 \rangle_0 / \langle s^2 \rangle_0^2$ within a specified percent of the values at the asymptotic limits as $N \to \infty$, you can use

$$\left(\frac{\partial \ln (\langle r^4 \rangle_0 / \langle r^2 \rangle_0^2)}{\partial (1/N)} \right)_{1/N=0} = -\frac{2}{5}$$

$$\left(\frac{\partial \ln (\langle s^4 \rangle_0 / \langle s^2 \rangle_0^2)}{\partial (1/N)} \right)_{1/N=0} = -\frac{12}{19}$$

which follow from Eqs. (I-13) and I-14). The results show the initial slope (in "ratio" versus $1/N$) is more different from zero for the ratio involving s^2 than for the one involving r^2. Hence the percent deviation is larger for this ratio at finite N than for the analogous ratio involving r^2.

I-2. The width of a distribution for a variable x is measured by $\langle x^2 \rangle / \langle x \rangle^2$. Application when x is either r^2 or s^2 for an infinitely long freely jointed chain yields the following, via Eqs. (I-13) and (I-14):

$$\frac{\langle r^4 \rangle_0}{\langle r^2 \rangle_0^2} = \frac{5}{3} > \frac{\langle s^4 \rangle_0}{\langle s^2 \rangle_0^2} = \frac{19}{15}$$

The distribution function for r^2 is broader than the distribution function for s^2.

I-3. From the answer to Problem I-1, $\langle r^4 \rangle_0 / \langle r^2 \rangle_0^2$ and $\langle s^4 \rangle_0 / \langle s^2 \rangle_0^2$ approach their asymptotic limits from below. Both distribution functions are infinitely sharp (δ functions) when $N = 1$. They broaden as N increases.

I-4. Since conformations with very large values of r^2 tend to have large values of s^2, there is a positive correlation of r^2 with s^2, producing

$\rho > 0$. But the correlation is not perfect, as is apparent from a consideration of cyclic conformations, which must have $r^2 \sim 0$, but that can have a rather wide range of s^2 ($0 < s^2 < N^2L^2/48$). The imperfection in the correlation produces $\rho < 1$.

I-5. Calculation of $\langle r^2 \rangle_0 = C_\infty n l^2$, with $n = 10^6/14$ and $l = 153$ pm, yields $\langle r^2 \rangle_0 = 1.12 \times 10^4$ nm^2. $r_{max} = nl \sin \theta/2 = 9.1 \times 10^3$ nm. $L = 1.23$ nm, $N = 7.4 \times 10^3$, $a = L/2 = 0.62$ nm.

I-6. Using $\theta_c = \pi - \theta$, we obtain

$$C_n = \frac{1 + \cos \theta_c}{1 - \cos \theta_c} - \frac{2 \cos \theta_c (1 - \cos^n \theta_c)}{n(1 - \cos \theta_c)^2}$$

$$C_\infty = \frac{1 + \cos \theta_c}{1 - \cos \theta_c}$$

I-7. Since $\langle \cos \phi \rangle = 0$, C_∞ is temperature-independent, with the value given by Eq. (I-20).

I-8. Now $\langle \cos \phi \rangle < 0$, and the numeric result depends on T. C_∞ is given by Eq. (I-29), and it depends on T.

I-9. $A_2 = b - a/RT$, which is zero when $T = a/bR$, positive at larger T (where b dominates), and negative at smaller T (where the term in a dominates). The Boyle temperature is a/bR.

I-10. Many solutions are possible. One set employs $\theta < 90°$ and two pairs of symmetrically placed rotational isomeric states at $\pm\phi_a$ and $\pm\phi_b$, with $0 < \phi_a < \phi_b < 90°$, and $E_{\phi_a} > E_{\phi_b}$.

I-11. Chains A, B, and C have $\langle \cos \phi \rangle = 0$ at all T. Chains D, F, and G have the same temperature coefficient, but different C_∞ at 300 K. Chains D and E have temperature coefficients of opposite sign, and D is more extended than E.

Chain	C_∞	$10^3 (\partial \ln \langle r^2 \rangle_0 / \partial T)_\infty$, deg^{-1}
A, B, C	2.198	0
D	4.000	−2.18
E	1.390	1.27
F	3.091	−2.18
G	5.459	−2.18

I-12. With $\theta = 112°$, Eq. (I-29) requires $\langle \cos \phi \rangle = -0.506$ if $C_\infty = 6.7$. This value of $\langle \cos \phi \rangle$ requires $E_g = 4.83$ kJ mol^{-1} at 140° C. But then

$(\partial \ln \langle r^2 \rangle_0 / \partial T)_\infty = -3.0 \times 10^{-3}$ deg^{-1}, in conflict with experiment. This simple model for a chain cannot account simultaneously for C_∞ and $(\partial \ln \langle r^2 \rangle_0 / \partial T)_\infty$ for polyethylene.

Chapter II

II-1. $87° < \theta < 93°$.

II-2. $C_\infty = 6.2$.

II-3. $V_E \propto N^{3/2}$.

II-4. No.

II-5. $(\alpha_\mu^2 - 1)(\alpha^2 - 1)^{-1} \rightarrow 0$ as $N \rightarrow \infty$, but we also have $\alpha^2 \rightarrow \infty$ in this limit if the chain is in a good solvent. Hence α_μ^2 is not necessarily equal to 1; it may increase with N, provided it approaches ∞ more slowly than α^2.

II-6. $\bar{\alpha} = 4x/3$, and $\hat{\alpha} = \text{diag} (2x/3, -x/3, -x/3)$.

II-7. $\bar{\alpha}$ and $\langle \gamma^2 \rangle$ must be positive.

II-8. 2-Phenylpropane cannot form an intramolecular excimer, and hence its emission in dilute solution must come from a single excited chromophore ("monomer" emission). The 2,4-diphenylpentanes contain two identical chromophores and, in principle, could form an intramolecular excimer in dilute solution. Apparently the geometry for an excimer is more easily accessible for the *meso* compound that for the *racemic* compound, because the *meso* compound shows the stronger emission to the red of the "monomer" emission. (Chapter VIII treats *meso* and *racemo* diads.)

Chapter III

III-1. Ethane. It is the only molecule for which all rotational isomeric states have the same shape, and the same energy at the minimum. Each rotational isomer has a population of $\frac{1}{3}$ at all temperatures.

III-2. The formalism is unchanged, with $p_{g^+} = p_{g^-} = (1 - p_t)/2 = \sigma/(1 + 2\sigma)$, but the value of σ will be different.

III-3. As a first approximation, the formalism is unchanged, but the values of σ and ω are different. Further discrimination between the nine conformations can be obtained by introduction of statistical weights for two additional second-order interactions: ψ for $g^+ g^+$ and $g^- g^-$, and τ for tt. The values of ψ and τ are very close to 1 in n-pentane, but the interaction of the Cl—C dipoles may cause them to differ more strongly from 1 in 1,3-dichloropropane.

III-4. The estimates from Fig. III-4 give about 0.54, 0.49, and 0.46 for p_t at 300, 400, and 500 K, respectively. These numbers are close to those estimated from Table III-2, with $\sigma = \exp(-E_\sigma/kT)$ and $E_\sigma = 2.1$ kJ mol^{-1}. This value of E_σ specifies $(1 + 2\sigma)^{-1}$ of 0.54, 0.48, and 0.45 at 300, 400, and 500 K.

III-5. In the following table, the third line is the sum over all conformations containing *gauche* states of the product of the statistical weight and the number of *gauche* bonds, the fourth line is the average number of *gauche* bonds, and the last line is the fraction of bonds in *gauche* states.

	n-Butane	*n*-Pentane
Z	$1 + 2\sigma$	$1 + 4\sigma + 2\sigma^2(1 + \omega)$
$\partial Z/\partial\sigma$	2	$4 + 4\sigma(1 + \omega)$
$\sigma(\partial Z/\partial\sigma)$	2σ	$4\sigma + 4\sigma^2(1 + \omega)$
$(\sigma/Z)(\partial Z/\partial\sigma)$	$2\sigma/Z$	$[4\sigma + 4\sigma^2(1 + \omega)]/Z$
$(n - 2)^{-1}(\sigma/Z)(\partial Z/\partial\sigma)$	$2\sigma/Z$	$[4\sigma + 4\sigma^2(1 + \omega)]/2Z$

III-6. $(n - 3)^{-1}\partial \ln Z/\partial \ln \omega$.

III-7. $(n - 3)^{-1}\partial \ln Z/\partial \ln \psi$, with $Z = 1 + 4\sigma + 2\sigma^2\omega + 2\sigma^2\psi$, with the last term arising from the g^+g^+ and g^-g^- states. The derivative yields $2\sigma^2\psi/Z$. The appropriate numeric assignment is $\psi = 1$.

III-8. The energies of the new c, s^+, and s^- must be assigned so high that their population is negligible. Hence their inclusion is a complication that will not significantly change the results obtained from the simpler three-state model.

III-9. The population of ttt in *n*-hexane is smaller than the population of tt in *n*-pentane. Both populations are $1/Z$, and Z is larger for *n*-hexane than for *n*-pentane.

III-10. Inclusion of the new statistical weight, ω', causes the $g^\pm g^\mp g^\pm$ conformations to be weighted by $\sigma^3\omega^2\omega'$, instead of by $\sigma^3\omega^2$. However, $\sigma^3\omega^2$ is only 0.0002 at 300 K. Since $\omega' \ll 1$, its inclusion will make this number smaller ... but the number is already so small that the $g^\pm g^\mp g^\pm$ states will make a negligible contribution to the ensemble, in terms of first- and second-order interactions alone. Inclusion of ω' will have a trivial effect on the populations calculated for the rotational isomeric states.

Chapter IV

IV-1. For both limits, $z_i = 3$ when it is evaluated as the sum over three discrete rotational isomeric states with $\phi = \pm 60°$, $180°$. The integral yields $z_i \rightarrow 2\pi$ as $\Delta E/kT \rightarrow 0$, and smaller values of z_i as $\Delta E/kT$ increases. Both the integral and the summation correctly capture the fact that any $120°$ range for ϕ is weighted the same as any other $120°$ range, for any value of $\Delta E/kT$.

$$\frac{1}{z_{i,int}} \int_{\phi_0 - \pi/3}^{\phi_0 + \pi/3} \exp\left[-\left(\frac{\Delta E}{2}\right)(1 + \cos 3\phi)\right] d\phi$$

$$= \frac{1}{z_{i,sum}} \exp\left(-\frac{\Delta E_{\phi_0}}{kT}\right) = \frac{1}{3}$$

IV-2. The 3^3 conformations of n-hexane have

$$Z = 1 + 6\sigma + 8\sigma^2 + 2\sigma^3 + 4\sigma^2\omega + 4\sigma^3\omega + 2\sigma^3\omega^2$$

IV-4. The conformation partition function is given by

$$Z = [1 \quad \sigma \quad \sigma]\begin{bmatrix} \tau & \sigma & \sigma \\ 1 & \sigma\psi & \sigma \\ 1 & \sigma & \sigma\psi \end{bmatrix}^{n-3}\begin{bmatrix} 1 \\ 1 \\ 1 \end{bmatrix}$$

with $\sigma < 1, \tau > 1, \psi < 1$. Statistical weights for second-order interactions should not occur in U_2, because they would involve the interaction of A_3 with a nonexistent A_{-1}. Hence $\tau = 1$ in U_2, so that $U_1 U_2 = [1 \quad \sigma \quad \sigma]$.

IV-5. $U_2 = [1 \quad 2\sigma]$ and

$$U_i = \begin{bmatrix} \tau & 2\sigma \\ 1 & \sigma(1+\psi) \end{bmatrix}, \qquad 2 < i < n$$

IV-6. Calculation of Z yields 1.948×10^{26} at 410 K and 6.878×10^{26} at 430 K. These points yield $\partial \ln Z/\partial T = 0.0631$ K^{-1}, which give $\langle E \rangle - E_0 = 92.5$ kJ mol^{-1} at 420 K via Eq. (IV-9).

IV-7. When $\omega = 1$, $\lambda_2 = 0$. Equation (IV-42) reduces to Eq (IV-45) with this assignment for ω, and the constant in Eq. (IV-45) is $-2\ln(1+2\sigma)$. In the other four combinations of statistical weights, where $\omega < 1$, the Z from Eqs. (IV-42) and Eq. (IV-44) agree to within at least three significant figures when $n = 5$, and the agreement extends to at least four significant figures when $n = 10$. The constant in Eq. (IV-45) has the value $\ln[(1 - \lambda_2)/\lambda_1(\lambda_1 - \lambda_2)]$.

IV-8. At each n, Z is the sum of the statistical weights for 3^{n-2} conformations. Each statistical weight is of the form $\sigma^{n_g}\omega^{n_\pm}$, where n_g denotes the number of bonds in *gauche* states and $n_\pm$ denotes the number of pairs of consecutive bonds that occupy *gauche* states of opposite sign:

$$Z = \sum \sigma^{n_g}\omega^{n_\pm}$$

$$\langle n_g \rangle = \frac{\partial \ln Z}{\partial \ln \sigma} = \frac{\sum n_g \sigma^{n_g}\omega^{n_\pm}}{\sum \sigma^{n_g}\omega^{n_\pm}}$$

$$p_g = \frac{\langle n_g \rangle}{n - 2}$$

IV-9. The statistical weight of a specified conformation is written as a product of $n - 2$ statistical weights, u_i:

$$w = \prod_{i=2}^{n-1} u_i$$

$u_i = 1$ if bond i is in a *trans* state, and $u_i < 1$ if bond i is in a *gauche* state. Hence $w = 1$ if all bonds are in *trans* states, and $w < 1$ if one or more *gauche* states are present. The population of the all-*trans* state is $1/Z$ in the unperturbed ensemble. See Table IV-2 for values of Z. The 1-kg sample is unlikely to contain any molecule in the all-*t* conformation if n is larger than ~95.

IV-10.

$$\begin{bmatrix} \tau & 2\sigma \\ 1 & \sigma(1 + \omega) \end{bmatrix}$$

IV-11. $p_{tt} = (\partial \ln \lambda_1/\partial \ln \tau)$, with $\lambda_1, \lambda_2 = (1/2)[\tau + \sigma(1 + \omega) \pm \sqrt{([\tau - \sigma(1 + \omega)]^2 + 8\sigma)}]$:

$$p_{tt} = \left(\frac{\tau}{\lambda_1}\right)\frac{\tau - \lambda_2}{\lambda_1 - \lambda_2}$$

IV-12. For a very long chain, the probability of the state(s) that index the second column increases as B increases. The sharpness of the change in probability is controlled by A; the smaller A, the sharper the change in probability.

Chapter V

V-2. If the statistical weight matrix is

$$
\mathbf{U}_i = \begin{bmatrix} 1 & \sigma & \sigma \\ 1 & \sigma & \sigma \\ 1 & \sigma & \sigma \end{bmatrix}_i
$$

both Eqs. (V-1) and (V-5) yield

$$
p_{g^+} = p_{g^-} = \frac{1 - p_t}{2} = \frac{\sigma}{1 + 2\sigma}
$$

V-3. The conformation partition functions was evaluated in Problem IV-2, and Table V-1 contains the $Z_{t;i}$. The probabilities for trans placements are

$$
\begin{aligned}
p_{t;2} = p_{t;4} &= Z^{-1}(1 + 4\sigma + 2\sigma^2 + 2\sigma^2\omega) \\
p_{t;3} &= Z^{-1}(1 + 4\sigma + 4\sigma^2) \\
p_t &= (3Z)^{-1}(3 + 12\sigma + 8\sigma^2 + 4\sigma^2\omega)
\end{aligned}
$$

Using $\sigma = 0.54$ and $\omega = 0.088$ (as appropriate for $T \sim 140°$ C), $p_{t;2} = p_{t;4} = 0.538$, $p_{t;3} = 0.614$, $p_t = 0.563$.

V-4. "?" does not have a simple description.

V-5. (a) No; (b) no; (c) no; (d) yes; (e) yes; (f) no; (g) no; (h) no.

V-6. $p_t(p_{g^+t} + p_{g^-t})^{-1}$ is 2.7 at 300 K and 2.4 at 450 K.

V-7. The temperature coefficents at ~ 400 K, in 10^{-3} deg^{-1}, are

$$
\frac{\partial \ln p_{tt}}{\partial T} = -1.2
$$

$$
\frac{\partial \ln p_{g^+t}}{\partial T} = 0.22
$$

$$
\frac{\partial \ln p_{g^+g^+}}{\partial T} = 1.7
$$

$$
\frac{\partial \ln p_{g^+g^-}}{\partial T} = 8.0
$$

V-8. $\psi = 0.74$, $\tau = 1.36$, $\omega = 0.064$

V-9. For that ~20% of the ensemble in which bond i in a long polyethylene chain is in the g^+ state, the population of the g^- state at bond $i+j$ is

$$p_{g^-;i+j} = [0 \quad 1 \quad 0]\mathbf{q}'^j_{\xi\eta} \begin{bmatrix} 0 \\ 0 \\ 1 \end{bmatrix}$$

Using Eq. (V-38) for $\mathbf{q}_{\xi\eta}$, the values for $p_{g^-;i+j}$ are 0.026, 0.173, 0.192, 0.200, and 0.201 for $j = 1, 2, 3, 4,$ and 5.

V-10. The number of distinct elements decreases to three as $i-j$ increases, provided neither i nor j is near an end. The decrease arises because the distinction between the probabilities of g pairs of the same sign, and of opposite sign, is lost as the number of intervening bonds increases. One element in $\mathbf{p}_{(\xi;j)(\eta;i)}$, $1 \ll j \ll i \ll n$, is $p_{t;j}p_{t;i}$, four elements have the value of $p_{t;j}p_{g^+;i}$, and four elements have the value of $p_{g^+;j}p_{g^-;i}$.

Chapter VI

VI-1. If $\mathbf{v}^{i+1} = [1 \quad 0 \quad 0]^T$, then

$$\mathbf{v}^i = \begin{bmatrix} -\cos\theta \\ -\sin\theta\cos\phi \\ -\sin\theta\sin\phi \end{bmatrix}$$

If $\mathbf{v}^{i+1} = [0 \quad 1 \quad 0]^T$, then

$$\mathbf{v}^i = \begin{bmatrix} \sin\theta \\ -\cos\theta\cos\phi \\ -\cos\theta\sin\phi \end{bmatrix}$$

If $\mathbf{v}^{i+1} = [0 \quad 0 \quad 1]^T$, then

$$\mathbf{v}^i = \begin{bmatrix} 0 \\ -\sin\phi \\ \cos\phi \end{bmatrix}$$

VI-2. Using the supplement of the bond angle ($\cos \theta_c = -\cos \theta$ and $\sin \theta_c = \sin \theta$)

$$\mathbf{T} = \begin{bmatrix} \cos \theta_c & \sin \theta_c & 0 \\ -\sin \theta_c \cos \phi & \cos \theta_c \cos \phi & -\sin \phi \\ -\sin \theta_c \sin \phi & \cos \theta_c \sin \phi & \cos \phi \end{bmatrix}$$

Using $\phi = 0°$ at *trans*, rather than *cis*

$$\mathbf{T} = \begin{bmatrix} -\cos \theta & \sin \theta & 0 \\ \sin \theta \cos \phi & \cos \theta \cos \phi & \sin \phi \\ \sin \theta \sin \phi & \cos \theta \sin \phi & -\cos \phi \end{bmatrix}$$

Using $\phi = 0°$ at *trans*, rather than *cis*, and also the supplement of the bond angle

$$\mathbf{T} = \begin{bmatrix} \cos \theta_c & \sin \theta_c & 0 \\ \sin \theta_c \cos \phi & -\cos \theta_c \cos \phi & \sin \phi \\ \sin \theta_c \sin \phi & -\cos \theta_c \sin \phi & -\cos \phi \end{bmatrix}$$

VI-3. The simple form is $\mathbf{I}_3$.

VI-4. In nanometers,

$$\mathbf{r}_{01} = \mathbf{l}_1 = \begin{bmatrix} 1 \\ 0 \\ 0 \end{bmatrix} \quad \mathbf{r}_{02} = \begin{bmatrix} 1 \\ 2 \\ 0 \end{bmatrix} \quad \mathbf{r}_{03} = \begin{bmatrix} 4 \\ 2 \\ 0 \end{bmatrix} \quad \mathbf{r}_{04} = \begin{bmatrix} 4 \\ 4 \\ 0 \end{bmatrix} \quad \mathbf{r}_{05} = \mathbf{r} = \begin{bmatrix} 3 \\ 4 \\ 0 \end{bmatrix}$$

VI-5. In square nanometers, the squares of the lengths of the bonds give $r_{01}^2 = 1$, $r_{12}^2 = 4$, $r_{23}^2 = 9$, $r_{34}^2 = 4$, $r_{45}^2 = 1$. The other r_{ij}^2 are $r_{02}^2 = 5$, $r_{03}^2 = 20$, $r_{04}^2 = 32$, $r_{05}^2 = r^2 = 25$, $r_{13}^2 = 13$, $r_{14}^2 = 25$, $r_{15}^2 = 20$, $r_{24}^2 = 13$, $r_{25}^2 = 8$, $r_{35}^2 = 5$. The squared radius of gyration is $s^2 = 5.14$.

VI-6. $s^2 = m_0 m_n (m_0 + m_n)^{-2} r_{0n}^2$, which is $r_{0n}^2/4$ if $m_0 = m_n$.

VI-8. For $\omega = 0$ and 1, respectively, the results in picometers as $n \to \infty$ are

$$\langle \mathbf{r} \rangle_0 = \begin{bmatrix} 611 \\ 517 \\ 0 \end{bmatrix} \quad \langle \mathbf{r} \rangle_0 = \begin{bmatrix} 340 \\ 291 \\ 0 \end{bmatrix}$$

VI-9. The result is $C_n = 1$, using

$$\langle T \rangle = \begin{bmatrix} 0 & \pi/4 & 0 \\ 0 & 0 & 0 \\ 0 & 0 & 0 \end{bmatrix}$$

VI-10. All of the terms containing ϕ become null after integration when rotation is free, leaving

$$\langle T \rangle = \begin{bmatrix} -\cos\theta & \sin\theta & 0 \\ 0 & 0 & 0 \\ 0 & 0 & 0 \end{bmatrix}$$

VI-11. Integration over the accessible values of ϕ yields $\langle \sin\phi \rangle = 0$ (because the torsion potential is symmetric), and $\langle \cos\phi \rangle = 0$, independent of $\Delta\phi$. The average transformation matrix is the same as in Problem VI-10. C_∞ is independent of $\Delta\phi$ because the three square wells are equally spaced, of the same width, and equally weighted.

VI-12. C_∞ depends on $\Delta\phi$ if each of the *gauche* states is weighted by σ ($\sigma \neq 1$), while the *trans* state is weighted by 1. $\langle \sin\phi \rangle$ is still zero, but

$$\langle \cos\phi \rangle = \frac{(\sigma - 1)\sin\Delta\phi}{(1 + 2\sigma)\Delta\phi}$$

which is nonzero if $\sigma \neq 1$. C_∞ is given by Eq. (I-29), using $\cos\theta$ and $\langle \cos\phi \rangle$. The sensitivity of C_∞ to $\Delta\phi$ depends on σ:

$$\frac{\partial \ln C_\infty}{\partial \Delta\phi} = \frac{\frac{2(\sigma-1)}{\Delta\phi(1+2\sigma)}\left(\frac{\sin\Delta\phi}{\Delta\phi} - \cos\Delta\phi \right)}{1 - \left(\frac{\sigma-1}{1+2\sigma}\frac{\sin\Delta\phi}{\Delta\phi} \right)^2}$$

Some numerical results for $\partial \ln C_\infty/\partial\Delta\phi$ are given in the following table.

$$\sigma$$

$\Delta\phi$	0	0.1	0.25	0.5	1	2.5	∞
60°	−1.98	−0.76	−0.38	−0.16	0	0.16	0.38
45°	−2.60	−0.68	−0.31	−0.13	0	0.13	0.31
30°	−4.04	−0.55	−0.23	−0.09	0	0.09	0.23
15°	−7.66	−0.29	−0.11	−0.05	0	0.05	0.11
10°	−11.47	−0.20	−0.08	−0.03	0	0.03	0.08
5°	−22.92	−0.10	−0.04	−0.02	0	0.02	0.04

An increase in $\Delta\phi$ will produce a decrease in C_∞ if $\sigma < 1$, and an increase in C_∞ if $\sigma > 1$. The characteristic ratio is most sensitive to $\Delta\phi$ when both σ and $\Delta\phi$ are small:

$$\lim_{\sigma \to 0} C_\infty = \left(\frac{1 - \cos\theta}{1 + \cos\theta} \right) \left(\frac{1 + \frac{\sin\Delta\phi}{\Delta\phi}}{1 - \frac{\sin\Delta\phi}{\Delta\phi}} \right)$$

$$\lim_{\sigma \to 0, \Delta\phi \to 0} C_\infty = \infty$$

In the limit where $\sigma \to 0$, a nonzero $\Delta\phi$ is required if C_∞ is to be finite. Hence $\partial \ln C_\infty / \partial \Delta\phi = -\infty$ in this limit.

In the other limit, where $\sigma \to \infty$, a finite C_∞ is retained for any value of $\Delta\phi$:

$$\lim_{\sigma \to 0} C_\infty = \left(\frac{1 - \cos\theta}{1 + \cos\theta} \right) \left(\frac{2 + \frac{\sin\Delta\phi}{\Delta\phi}}{2 - \frac{\sin\Delta\phi}{\Delta\phi}} \right)$$

$$\lim_{\sigma \to \infty, \Delta\phi \to 0} C_\infty = 3 \left(\frac{1 - \cos\theta}{1 + \cos\theta} \right)$$

VI-14. The discrete enumeration becomes too tedious, even for a computer, when n increases to 20–30. The literature contains few results for discrete enumeration when $n = 30$ and there are three rotational isomeric states. The number of conformations is then $3^{28} \sim 2.28767 \times 10^{13}$.

VI-15. For $\langle r \rangle_{0,\text{odd}}$ use Eq. (VI-53) with $l_i = 153$ pm in each $\mathcal{A}_i$ for which i is odd, and $l_i = 0$ in each $\mathcal{A}_i$ for which i is even. And for $\langle r \rangle_{0,\text{even}}$ use Eq. (VI-53) with $l_i = 153$ pm in each $\mathcal{A}_i$ for which i is even, and $l_i = 0$ in each $\mathcal{A}_i$ for which i is odd. As $n \to \infty$, the results in picometers are

$$\langle \mathbf{r} \rangle_{0,\text{odd}} = \begin{bmatrix} 350 \\ 171 \\ 0 \end{bmatrix} \qquad \langle \mathbf{r} \rangle_{0,\text{even}} = \begin{bmatrix} 223 \\ 315 \\ 0 \end{bmatrix}$$

The contributions from the two sets of bonds are nearly of the same magnitude, but they are in different directions. The contributions from the odd bonds are primarily along the x axis, while the contributions from the even bonds are primarily along the y axis. The contributions become more nearly similar in their directions if we eliminate the first and second bond:

$$\langle \mathbf{r} \rangle_{0,\text{odd}} - \mathbf{l}_1 = \begin{bmatrix} 197 \\ 171 \\ 0 \end{bmatrix} \qquad \langle \mathbf{r} \rangle_{0,\text{even}} - \mathbf{T}_1\mathbf{l}_2 = \begin{bmatrix} 166 \\ 173 \\ 0 \end{bmatrix}$$

VI-16. The differences are too small for detection by experiment. With $\sigma = 0.54$, $\omega = 0.088$, $\theta = 112°$, and g placements $\pm 120°$ from the t placements, $\langle s^2 \rangle_0 / l^2 = 1132$ for 1001 atoms, 1133 for 501 atoms, 1143 for 101 atoms.

VI-17. Four significant figures are needed to see a difference. With $\sigma = 0.54$, $\omega = 0.088$, $\theta = 112°$, and g placements $\pm 120°$ from the t placements, the characteristic ratios for the subchains of 100 bonds are 6.534 for the subchain at the end, and 6.538 for the subchain in the middle of the chain of 300 bonds. The same parameters give $C_n = 6.531$ for the chain of 100 bonds.

VI-18. $\langle \mathbf{r} \rangle_0$.

VI-19. The result for $\lim_{n \to \infty} \langle \mathbf{r}_{jn} \rangle_0$ will depend on j, for very small j, if the chain is subject to end effects (if $p_{\eta;i}$ and $p_{\xi\eta;i}$ depend on i at small i). In general, such a dependence will be obtained if the bonds are interdependent. The $\lim_{n \to \infty} \langle \mathbf{r}_{jn} \rangle_0$ is sensitive to the assignment of "ϕ_1" as 180° in $\mathbf{T}_1$ only if $j = 0$; at larger values of j, $\mathbf{T}_1$ is not used.

On the other hand, $\lim_{n \to \infty} [\langle r_{jn}^2 \rangle_0 / (n - j)l^2]$ is equal to C_∞ for any finite j.

Chapter VII

VII-1. (a) Sequences of *trans* states; (b) helices with *gauche* states of the same sign, (c) strings of *gauche* placements that alternate in sign.

VII-2. Using the interaction energies in Table VII-1 in Eqs. (V-51) and (V-46) shows p_t is most sensitive to temperature in polydimethylsilylene. Calculations at 400 K (and 450 K) yield p_t of 0.601 (and 0.585) for polyethylene, 0.175 (and 0.191) for polysilane, and 0.219 (and 0.254) for polydimethylsilylene.

VII-3. For calculations at 300 K:

(a) The model for polyethylene, using $\theta = 112°$, $\phi_g = \pm60°$, $\sigma = 0.43$, and $\omega = 0.034$, yields $C_\infty = 7.97$. Changing one parameter at a time to the indicated values yields the following C_∞: $\theta = 116°, C_\infty = 9.38$; $\phi_g = \pm65°, C_\infty = 8.71$; $\sigma = 0.135, C_\infty = 15.7$. Every change increases C_∞, but the largest increase comes from decreasing the weight of the g states.

(b) The model for polysilane, using $\theta = 109.4°$, $\phi_g = \pm55°$, $\sigma = 1.68$, $\psi = 1.98$, and $\omega = 1$, yields $C_\infty = 1.42$. Changing one parameter at a time to the indicated values yields the following C_∞: $\theta = 115.4°, C_\infty = 1.78$; $\sigma = 1.17, C_\infty = 1.60$; $\psi = 3.75, C_\infty = 2.08$; $\omega = 0, C_\infty = 5.96$. Every change increases C_∞, but the largest increase comes from suppression of the $g^\pm g^\mp$ states.

(c) The model for polymeric sulfur, using $\theta = 106°$, $\phi_g = \pm90°$, and $\omega = 1.55$, yields $C_\infty = 1.14$. Changing one parameter at a time to the indicated values yields the following C_∞: $\theta = 104°, C_\infty = 1.06$; $\omega = 1.98, C_\infty = 0.89$. Both changes decrease C_∞ slightly. The larger decrease is from the change in ω.

VII-4. Using $\sigma = 0.17$ and $\omega = 0.25$, C_∞ is 37. If ω is reassigned as 1, C_∞ increases to 63.

VII-5. Replacement of θ_{ABA} and θ_{BAB} by their average causes an increase in C_∞ at 300 K of 21% for polydimethylsilmethylene and polyisobutylene, 78% for polydimethylsiloxane, and 91% for polyphosphate.

VII-6. The interchange of θ_{ABA} and θ_{BAB} has no effect on C_∞ for polyphosphate, because $\mathbf{U}_{AB} = \mathbf{U}_{BA}$ for this polymer. For the other three polymers, where $\mathbf{U}_{AB} \neq \mathbf{U}_{BA}$, the interchange increases C_∞ by 8% for polydimethylsiloxane, 34% for polydimethylsilmethylene, and 46% for polyisobutylene.

VII-7. For calculations at 300 K, performed with $E_\omega = \infty$ and with finite E_ω, the change in C_∞ is ~2% when $E_\omega = 9$ kJ mol^{-1} in polyethylene and $E_\omega = 13$ kJ mol^{-1} in polysildimethylene. If the ω values in both statistical weight matrices are assigned the same values, the E_ω for the same change are 15 kJ mol^{-1} in polyoxymethylene, and 9 kJ mol^{-1} in polyphosphate. In all four polymers, C_∞ decreases when ω rises above zero, because of the introduction of $g^\pm g^\mp$ states. When E_ω is large, retention of its real value, rather than the approximation $E_\omega \sim \infty$, may be more important for $\partial \ln C_\infty/\partial T$ than for C_∞.

VII-9. (a) three σ values and three ω values; (b) three σ values and three ω values; (c) three σ values and four ω values.

VII-11. The *trans* state at the $CH_2\!-\!CH_2$ bond does not generate "pentane-effect" second-order interactions, and therefore it should not be split. A similar statement applies to the $CH_2\!-\!C(CH_3)_2$ bond. Approximation for the $CH_2\!-\!CH_2$ bond,

$$\mathbf{U} = \begin{bmatrix} 1 & x & x \\ 1 & x & x \\ 1 & x & x \end{bmatrix}, \qquad x < 1$$

Approximation for the $CH_2\!-\!C(CH_3)_2$ bond,

$$\mathbf{U} = \begin{bmatrix} 1 & 1 & 1 \\ y & y & y \\ y & y & y \end{bmatrix}, \qquad y < 1$$

Approximation for the $C(CH_3)_2\!-\!CH_2$ bond,

$$\mathbf{U} = \begin{bmatrix} 1 & 1 & 1 \\ 1 & 1 & \omega \\ 1 & \omega & 1 \end{bmatrix}, \qquad \omega < 1$$

VII-13.

$$\begin{bmatrix} 1 & 1 \\ 1 & 1 \end{bmatrix} \begin{bmatrix} 0 & 1+\xi \\ 1+\xi & 0 \end{bmatrix}$$

VII-14. All of the chains described in this chapter have $\langle z \rangle_0 = 0$ due to the symmetry of the torsion potential energy functions. For most chains, $\langle x \rangle_0 > \langle y \rangle_0 > 0$. Polyoxyethylene has $\langle y \rangle_0 > \langle x \rangle_0$ if an O—C bond is indexed as bond 1 (see Figure 3 of Abe, A.; Kennedy, J. W.; Flory, P. J. *J. Polym. Sci. Polym. Phys. Ed.* **1976**, *14*, 1337).

VII-15. In the sequence they are used, the four symmetric statistical weight matrices are

$$\mathbf{U}_{OC} = \begin{bmatrix} 1 & \sigma_{CC} & \sigma_{CC} \\ \sigma_{CC} & \sigma_{CC}^2 & \sigma_{CC}^2 \omega'_{CC} \\ \sigma_{CC} & \sigma_{CC}^2 \omega'_{CC} & \sigma_{CC}^2 \end{bmatrix}$$

$$\mathbf{U_{CC}} = \begin{bmatrix} 1 & 1 & 1 \\ 1 & 1 & \omega_{CC} \\ 1 & \omega_{CC} & 1 \end{bmatrix}$$

$$\mathbf{U_{CC}} = \begin{bmatrix} 1 & \sigma_{OC} & \sigma_{OC} \\ \sigma_{OC} & \sigma_{OC}^2 & \sigma_{OC}^2\omega_{OO} \\ \sigma_{OC} & \sigma_{OC}^2\omega_{OO} & \sigma_{OC}^2 \end{bmatrix}$$

$$\mathbf{U_{CO}} = \begin{bmatrix} 1 & 1 & 1 \\ 1 & 1 & \omega_{CC} \\ 1 & \omega_{CC} & 1 \end{bmatrix}$$

The second and fourth symmetric statistical weight matrices are identical.

VII-16. The preferred state at bond 2, the first rotatable bond, is g if the chain starts as O—C, but it is t if the chain starts as C—C or C—O. The vectors are depicted in Figure 3 of Abe, A.; Kennedy, J. W.; Flory, P. J. *J. Polym. Sci., Polym. Phys. Ed.* **1976**, *14*, 1337.

Chapter VIII

VIII-1. The expressions for Z are different:

$$Z_{meso} = 2\eta + 2\eta\tau\omega_{CX} + 2\tau\omega_{CX} + \omega + \eta^2\omega_{XX} + \eta\tau^2\omega\,\omega_{XX}$$

$$Z_{racemic} = 1 + \eta^2 + 2\eta\,\omega_{CX} + 2\eta\tau\omega_{XX} + 2\tau\omega + \tau^2\omega_{CX}^2$$

VIII-3. The isotactic chains prefer tg helices, and the syndiotactic chains prefer sequences of t states, with occasional g interruptions at C^α—C—C^α.

VIII-4. The statistical weight denoted by ω_{XX} should be increased in the statistical weight matrix for the C—C_x^α bond (to enhance the attraction of rings in units $x - 1$ and x) and in the statistical weight matrix for the C—$C\alpha_{x+1}$ bond (to enhance the attraction of rings in units x and $x + 1$). The normal value for ω_{XX} is restored when the excited state decays to the ground state.

VIII-7. The calculated C_∞ for isotactic and syndiotactic chains at 300 K are: 13 and 17 for polystyrene, 4.9 and 7.7 for poly(2-vinylpyridine), and ~15 and ~39 for poly(N-vinylcarbazole).

VIII-9. This behavior is seen with the Flory–Williams model for poly(vinyl chloride), Table VIII-2, because the strongly negative E_η, in conjunction with positive energies for the other short-range interactions, causes a strong preference for sequences of t states in the syndiotactic chain.

VIII-10. This behavior is seen with the model for polystyrene developed by Rapold and Suter, Table VIII-3.

VIII-11. C_∞ is weakly dependent on p_m in poly(methyl methacrylate).

VIII-12. This behavior is seen with the model for polystyrene developed by Rapold and Suter, Table VIII-3.

VIII-13. There are three nonzero components for $\langle \mathbf{r} \rangle_0$ for long chains. The asymmetric rotation potential energy function allows the z component to be nonzero.

Chapter IX

IX-1. The first-order interactions are the same as at the equivalent bonds in the homopolymers. As a first approximation, the higher-order interactions with the statistical weight β can be introduced using

$$
[\zeta \quad 1 \quad 1]
\begin{bmatrix} 1 & 1 & 1 \\ 1 & 1 & 1 \\ 1 & 1 & 1 \end{bmatrix}
\begin{bmatrix} \rho & 1 & 1 \\ \rho\beta & 1 & 1 \\ \rho\beta & 1 & 1 \end{bmatrix}
$$

for the C—CH$_2$—CH$_2$—CH bonds between *cis* and *trans* units, and

$$
[\rho \quad 1 \quad 1]
\begin{bmatrix} 1 & \beta & \beta \\ 1 & 1 & 1 \\ 1 & 1 & 1 \end{bmatrix}
\begin{bmatrix} \zeta & 1 & 1 \\ \zeta & 1 & 1 \\ \zeta & 1 & 1 \end{bmatrix}
$$

for the C—CH$_2$—CH$_2$—CH bonds between *trans* and *cis* units.

IX-2. The value of C_∞ is 1.93, and the average transformation matrix is

$$
\begin{bmatrix} 0.31 & -0.13 & 0 \\ -0.07 & 0.03 & 0 \\ 0 & 0 & 0 \end{bmatrix}
$$

IX-3. The value of C_∞ is only 1.7, and the average transformation matrix is

$$
\begin{bmatrix}
0.21 & -0.085 & 0 \\
-0.50 & 0.20 & 0 \\
0.75 & -0.31 & 0
\end{bmatrix}
$$

IX-5. The conversion factor is $(l^2_{NC^\alpha} + l^2_{C^\alpha C'} + l^2_{C'N})/l^2_p$.

IX-9. $\langle T \rangle$ is constructed with $\langle \cos \phi \rangle = \langle \cos \psi \rangle = 0$, $\langle \sin \phi \rangle = -2/\pi$, and $\langle \sin \psi \rangle = (2/\pi)X$, where $X = (1-y)(1+y)^{-1}$, with $y = \exp(-E/kT)$.

$$
\langle T \rangle =
\begin{bmatrix}
0.308 + 0.035X & 0.126 - 0.086X & 0.582X \\
-0.072 + 0.149X & -0.030 - 0.365X & -0.137X \\
0.554 & 0.226 & -0.138X
\end{bmatrix}
$$

For $y = 0$, 1, and ∞, the values of C_∞ are 4.37, 1.85, and 0.80, respectively. The largest C_∞ is obtained with the heaviest weighting of the upper left quadrant of the conformational energy surface.

IX-11. The hydrogen atom bonded to the amide nitrogen in poly(L-alanine), which has no counterpart in poly(L-lactic acid), contributes to the short-range interactions that produce differences in the conformational energy surfaces.

IX-12. With $\langle \sin \phi \rangle = \langle \sin \psi \rangle = 0$, $\langle \cos \phi \rangle = \cos \phi$, and $\langle \cos \psi \rangle = \cos \psi$, the elements of $\langle T \rangle$, expressed as a column in reading order, are

$$
\begin{bmatrix}
0.308 - 0.199 \cos \phi - 0.346 \cos \psi - 0.030 \cos \phi \cos \psi \\
-0.126 + 0.081 \cos \phi - 0.847 \cos \psi - 0.072 \cos \phi \cos \psi \\
0 \\
-0.072 - 0.847 \cos \phi + 0.081 \cos \psi - 0.126 \cos \phi \cos \psi \\
0.030 + 0.346 \cos \phi + 0.199 \cos \psi - 0.308 \cos \phi \cos \psi \\
0 \\
0 \\
0 \\
-\cos \phi \cos \psi
\end{bmatrix}
$$

The chain approaches full extension as $\phi \rightarrow 180°$, $\psi \rightarrow 180°$.

Chapter X

X-1. The transformation matrices are related by

$$\mathbf{T}_{\text{left-handed}} = \text{diag}\,(1, 1, -1)\mathbf{T}_{\text{right-handed}}\,\text{diag}\,(1, 1, -1).$$

$$\mathbf{T}_{\text{right-handed}} = \begin{bmatrix} 0.425 & 0.547 & -0.721 \\ -0.422 & 0.825 & 0.377 \\ 0.801 & 0.144 & 0.581 \end{bmatrix}$$

X-2. The same conformational partition function (see Lifson, S.; Roig, A. *J. Chem. Phys.* **1961**, *34*, 1963, who use a different notation) is obtained with an initial row vector of $[1\ \ 0\ \ 0]$, a final column vector of $[1\ \ 1\ \ 0]^T$, and

$$\mathbf{U}_i = \begin{array}{c} \\ c(c\text{ or }h) \\ hc \\ hh \end{array} \begin{array}{c} c(c\text{ or }h) \qquad hc \qquad hh \\ \begin{bmatrix} 1 & 0 & \hat{\sigma}^{1/2}s \\ 1 & 0 & 0 \\ 0 & \hat{\sigma}^{1/2}s & s \end{bmatrix}_i \end{array}$$

X-3. With rows and columns indexed in the order c, h_r, and h_l,

$$\mathbf{U}_i = \begin{bmatrix} 1 & \hat{\sigma}s_r & \hat{\sigma}s_l \\ 1 & s_r & \hat{\sigma}s_l \\ 1 & \hat{\sigma}s_r & s_l \end{bmatrix}_i$$

The population of the left-handed helix is never higher than 0.006 as $n \to \infty$, although it can be higher at small n.

X-4. The differences will be most apparent when the average number of residues in an ordered region is small. This condition is obtained when $\hat{\sigma}$ is large.

X-5. 1.3.

X-6. The three components (in nm), for several values of s and p_h, are given in the table below. The ends of the chain remain disordered until very large values of s. For this reason, most of the change in $\mathbf{r}$ does not occur until very large values of s, where the chain ends finally join the helix.

s	p_h	r_x	r_y	r_z
0.9	0.01	2.0	0.1	1.0
0.99	0.23	2.0	0.0	1.0
1.00	0.40	2.1	−0.0	1.1
1.01	0.59	2.5	−0.1	1.2
1.05	0.89	5.0	−1.2	2.6
1.10	0.95	8.0	−2.9	4.2
1.8	0.995	19.2	−22.1	8.4
6.0	0.999	26.5	−52.7	4.8
∞	1	30.8	−70.1	0.2

X-7. Using $\hat{\sigma} \ll 1$, we obtain

$$Z = \begin{bmatrix} 1 & 1 \end{bmatrix} \begin{bmatrix} 1 & \hat{\sigma} \\ \hat{\sigma} & 1 \end{bmatrix}^{x-1} \begin{bmatrix} 1 \\ 1 \end{bmatrix}$$

X-8. The conformational partition function might be written as

$$Z = \begin{bmatrix} 1 & s \end{bmatrix} \begin{bmatrix} 1 & \hat{\sigma}s \\ \hat{\sigma} & s \end{bmatrix}^{x-1} \begin{bmatrix} 1 \\ 1 \end{bmatrix}$$

At 300 K, this conformational partition function, and the conformational energies stated in the problem, predict that 72% of the units should be h^-, and the average number of units in a string of h^- units is 981.

X-9. With rows and columns indexed in the order $c(c$ or $h)$, hc, hh:

$$\mathbf{U}_i = \begin{bmatrix} 1 & 0 & \hat{\sigma}^{1/2}s \\ 1 & 0 & 0 \\ 0 & \hat{\sigma}^{1/2}s & s \end{bmatrix}_i$$

Chapter XI

XI-1. The oxygen atom is involved directly in certain of the first- and second-order interactions that appear in $\mathbf{U}_b$. Specifically, it appears in the $\mathbf{D}$ of $\mathbf{DQ}_d^T\mathbf{DQ}_d$, and in the $\mathbf{V}$ of $\mathbf{Q}_d^T\mathbf{V}$ and $\mathbf{VQ}_d$, for $\mathbf{U}_{b,d}$ (see Table XI-1). The shorter bond length of C—O, rather than C—C, will affect the assignments of the statistical weights in $U_{i;3}, i = 2$–5. The first-order interaction in $U_{4;2}$, and the second-order interaction in $U_{5;2}$, involve O with CH_2.

XI-4. Only two of the 27 conformations for the three bonds to the atom at the trifunctional branch point are not weighted by τ_1 or ω. At a d pseudoasymmetric center, these conformations are tg^+t and g^-tg^-.

XI-5. The final result for U_b, presented in sequence as the first, second, and third blocks, each of size 3×3^2, is

	ttt	ttg^+	ttg^-	tg^+t	tg^+g^+	tg^+g^-	tg^-t	tg^-g^+	tg^-g^-
t	$\tau_2\omega^3$	$\tau_2\psi^2\omega$	$\tau_2\psi\omega^2$	$\tau_2\psi\omega^2$	$\tau_2\psi^2\omega$	$\tau_2^2\omega^3$	$\tau_2\psi^2\omega$	$\tau_2\psi^4$	$\tau_2\psi^2\omega$
g^+	$\tau_2\omega^3$	$\psi\omega^2$	$\tau_2\psi^2\omega$	$\tau_2\psi\omega$	$\psi\omega$	$\tau_2\psi\omega$	$\tau_2\psi\omega$	$\psi^2\omega$	ψ^2
g^-	$\tau_2\omega^3$	$\tau_2\psi\omega$	$\tau_2\psi\omega$	$\psi^2\omega$	ψ^2	$\tau_2\psi\omega$	$\psi\omega^2$	$\psi^2\omega$	$\psi\omega$

	g^+tt	g^+tg^+	g^+tg^-	g^+g^+t	$g^+g^+g^+$	$g^+g^+g^-$	g^+g^-t	$g^+g^-g^+$	$g^+g^-g^-$
t	$\psi\omega$	ψ^2	$\psi^2\omega$	$\psi\omega$	$\psi\omega$	$\tau_2\omega^3$	$\tau_2\psi\omega$	$\tau_2\psi^2\omega$	$\tau_2\psi\omega^2$
g^+	$\tau_2\psi^2\omega$	$\psi^2\omega$	ψ^4	$\tau_2\psi^2$	$\psi\omega$	$\tau_2\psi^2\omega$	$\tau_2^2\psi\omega$	$\tau_2\psi\omega^2$	$\tau_2\psi^2\omega$
g^-	$\tau_2\psi\omega^2$	$\tau_2\psi\omega$	$\tau_2\psi^3\omega$	$\psi^2\omega$	$\psi\omega$	$\tau_2\psi\omega^2$	$\tau_2\omega^3$	$\tau_2\omega^3$	$\tau_2\omega^3$

	g^-tt	g^-tg^+	g^-tg^-	g^-g^+t	$g^-g^+g^+$	$g^-g^+g^-$	g^-g^-t	$g^-g^-g^+$	$g^-g^-g^-$
t	$\psi\omega$	$\tau_2\psi\omega$	$\psi\omega$	$\psi^2\omega$	$\tau_2\psi\omega^2$	$\tau_2\omega^3$	ψ^2	$\tau_2\psi^2\omega$	$\psi\omega$
g^+	$\tau_2\psi\omega^2$	$\tau_2\omega^3$	$\psi^2\omega$	$\tau_2\psi^2\omega$	$\tau_2\omega^3$	$\tau_2\psi\omega^2$	$\tau_2\psi\omega$	$\tau_2\omega^3$	$\psi\omega$
g^-	$\tau_2\psi^2\omega$	$\tau_2^2\psi\omega$	$\tau_2\psi^2$	ψ^4	$\tau_2\psi^2\omega$	$\tau_2\psi^2\omega$	$\psi^2\omega$	$\tau_2\psi\omega^2$	$\psi\omega$

XI-6. Using $\sigma = \tau_1 = 0.54$ and $\omega = 0.088$, with all other statistical weights equal to 1, the probabilities for *trans* placements at the last two bonds in branch 1 are 0.40 and 0.73 if the branch point is trifunctional, and they are both 0.33 if the branch point is tetrafunctional.

XI-10. If τ_X is to be greater than 1, then $\omega_{X_2} > 1$. In most structures, the energy is positive in the second-order interaction associated with two successive *gauche* placements of opposite sign, which produces the expectation that we will usually have $\omega_{X_2} < 1$ and $\tau_X < 1$.

Chapter XII

XII-3. A first approach might employ four σ values and eight ω values, in eleven statistical weight matrices.

Fragment	σ	ω
O–C–O–C–O	σ_1	ω_1
C–C–O–C–O	σ_1	ω_2
O–C–C–C–O	σ_2	ω_3
C–C–C–C–O	σ_2	ω_4
C–O–C–O–C	σ_1	ω_5
C–C–C–O–C	σ_3	ω_6
O–C–O–C–C	σ_3	ω_2
C–C–O–C–C	σ_3	ω_7
C–O–C–C–C	σ_2	ω_6
O–C–C–C–C	σ_4	ω_3
C–C–C–C–C	σ_4	ω_8

XII-9. Using the average transformation matrix in the text, the C_∞ are 2.02 if each chain contains either L or D units, 1.79 if the units are placed randomly, and 1.62 if there is perfect alternation of the units.

XII-10. The values of C_∞ are 2.15, 2.44, 3.11, 4.63, and 9.44 when the fraction of L-alanyl residues is 0, 0.25, 0.5, 0.75, and 1, respectively.

XII-11. C_∞ is 3.11 for the random copolymer, and 2.21 for the alternating copolymer.

XII-12. 0.76%.

XII-14. C_{1000} is 4.40 for the tapered copolymer, 1.87 for the alternating copolymer, and 3.03 for the random copolymer.

XII-15. C_{200} is 5.90, 3.51, 3.05, 3.51, and 5.90 for the five subchains. The value of C_{200} for the middle subchain is very close to the expectation of 3.01 for a random copolymer.

Chapter XIII

XIII-1. $(3^p + 1)/2$.

XIII-2. Using $\mathbf{T}_1$ as defined in Eq. (VI-4), the reduced form of $\mathbf{A}_1^{\times 2}$ is

$$
\begin{bmatrix}
\cos^2\theta & -\cos\theta\sin\theta & 0 & -l\cos\theta & \sin^2\theta & 0 & l\sin\theta & 0 & 0 & l^2 \\
-2\cos\theta\sin\theta & \sin^2\theta - \cos^2\theta & 0 & l\sin\theta & 2\cos\theta\sin\theta & 0 & l\cos\theta & 0 & 0 & 0 \\
0 & 0 & \cos\theta & 0 & 0 & -\sin\theta & 0 & 0 & -l & 0 \\
\sin^2\theta & \cos\theta\sin\theta & 0 & 0 & \cos^2\theta & 0 & 0 & 0 & 0 & 0 \\
0 & 0 & -\sin\theta & 0 & 0 & -\cos\theta & 0 & 0 & 0 & 0 \\
0 & 0 & 0 & 0 & 0 & 0 & 0 & 1 & 0 & 0
\end{bmatrix}
$$

and the reduced for of $\mathbf{A}_n^{\times 2}$ is

$$
\begin{bmatrix}
l^2 \\
0 \\
0 \\
2l \\
0 \\
0 \\
0 \\
0 \\
0 \\
1
\end{bmatrix}
$$

XIII-8. Terms that are always identical for the two directions of indexing: $\langle L_1^2 \rangle_0$, $\langle L_2^2 \rangle_0 / \langle L_1^2 \rangle_0$, $\langle L_1^2 \rangle_0 + \langle L_2^2 \rangle_0 + \langle L_3^2 \rangle_0$, $P_1^2 + P_2^2 + P_3^2$.

Terms that become identical in the limit where $n \to \infty$: $\langle S \rangle_0$, P_1^2, P_2^2 / P_1^2.

XIII-9. The principal moments of $\langle S \rangle_0$ will be nearly equal if the terminal bond in the flexible block is indexed as bond 1. One principal moment will be much larger than the other two if the terminal bond in the rigid block is indexed as bond 1.

XIII-10. Decrease.

Chapter XIV

XIV-1. $\lim_{n \to \infty} \langle \mu^2 \rangle / 2m^2 = 1$.

XIV-2. For the **CClH—CH$_2$** bond.

APPENDIX C
Program for Calculation of C_∞ and D_∞

```
 1      program V2_RIS
 2 C
 3 C---
 4 C Computes the exact average (V**2) of a vector V that is the sum of
 5 C contributions V(i), rigidly connected to the skeletal bonds of polymer
 6 C chains of regular constitution and configuration.
 7 C
 8 C The program is capable of handling any number of bonds/unit, but not
 9 C more than NU rotational isomeric states/bond (NU is a parameter). The
10 C chain length increases step-wise by a factor of 2; the maximum length is
11 C selected by the number of steps ("loops") LMAX [max. X = 2**(LMAX-1)].
12 C
13 C Matrices are normalized in each step to prevent numerical overflow or
14 C underflow and there is no strict upper bound to the chain length.
15 C Roundoff error, however, can corrupt the results for very long chains;
16 C adjustment of the quantity EPSILON (a parameter in subroutine NRMLZ)
17 C allows for maximum (machine-dependent!) accuracy without underflow
18 C errors (EPSILON = 10**(-24) is a reasonable setting for most computers).
19 C
20 C (C) Copyright 1993 : Wayne L. Mattice & Ulrich W. Suter
21 C---
22      integer   I, II, IN, J, JJ, JN, L, LMAX, M, M1, M5, MAXL, MOLD, N,
23    &           N5, NBPU, NOLD, NU, NUN, NUNIT,
24    &           LNBLNK
25      parameter (NU=6)
26      character TITLE*80
27      real*8    PHI(NU), THETA(NU), U(NU,NU), UI(NU,NU), U1(NU), UX(NU),
28    &           VEC(3), VECN(3), VEC1(3),
29    &           G(5,5), GG(5*NU,5*NU), GGI(5*NU,5*NU), GG1 (5*NU),
30    &           GGX(5*NU), TMP(5*NU,5*NU),
31    &           C, CINF, COLD, ENLSQ, FAC, FG, GMAX, LNZ, SUMVEC, UMAX,
32    &           VEC1N, VM2, XINV, Z, ZFAC
33      data      U1 / NU*0. /
34      data      UX / NU*1. /
35 C---
36 C Format statements too long to fit into the write statements themselves:
37 C
38    1 format ('1Characteristic Ratio of (Vector**2)0 for',
39    &         ' Chains with Regular Constitution',//,
```

```
40    &          1x,' +++ ',a,' +++',///,
41    &             ' Number of bonds per unit =',i3,/,
42    &             ' Longest chain :',i14,' units')
43   2 format (//,' Bond',i3,' : ',i3,' x',i3,' U-matrix',10x,
44    &            'associated vector =',3f8.4,/)
45   3 format (1h1,//,1x,' +++ ',a,' +++',///,
46    &             ' Dimension of U(unit) =',i3,' x',i3,/)
47   4 format (//,10x,'X',5x,'1/X',8x,'ln(Z)',14x,'(V**2)',6x,'C(X)',
48    &            7x,'est. C(inf)',/)
49 C---
50 C Initialize matrices to unit-matrices for later convenience
51 C
52       U1(1) = 1.
53    50 do 70 I = 1, NU
54          do 60 J=1,NU
55    60       U(I,J) = 0.
56    70    U(I,I) = 1.
57       do 90 I=1,5*NU
58          do 80 J=1,5*NU
59    80       GG(I,J) = 0.
60    90    GG(I,I) = 1.
61 C
62 C Read running TITLE for all pages
63 C
64       read (*,'(a)',end=900) TITLE
65 C
66 C Read # of bonds/unit & # of chain length loops [max.X = 2**(LMAX-1)].
67 C Defaults are: NBPU : none, LMAX : 10
68 C
69       SUMVEC = 0.
70       read (*,*,end=900) NBPU, LMAX
71       if (LMAX.1e.0) LMAX = 10
72       MAXL = 2**(LMAX-1)
73       write (*,1) TITLE(1:LNBLNK (TITLE)), NBPU, MAXL
74 C
75 C Read U-matrices and vectors associated with the bonds, then
76 C torsion angles and bond angles for each unit.   Running sum for V**2.
77 C
78       do 300 NUM=1,NBPU
79          read (*,*,end=900) M, N, (VEC(I),I=1,3)
80          SUMVEC = SUMVEC + VEC(1)**2 + VEC(2)**2 + VEC(3)**2
81          if (NUN.eq.1) then
82             M1 = M
83             MOLD = M
84             NOLD = M
85          endif
86          write (*,2) NUN, M, N, (VEC(I), I=1, 3)
87          if (M.ne.NOLD) go to 800
88          if ((NUN.eq.NBPU).and.(N.ne.M1)) go to 800
89          if (NUN.eq.1) then
90             VECN(1) = VEC(1)
91             VECN(2) = VEC(2)
92             VECN(3) = VEC(3)
93          endif
```

```
 94          if (NUN.eq.NBPU) then
 95              VEC1(1) = VEC(1)
 96              VEC1(2) = VEC(2)
 97              VEC1(3) = VEC(3)
 98          endif
 99          do 130 I=1,M
100              read (*,*,end=900) (UI(I,J),J=1,N)
101   130        write (*,'(1x,10f12.6)') (UI(I,J),J=1,N)
102          write (*,'(/,a)') ' Torsion angles'
103          read (*,*,end=900) (PHI(I),I=1,N)
104          write (*,'(1x,10f12.1)') (PHI(I),I=1,N)
105          write (*,'(/,a)') ' Bond angles'
106          read (*,*,end=900) (THETA(I),I=1,N)
107 C
108 C - default for any THETA-angle beyond the first: take the previously
109 C   defined one (if THETA is the same for all PHI's of a bond, simply
110 C   add zero's to complete the number required by the input statement).
111 C
112          do 140 I=2,N
113   140        if (THETA(I).lt.1.) THETA(I) = THETA(I-1)
114          write (*,'(1x,10f12.1)') (THETA(I),I=1,N)
115 C
116 C Now compute U(unit) and GG(unit) (=Script-G(unit)) as running product.
117 C First U
118 C
119          call MATMAT (U,UI,U,NU,MOLD,M,N,TMP,5*NU)
120 C
121 C Now Script-G (according to equation (VI-51), with A replaced by G)
122 C
123          M5 = 5*(M-1) + 1
124          N5 = 5*(N-1) + 1
125          do 160 J=1,N5,5
126              JN = 1 + J/5
127              call GMAT (THETA(JN),PHI(JN),VEC,G)
128              do 160 I=1,M5,5
129                  IN = 1 + I/5
130                  do 160 II=1,5
131                      do 160 JJ=1,5
132   160                    GGI (II+I-1,JJ+J-1) = G(II,JJ)*UI(IN,JN)
133          call MATMAT (GG,GGI,GG,5*NU,5*MOLD,5*M,5*N,TMP,5*NU)
134 C
135 C Now compute Script-G for first and last bond (GG1 and GGX)
136 C according to equations (VI-50) and (VI-52) with A replaced by G
137 C (the first bond of the chain is taken as the last bond of the repeat unit
138 C occurring once before the repetition starts, the last bond as the first
139 C bond of the repeat unit, taken once after the repetition ends - hence,
140 C the chain-length will be equal to the [number of units]*NBPU + 2).
141 C
```

```
142              if (MOLD.lt.M) MOLD = M
143              NOLD = N
144              if (NUN.le.1) then
145                 do 170 I=1,M5,5
146                    do 170 II=1,5
147     170                GGX(I+II-1) = G(II,5)
148              endif
149              if (NUN.ge.NBPU) then
150                 do 190 I=1,5
151     190            GG1 (I) = G(1,I)
152                 do 200 I=6,5*NU
153     200            GG1(I) = 0.
154              endif
155     300      continue
156 C
157 C Ready for looping over chain length.
158 C First write summery of setup, then go ...
159 C
160         write (*,3) TITLE(1:LNBLNK(TITLE)), M1, N
161         do 350 I=1,N
162     350   write (*,'(1x,10f12.6)') (U(I,J),J=1,N)
163         write (*,'(/,1x,a,f10.4,/)')
164       &         'Sum of mean-square length of vectors / unit =',SUMVEC
165         write (*,4)
166         NUNIT = 1
167         FAC = 0
168         ZFAC = 0.
169         VEC1N = VEC1(1)**2 + VEC1(2)**2 + VEC1(3)**2
170       &        + VECN(1)**2 + VECN(2)**2 + VECN(3)**2
171 C
172 C Now loop
173 C
174         do 500 L=1,LMAX
175 C
176 C -   calculate Z, Z(V**2), (V**2), V[av]**2 (V[av] is the rms average
177 C     of V(i) over the entire chain), and C(X)=(V**2)/(N*V[av]**2).
178 C
179            call VMATV (U1,U,UX,NU,N,Z,TMP)
180            call VMATV (GG1,GG,GGX,5*NU,5*N,FG,TMP)
181            LNZ = ZFAC + log(Z)
182            VM2 = exp(FAC)*FG/Z
183            ENLSQ = real (NUNIT)*SUMVEC + VEC1N
184            C = VM2/ENLSQ
185            XINV = 1./real(NUNIT)
186 C
187 C -   extrapolate and write results.
188 C
```

```
189             if (L.ge.2) then
190                CINF = 2*C - COLD
191                write (*,'(i11,2e12.4,f17.2,f11.4,f15.4)')
192      &                NUNIT, XINV, LNZ, VM2, C, CINF
193             else
194                write (*,'(i11,2e12.4,f17.2,f11.4)')
195      &                NUNIT, XINV, LNZ, VM2, C
196             endif
197             if (L.lt.LMAX) then
198                COLD = C
199 C -   prevent overflow and underflow by conditioning U and GG.
200                call NRMLZ (U,NU,N,UMAX)
201                call NRMLZ (GG,5*NU,5*N,GMAX)
202                FAC = FAC + log(GMAX) - log(UMAX)
203                ZFAC = ZFAC + log(UMAX)
204 C -   square U and GG for double chain length.
205                call MATMAT (U,U,U,NU,N,N,N,TMP,5*NU)
206                call MATMAT (GG,GG,GG,5*NU,5*N,5*N,5*N,TMP,5*NU)
207                NUNIT = 2*NUNIT
208                FAC = 2*FAC
209                ZFAC = 2*ZFAC
210             endif
211  500        continue
212 C and restart.
213             go to 50
214 C That's it.
215  800 write (*,'(/,a)') ' ***** WRONG DIMENSIONS *****'
216  900 write (*,'(///,1x,a)') '***** end of run *****'
217          end
```

```
1          subroutine GMAT (THETA,PHI,V,G)
2 C---
3 C Sets up the generator matrix G for V**2, given scalar values of
4 C THETA(i), PHI(i), and the vector V(i), according to equation (VI-21).
5 C---
6          real*8    G(5,*), V(*), T(3,3),
7        &           PHI, THETA
8          integer  I,J
9 C---
10         do 100 I=1,5
11            do 100 J = 1,5
12   100        G(I,J) = 0.
13 C
14 C Transformation matrix T according to equation (VI-5).
15 C
16         call TMAT (THETA,PHI,T)
17 C
18         do 110 I=2,4
19            do 110 J=2,4
20   110        G(I,J) = T(I-1,J-1)
21         G(1,1) = 1.
22         G(1,2) = 2.*(V(1)*T(1,1) + V(2)*T(2,1) + V(3)*T(3,1))
23         G(1,3) = 2.*(V(1)*T(1,2) + V(2)*T(2,2) + V(3)*T(3,2))
24         G(1,4) = 2.*(V(1)*T(1,3) + V(2)*T(2,3) + V(3)*T(3,3))
25         G(1,5) = V(1)**2 + V(2)**2 + V(3)**2
26         G(2,5) = V(1)
27         G(3,5) = V(2)
28         G(4,5) = V(3)
29         G(5,5) = 1.
30         return
31         end
```

```
1          integer function LNBLNK (STRING)
2 C---
3 C returns the location of the last non-blank character in string STRING.
4 C---
5          character STRING*(*)
6          integer I
7 C---
8          do 100 I=len(STRING),1,-1
9    100     if (STRING(I:I).ne.' ') go to 200
10         I = 0
11   200 LNBLNK = I
12         return
13         end
```

```
1          subroutine MATMAT (A,B,AB,MM,MA,MB,NB,TMP,MTMP)
2 C---
3 C Multiplies matrix A (MAxMB) with matrix B (MBxNB) and stores
4 C the result in matrix AB (MAxNB). AB can be one of A or B.
5 C The first dimension of the arrays containing A, B, and AB in the
6 C calling program's defining statement is MM.
7 C The first dimension of the array containing TMP in the
8 C calling program's defining statement is MTMP.
9 C---
10         real*8     A(MM,*), B(MM,*), AB(MM,*), TMP (MTMP,*),
11      &             X
12         integer    I, J, K, MA, MB, MM, NB
13 C---
14         do 100 I=1,MA
15            do 100 J=1,NB
16               X = 0.
17               do 90 K=1,MB
18    90            X = X + A(I,K)*B(K,J)
19   100          TMP(I,J) = X
20 C
21         do 200 I=1,MA
22            do 200 J=1,NB
23   200          AB(I,J) = TMP(I,J)
24         return
25         end
```

```
1          subroutine NRMLZ (A,MM,N,FAC)
2 C---
3 C Divides every element of square matrix A (NxN) by that with the largest
4 C magnitude, FAC.  Set elements that are smaller than EPSILON in magnitude
5 C to true zero (EPSILON is a parameter).
6 C The first dimension of the array containing A in the
7 C calling program's defining statement is MM.
8 C---
9          real*8     A(MM,*),
10      &             EPSILON, FAC
11         integer    I, J, MM, N
12         parameter (EPSILON=1.0d-24)
13 C---
14         FAC = 0.
15         do 100 I=1,N
16            do 100 J=1,N
17   100          FAC = max(FAC,abs(A(I,J)))
18 C
19         do 200 I=1,N
20            do 200 J=1,N
21               A(I,J) = A(I,J)/FAC
22   200          if (abs(A(I,J)).lt.EPSILON)  A(I,J) = 0.
23         return
24         end
```

```
 1          subroutine TMAT (THETA,PHI,T)
 2 C---
 3 C Sets up the transformation matrix T according to equation (VI-5).
 4 C---
 5          real*8     T(3,*),
 6      &              COSTHT, COSPHI, PHI, SINTHT, SINPHI, THETA
 7          integer    I, J
 8 C---
 9          COSTHT = cos (THETA*3.1415926535/180.)
10          SINTHT = sin (THETA*3.1415926535/180.)
11          COSPHI = cos (PHI*3.1415926535/180.)
12          SINPHI = sin (PHI*3.1415926535/180.)
13          T(1,1) = - COSTHT
14          T(1,2) =   SINTHT
15          T(1,3) =   0.
16          T(2,1) = - SINTHT * COSPHI
17          T(2,2) = - COSTHT * COSPHI
18          T(2,3) = - SINPHI
19          T(3,1) = - SINTHT * SINPHI
20          T(3,2) = - COSTHT * SINPHI
21          T(3,3) =   COSPHI
22          return
23          end
```

```
 1          subroutine VMATV (VL,A,VR,MM,M,RESULT,TMP)
 2 C---
 3 C Forms the bilinear product from the vector VL,
 4 C the square matrix A (MxM), and the vector VR:
 5 C              RESULT = VL(transpose) A VR
 6 C The first dimension of the array containing A in the
 7 C calling program's defining statement is MM.
 8 C---
 9          real*8     A(MM,*), TMP(*), VL(*), R(*),
10      &              RESULT, X
11          integer    I, K, MM, M
12 C---
13          do 100 I=1,M
14             X = 0.
15             do 90 K=1,M
16     90         X = X + VL(K)*A(K,I)
17    100         TMP(I) = X
18          X = 0.
19 C
20          do 200 I=1,M
21    200     X = X + TMP(I)*VR(I)
22          RESULT = X
23          return
24          end
```

NAME INDEX

SUBJECT INDEX

Entries in italics denote problems.